Applications of Ac
Models

Applications of Activated Sludge Models

Damir Brdjanovic
Sebastiaan C.F. Meijer
Carlos M. Lopez-Vazquez
Christine M. Hooijmans
Mark C. M. van Loosdrecht

IWA PUBLISHING

Published by **IWA Publishing**
Alliance House
12 Caxton Street
London SW1H 0QS, UK
Telephone: +44 (0)20 7654 5500
Fax: +44 (0)20 654 5555
Email: publications@iwap.co.uk
Web: www.iwapublishing.com

First published 2015

British Library Cataloguing in Publication Data
A CIP catalogue record for this book is available from the British Library

Library of Congress Cataloguing- in-Publication Data
A catalogue record for this book is available from the Library of Congress

Front page: Peter Stroo
Graphic design: Hans Emais

ISBN: 9781780404639 (Hardback)
ISBN: 9781780404660 (eBook)

About the editors

Prof. Damir Brdjanovic is Professor of Sanitary Engineering and Head of Environmental Engineering and Water Technology Department at UNESCO-IHE. He also holds a position of Guest Professor at Delft University of Technology in the Environmental Biotechnology group of Prof. Van Loosdrecht. In 1998 he was awarded a PhD degree on the topic "Modelling biological phosphorus removal (BPR) in activated sludge systems". Having 25 years of academic and consultancy expertise in the wastewater field, Prof. Brdjanovic is one of the pioneers of practical applications of ASM BPR models in sewage treatment plants in developing countries and countries in transition, and in industrial effluent treatment plants. He started with modelling two decades ago, when he assisted the late Prof. Marais of the University of Cape Town in delivering early courses on 'Activated Sludge Modelling' at UNESCO-IHE. Since then, he has been involved in teaching in this advanced course, currently entitled 'Modelling sanitation systems', which is carried out in cooperation with Prof. van Loosdrecht and Dr Meijer. Prof. Brdjanovic has a sound publication record and is author and editor of several books in the wastewater treatment and sanitation field.

Dr. Sebastiaan Meijer is environmental engineer and owner of the company Yuniko B.V. operating in the field of ICT, wastewater, natural resources and green energy recovery. He earned his PhD degree for his research publication "Theoretical and practical aspects of modelling activated sludge processes" at Delft University of Technology working with professors Heijnen and van Loosdrecht. He has been working for 18 years as a researcher and consultant for international firms, institutions and governments on the development and improvement of wastewater treatment. He has carried out over 50 full-scale modelling assignments, ranging from classical design examples up to advanced and state-of-the-art applications of modelling and process control. Since 1998, Dr. Meijer is teaching wastewater engineering at Delft University of Technology and UNESCO-IHE. He has a sound publication record and is recognized as a leading international specialist in the field of wastewater treatment and model-based design.

Carlos Manuel Lopez Vazquez is Associate Professor in Wastewater Treatment Technology at UNESCO-IHE institute for Water Education. In 2009, he received his doctoral degree on Environmental Biotechnology (*cum laude*) from Delft University of Technology and UNESCO-IHE Institute for Water Education on the topic "The Competition between Polyphosphate-Accumulating Organisms and Glycogen-Accumulating Organisms: Temperature Effects and Modelling". During his professional career, he has taken part in different advisory and consultancy projects for both public and private sectors concerning municipal and industrial wastewater treatment systems. After working for a couple of years in the Water R&D Department of Nalco Europe on industrial (waste)water treatment applications, he re-joined UNESCO-IHE's Sanitary Engineering Chair Group in 2009. Since then, he is involved in education, capacity building and research projects guiding several MSc and PhD students and co-authoring several peer-reviewed publications. By applying mathematical modelling as an essential tool, he has a special focus on the development and transfer of innovative and cost-effective wastewater treatment technologies to developing countries, countries in transition and industrial applications.

Tineke Hooijmans is Associate Professor of Sanitary Engineering at UNESCO-IHE in Delft. She obtained her PhD in 1990 at Delft University of Technology entitled "Diffusion coupled with bioconversion in immobilized systems - use of an oxygen micro-sensor". During her PhD research she became interested in the applicability of modeling to describe the behavior of micro-organisms combined with mass transfer phenomena. Since then she worked on nutrient removal and modeling of wastewater treatment processes and systems, supervising many PhD and MSc students. Since 2002 she organizes, in cooperation with the TU Delft, a three-week course on Modelling of Wastewater Treatment and Processes and Plants. Since 2011, she also coordinates the online course on Modelling Sanitation Systems.

Prof Mark van Loosdrecht is a well-renown scientist recognised for his significant contributions to the study of reducing energy consumption and the footprint of wastewater treatment plants through his patented and award winning technologies Sharon®, Anammox® and Nereda®. His main work focuses on the use of microbial cultures within the environmental process-engineering field, with a special emphasis on nutrient removal, biofilm and biofouling. Currently a full Professor and the Group Leader of Environmental Technology at TU Delft, Prof Loosdrecht was trained on environmental engineering, where his focus on microbiology and biotechnology engineering has led to significant contributions to advance the scientific knowledge in these areas. A fellow of the Royal Dutch Academy of Arts and Sciences (KNAW), the Dutch Academy of Engineering (AcTI) and the International Water Association (IWA), Prof Loosdrecht has won numerous prestigious awards. He is also a scientific advisor at KWR Water Cycle Research Institute where he conducts wastewater and water cycle research in his continuing efforts to identify more sustainable solutions to solve global water problems. His other research interests include granular sludge systems, microbial storage polymers, wastewater treatment, gas treatment, soil treatment, microbial conversion of inorganic compounds, production of chemicals from waste, and modelling. He also led the development of the metabolic model for biological phosphorus removal (TU Delft bio-P model). Apart from his other achievements, he has published over 450 papers, supervised 65 PhD students so far and is honorary professor at University of Queensland, Australia. He is also currently the Editor-in-Chief for Water Research and Advisor to IWA Publishing.

My (and others) most unforgettable character – A tribute to Professor GvR Marais

George A. Ekama

After serving as an apprentice fitter and tuner on the South African Railways and lecturer at Port Elizabeth technical college, Gerrit van Rooyen Marais commenced university studies at the age of 26. In 1955 he obtained a BSc degree in civil engineering from the University of Cape Town. After two years as a research engineer in the mining industry working on the slip failure of slimes dams, for which he obtained a MSc (Eng) from the Witwatersrand University, he spent six years with the Zambian (then Northern Rhodesia) Housing Board doing research in low-cost housing, water supply and sanitation. During this time, he did a postgraduate study at the Imperial College, London and was awarded a DIC in Public Health Engineering. The research that he conducted during this time resulted in 19 publications in the low-cost water supply and sanitation area. One of these papers describes how a robust dissolved oxygen (Mackereth) probe and meter can be "homemade" and used for measuring dissolved oxygen concentration in oxidation ponds, for which he was awarded a prize for the most innovative short paper by the South African Institute of Civil Engineers in 1967. The quality of the research from this first phase of his career as a sanitary engineer was such that in 1970 he was appointed as a member of the Expert Committee on Environmental Sanitation of the World Health Organisation, a position he held until 1987, and in 1972 he was invited to the United States of America by the American Society of Sanitary Engineering as Distinguished Foreign Professor.

After a brief spell as senior lecturer in geotechnical engineering at the Witwatersrand University, the second and most successful phase of his career in Sanitary Engineering commenced in 1967 when he was appointed to the chair of Water Resources and Public Health Engineering in the Department of Civil Engineering at the University of Cape Town, a position he held until his retirement in 1992. It was early in this phase that I met "Prof" as a Masters and PhD student. His main research interests during this time were in municipal water and wastewater treatment covering topics as water chemistry and water conditioning, anaerobic digestion, activated sludge systems, biological nitrogen and phosphorus removal, filamentous bulking control in activated sludge systems and industrial wastewater treatment in aerobic and anaerobic systems. As a result of his indefatigable energy and research activity, he was elected fellow of the University of Cape Town for distinguished research in 1977 and in 1984 was awarded a DSc (Eng) degree by the University of Cape Town. On two occasions he was appointed short-term consultant by the World Health Organization, in 1976 to Brazil and in 1980 to Colombia to advice on wastewater treatment problems in the cities of San Paulo and Bogota. In 1983, he was appointed by the International Water Association (IWA, then IAWPRC) together with four other research engineers to serve on the international task group for modelling biological wastewater treatment plants. In 1984, at the invitation of Professor Clifford Randall of the Virginia Polytechnic and State University, Professor Marais was a major contributor to a short course in Richmond, Virginia on the design of biological nitrification denitrification (ND) biological excess phosphorus removal (BEPR) systems for practicing engineers and wastewater treatment plant operators and state legislators to promote biological nutrient removal (BNR) technology in Virginia and surrounding East Coast states to protect the Chesapeake Bay. On many occasions he was invited plenary speaker at international conferences to present state of the art review papers on the field of biological nutrient removal, inter alia Italy, Australia and Japan. In recognition of his contribution to the field of environmental engineering, Professor Marais was invited to join the Royal Society of South Africa, one of very few engineers who at that time won this distinction.

In 1970 when Professor Marais commenced research into the activated sludge system, the design criteria for the system were very limited, empirical and inadequate to deliver effluent quality suitable for discharge under standards promulgated by the legislation agencies to limit eutrophication. With time the standards became more stringent, in particular for certain catchments were high levels of N and P removal were required. From the outset of this research, Professor Marais took the viewpoint that the various activated sludge systems operate on the same basic bioprocess principles and all systems should be amenable to description and simulation by one basic biological kinetic model. With this approach, he commenced with his postgraduate students a wide ranging long-term experimental and theoretical modelling research programme supported by the Water Research Commission, in two parallel but complementary projects: (1) the experimental, to provide data, firstly to meet short-term need for design criteria for construction of BNR plants, and secondly to meet the long-term need for validation data of simulation kinetic models, and (2) the theoretical, to develop an integrated kinetic model for the activated sludge system in most of its variations. In experimental project, in the period from 1972 to 1980, practically all the known forms of activated sludge system comprising single and multi-reactor configurations including aerobic, anoxic and anaerobic zones for BNR had been operated and studied. To eliminate the confounding and adverse effect of the nitrate on the EBPR, the UCT configuration was developed in 1980. Many other configurations followed, for example, the Johannesburg system and the biological chemical flexible system (BCFS). Today, BNR activated sludge plants are the workhorse for municipal wastewater treatment. The practical experience gained from the wide-ranging experimental investigation was compiled in 1984 by Professor Marais and his colleagues into a design guide, colloquially called "The Green Book", which has enjoyed wide circulation internationally and has proven very useful to design engineers and plant operators, particularly in the 1980s and 1990s.

In the theoretical part of the programme, once fully aerobic systems for organics removal by ordinary heterotrophic organisms (OHO) in its various forms were satisfactory understood and could be accurately simulated with an aerobic general kinetic model, other biological processes such as nitrification by autotrophic organisms (ANO) and denitrification by facultative OHO were integrated into the kinetic model. By 1981, a general kinetic model incorporating nitrification and denitrification (ND) was developed capable of accurately simulating multi-reactor systems comprising un-aerated reactors anywhere in the configuration and the cyclic flow and load conditions from sludge ages 3 to 30 days and temperatures 12 to 24°C. When Professor Marais joined IAWPC Task Group on modelling biological wastewater treatment systems in 1985, this model was accepted as the basis from which to develop the Activated Sludge Model No 1 (ASM1). The ASM1 model has been widely distributed around the world in the form of WWTP simulation programmes written by software development companies and has been found very useful by design engineers and plant operators.

In the 8 years after 1984 to Professor Marais retirement in 1992, the experimental and modelling research projects focussed on integrating the enhanced biological phosphorus removal (EBPR) into the model. Experimentally, the understanding of EBPR progressed from the parametric and empirical to biochemical, kinetic and mechanistic. By 1988, enhanced cultures of the organisms that mediate the EBPR - the phosphorus accumulating organisms (PAO) - were developed, from which their behaviour and kinetics were delineated. The enhanced culture PAO model was integrated into the ASM1 model by 1991 to produce the first mixed culture (OHO, ANO, PAO) activated sludge model for design and operation of biological N and P removal activated sludge systems (UCTPHO). Later the IWA Task Group used this model to develop ASM2. In honour of his outstanding contribution to water management and science, Professor Marais was awarded the International Association for Water Quality (IAWQ, Now IWA) Karl Imhoff - Pierre Koch medal at the 17th IAWQ biennial conference in Budapest in 1994. On the many occasions when Prof Marais was honoured for his tireless contribution to wastewater treatment both nationally and internationally, he always acknowledged his postgraduate students, 19 PhD and 67 master students who received their degrees and published their research under his guidance.

Over the year preceding his 75th birthday, I invited the graduates from the Water Research Group he founded at UCT to write some recollections and reflections of him. Their stories make amazing reading and describe much more elegantly and personally the unforgettable character Prof Marais really was than the professional achievements above. I collated the responses in a book, which I gave him on the day of his 75th birthday (20 Dec., 2002). I include a selection of these below.

"As with any student, the minimum effort is expended right up until the last moment or final exam when one can cram in as much information as possible to achieve a pass. I think Prof was always aware of this and one of his routines was always to pop into the lab wanting to clean his pipe with the compressed air hose and acetone and upon seeing a student waiting for some cooking or boiling cycle to finish, would ask-"Where are your plots?" In most cases we were always at least a week behind in plotting the data. Prof would then recite a story about a certain Roman Emperor who, after winning many battles found the site he had always wanted to build his palace. Upon completion he was standing on the paved road leading up to his palace talking to the chief gardener, who was responsible for all the planting, landscaping, etc. The Emperor described a tree canopy covering the approach road and told the gardener that he wanted a particular type of tree planted on either side of the road. The gardener politely pointed out to the Emperor that this particular type of tree took hundreds of years to reach the majestic maturity described by the Emperor. The Emperor then looked at the gardener and said- "then there is no time to start like the present is there". After this punch line Prof would shuffle off leaving the student wondering if he had been reprimanded or just told and interesting bit of Roman history. Once again the manner in which this was done was such that you felt that you had better get up to date and stay up to date for fear of another story. (Ian Lilley, MSc, 1991).

"My first memory of "Prof" was of him in the position of Senior Lecturer at Wits University in the field of foundation engineering, and me as a final year Civil Engineering student. It was the day of thesis allocation - effected with much pomp and ceremony by the academics of the department selecting particular students in fields of research investigation to be effected under their supervision. Eventually one student remained unselected in a silent auditorium - yours truly. In retrospect there were a thousand reasons why any academic would have run the proverbial mile before volunteering to supervise me in a thesis project. However, after minutes of painful silence my lot was saved and Prof stepped forward ending the embarrassment. Oddly enough (given my instability at the time), a reasonable piece of work emerged on "the application of quasi linear differential equations to slope stability problems". This had two important outcomes for me - first, I had my introduction into the excitement of research, and second Prof made a verbal offer of future research supervision. Shortly after this he was offered the chair in Water Quality Engineering at UCT and left Johannesburg. After graduating, it did not take long for me (and my employers) to discover that I had zero interest or ability in a consulting engineering career. Thus it was that I applied to do research under Prof at the University of Cape Town. I have little doubt that it was similar acts of kindness on Prof's part that resulted in the majority of his students entering into the world of research. All of us greenhorns or worse at the start of his guidance -

but all international researchers/engineers at the end. I know I speak for all Prof's students in saying that research under his supervision was a special experience - there was science and the history of science, Shakespeare, historical politics of Europe and the never to be forgotten monologues around the subtleties of probability and statistical theory (still a mystery to me). He ruled our research/teaching world as a compassionate dictator, with a string of homilies: "that's not good, it's not even bad", "good enough isn't good enough", "research involves starting, finishing and publishing - leave one out and you have done nothing". But underlying his bonhomie was a steel core of Calvinism - he had failed unless each student working under him realised their full potential. But the Calvinism terrified us all for a good reason. If it happened as it usually did in working with him or under him that a scientific breakthrough occurred the jubilation was always tempered with our knowledge that there was to be lot of suffering in giving birth to our creation. However, as with child birth, the pains were soon forgotten. Prof - happy birthday - however, I am sure that as you read what I have written above you are shaking your head and muttering "Dick how many times have I told you that when you write you do so for the reader NOT the writer". However, in this case it is for me in that I can say publically, thank you very much for all your help, guidance and kindnesses in our many years of working together (Dick Loewenthal, MSc, 1976, PhD 1983, lifelong colleague).

"There are two ways to shorten a weekend. The first is to drink to excess on a Friday night and only wake up after midday on Saturday. The second is to be caught by Prof in the lab on a Saturday morning and be asked to explain your results from the previous six months. Students who had been working in the lab for only one month were not exempt, for if they only had a small amount of data to present, they had the privilege of listening to a discourse on the history of the British monarchy from 1715 to 1810." (Tim Casey, PhD 1993).

"Prof, I was always aware of your international stature but moving to Australia really brought this into a sharper focus. Before I elaborate, let me recall a comment which you made when you were sitting in the staff tea room at UCT. I remember you lighting your pipe with a match and through the clouds of smoke you remarked on the fact that after so many years of work all you could point to as evidence of your endeavour were piles of paper. I felt that your comment was overly modest but only in Australia could I see that it was also inaccurate. The true evidence of your work can be found outside of the confines of the university campus. Within days of arriving in Australia I was joined by Barry Rabinowitz of Reid Crowther in Vancouver BC. Ian Law, another of your students, joined us to work on a project in Tasmania. Subsequently I worked with Richard Moosbrugger out from Austria for a project at a milk factory wastewater plant in New Zealand. Peter Dold from Canada & I keep contact and have had dealings on the same project in Adelaide. I speak to Tim Casey and Wayne Bagg in Perth. A Victorian client remarked that if I was from UCT then surely I must live and breathe wastewater treatment. Australian educated engineers use the process terminology developed through your research, in the same way in which we might use terms like enthalpy or entropy, attaching a familiarity to it that implies entrenched and long established use. The Modified UCT process has been vigorously advocated by some of these same Australian engineers and installations can be found in many different parts of Australia." (John Messenger, PhD 1991).

"Like every quality research lab, there was a wonderful mix of the most curious characters in the WR Lab., united in their respect for you. The great tutor, Taliep, made his Lakay Gallery - a collection of your quotes: e.g. 'You must think until your head gets hurt' or 'You must work until you get dark rings under your eyes'. Every piece of writing went through an astonishing number of drafts, due to the critical reception each revision received from you. 'You must write it over and over until you get it right!' was the mantra I received from the Guru of Water Resources, who was usually found meditating in his cave-office, half-hidden behind clouds of pipe smoke

and enormous piles of ash-covered papers. You knew too much classical Greek, too much Lord Macaulay, and wrote too well in English for me to ever convince you that my version was in any way superior to yours. Of all the treasured memories, the one that makes me smile the most is the occasion we had a site visit from the Water Research Commission. For some reason you were extremely nervous about your presentation, which was due at 7pm. All day you kept shuttling between your office and the WR Lab., checking that all was ready and making additional demands for material for your presentation. George Ekama was sitting in his mezzanine office working like a dervish trying to get a computer simulation finished in time, but he had severe problems with his program – then run with punch cards on the central computer. Try as he might, it wasn't working. You, in the meantime, were getting more and more agitated. Eventually at about 5 pm you burst into the lab to find that George still hadn't solved the problem. Standing in the middle of the lab, you barked up at the mezzanine: 'EKAMA – YOU'RE FIRED! YOU'VE FAILED IN YOUR FIRST CRISIS!' The rest of us unsympathetically fell about in silent laughter. Needless to say, the unfortunate George solved the problem in time, your presentation was flawless, and you did not mention his sacking again." (Kevin Martin, MSc 1975).

A little more than a year after his 75th birthday, Prof Marais was diagnosed with inoperable lung cancer. After enduring radio and chemo-therapy, he enjoyed another year life. The cancer returned in July 2005, and after a brave and stoic struggle, he succumbed on the 7th December, 2005.

George A. Ekama
Cape Town, 7 December 2012.

Professor Marais paging through the collection of students tributes on his 75th birthday.

Foreword

Models for activated sludge processes have been made for more than 50 years. They were first developed as scientific tools to organise insights into the various processes involved and were mainly used by scientists and sometimes by students. These early models had little flexibility and were not meant for practical purposes. Also, each research group had their own model with varying degrees of complexity and different elements, and so initially there was very little progress made in their development. However, in 1981 Professor Poul Harremoës from the Technical University of Denmark got the idea to combine all these models and to get international agreement on a common framework in order to accelerate their development. Being at that time the President of IAWPRC (The International Association of Water Pollution, Research and Control) he was able to organise the establishment of an international task group to realize this idea. I was lucky enough to be employed by Poul Harremoës and he asked me to come up with suggestions for members for such a group, which should involve the significant players in this topic area. At that time they were Professor Gerrit Marais of Cape Town University in South Africa, Professor Les Grady of Clemson University in the US and Professor Willy Gujer of EAWAG in Switzerland. We wanted to include a representative of Asia in the group as well and the choice fell on Professor Tom Matsuo of Tokyo University. Professor Harremöes thought that I would be a good chairman, since I did not have my own models to defend. Thus the Task Group on Mathematical Modelling for Design and Operation of Activated Sludge Processes was created in 1982 and supported by IAWPRC. Many people believed that this group could not agree on anything because of the strong personalities involved, but after three years of work we presented the outline of the IAWPRC Activated Sludge Model. At the biennial IAWPRC conference in Rio de Janeiro in 1986 there was reservation from some delegates due to the IAWPRC name being involved, so I changed the name to Activated Sludge Model No.1. The appendix No.1 was intended to reduce the criticism of the first suggested name. That decision turned out to be a very good one because it allowed us to come up with new model numbers in the following years. Later on in his life Poul Harremoës said that when he travelled and presented himself as the President of IAWPRC, many people answered: 'Ah, that's the organisation with the model'.

The first model (ASM1) was soon grasped by engineering companies for the design of nitrogen removal processes, which at that time was not straightforward. Since then the models have been used and developed for many other purposes: biological phosphorus removal, sludge bulking, anaerobic processes, etc. As a teaching tool the models have been excellent and allowed the students and treatment plant operators to 'operate' a treatment plant model without interrupting the function of the plant.

Until his untimely death on November 26, 2003, Professor Poul Harremoës was able to enjoy the development he had initiated with such remarkable vision 20 years earlier.

Mogens Henze
Copenhagen, 25 November 2012

Preface

Modelling is an important activity in the development of science. Modelling not only requires the explicit and quantitative formulation of theoretical concepts, it also allows transfer of complicated knowledge between scientific disciplines as well as between theoretical and practical applications. For 25 years, activated sludge models have played a crucial role in the development of the activated sludge process. These models are not typically academic; they do not aim to include every potential sub-processes involved in the activated sludge process. Instead, they are formulated with the minimum complexity needed to describe the relevant features of the process in practice. They also provide a systemized platform for the description of environmental biotechnological models in general, through the use of standardized notation and a matrix presentation. Over the years, wastewater research in Delft has benefitted greatly from the development of activated sludge models. On one hand, modelling has been expanded through the development of novel theoretical concepts and their application in new fields. On the other hand, models have been used for practical projects. This book has been prepared to celebrate the 25th anniversary of Activated Sludge Model Nr. 1 (ASM1), and in tribute to the activated sludge modelling pioneer, the late professor Marrais. It presents fifteen practical applications of the activated sludge model and their development, applied to plant optimization, the extension, upgrading, retrofitting and troubleshooting of wastewater treatment plants, carried out by the members of the Delft modelling group over the last two decades. These applications present a good overview of the potential of modelling, and can be used as examples in courses. We trust the cases will inspire future engineers to use model as central tools in their work on improving the wastewater treatment technology through innovation and optimization.

Editors
Delft, 3 January 2015

Contents

Chapter 1 1
Introduction to modelling of activated sludge processes
van Loosdrecht M.C.M., Ekama G.A., Wentzel M.C., Hooijmans C.M., Lopez-Vazquez, C.M., Meijer S.C.F., and Brdjanovic D.

Chapter 2 69
WWTP Holten, The Netherlands: Model development and full-scale testing
Van Veldhuizen H. M., van Loosdrecht M.C.M. and Heijnen J.J.

Chapter 3 81
WWTP Haarlem Waarderpolder, The Netherlands: Model evaluation of a full-scale bio-P side-stream process
Brdjanovic D., van Loosdrecht M.C.M., Versteeg P., Hooijmans C.M., Alaerts G.J. and Heijnen J.J.

Chapter 4 115
WWTP Katwoude, The Netherlands: Development of wastewater treatment data verification techniques
Meijer S.C.F., van der Spoel H., Susanti S., Heijnen J. J. and van Loosdrecht M.C.M.

Chapter 5 129
WWTP Hardenberg, The Netherlands: Modelling full-scale start-up of the BCFS® process
Meijer S.C.F., van Loosdrecht M.C.M., Heijnen J.J.

Chapter 6 165
WWTP Shell Godorf, Germany: Optimization of oil refinery wastewater treatment
Pinzon A., Brdjanovic D, Moussa M., Lopez-Vazquez C.M., Meijer S.C.F., Van Straaten H., Janssen A., Van Loosdrecht M.C.M., and Amy G.

Chapter 7 183
WWTP Walcheren, The Netherlands: Model evaluation of a novel upgrading concept for N removal
Salem S., Berends D., Van Loosdrecht M.C.M. and Heijnen J.J.

Chapter 8 191
WWTP Anjana, India: Coupling models for integrated and plant wide modelling
Lopez-Vazquez C.M., Moussa M.S., Mithaiwala M., van Loosdrecht M.C.M., and Brdjanovic D.

Chapter 9 209
WWTP Ecco, The Netherlands: Modelling nitrogen removal from tannery wastewater
Moussa M.S., Rojas A.R., Hooijmans C.M., Gijzen H.J. and van Loosdrecht M.C.M.

Chapter 10 221
WWTP Sarajevo, Bosnia and Herzegovina: Use of modelling for cost-effective reconstruction of urban wastewater infrastructure
Hodzic A., Vojinovic Z., Seyoum S.D., Pathirana A., Drnda N., Meijer S.C.F., and Brdjanovic D.

Chapter 11 241
WWTP Illidge Rd., Sint Maarten: Use of modelling for cost-effective design of wastewater treatment
Lake, O., Lopez-Vazquez, C.M., Hooijmans, C.M. and Brdjanovic, D.

Chapter 12 259
WWTP Dulces Nombres, Mexico: Model-based evaluation of full-scale plant hydraulics
Fall C., Flores-Alamo N., Esparza-Soto M. and Hooijmans C.M.

Chapter 13 267
WWTP Varaždin, Croatia: Use of models for cost-effective planning of plant retrofit and upgrade scenarios
Meijer S.C.F., Brdjanovic D, Šikić T., Matošić M., Deduš B. and Širac S.

Chapter 14 339
WWTP Amsterdam-West, The Netherlands: Use of modelling to explain deterioration of effluent quality under wet weather conditions
Meijer S.C.F., and Piekema P.

Chapter 15 357
WWTP Houtrust, The Netherlands: High performance plant upgrade using big-data and reconciliation techniques
Meijer S.C.F., Van Kempen R.N.A. and Appeldoorn K.J.

Chapter 16 411
WWTP UPM, Uruguay: Modelling pulp mill wastewater treatment
Bentancur S., Lopez-Vazquez C.M., Garcia H.H., Travers D. and Brdjanovic D.

Chapter 17 437
The past, present and future of wastewater treatment modelling
Lopez-Vazquez C.M., Hooijmans C.M., Meijer S.C.F., Brdjanovic D., and van Loosdrecht M.C.M.

Chapter 1

Introduction to modelling of activated sludge processes

Van Loosdrecht M.C.M., Ekama G.A., Wentzel M.C., Hooijmans C.M., Lopez-Vazquez, C.M., Meijer S.C.F., and Brdjanovic D.

This chapter is based on the chapter 'Modeling Activated Sludge Processes' by Mark C.M. van Loosdrecht, George A. Ekama, Mark C. Wentzel, Damir Brdjanovic and Christine M. Hooijmans in 'Biological Wastewater Treatment: Principles, Modeling and Design' edited by M. Henze, M.C.M. van Loosdrecht, G.A. Ekama and D. Brdjanovic. ISBN: 9781843391883. IWA Publishing, London, UK. (2009), and chapter 'The Role of Modelling in Modern Wastewater Management' in 'A Practical Guide to Activated Sludge Modelling' by Meijer S.C.F. and Brdjanovic D. ISBN: 9789073445260, UNESCO-IHE (2012).

1.1 What is a model?

A model can be defined as a purposeful representation or description (often simplified) of a system of interest (Wentzel and Ekama, 1997). This consequently means that the model will never exactly reflect the reality. So, the question 'does this model describe a wastewater treatment plant?' is meaningless, unless it defines what part(s) of the treatment plant the model should describe in the first place. One never develops a model that describes every single organism, every molecule of water or every detail of the process. Models are used as a simplification of reality in such a way that they describe that part of reality that is relevant to understand and to deal with. It is also important to note that a mathematical model can only be successful if it fulfils people's expectations. There are two aspects that are extremely relevant in modeling: the aspect of time and that of scale. In general, processes can be separated into three groups from the perspective of time. Processes can be in a so-called frozen state, dynamic state, steady state or equilibrium. Models are usually made to describe the dynamic state, the state in which variations occur as a function of time. When a process is in a frozen state it means that the process will change over time, but not in the time interval that one is interested in. For example, if the daily dynamics of the wastewater treatment plant is of concern, the concentration of ammonia in the effluent will vary over time; the concentration of nitrate will vary in the reactors, etc. However, within one day, the ammonia and nitrate concentrations in the sludge digester, which can be part of the total activated sludge model, will not change. Usually the hydraulic retention time is 30 days, resulting in a characteristic time of change in this digester being in the order of two to three weeks. Consequently, one can consider the processes taking place in the digester as being in a kind of frozen state. There are hardly any daily variations in the processes in comparison with those taking place in the wastewater treatment line. On the other hand, there are processes that occur so fast that they are in steady state or in an equilibrium condition. These processes occur so rapidly that the speed of change greatly exceeds the dynamics that one is interested in. The changes that are of usual concern in wastewater treatment are, for example, changes in the ammonia concentration, which have a rate value in the order of magnitude of hours. The changes that are relevant to process control have a rate value in the order of magnitude of minutes. However, if one considers the chemical precipitation processes, these will occur more or less instantaneously (in a few seconds). These rapid processes do not have to be described in a dynamic way because they happen so fast that one can assume they are in equilibrium condition or completed. Therefore, one of the first considerations when making a model is to consider which processes are of interest, followed by the determination of the relevant timeframe, an assessment of the dynamics within the process, and finally an accurate description of those processes that are time-varying. The other processes, that are in a frozen or steady state, are not of primary importance as they can be introduced in a much quicker, simpler way within the model, or even omitted. This is because they can be considered as continuous processes with stable concentrations under certain conditions (as in digesters). So, the aspect of time is the first major issue in trying to simplify the reality. The recommended approach is to consider the time constants and select those processes that have the dynamics in the order of time constants that one is interested in. For wastewater treatment this usually means hourly or daily dynamics, sometimes yearly dynamics. In the latter example, of course, digestion will become important as over the year the performance of a digester will change because the sludge production will vary during the year.

The second relevant issue for modeling is space resolution. One can theoretically make a model that describes every square inch of an activated sludge plant. But the question is whether one is interested in such a detailed description in the first place. The answer depends once again on the purpose of the model. In general, in wastewater treatment practice, the reactor size is in the order of tens of meters. In order to describe the concentration gradients of relevant components in the reactor, of which oxygen is the most sensitive one, usually a scale size of a few meters is needed. On a different scale, there is a gradient of concentrations inside the activated sludge floc that theoretically can be described by a model. However, in standard activated sludge modeling it is neglected, as not being relevant enough to be taken into account. This consequently means that activated sludge models are usually not designed to describe the system at the length scale of an activated sludge floc but at the length scale of a reactor.

The next step in modeling is to look at the relevant level of detail of a microbial model. Typical traditional wastewater treatment design methods are based on the 'black-box approach' focusing on plant influent and effluent characteristics, while nothing or very little is known about what is happening inside the wastewater treatment plant. Traditional design parameters, such as the F/M ratio (sludge loading rate) are not based on the understanding of the processes within a wastewater treatment plant. However, one can design a plant reasonably well by applying a proper loading rate, without really knowing what processes are taking place in the plant. So the black-box model can work well in reality (Figure 1.1).

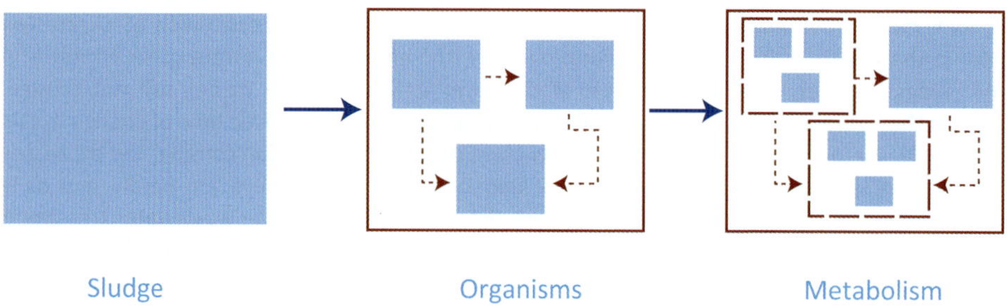

Figure 1.1 Schematic representation of the step-wise refinement of a model (Smolders *et al.*, 1995).

The black-box model is not by definition wrong or unscientific, but the application of the black-box approach very much depends on the purpose of the model. If the purpose is to design a wastewater treatment plant, experience has shown that the F/M ratio can be a very good basic approach for the design, despite the fact that it does not give information about the sludge composition. One can refine this approach and move towards grey-box models, as was the case in the Activated Sludge Model No.1 (ASM1, Henze *et al.*, 1987), No. 2 (ASM2, Henze *et al.*, 1995), and No. 2d (ASM2d, Henze *et al.*, 1999). Here the sludge was split up into relevant fractions: an inert organic matter fraction, a fraction of nitrifying bacteria, heterotrophic bacteria, denitrifying bacteria, and a fraction of phosphate-removing bacteria. Different functional aspects of the sludge were specified for a population-based model where selected microbial communities are defined inside the activated sludge and as such incorporated in the model.

Furthermore, the metabolism of the organisms and the metabolic routes inside the organisms can also be described. With such an increase in information, the approach becomes close to glass-box modeling (such as the Activated Sludge Model No.3: ASM3, Gujer *et al.*, 1999, and the TU Delft EBPR model: TUDP model, Van Veldhuizen *et al.*, 1999). This results in a larger and more complex model. The challenge here is to figure out for each process what the adequate level of description is. The question is: does the increase in complexity also increase the quality of the model (outputs), in other words, does it provide a better description of the wastewater treatment plant? For example, it has been shown that with an increased level of detail in the description of nitrification, a marginal improvement of model performance can be obtained while in the case of phosphate removal, significant improvement can be gained by including a metabolic description. Therefore, the preference for black-, grey-, or glass-box modeling depends largely on the purpose and application of the model. This is the point where modeling often goes wrong as modellers neglect the purpose and make the model itself the purpose of modeling.

Of course, one can attempt to proceed to the next level of complexity by including fundamentals of microbial genetics and genetic change. Technically and in principle this is possible, but again, this must happen depending on what the purpose and use of the model are. If the model is made too complex and with too many parameters in relation to what one ultimately wants to describe, then such an approach can generally be considered a waste of time and effort. There is also no absolute need for a model that describes the exact reality. How far reality should be matched depends again on the purpose. For example, if one wants to get an idea about N_2O emission from wastewater treatment plants, maybe three of four theories can be created and incorporated into the model. At this point it is of prime interest to look at the simulation results of different models in terms of trends and to what extent these trends reflect reality. At this stage one only has to focus on trends; good calibration, exact fitting and accurate knowledge of parameter values are not necessary. On the contrary, as an example, if one needs to predict the performance of a plant to satisfy legislation that requires every sample taken from the effluent to be below 1 mgNH_4/L, then the accuracy of the parameters put in the model must be much higher. One has to guarantee that the model prediction must be below 1 mgNH_4/L exactly. These two examples once again show that the model should always be judged in relation to the purpose of its use.

Two extremes in types of mathematical models can be identified: empirical and mechanistic models. An empirical model is based on recognition of the parameters that seem to be essential to describe the behavioural pattern of interest, and linking these through empirical relationships established by observation. The mechanisms and/or processes operating in the system are not known or are often ignored: a classical black-box approach. In contrast, a mechanistic model is based on some conceptualization of the biological/physical mechanisms operating in the system, i.e. is based on a conceptual idea (or model). The complexity of this mechanistic model will depend on the degree of understanding of the biological and chemical processes occurring in the system. Because mechanistic models have a conceptual basis, they are often more reliable than empirically-based models. Because of their black-box approach, the empirical models have an application strictly limited by boundaries (e.g. wastewater characteristics, system parameters) within which the model was developed; only interpolation is possible. Being conceptually based, the mechanistic models have greater sureness in application outside the boundaries within which the model was developed; both interpolation and extrapolation are possible. However, ultimately all models are only our rationalization of behavioural patterns or processes we conceive to be of interest. Because of this rationalization, any model needs to be rigorously calibrated and adequately verified by appropriate tests. Also, the conditions within which the model is expected to operate successfully need to be firmly delineated. For the empirical models these are strict conditions within which the model was developed, and for the mechanistic models these are the conditions under which the conceptualized behaviour is expected to remain valid. It is evident from the discussion above that the mechanistically based models have greater potential for application to wastewater treatment plants, and attention will be focused on these models.

Processes operating in a system and the compounds on which these act must be identified to set up a conceptual model on which the mechanistic mathematical model is based. The various interactions between the processes, and between the processes and compounds, are delineated descriptively. In order to develop the mechanistic model from the conceptual model, the process rates and stoichiometric interactions with the compounds are formulated mathematically. The mathematical equivalent of the mechanistic model will very probably not include all the processes and compounds that are present in the system; only those perceived to be of significance for fulfilling the objectives set for the model need be included. The art of constructing conceptual and mechanistic models is in eliminating those processes and compounds that contribute little or nothing to fulfilling the objectives set for the model. It is a waste of time and effort to develop a complicated model when it is unlikely that a model can be developed that describes a phenomenon completely. Theoretically a complete description should include aspects down to the most fundamental level. The level of organization is usually set by the objectives for the model. For example, in modeling biological behaviour in wastewater treatment systems, we cannot directly implicate biochemical control mechanisms (such as ADP/ATP and NAD/NADH ratios), or even the behaviour of specific microorganism species. The mixed liquor in the activated sludge system contains a wide diversity of different microorganism species for which identification and enumeration techniques have recently become available. These techniques, however, are time and labour consuming. Instead, microorganisms that fulfil a particular function in the activated sludge system (e.g. aerobic degradation of organics or nitrification) are grouped together as a single entity, which is called a 'surrogate' organism. This surrogate organism is assigned a set of unique characteristics that reflect the behaviour of the group, but might not reflect the characteristics of any individual organism or species of organisms in the group. To illustrate, this approach is equivalent to modeling the 'macroscopic' behaviour of a forest of trees as opposed to the 'microscopic' behaviour of each individual tree or species of trees that makes up the forest. In considering the behaviour of the forest, a parameter that could be modelled, for example, is carbon dioxide (CO_2) production. The forest as an entity will have defined CO_2 production and consumption rates. Individual tree species within the forest, or even every single tree, might have specific CO_2 production and consumption rates that deviate significantly from those of the forest entity. However, the effect achieved by modeling the forest as an entity will closely equal the net effect of modeling the cumulative contribution of each individual tree or tree species. The great advantage in modeling the forest as an entity over modeling the individuals is that considerably less information is required to develop the model and to calibrate it. In modeling biological wastewater treatment systems, the utilization of substrate by organisms is a typical example: Monod's Equation (Monod, 1949) is used to relate the specific growth rate of the surrogate organism to the surrounding substrate concentration, whereas the organisms making up the surrogate group might have different specific growth rates or might respond differently to the various substrates present in the influent wastewater. Thus, for modeling wastewater treatment systems the organizational level that is modelled is the mass behaviour of a population or group of selected microorganisms. In the models developed for activated sludge systems, this is the principle organism groups, their functions and the zones in which these functions are performed.

The parameters at that level that need to be included in the mathematical model depend greatly on the objectives for the model accepting the level of organization described above. For mathematical modeling of wastewater treatment systems two different kinds of mathematical models are generally developed: steady state and dynamic models. Steady state models have constant flows and loads and tend to be relatively simple. This simplicity makes these models useful for design. In these models complete descriptions of system parameters are not required. They are oriented towards determining the more important system design parameters. The dynamic models have varying flows and loads and accordingly include time as a parameter. Dynamic models are more complex than the steady state ones. The dynamic models are useful in predicting the time-dependent system response of an existing or proposed system. Their complexity means that the system parameters have to be completely defined for its application. For this reason the use of dynamic models for design is restricted. Often the steady state design and dynamic kinetic models evolve interactively. The dynamic kinetic models can provide guidance for the development of the steady state design models; they help to identify the design parameters that have a major influence on the system response and help to eliminate those processes that are not of major importance at steady state. For the dynamic models, with their greater complexity, only those parameters that seem to be of importance are considered for inclusion in the model. For activated sludge systems, when selecting the level of organization at the surrogate organism or mass behaviour of populations, until recently the dynamic models have been structured to consider only the net effects as present in the bulk liquid. For example, when using Monod's Equation the kinetic rate is determined by the bulk liquid soluble COD and surrogate organism concentrations.

However, with the extensions of the models to include enhanced biological phosphorus removal (EBPR), parameters internal to the surrogate biomass have had to be included, e.g. poly-β-hydroxyalkanoate (PHA), glycogen and polyphosphate. With this development, although the model might be at a selected level of organization, information on processes and behaviour from lower levels of organization is often essential, particularly to identify the key processes that control the response of the system. Usually information from lower levels of organization is of a microbiological and/or biochemical nature; the more complete this information is, the more reliable is the model. To make use of this information, 'model' organisms that are part of the 'surrogate' are identified and the known microbiological and biochemical characteristics of the organism are used to obtain a greater understanding of the surrogate. More recently the surrogate organism approach to modeling has been found to be inadequate to completely describe some behavioural patterns observed in activated sludge systems; for example the selector effect (Gabb *et al.*, 1991), substrate utilization inhibition on transfer from anoxic to aerobic zones (Casey *et al.*, 1994), and generation of nitrogen intermediates in denitrification (Casey *et al.*, 1994). To describe these and similar observations, it has been found that a lower level of organization needs to be selected: the synthesis and activity of certain key enzymes and the processes they mediate need to be modelled (Wild *et al.*, 1994). Modeling at this level of organization has been termed modeling with structured biomass. Detailed microbiological and biochemical information is required for this modeling approach (Wentzel and Ekama, 1997).

It should be noted that there is an essential difference between an activated sludge model and a wastewater treatment (plant) model. The latter term is used to indicate the ensemble of activated sludge model, hydraulic model, oxygen transfer model and sedimentation tank model, all needed to describe an actual wastewater treatment at a full-scale installation (Gernaey *et al.*, 2004). The wastewater treatment plant model should be furthermore distinguished from a plant-wide model, which combines wastewater treatment models and sludge treatment models.

1.2 Modeling basics

1.2.1 Model building

Many different types of models exist; these can be broadly categorized into (*i*) physical, (*ii*) verbal or conceptual and (*iii*) mathematical models. The physical model is a spatially-scaled representation of the system. For example, the laboratory- and pilot-scale experiments used by scientists and engineers to investigate system response and behaviour are physical models. The verbal or conceptual model provides a qualitative description of the system and is usually developed from detailed observations; these models can be presented as schematic diagrams (e.g. flow diagrams) or as a series of narrative statements. Preparation of a mechanistic (verbal) model is the most important but also the most complex part of model building. The mathematical model provides a quantitative description of the system. With mathematical models the rates of the processes acting in the system and their stoichiometric interaction with the compounds are formulated mathematically. The mathematical formulations need to be incorporated in a solving procedure that takes account of the physical constraints and characteristics imposed by the system in which the processes take place, e.g. temperature and mixing conditions. Mathematical models are seldom developed in isolation, but usually evolve interactively from a conceptual model that might to some degree be based on observations made in a physical model, e.g. laboratory- or pilot-scale experiments (Wentzel and Ekama, 1997).

Methodology in research which combines verbal, mathematical and physical models (Figure 1.2) is very helpful to rapidly progress and evaluate new systems. A number of factors should be considered regarding activated sludge modeling and simulation, and a step-wise approach is needed to progress from the model purpose definition to the point where a wastewater treatment plant model is available for simulations. The following main steps can be distinguished in this process (Coen *et al.*, 1996; Petersen *et al.*, 2002; Hulsbeek *et al.*, 2002):

- Definition of the model purpose or the objectives of the simulation study;
- Model selection: choice of the models needed to describe the different plant units to be considered in the simulation, i.e. selection of the activated sludge model, the sedimentation model, etc.;
- Hydraulics, i.e. determination of the hydraulic models for the plant or plant tanks;
- Wastewater and biomass characterization, including biomass sedimentation characteristics;
- Calibration of the activated sludge model parameters;
- Model verification;
- Scenario evaluations.

The methodology is illustrated in detail by Petersen *et al.* (2002).

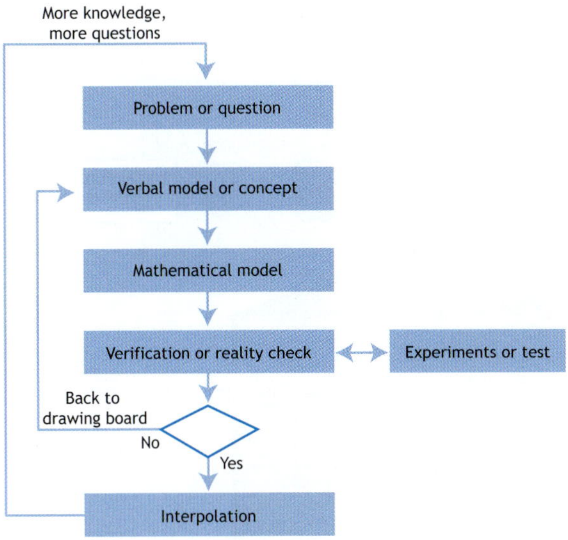

Figure 1.2 Model-building process.

1.2.2 General model set-up

Balance equations form the basis of any model description. These equations describe the change in concentration in a reactor over time as the result of chemical and biological conversions and of transport processes. In steady state the change of the concentration as a function of time becomes zero. Transport and conversion processes are two different parts of the model (of a physical and chemical-biological nature, respectively). The biological processes are only dependent on the concentration in a reactor at the place where the conversion takes place. In essence the conversion processes are therefore independent of the type of reactor or the size of the reactor (microorganisms do not know in which type of reactor they are: concrete or steel, plug-flow or fully-mixed, activated sludge or biofilm etc.). Therefore the biological and chemical conversions are called micro-kinetics and can be easily studied in the laboratory and will not change at a full-scale installation. This part of a process model is therefore universal and can be formulated as a general activated sludge model, like the ASM model family. The local concentrations in a reactor depend on the transport of the reacting compounds in the reactor system or treatment plant. When comparing full-scale systems, the difference lies effectively in these transport processes. The advantage of transport processes (such as convective flow, mixing, aeration) is that they have already been thoroughly studied and described by general rules. They can therefore be predicted relatively accurately for999 different types and scales of processes. One can study biology and chemistry in the laboratory (for instance the effect of temperature, concentration and pressure on microorganisms) and then use physical transport models to predict what is going to happen at full-scale. Acknowledging the fact that microbes will not undergo changes between laboratory and full-scale conditions, as opposed to transport processes, helps to understand the processes and their integration in the mathematical model. Such integration allows the models to be used in the process design (selection of bio-reactors, types, stability, optimization, automation and control, scale up etc.).

The components of a full wastewater treatment model are given schematically in Figure 1.3.

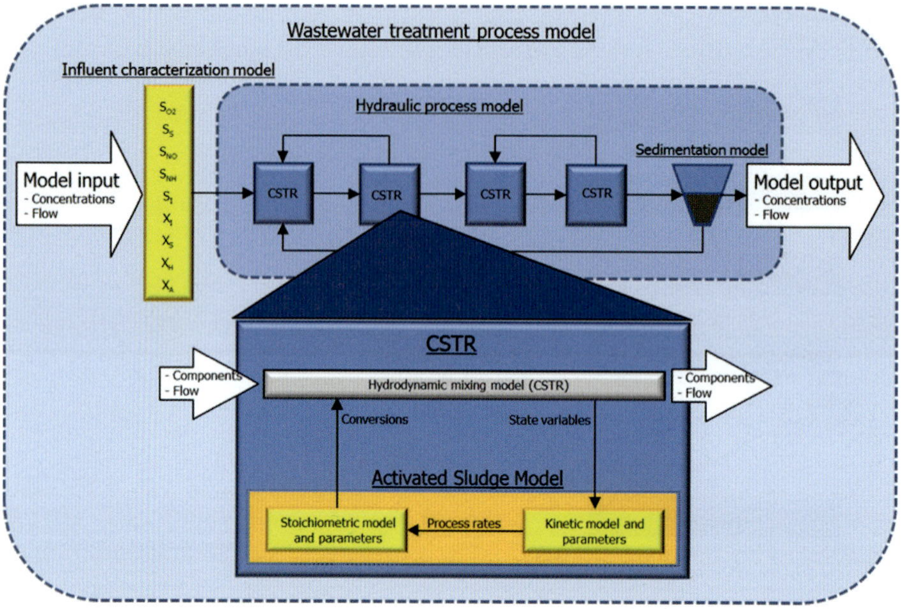

Figure 1.3 Schematic representation of a complete wastewater treatment plant model (Meijer, 2004).

First, measurable wastewater parameters have to be transformed to an influent vector with the concentrations of the different model compounds. The wastewater treatment plant is modelled hydraulically, describing the different zones/reactor compartments of the plant, including the settler. Each reactor compartment is modelled individually for its mixing and mass transfer (e.g. aeration) characteristics. Usually a completely mixed tank reactor is used. A mass balance equation is applied to each reactor. In such a mass balance equation the bioconversion model is included. In the overall model all the compartments are coupled by the state vector including the concentrations and flow rates of the links between the compartments. This overall model is usually solved numerically to show the concentrations of all compounds as a function of time for each compound included in the model. Therefore effectively we can speak of four models: the process model, the hydraulic model, the reactor/compartment model and finally the activated sludge model. Only this last model is general. The mass balance equation in steady state in mathematical terms reads:

$$\frac{\delta(S_{in}\cdot Q_{in})}{\delta t}=\frac{\delta(S_{out}\cdot Q_{out})}{\delta t}+\left(\alpha\cdot q\cdot X\cdot V\right)+\left(k_l A\cdot\left(S_{max}-S\right)\right) \quad (1.1)$$

In which:

α	stoichiometry	S_{max}	saturation concentration ($gCOD/m^3$)
A	surface area (m^2)	S	concentration in liquid ($gCOD/m^3$)
k_l	external mass transfer coefficient (m/h)	S_{in}	concentration in influent ($gCOD/m^3$)
q	specific conversion rate (1/h)	S_{out}	concentration in effluent ($gCOD/m^3$)
Q_{in}	influent flow (m^3/h)	t	time (h)
Q_{out}	effluent flow (m^3/h)	V	volume (m^3)
		X	biomass concentration ($gCOD/m^3$)

Effectively it states that a compound entering a reactor either leaves with the effluent is converted in the reactor or is exchanged with the gas phase in the compartment. Each term on balance has the dimension of mass over time. It is useful to realize that in order to analyse a complex system it is better to work in these dimensions than with concentration terms.

1.2.3 Stoichiometry

From the system definition one takes only those compounds of the system that are considered important and/or make a significant part of the total system mass (being at least a small percentage of it). For example, in the case of nitrification the nitrite concentration will remain very low or close to the detection limit in most plants, so from the perspective of mass balance there is no need to take nitrite into account. In anaerobic digestion, similarly, there is no need to take hydrogen into account, as the hydrogen content of gas is very low, as almost everything ends up as methane. Such intermediates will only be specified if considered important, for example, when there is nitrite or hydrogen accumulation. Nitrite is not included in the nitrification process in ASM1, while in the Anaerobic Digestion Model (ADM1, Batstone *et al.*, 2000) hydrogen is included as it plays an important role in the stability of the anaerobic system. ASM models are specifically designed for applications at lower temperatures (5 to 20°C), under which no significant accumulation of nitrite is expected to take place. Nitrite will only accumulate at higher temperatures or in the case of unusual toxic events. Thus nitrite is left out of the model. Similarly, in denitrification, only a small amount of nitrate turnover is in the form of N_2O, so from the perspective of describing the N removal it is not relevant to include the contribution of N_2O. However, if the plant has to fulfil N_2O gas limits, then it becomes important to take it into account. Again, it depends on the purpose of the application of the model. Besides the determination of relevant compounds and processes, defining relevant balances is essential. For each conserved balance the number of atoms of a compound entering the plant is equal to those leaving. Examples of conserved balances are nitrogen, phosphorus, COD or alkalinity conversions. Using balance equations, unknown stoichiometry coefficients can be calculated. This substantially reduces the information required for modeling as the approach allows a number of unknown values to be calculated. The use of BOD measurement as the characteristic of wastewater is declining and, instead, modern approaches rely on COD. BOD-based design is associated with a black-box approach, which cannot be used for balancing as it is not conserved, and depends on many factors (e.g. reaction time, temperature). In reality, it is still mostly used to link the output of ASM regarding the effluent impact on the receiving waters (where BOD is still a relevant indicator of water quality). In contrast, the COD balance is conserved because COD is by definition the amount of electrons which are transferred to oxygen in order to oxidize all the organic matter in the system to CO_2 and water. That is why the modeling is based on COD rather than on BOD.

Stoichiometry can be determined based on relevant compounds involved in the reaction and use of balances to calculate those relevant coefficients. For example, in the reaction for heterotrophic growth, the relevant compounds are organic matter, oxygen, ammonia, alkalinity, biomass, carbon dioxide and water. At this stage of equation development it is not necessary to determine which compound is utilized and which produced as it will simply receive a negative or positive sign, in other words, it is not important on which side of the equation the parameter is. The next step is to choose one coefficient as 1 and to use balances to calculate all the other (relevant) coefficients. In the example below, 5 balance equations can be made (for carbon, oxygen, hydrogen, nitrogen and charge) and there are 7 unknown coefficients. One of these coefficients can be made equal to 1 and there is only one coefficient that we need to know, for example the amount of oxygen consumed per COD converted or the amount of biomass produced per substrate COD mass utilized (yield coefficients).

This is a general approach for setting up the stoichiometry of the reaction for any biological process e.g. if organics (or COD) are utilized aerobically (with O_2).

$$?COD + ?O_2 + ?NH_4^+ + ?HCO_3^- \rightarrow ?Biomass\text{-}COD + ?CO_2 + ?H_2O \quad (1.2)$$

In wastewater treatment systems we are usually not interested in the CO_2 and H_2O conversion and the COD balance is used to replace one of the elemental balances. If one sets the coefficient for substrate-COD as 1, and if the yield coefficient for the biomass is known, Eq.3.2 becomes:

$$1COD + (1-Y_H)O_2 + f_N Y_H NH_4^+ + f_N Y_H HCO_3^- \rightarrow Y_H\, Biomass\text{-}COD \quad (1.3)$$

where:

Y_H heterotrophic yield (gBiomass-COD /gSubstrate COD), and

f_N fraction of nitrogen in biomass (gN/gBiomass-COD)

To derive this equation we effectively used the COD, nitrogen and charge balance. The COD balance states that oxygen consumption and biomass production are always linked; it is impossible to save oxygen and produce less sludge as the substrate (COD) is either oxidized by oxygen or becomes sludge. From the N balance the amount of ammonia needed can be calculated, and from the charge balance the amount of bicarbonate (alkalinity) can be determined, etc. The stoichiometric reaction can be written down as a function of the yield coefficient and, in this particular example, the amount of nitrogen inside the biomass. The stoichiometric coefficients for each compound are included in the model matrix (Table 1.1).

1.2.4 Kinetics

Each reaction has its own rate equation. The rate equation specifies the rate of conversion of the compound with the stoichiometric yield coefficient of 1. The conversion rate of the other compounds follows from multiplying each yield coefficient with the rate equation. The model can be based either on substrate-based kinetics (the substrate stoichiometric coefficient is equal to 1) or growth-based kinetics (the biomass stoichiometric coefficient is equal to 1). It is not advisable to use both at the same time in one model. In ASM1 the rates are described based on the growth rate; the biomass coefficient is therefore set as 1. In ASM, a saturation equation is used as the standard rate equation. Saturation (Monod) kinetics includes two main parameters, the maximum rate parameter and affinity or saturation constant (K value, defined as the concentration at half the maximum rate). The saturation term S/(K+S) can have a value between 0 and 1, and can have a different function in the model. Several affinity terms reflect a real value, e.g. the oxygen affinity term is an observed parameter. However, in some cases the saturation term is only a switching term. For example, a switching function is used in the model to stop the growth process when there is no ammonia present (Eq. 1.4). The affinity constant for ammonia is effectively very low and hardly measurable, so the sole purpose of the coefficient placed in the equation is to guarantee there is no growth anymore if the ammonia is fully consumed. This consequently means that one does not need to calibrate this value. How to distinguish between real measurable parameters and switching functions is a bit vague and inexplicit in activated sludge modeling. Therefore it is important to realize whether the K values are there as real model parameters or as a switching function to stop the process when the relevant compound is not present.

$$\mu = \mu^{max} \frac{S}{K_S + S} \cdot \frac{S_O}{K_O + S_O} \cdot \frac{S_N}{K_N + S_N} \cdots \quad (1.4)$$

To describe inhibition kinetics, a similar approach is applied, but now the affinity constant is called the inhibition constant, and consequently it is possible to define an inhibition term (eq. 1.5), which again has a value between 0 and 1. The inhibition constant is equal to the substrate concentration at which a 50% decrease in the rate is observed. There are also much more complex inhibition terms, but in ASM this is the usually applied term, especially for the substrate inhibition.

$$1 - \frac{S}{K_S + S} = \frac{K_S}{K_S + S} \tag{1.5}$$

It is important to note that multiplying so many factors causes deviation because these factors are never exactly 1. If one multiplies the two factors with the value of 0.9 with the value of the third factor that is 0.5, the result will be 0.4, while the real value should be 0.5 because this is the limiting factor. This consequently means a 20% lower rate value. Therefore it is better to use a logical operator in the model and choose the minimum factor among the terms (eq. 1.7) instead of multiplying these factors (eq. 1.6) as it seems that it gives a better approximation of the reality.

$$\mu = \mu^{max} \cdot \frac{S}{K_S + S} \cdot \frac{S_O}{K_O + S_O} \cdot \frac{S_{NH}}{K_{NH} + S_{NH}} \cdot \frac{S_{KI}}{K_I + S_{KI}} \tag{1.6}$$

$$\mu = \mu^{max} \cdot MIN\left(\frac{S}{K_S + S}; \frac{S_O}{K_O + S_O}; \frac{S_{NH}}{K_{NH} + S_{NH}}; \frac{S_{KI}}{K_I + S_{KI}} \right) \tag{1.7}$$

The reason that Eq. 1.6 is used is partly an inherited habit (at the time of early model development in the 1970s, computing logical operators by integral differential equations was difficult and extremely time-consuming and thus not applied). It does not matter so much which equation is used for activated sludge modeling; the point is to understand the reasons for the choices at different stages of the model development.

1.2.5 Transport

A typical wastewater treatment model has several transport terms, often being time-dependent (Figure 1.3). The model input is the time variable flow and composition of the wastewater. The process is described in a hydraulic model, representing the hydraulics of the full-scale plant. An example is given in Figure 1.4.

Within each zone one then has to observe whether e.g. in an aerated tank a gradient exists in the oxygen concentration. As long as the oxygen concentration is always well above the saturation coefficient for oxygen used in the kinetic equations, there is no direct need to describe the changes in concentration in the aerobic zone, and the tank can be considered fully mixed. If the concentration of the reacting compounds becomes close to or lower than the saturation constants, the hydraulic model should be such that the change in concentrations is well described. In general, this means using a plug flow model or describing the system as a series of tanks. In the case that the observed concentration of e.g. ammonia in the aeration tank is around 4 mg/L across the tank, it can be considered as a fully mixed tank and be represented in the hydraulic scheme of the plant as a single reactor. However, if the observed ammonia concentration changes from 4 mg/L at the inlet to the aeration tank to 0 mg/L at the outlet, this indicates a strong concentration gradient within the tank, and consequently it is much better to model it as a multi-compartmental tank with a number of fully-mixed smaller reactors in series.

A second aspect taken into consideration is the transfer of compounds between gas and liquid (e.g. oxygen) in aerated reactors or between biofilm and liquid.

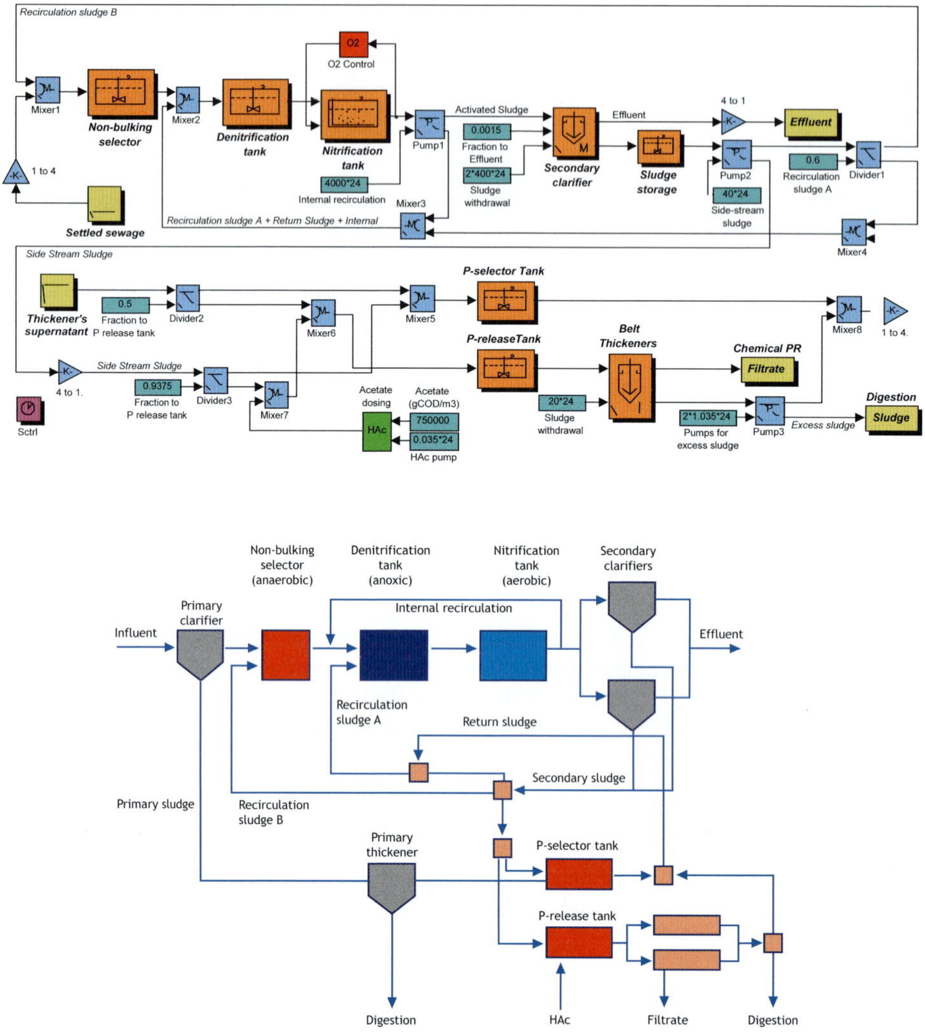

Figure 1.4 Hydraulic process scheme of the PhoStrip® process at WWTP Haarlem Waarderpolder in the Netherlands and its representation in the modeling simulator SIMBA (Brdjanovic *et al.*, 2000).

The main problem is associated with making the hydraulic model of a wastewater treatment plant. A rigorous solution would be to make a full computational fluid dynamics model of the plant, which exactly describes the flow in the reactors. However, in general the details obtained in this way for the flow are far too large for most conversion models. Since we are mainly interested in the bioconversion we need to adequately describe changes in concentrations in the treatment plant. Measuring several relevant compounds can help to define the hydraulic model. For activated sludge models these compounds are in general oxygen, ammonium, and nitrate and, for phosphate-removing systems, phosphate. As a first step a clear division can be made between aerobic and anoxic or anaerobic zones in a treatment plant.

1.2.6 The matrix notation

The balance equation (Eq. 1.1) can be described for each individual compound. The large number of relevant compounds and conversions make activated sludge modeling complex. A large number of balance equations need to be formulated resulting in a loss of overview. The IAWQ Task Group (Henze *et al.*, 1987) on 'Mathematical Modeling of Wastewater Treatment' has recommended the matrix method for model presentation. This format facilitates a clear and unambiguous presentation of the compounds and processes and their interaction on a single page. In addition, the matrix format allows an easy comparison of different models, and facilitates transforming the model into a computer program. The matrix is represented by a number of columns and rows; one column for each compound and one row for each process. A simplified example is given in Table 1.1.

Table 1.1 Example of a simple stoichiometric matrix for activated sludge modeling (Henze *et al.*, 1987).

Components i	1: S_O	2: S_S	3: X_H	Process rate equation ρ_j
List of processes j				
Aerobic growth	$-\frac{1}{Y_H}+1$	$-\frac{1}{Y_H}$	+1	$\mu_H^{max} \cdot \frac{S_S}{K_S+S_S} \cdot X_H$
Lysis		+1	-1	$b_H \cdot X_H$
Observed transformation rate r_i	$r_i=\sum_j \nu_{j,i} \cdot \rho_j \; [M_iL^{-3}T^{-1}]$			
Definition of stoichiometric parameters: Y_H Heterotrophic yield coefficient $[M_H M_S^{-1}]$	Dissolved oxygen (O_2)	Dissolved organic substrate (COD)	Heterotrophic biomass (COD)	Definition of kinetic parameters: μ_H^{max} Maximum specific growth rate $[T^{-1}]$ K_S Saturation coefficient for substrate $[M_{COD}L^{-3}]$ b_H rate constant for decay $[T^{-1}]$

The first step in setting up the matrix is to identify the compounds that are of relevance to the model. The compounds are presented as symbols listed at the head of the appropriate column including a row with the dimensions.

The second step in setting up a matrix is to identify the biological processes occurring in the system. These are conversions or transformations that affect the compounds considered in the model and are itemized one below the other down the left-hand side of the matrix. The process rates are formulated mathematically and are listed on the right-hand side of the stoichiometry matrix in line with the respective process. The stoichiometric coefficient for conversion from one compound to another is inserted along each process row so that each compound column lists the stoichiometric coefficients for the processes that influence that compound. If the stoichiometric coefficient equals zero it is generally not shown in the printed matrix to maintain clarity. The sign convention used in the matrix for each compound *per se* is 'negative for consumption' and 'positive for production'. In this convention the process rates always have a positive sign. Note that oxygen is negative COD because oxygen accepts electrons: electrons are passed from the substrate to oxygen to form water. Care must be taken of the units used in the rate equations for the processes. The stoichiometric coefficients are greatly simplified by working in consistent units.

In the example presented in Table 1.1 the compounds are expressed as COD equivalents. Provided that consistent units are used, continuity can be checked from the stoichiometric parameters by moving across any row of the matrix; the sum of the stoichiometric coefficients must be zero. This matrix forms a succinct summary of the complex interactions between compounds and processes. It allows alterations in processes, compounds, stoichiometry and kinetics to be readily incorporated. The matrix shows two important process aspects: the reaction equation for each process is represented in the different rows. In the columns of each compound one directly observes in which conversions the compound is involved. By multiplying the stoichiometric factors with their respective rate equations one gets the total conversion equation for each compound. For convenience two extra aspects can be added to the matrix description (Table 1.2).

Table 1.2 Example of a stoichiometric matrix for activated sludge modeling (adapted from Gujer and Larsen, 1995).

Component	Oxygen	Inert	Substrate	Ammonia	Alkalinity	Biomass	Inert	Substrate	TSS	Rate
Symbol	S_O	S_I	S_S	S_{NH}	S_{HCO}	X_H	X_I	X_S	X_{TSS}	
Unit	gO_2	gCOD	gCOD	gN	mole	gCOD	gCOD	gCOD	gTSS	
Process	STOCIHIOMETRY MATRIX									
Hydrolysis			1					-1	-0.75	r_1
Aerobic growth	-0.5		-1.5	-0.08	-0.005714	1			0.9	r_2
Lysis				0.07	0.005	-1	0.2	0.8	-0.12	r_3
Conservatives	COMPOSITION MATRIX									
ThOD-COD	-1	1	1	0		1	1	1		
N		0.02		1		0.08	0.05	0		
Charge				0.071429	-1					
Observables										
TSS						0.9	0.9	0.75		

The first aspect is a matrix with the composition in terms of conserved balances, in this case the COD, N and charge balance. Biomass is expressed in the stoichiometric matrix in terms of COD, but it also contains nitrogen. In the composition matrix this is included. Since the composition matrix and the stoichiometry matrix contain all the conserved balances, therefore multiplication of the two matrices leads to zero. Secondly we are generally interested not only in the compounds expressed in the dimensions as used in the model, but also in their measured or observed units. For instance the amount of sludge is usually measured as gTSS and not gCOD. The observed matrix contains these conversion numbers between e.g. gCOD and gTSS. Other potentially interesting observed quantities are Kjeldahl nitrogen, VSS or BOD.

1.3 The stepwise development of biokinetic model: ASM 1

Model development is a step-wise, bottom-up process in which only the strictly necessary processes for the pre-defined purpose of modeling are included. Starting simple and increasing complexity when needed is the general governing principle in model development. In general, activated sludge models from the ASM family are developed to describe the oxygen uptake rate and sludge production (coupled with COD balance), and N and P conversions at domestic wastewater treatment plants. However, despite the fact that they are designed for practical (and therefore not academic) purposes, they are not sanitation models as they do not describe the removal of pathogens. Probably the best way to describe the stepwise activated sludge model development is the original approach of Ekama and Marais (1978), later depicted by Dold *et al.* (1980), and further elaborated on in Gujer and Henze (1991). The outcome of this approach is the model which comes close to ASM1 and as such is briefly described here. The experimental system used in this approach comprised a completely mixed activated sludge system using settled domestic wastewater, and basic influent and sludge characterization and operational conditions are listed in Table 1.3.

Table 1.3 Experimental system summary data according to Ekama and Marais (1978).

Parameter	Value
Feeding regime	12 h/d on between 02 and 14 hrs
Flow	18 L/d
Reactor volume	6.73 L
Processes	COD removal and nitrification (fully aerobic)
Biomass content in the reactor	1,375 mgVSS/L or 2,090 mgCOD/L
Sludge retention time (SRT)	2.5 d
Operating temperature	20.4°C
Influent COD concentration	570 mgCOD/L
Influent TKN concentration	46.8 mgN/L

The objective of the study was to use the model to correctly describe the biomass content in the system, oxygen uptake by the biomass and nitrogen conversion. To begin with, one can use a very simple model consisting of only three relevant components (dissolved oxygen S_O, dissolved organic substrate S_S, and heterotrophic biomass X_H) and two relevant conversion processes (aerobic biomass growth and lysis). With an increase in SRT, the biomass (live organisms) as a fraction of the sludge mass (VSS) in the system decreases. To describe this, lysis process or death regeneration was used i.e. disintegration of death cells resulting in generation of soluble biodegradable substrate available for generation of new biomass. Lysis of heterotrophic biomass here summarizes all the processes which lead to a loss of biomass (decay, lysis, endogenous respiration, predation etc). Maintenance or endogenous decay could have also been used here to describe the reduction in biomass. For the aerobic growth process all three components are relevant; dissolved oxygen and organic substrate are utilized by the biomass under aerobic conditions (thus negative coefficient) to produce biomass (positive coefficient). In general, a matrix can be simplified if one can choose to freely assign one stoichiometric coefficient to each process with a value of +1 or -1. The choice of the yield coefficient Y_H (0.67 gCOD/gCOD) together with the COD conservation equation is sufficient to determine all the stoichiometric coefficients for aerobic growth (Figure 1.5).

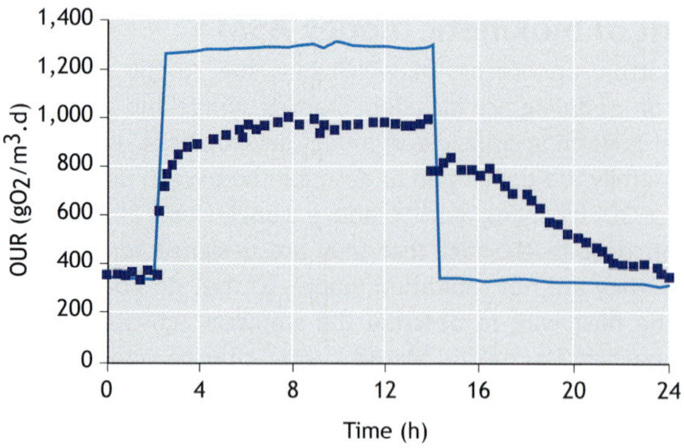

Figure 1.5 Model A: comparison of experimentally observed values (data points) with theoretically predicted oxygen uptake rate (continuous line) (adapted from Ekama and Marais, 1978; Gujer and Henze, 1991).

For both processes one can define the rate; for aerobic growth it is a product of maximum specific growth rate, the affinity of substrate and the biomass concentration (assuming that the oxygen is not limiting the growth). For lysis, it is a 1st order process in which biomass falls apart in proportion to the biomass concentration present and the constant of proportionality is called the rate constant for decay. Substituting the coefficients in the biokinetic model gives the matrix for model A (Table 1.4).

Table 1.4 Matrix presentation of the model A.

Component	S_O	S_S	X_H	Rate
Growth	*-0.5*	*-1.5*	*1.0*	$\mu_H^{max} \cdot \frac{S_S}{K_S+S_S} \cdot \frac{S_O}{K_{O,H}+S_O} \cdot X_H$
Lysis		*+1.0*	*-1.0*	$b_H \cdot X_H$

If this model is used to compare the experimentally observed oxygen uptake rate (OUR), it can be seen that experimental observations strongly deviate from the model predictions, except for the period between 0 and 2 hours and at the very end of the experiment (endogenous respiration). In general, if the model predictions are deviating considering the levels of parameter of interest, it can be relatively straightforward to adjust by changing the value of the selected model parameter(s). However, if the model predictions in terms of trends and shapes are wrong, very likely the relevant process or processes are overlooked. In this particular experiment, the difference between the total oxygen consumption observed over 24 hrs and the one predicted by the model seems to be quite close. However, it was the deviation between the model prediction and experimental results that led Dold *et al.* (1980) to suggest splitting the degradation of organic matter in wastewater into two processes (fractions): the relatively quick process of biodegradation of part of COD comprising of organics such as VFAs and glucose, and the relatively slow process of COD degradation (cellulose, starch, proteins etc). This fractionation of biodegradable COD to readily and slowly biodegradable COD (RBCOD and SBCOD, respectively) was triggered by the experimental observation of the OUR profile which showed a very sharp drop almost immediately after feeding stopped (14 hrs), followed by a slow decrease observed until the end of the experiment where it reached the steady value observed during the first two hours of the test. Therefore, the lysis process was reasonably well described by the model. It was furthermore concluded that the SBCOD is converted into RBCOD by the relatively slow process of hydrolysis (Figure 1.5). This implies that there is a need to extend model A by introducing two types of substrate (RBCOD and SBCOD) and one additional process (hydrolysis). In ASM1 it was also assumed that slowly biodegradable substrate consists fully of particulate substrate (X_S), which is not necessarily true, but in ASM1 it is accepted as such. A distinction between soluble (S) and particulate (X) material is necessary in order to determine which compound will settle in the clarifier and which will leave the system with the effluent. The introduction of a SBCOD fraction (X_S) did not affect the heterotrophic growth process as it is assumed that growth is not directly based on SBCOD. Also, the lysis processes were adjusted by the assumption that the lysis products are slowly biodegradable and, as such, are added to the pool of X_S. These products are made available for aerobic heterotrophic growth by hydrolysis. This means that there are two types of particulate substrate: one derived from the influent wastewater, and the second generated by the biomass decay. In some cases these are lumped together (as in this case), while in some models they are taken into account separately. However, both options result in practically no net difference. Furthermore, hydrolysable material X_S is assumed to adsorb onto heterotrophic biomass X_H resulting into a kind of Lagrangian kinetic expression as the rate equation for hydrolysis. So it is the amount of substrate per biomass which is important here (rate limiting) and not the substrate concentration with respect to the bulk liquid as in the case for the RBCOD. By implementing this, Model B was formed (Table 1.5).

Table 1.5 Matrix presentation of Model B

Component	S_O	S_S	X_H	X_S	Rate
Growth	*-0.5*	*-1.5*	*1.0*		$\mu_H^{max} \cdot \frac{S_S}{K_S+S_S} \cdot \frac{S_O}{K_{O,H}+S_O} \cdot X_H$
Lysis			*-1.0*	*+1.0*	$b_H \cdot X_H$
Hydrolysis		*+1.0*		*-1.0*	$k_H \cdot \frac{(X_S/X_H)}{K_x+(X_S/X_H)} \cdot X_H$

Model B describes the OUR competently (Figure 1.6). However, the predicted activated sludge concentration was 22% less than the measured concentration. This indicated the need to increase the sludge production by introduction of an influent non-biodegradable fraction of COD (being inert and also particulate organic matter which accumulates in the reactor: X_I). The term 'non-biodegradable' is in the context of wastewater treatment used for the compound which is not degraded by microorganisms during its retention in the treatment system.

Materials such as plastics, wood-based and fibrous materials, nails, and hair are all organic and strictly considered biodegradable, but not in wastewater treatment systems. Even a compound such as cellulose is considered non-biodegradable in high-loaded wastewater treatment plants but biodegradable in low-loaded systems. Besides inert particulate material derived from the influent wastewater, there is also the second component generated by the biomass decay. The latter arises from the fact that cell walls are very slowly biodegradable COD which is considered non-biodegradable, resulting in the experimentally-determined division of lysis process products of 92% being X_s (biodegradable) and 8% X_I (non-biodegradable). Consequently the rates in Model B are not changed given the fact that the OUR profile was described well. Inclusion of X_I resulted in the new Model C shown in Table 1.6.

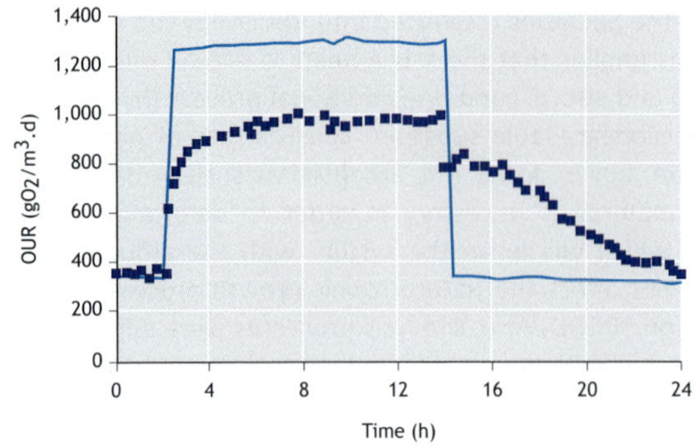

Figure 1.6 Model B: comparison of experimentally observed values (data points) with theoretically predicted oxygen uptake rate (continuous line) (adapted from Ekama and Marais 1978; Gujer and Henze 1991).

Table 1.6 Matrix presentation of Model C.

Component	S_O	S_S	X_H	X_S	X_I	Rate
Growth	*-0.5*	*-1.5*	*1.0*			$\mu_H^{max} \cdot \frac{S_S}{K_S+S_S} \cdot \frac{S_O}{K_{O,H}+S_O} \cdot X_H$
Lysis			*-1.0*	*+0.92*	*+0.8*	$b_H \cdot X_H$
Hydrolysis		*+1.0*		*-1.0*		$k_H \cdot \frac{(X_S/X_H)}{K_x+(X_S/X_H)} \cdot X_H$

The observed biomass concentration in the reactor was predicted well by Model C, however, due to higher sludge (COD) production/removal from the system, oxygen consumption was significantly underestimated by the model (despite the fact that the general OUR profile was well matched, Figure 1.7).

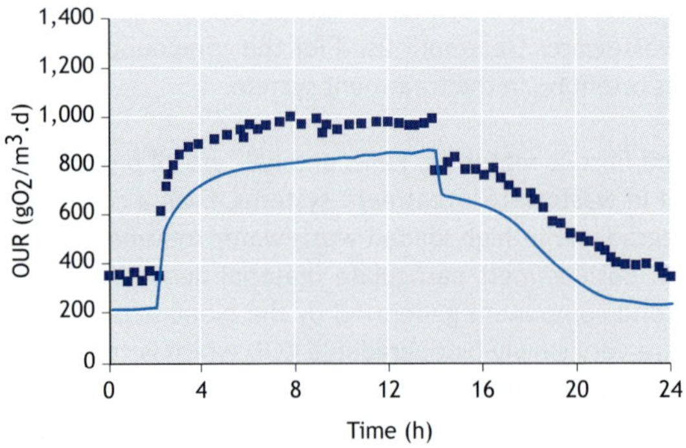

Figure 1.7 Model C: comparison of experimentally observed values (points) with theoretically predicted oxygen uptake rate (continuous line) (adapted from Ekama and Marais, 1978; Gujer and Henze, 1991).

This is to be expected as oxygen consumption and sludge production are linked through the COD balance, and consequently, an increase in sludge production will cause a decrease in oxygen demand. It was not possible to correctly predict both oxygen consumption and sludge production using Model C. From the experimental observations (results not shown) it was clear that the effluent from the treatment plant contained nitrate which implied that nitrification should be included. By inclusion of nitrification the model had to be extended by addition of three materials and two processes, namely ammonium (NH_4^+, S_{NH}), nitrate (NO_3^-, S_{NO}) and nitrifying autotrophic biomass (X_A), and aerobic (nitrifier) growth and nitrifier lysis. Again, it was necessary to evaluate the influence of each of the additional materials on the existing reactions. Ammonia is not only used in nitrification, but also for cell growth, therefore it was necessary to add a stoichiometric factor for ammonia in the growth relation. If the biomass contains 8% nitrogen (0.08 mgN/mgCOD), then the factor becomes 0.08. Furthermore, it was assumed that in the lysis process nitrogen remains within the biomass. However, in the hydrolysis process the biomass SBCOD is degraded (e.g. proteins into amino-acids) resulting in a release of ammonia. This has as a consequence that, besides one unit of substrate, also 0.08 units of ammonia is produced. This also happens with the influent SBCOD – as the protein part of it is hydrolysed, ammonia is released – the proteins are measured in the influent as organic N, i.e. the difference between the TKN and FSA. So for nitrification, a certain amount of oxygen and ammonia are consumed, and nitrate and biomass are produced. The amount of ammonia consumed is not exactly the same as the amount of nitrate produced, due to dual use of ammonia, namely (*i*) for energy generation in the nitrification process, and (*ii*) as nitrogen source for the heterotrophic biomass growth. The difference between the ammonia consumed and the nitrate formed is 0.08, representing the nitrogen content of the heterotrophic biomass. The overall N balance in this case will fit ($4.25 = 4.17 + 0.08 \times 1.0$). Furthermore, the lysis process for the autotrophs was assumed to be the same as for the heterotrophs, by which particular substrate and a small amount of inert substrate is produced. The process rate for lysis for autotrophs is generated analogously to heterotrophs with saturation terms for ammonia and oxygen. By the inclusion of additional three materials and two processes, Model D was created which satisfied both the COD and N balance. By having such a model it was possible to split the total oxygen consumption into oxygen consumption for ammonium oxidation and oxygen consumption for COD degradation. This shows the added value of using the model by providing insight into where the oxygen is used and for which processes (so-called process analysis). The results of simulation of the OUR by Model D are presented in Figure 1.8.

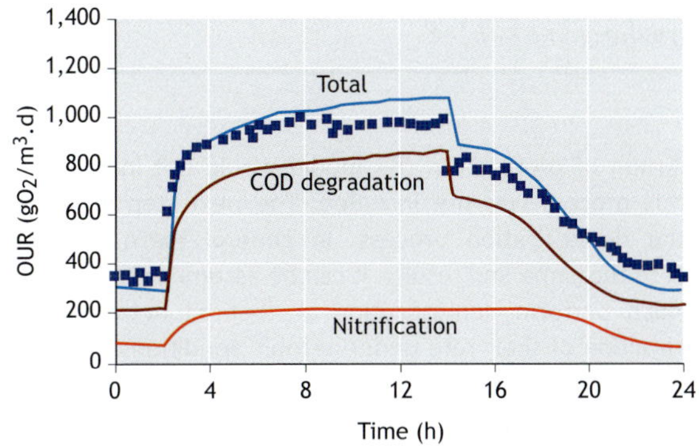

Figure 1.8 Model D: comparison of experimentally observed values (data points) with theoretically predicted total oxygen uptake rate (continuous line). Oxygen uptake for COD degradation and nitrification are separated out (adapted from Ekama and Marais, 1978; Gujer and Henze, 1991).

There were a few dilemmas at this stage of the model development such as 'is the match sufficiently accurate?' or 'is deviation, for example, in OUR of 5 to 10% at 14.00h acceptable?'. The answer to this depends entirely on the quality of the experimental data. If the COD and N balances of the data are exactly 100%, then it may be worth pursuing refinements in the model to get a better prediction because the data is reliable and accurate. If the COD and N balances are not 100% accurate but in the 95-105% range, then there is not much sense in making the model much more accurate. Making models is relatively easy – getting reliable and accurate mass balanced experimental data is the most difficult part of developing models for wastewater treatment systems. If the equipment results in an inaccuracy of 5-10%, it makes no sense to make the model more accurate. One should not forget that a model can only be successful if it fulfils the expectations the modeller has of it. If the purpose of the model is to describe correctly general trends, there is no need for further refinements. Of course, fine-tuning and calibration for a more accurate fit are possible, but this increases the model's complexity. Here, it was decided not to add extra compounds or processes, as the last part of calibration can be done by the straightforward shift of some model parameters. In general, the model simulations showed that the measured line and model line are fitting rather well: OUR, sludge production and nitrification (data not shown, see Dold *et al.*, 1980; Gujer and Henze, 1991) are predicted correctly. For the pre-defined goal, the model is considered correct, but it does not necessarily mean that the assumptions used are correct. Indeed, by using those assumptions, a mathematical description sufficiently appropriate for the purpose of its use is obtained. However, Model D omitted some processes which in reality play an important role, such as protozoa activity (Table 1.7).

Table 1.7 Matrix presentation of Model D.

Component	S_O	S_S	S_{NH}	S_{NO}	X_H	X_S	X_I	X_A	Rate
Growth	-0.5	-1.5	-0.08		+1.0				$\mu_H^{max} \cdot \frac{S_S}{K_S+S_S} \cdot \frac{S_O}{K_{O,H}+S_O} \cdot \frac{S_{NH}}{K_{N,H}+S_{NH}} \cdot X_H$
Lysis					-1.0	+0.92	+0.08		$b_H \cdot X_H$
Hydrolysis		+1.0	+0.08			-1.0			$k_H \cdot \frac{(X_S/X_H)}{K_x+(X_S/X_H)} \cdot X_H$
Autotrophic growth	-18.0		-4.25	+4.17				+1.0	$\mu_A^{max} \cdot \frac{S_O}{K_{O,A}+S_O} \cdot \frac{S_{NH}}{K_{N,A}+S_{NH}} \cdot X_A$
Autotrophic lysis						+0.92	+0.08	-1.0	$b_A \cdot X_A$

In this case, it was evaluated that inclusion of protozoa in the model would not increase its descriptive power and consequently this process was not included. The next step in model development was the inclusion of the denitrification process. In general there are two approaches possible that ultimately lead to the same end results. It can be assumed that there is either a special group of bacteria which perform denitrification or that all heterotrophic microorganisms can denitrify, but at a fraction of their rate under aerobic conditions. In other words, there is either a specialized population that can utilize both oxygen and nitrate, and another part of the population that can only use oxygen, or all heterotrophs can denitrify but at a reduced rate, corrected by the η factor (reduction factor for growth rate under anoxic conditions).

Conceptually, these are different assumptions but mathematically they come down to the same equation. Since the last assumption simplifies the model, it has been chosen to use the stoichiometry for denitrification and the bacteria that are ordinary heterotrophic organisms. Although the reality is probably much more complex, it was demonstrated that in reality this simplified approach works satisfactorily. Another important aspect concerns differentiation between fractions of inert material and nitrogen which is one of the items that differs from one commercial model to another. As mentioned earlier, inert material can originate from influent or from degradation of biomass, and the end content of the degraded biomass, inert material, might be different in composition to that of inert material in the influent. This can also be taken into account in the model; the inert material can be either separated or lumped together. In principle, it is not strictly necessary to define these fractions separately, but sometimes it is done based on aesthetic reasons or specific purpose of the model application. Similar reasoning applies for nitrogen fractionation. The model development as described by Ekama and Marais (1978) is still considered valid and Model D extended with denitrification becomes close to the ASM1 (Henze *et al.*, 1987). For further details on ASM1 the reader is referred to Dold *et al.*, 1980; van Haandel *et al.*, 1981; Alexander *et al.*, 1983; Warner *et al.*, 1986; Henze *et al.*, 1987, 2000. One of the most important limitations of the ASM is that it does not describe the sludge bulking phenomenon. Therefore, if the ASM is used, for example to improve the nitrification process, it is necessary to check if the proposed changes create bulking sludge. Limited aeration, which is beneficial for nitrogen removal, will almost inevitably induce bulking sludge. Sludge bulking itself cannot be modelled in a way that is sufficiently reliable for implementation in commercial software packages despite some attempts described in the literature (Krebs, 1995). This consequently means that the model cannot be accurately applied to predict very low effluent concentrations when highly efficient processes are implemented. Furthermore, the analogous consideration is valid for denitrification process as well. On the other hand, even if the model is able to predict the concentration of ammonia at 0.5 mg/L, there are always some inaccuracies and imperfections in the analytical procedures for determining ammonia concentration, as well as in the sampling procedure and sample handling. Another ASM limitation is that it does not take into account removal of micro-pollutants such as metals, xenobiotics or oestrogenic endocrine disrupting compounds. This is partially due to the required increase in the complexity of the model and partially due to lack of knowledge on microorganisms and biochemical reactions involved in the conversion of these compounds. In some cases, as in modeling of the wastewater treatment at oil refineries, it is necessary to predict the phenol reduction. To support denitrification, methanol is often added under anoxic conditions and its conversion needs to be included. And there are cases when, for example, one is interested in sulphite reduction. In all these cases a new specialized organism must be included in the model, as the biomass included in the ASM1 does not convert these micro-pollutants. Examples of such extensions can be found in the literature, and nowadays some commercial software packages include methanol utilization, for example. In the case of other COD compounds such as volatile fatty acids (VFAs), the ordinary organisms which remove COD from the wastewater will convert these and therefore the model does not have to be extended for that. Beyond the ASM1 level, the model can be extended to take into account oxygen transfer, pH and alkalinity, anaerobic digestion, chemical phosphorus removal and precipitation, additional units (like settlers etc), side stream processes, gas phase etc. Again, whether the model needs to be extended depends on the purpose of the model.

1.4 ASM3

In essence, ASM3 describes the same processes as ASM1, although ASM3 was introduced to correct the deficiencies of ASM1. This is partly based on the observations from OUR tests with activated sludge which revealed the fact that bacteria rapidly take-up readily biodegradable COD and store it as internal substrate which will then be converted slowly (conversion of readily biodegradable COD into slowly biodegradable COD). When the acetate (defined substrate) is added to the activated sludge the observed OUR suggests the presence of two substrates; a rapid and slow degradation of substrate associated with OUR can be observed (Figure 1.9).

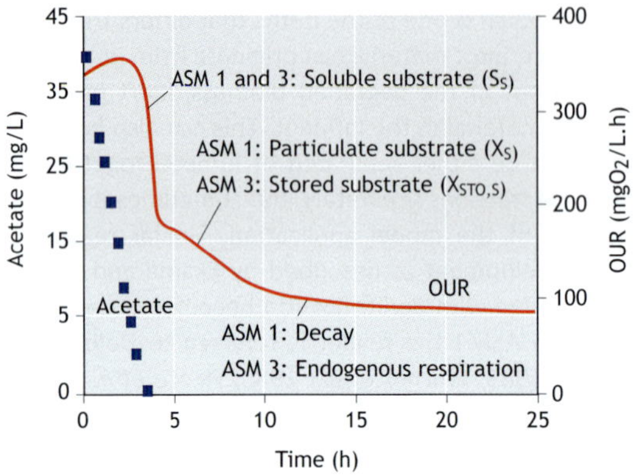

Figure 1.9 Differences between ASM1 and ASM3.

In ASM1 it appears as if two substrates are present (S_S and X_S) while in reality only acetate (S_S) is dosed. In order to describe the observed OUR by ASM1 in this case, it is necessary to define that the acetate is partly soluble and partly particulate, which is not recommended. This deficiency is solved by the introduction of a storage compound, $X_{STO,S}$, in ASM3. This means that substrate is taken up rapidly and stored, while growth occurs within the stored substrate. Both models will describe the observed OUR, but only ASM3 will accurately describe the uptake. However, there is no problem in using ASM1 for simulation of nitrogen removal systems because nitrification is a slow process, and thus enough time is available for biodegradation of slowly biodegradable COD.

The second reason to introduce ASM3 was that ASM1 proved to be rather successful for simulation of wastewater treatment plants and consequently too many started to believe that what was in ASM1 was 100% true and the reality. However, the storage mechanisms exhibited by the biomass show that what is in ASM1 is not all true, but close enough to reality to serve its purpose. Therefore, ASM3 has an added educational value because it demonstrates that there are different (but not necessarily better) ways to model the same treatment plant.

However, the most important reason to introduce ASM3 was the recognition of the importance of three rates of oxygen consumption in the process, namely: the rapid rate of oxygen consumption for degradation of RBCOD, the slow rate associated with degradation of SBCOD, and the even slower endogenous OUR (Figure 1.9). In contrast, in ASM1 there is only one oxygen-consuming process, so it is very difficult to perform calibration as one needs to calibrate other processes that indirectly influence the processes that consume oxygen.

The other problem is the cycling of the COD in the process, as in the decay process particulate COD is produced, hydrolyzed, and used for growth again. It means that if in the process one parameter is changed, it influences all the other processes due to the cycling, and it is difficult to use automated calibration as every parameter has influences on every process. In ASM3 this issue has been solved as the decay process has been replaced by endogenous respiration which eliminates the COD cycle (Figure 1.10).

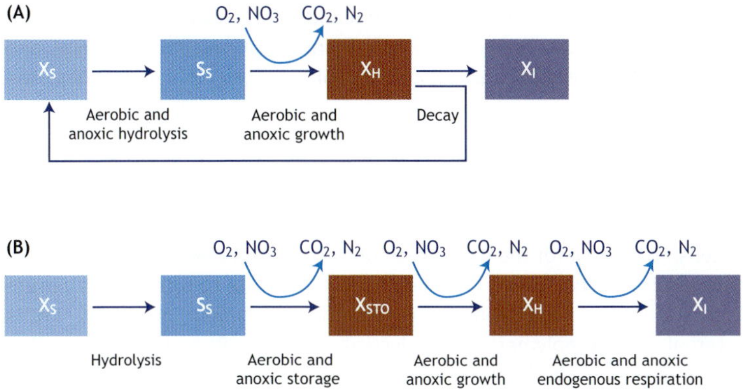

Figure 1.10 Degradation of COD in (A) ASM1 and (B) ASM3.

In other words, once the cells are produced, they start to oxidize themselves and by this means the biomass is reduced by the aerobic mineralization process (the classical endogenous respiration). While this has some conceptual controversy, e.g. why would an organism oxidize itself (i.e. go on a diet) when there is food around, it is useful to eliminate the bioprocess interaction from the substrate recycling of the death–regeneration model.

In addition, in ASM3 the oxygen consumption is divided into three processes (storage, growth and endogenous respiration) instead of having only one as in ASM1. ASM3 allows one of these three rates to be fitted if one knows which process to target, which directly links the measurements and calibration parameter. The fact that the $RBCOD_i$ is taken up and stored is irrelevant for most plants (and therefore also the choice between ASM1 and ASM3).

However, the only place where it really makes sense to use ASM3 is in plug flow reactors, such as selectors. If, for example, acetate must be removed in the aerobic selector to prevent sludge bulking, the design of the selector is governed by the time needed to take up the acetate and by the amount of oxygen needed for it. If ASM1 is used instead the oxygen requirements in the selector will be significantly overestimated. In reality a large proportion of acetate is stored inside the biomass, and once it is stored, there is no longer a problem with bulking sludge. If one wants to design the aerobic selector and include it in the model, then ASM3 is the best model to use.

The other preferred application of ASM3 is for the description of a pre-denitrifying nitrogen removal plant operating at a short SRT. Here, it makes a substantial difference whether or not readily or slowly biodegradable COD is present or whether COD is stored or not. In systems with a long SRT (10-20 days depending on temperature, which are more common in practice), a large part of the nitrate removal is effectively associated with the slowly biodegradable COD from the influent and death-regeneration in the pre-denitrification reactor and from death–regeneration only in the post denitrification reactor, so the sensitivity to the exact ratio between readily and slowly biodegradable COD is much less. The same applies for the differentiation between ASM1 and ASM3. In highly loaded systems endogenous respiration is less important and accumulation of COD in the form of storage polymers and the carry over in the aerated phase of a treatment plant might be significant.

In conclusion, ASM3 is recommended to be used for (*i*) simulation of highly loaded nitrification-denitrification systems with short anoxic retention times (volumes), (*ii*) selector modeling, and (*iii*) automatic calibration. Otherwise ASM1 should be equally successful in describing the activated sludge plant.

The consequence of introducing EBPR and Phosphorus Accumulating Organisms (PAO) into ASM is that the model becomes quite complex, as illustrated in Figure 1.14. The left side of the figure depicts the part of conversions carried out by nitrifiers and ordinary heterotrophs, while the right side shows the extension needed for the description of the complex physiology of PAO. The nitrifiers and ordinary heterotrophs use oxygen to oxidize their substrate to form CO_2 or nitrate and biomass. They have a rather simple physiology resulting in simple processes. PAO's physiology includes internal storage polymers (PHA, glycogen and poly-P) and their behaviour under anaerobic, anoxic and aerobic conditions is different. They also behave differently under aerobic conditions depending on whether the substrate is present or not. Obviously, there are lots of possible variations and inclusion of EBPR in the model substantially increases its complexity (the number of processes in ASM increases from 11 to 22). The situation becomes even more complex when Glycogen Accumulating Organisms (GAO) are also included. ASM2 and ASM2d are similar to ASM1 in assuming the cell to be a black box as opposed to using the metabolic approach to modeling which takes into account what is happening inside the cell.

1.5 The metabolic model

Why is it useful to use a metabolic model? In the standard model for heterotrophic growth there are seven relevant compounds (substrate, oxygen, charge, carbon dioxide, water, ammonia and biomass), five independent balances (carbon, hydrogen, oxygen, nitrogen and charge), and two degrees of freedom. If one knows one yield and one rate coefficient, it is possible to describe the whole system with one model. If one was to describe COD removal and nitrification at a metabolic level, it would not bring any advantage as the yield and rate coefficients would still be needed. Although the metabolic stoichiometry allows tracking, the C, H, O, N, P and charge flows through a system give more information from a modeling point of view, which makes the model more complex but not more accurate. All the rates are linked through conservation relations (stoichiometry) and, therefore, the choices between the process rate or growth rate, and substrate uptake rate or oxygen utilization rate, is not important (Figure 1.11).

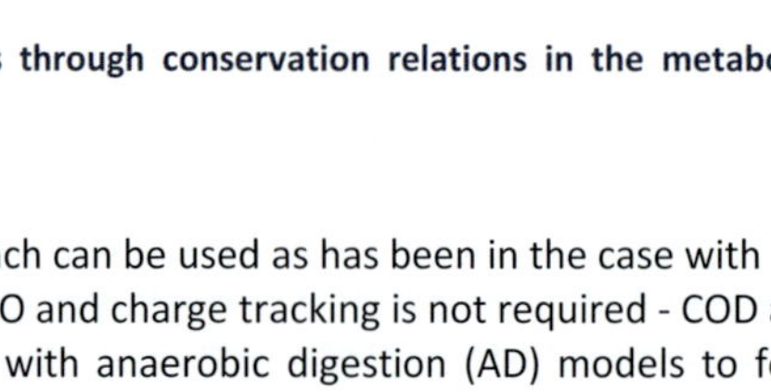

Figure 1.11 Linking of rates through conservation relations in the metabolic model for heterotrophic growth.

Thus the black box approach can be used as has been in the case with ASM1. So for the activated sludge system itself, C, H, O and charge tracking is not required - COD and N is enough, but when the ASMs are integrated with anaerobic digestion (AD) models to form plant-wide models, it becomes important because AD modeling requires C, H, O and charge tracking to predict gas production and composition and alkalinity generation (Brink *et al.*, 2007).

However, if one needs to describe the situation with heterotrophic growth and product (storage polymers) formation as for PAO in EBPR processes, the number of relevant compounds increases; each additional storage polymer brings an extra compound, but the amount of balances does not increase, which means that the degrees of freedom (unknown values) increase as a consequence of the increased number of unknown compounds. In this case one needs to know at least one yield and rate coefficient, and the choice of the process rate becomes important. For example, during aerobic conditions PAO use internally-stored PHA to produce the intermediate compound Acetyl-CoA, that is used further for biomass growth, glycogen formation and creation of energy required for these processes, and poly-P formation (Figure 1.12).

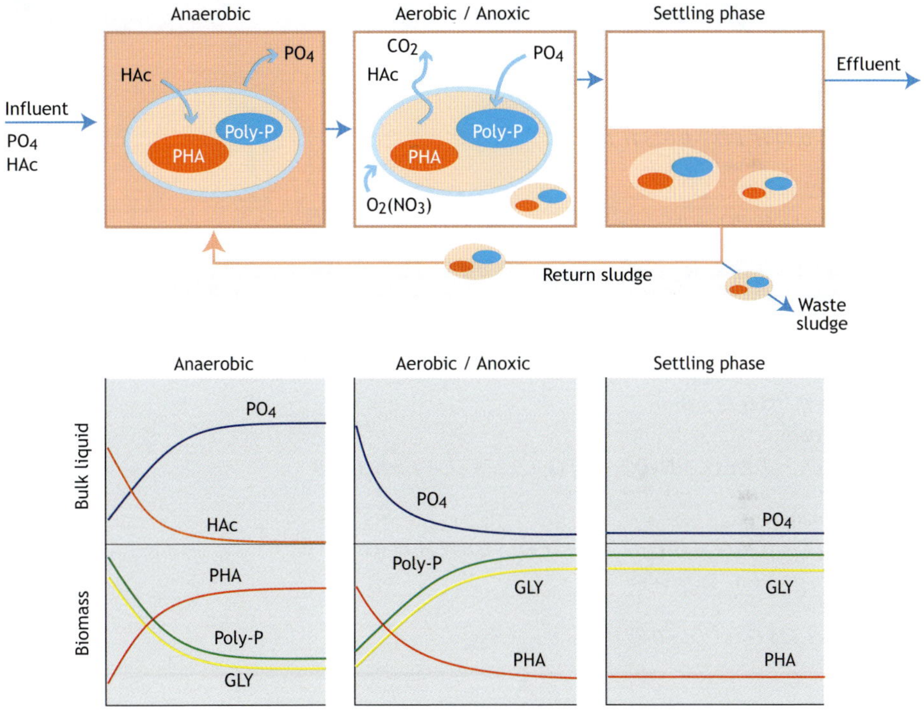

Figure 1.12 Schematic representation of the EBPR process. A typical EBPR process layout is shown with the typical concentration dynamics in the bulk liquid (top) and inside the biomass (bottom). EBPR depends on alternating anaerobic, anoxic and aerobic conditions. The process is cyclic, provided by the return activated sludge (RAS). Phosphorus (P) is removed from the system with the waste activated sludge (WAS). Provided by the storage polymers, PAO are capable of taking up substrate under anaerobic conditions, which is their competitive advantage in the bacterial suspension (Meijer, 2004).

Obviously, the introduction of storage compounds creates a more complicated network of processes. In the processes with extra storage polymers, extra yield coefficients will also be introduced. The efficiency of the conversion processes would however be the same for all the yields. Within a metabolic description one can link the macroscopic yields to the metabolic yield, which is the efficiency of energy (ATP) generation per unit of substrate oxidized. The substrate oxidation is related to electron transfer to oxygen or nitrate consumption. The yield coefficients are therefore all a function of this basic parameter (ATP produced per pair of electrons transferred) and the number of independent yield parameters is less in a metabolic description for these complex microorganisms.

In 1994, increasing knowledge of the cell-internal biochemistry of PAO resulted in the development of a metabolic model describing the anaerobic and aerobic phases of EBPR (Smolders *et al.*, 1994a,b; 1995a,b,c) (Table 1.8). The model was developed and validated using enriched PAO cultures cultivated on lab-scale anaerobic/aerobic (A/O) sequencing batch reactor (SBR) experiments.

Table 1.8 EBPR anaerobic metabolic reactions and derived stoichiometry.

Overall anaerobic stoichiometry (Smolders et al., 1994a), (in moles C and P)	
Acetate uptake	
$-\underset{\text{acetic acid}}{CH_2O} - 0.5\,\underset{\text{glycogen}}{CH_{10/6}O_{5/6}} - 0.44\,\underset{\text{poly P}}{HPO_3} + 1.33\,\underset{\text{PHA}}{CH_{1.5}O_{0.5}} +$ $+0.17\,CO_2 + 0.44\,\underset{\text{phosphate}}{H_3PO_4} - 0.023\,H_2O = 0$	(1.8)
Maintenance	
$-\underset{\text{poly P}}{HPO_3} - H_2O + \underset{\text{phosphate}}{H_3PO_4} = 0$	(1.9)
Overall aerobic stoichiometry (Smolders et al., 1994b), (in moles C, P and N)	
Biomass growth	
$-1.37\,\underset{\text{PHA}}{CH_{1.5}O_{0.5}} - 0.2\,NH_3 - 0.015\,\underset{\text{phosphate}}{H_3PO_4} - 0.42\,O_2 + \underset{\text{biomass}}{CH_{2.09}O_{0.54}N_{0.20}P_{0.015}} +$ $+0.37\,CO_2 - 0.305\,H_2O = 0$	(1.10)
Poly-P formation	
$-0.27\,\underset{\text{PHA}}{CH_{1.5}O_{0.5}} - 0.306\,O_2 - \underset{\text{phosphate}}{H_3PO_4} + \underset{\text{poly P}}{HPO_3} + 0.27\,CO_2 + 1.2\,H_2O = 0$	(1.11)
Glycogen formation	
$-1.12\,\underset{\text{PHA}}{CH_{1.5}O_{0.5}} - 0.26\,O_2 + \underset{\text{glycogen}}{CH_{10/6}O_{5/6}} + 0.12\,CO_2 + 0.007\,H_2O = 0$	(1.12)
Maintenance	
$-\underset{\text{PHA}}{CH_{1.5}O_{0.5}} - 1.125\,O_2 + CO_2 + 0.75\,H_2O = 0$	(1.13)

Initially, the model kinetics was chosen as simple as possible. Smolders *et al.* (1994b) proposed a kinetic structure in which the oxygen (or nitrate) consumption and PHA degradation are the net result of biomass growth (r_X), poly-P formation (r_{PP}), glycogen formation (r_{GLY}) and maintenance (m_O and m_{PHA}). This kinetic structure is expressed in the linear eq. 1.14 and 1.15, and led to the set of overall reactions previously shown in Table 1.8 (Eq. 1.8 to 1.13).

$$r_O = \frac{1}{Y_{X/O}} \cdot r_X + \frac{1}{Y_{PP/O}} \cdot r_{PP} + \frac{1}{Y_{GLY/O}} \cdot r_{GLY} + m_O \cdot C_X \tag{1.14}$$

$$r_{PHA} = \frac{1}{Y_{X/PHA}} \cdot r_X + \frac{1}{Y_{PP/PHA}} \cdot r_{PP} + \frac{1}{Y_{GLY/PHA}} \cdot r_{GLY} + m_{PHA} \cdot C_X \tag{1.15}$$

In the balances of eq. 1.14 and 1.15, C_X represents the PAO concentration. m_O and m_{PHA} are the maintenance rates. When expressed in COD, the maintenance rates relate according to $-m_O = m_{PHA}$. Yields are represented by $Y_{a/b}$, where 'a' is a product (namely either biomass 'X', poly-P 'PP', or glycogen 'GLY') from the aerobic utilization of 'b' as electron acceptor (oxygen 'O') and its correspondent electron donor (PHA) following from the metabolic reactions. Soon after, Kuba *et al.* (1996) proposed a metabolic model for denitrifying EBPR. The anoxic model was calibrated and validated on the basis of batch experiments and lab-scale anaerobic-anoxic (A2) SBR. Initially, the kinetic structure of this model was developed in accordance to Smolders *et al.* (1995a), but further suggestions were made to improve the model kinetics. In 1997, Murnleitner *et al.* combined the anaerobic, aerobic and anoxic models. Such an integrated model was tested on the basis of the original SBR experiments performed by Smolders *et al.* (1994a,b; 1995a,b) and Kuba *et al.* (1996). However, the tests showed that the kinetic structure proposed by Smolders *et al.* (Figure 1.13a) was not suitable to describe all SBR experiments at once only using one set of model parameter values. Therefore, Murnleitner *et al.* (1997) proposed a different kinetic structure, in which growth was the net result of PHA consumption and poly-P (PP) and glycogen formation (Figure 1.13b).

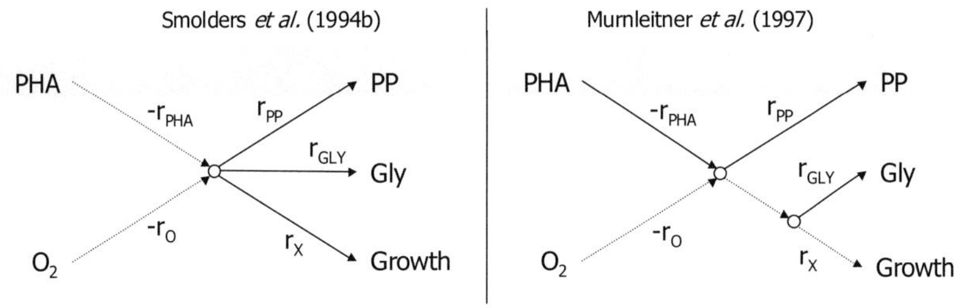

Figure 1.13a,b Two kinetic EBPR model concepts. Solid lines are modelled rates, dashed lines are rates resulting from the model. In the definition by Smolders *et al.* (1994b), growth is limited by the maximum growth rate. In the definition by Murnleitner *et al.* (1997), growth is the result of PHA degradation and storage of PP and glycogen.

To achieve a better description, Murnleitner *et al.* (1997) mathematically reformulated the model (A=B+C was reformulated as B=A-C) without changing its original stoichiometry (Eq. 1.16 and 1.17).

$$r_O = \frac{-1}{Y_{PHA/O}} \cdot r_{PHA} + \frac{1}{Y_{PP/O}} \cdot r_{PP} + \frac{1}{Y_{GLY/O}} \cdot r_{GLY} + m_O \cdot C_X \qquad (1.16)$$

$$r_X = Y_{X/PHA} \cdot r_{PHA} - \frac{Y_{X/PHA}}{Y_{PP/PHA}} \cdot r_{PP} - \frac{Y_{X/PHA}}{Y_{GLY/X}} \cdot r_{GLY} - Y_{X/PHA} \cdot m_{PHA} \cdot C_X \Rightarrow$$

$$r_X = \frac{1}{Y_{PHA/X}} \cdot r_{PHA} - \frac{1}{Y_{PP/X}} \cdot r_{PP} - \frac{1}{Y_{GLY/X}} \cdot r_{GLY} - m_X \cdot C_X \qquad (1.17)$$

Eq. 1.17 follows from eq. 1.14. Eq. 1.16 and 1.17 are derived from the columns of respectively oxygen (S_O) and PAO (X_{PAO}).

The kinetic structure proposed by Smolders *et al.* (1994b) defined the PHA degradation rate as a consequence of the PHA utilization rates for biomass growth, poly-P formation and glycogen formation (Figure 1.13a). Hereby, the growth of PAO was limited by the maximum growth rate. Alternatively, in the kinetic model proposed by Murnleitner *et al.* (1997), (Figure 1.13b), growth is determined by the net result of PHA degradation and poly-P and glycogen formation. From an ecological point of view, this structure, in which storage is preferred above growth, seems logical. In the competition with other micro-organisms, PAO rely on their storage ability. A rapid resupply of storage compounds is a primary condition for long term survival. In this formulation, the maximum growth rate is no longer an intrinsic property of PAO, but becomes dependent on environmental conditions and the maximum PHA storage capacity (Brdjanovic *et al.*, 1998). With the reformulated kinetic structure, Murnleitner *et al.* (1997) described all experiments performed by Smolders *et al.* (1994a,b; 1995a,b) and Kuba *et al.* (1996) with one set of model parameters. The proposed kinetic structure resulted in the overall model stoichiometry presented in Table 1.9 (Eq. 1.18 to 1.25). Nevertheless, one must underline that these reactions cannot be read separately, as they are merely the result of the mathematical reformulation of Eq. 1.16 and 1.17.

Table 1.9 Aerobic stoichiometry resulting from different kinetic structures. Eq. 1.18 to 1.21 result from the kinetic structure proposed by Smolders *et al.*, 1994b. Eq. 1.22 to 1.25 result from the improved kinetic structure proposed by Murnleitner *et al.*, 1997).

Overall aerobic stoichiometry (Smolders et al., 1994b), (in moles C, P and N)

Biomass growth

$$-1.37\,\underset{PHA}{CH_{1.5}O_{0.5}} - 0.2\,NH_3 - 0.015\,\underset{phosphate}{H_3PO_4} - 0.42\,O_2 + \underset{biomass}{CH_{2.09}O_{0.54}N_{0.20}P_{0.015}} + \\ +0.37\,CO_2 - 0.305\,H_2O = 0 \qquad (1.18)$$

Poly-P formation

$$-0.27\,\underset{PHA}{CH_{1.5}O_{0.5}} - 0.306\,O_2 - \underset{phosphate}{H_3PO_4} + \underset{poly\ P}{HPO_3} + 0.27\,CO_2 + 1.2\,H_2O = 0 \qquad (1.19)$$

Glycogen formation

$$-1.12\,\underset{PHA}{CH_{1.5}O_{0.5}} - 0.26\,O_2 + \underset{glycogen}{CH_{10/6}O_{5/6}} + 0.12\,CO_2 + 0.007\,H_2O = 0 \qquad (1.20)$$

Maintenance

$$-\underset{PHA}{CH_{1.5}O_{0.5}} - 1.125\,O_2 + CO_2 + 0.75\,H_2O = 0 \qquad (1.21)$$

Mathematically reformulated (Murnleitner et al., 1997), (in moles C, P and N)

PHA degradation

$$-\underset{PHA}{CH_{1.5}O_{0.5}} - 0.14\,NH_3 - 0.011\,\underset{phosphate}{H_3PO_4} - 0.32\,O_2 + 0.72\,\underset{biomass}{CH_{2.09}O_{0.54}N_{0.20}P_{0.015}} + \\ +0.28\,CO_2 - 0.23\,H_2O = 0 \qquad (1.22)$$

Poly-P formation

$$-0.19\,\underset{biomass}{CH_{2.09}O_{0.54}N_{0.20}P_{0.015}} - 0.218\,O_2 - 0.997\,\underset{phosphate}{H_3PO_4} + \underset{poly\ P}{HPO_3} + \\ +0.038\,NH_3 + 0.19\,CO_2 + 1.14\,H_2O = 0 \qquad (1.23)$$

Glycogen formation

$$-0.78\,\underset{biomass}{CH_{2.09}O_{0.54}N_{0.20}P_{0.015}} - 0.27\,H_2O - 0.22\,CO_2 + \underset{glycogen}{CH_{10/6}O_{5/6}} + 0.1\,O_2 + \\ +0.012\,\underset{phosphate}{H_3PO_4} + 0.156\,NH_3 = 0 \qquad (1.24)$$

Maintenance

$$-0.89\,\underset{biomass}{CH_{2.09}O_{0.54}N_{0.20}P_{0.015}} - O_2 + 0.013\,\underset{phosphate}{H_3PO_4} + 0.178\,NH_3 + 0.89\,CO_2 + 0.64\,H_2O = 0 \qquad (1.25)$$

From the metabolic reactions, an overall anaerobic, aerobic and anoxic stoichiometry was determined. A full description of the TUDP model is given by Meijer (2004) and de Kreuk *et al.* (2007). Overall, the formulation of an overall anaerobic reaction is unambiguous, as there is only one metabolic reaction. As such, by measuring the acetate uptake rate, all other rates are fixed. Concerning the aerobic and anoxic stoichiometry, five overall reactions (r_X, r_{PP}, r_{GLY}, r_{PHA} and m_{PHA}) are found but the system can be solved if four out of five rates are determined. In 1999, van Veldhuizen *et al.* integrated the metabolic EBPR model with the heterotrophic, hydrolytic and autotrophic processes from ASM2d (Henze *et al.*, 1999). With this model a full-scale MUCT process for COD, N and P removal was simulated (Chapter 2). That study showed that the TUDP model was capable of describing full-scale conditions, without significant adjustments. To strengthen the full-scale application of the model, a calibration protocol was developed and tested. Using the same model, Brdjanovic *et al.* (2000) simulated a full-scale side-stream P-removing process (Chapter 3). After calibrating glycogen formation, the model described the process without the need to further adjust other parameters. Since temperature plays a major role on microbial conversions, Brdjanovic *et al.* (1998) studied the effect of temperature on EBPR. Their results were incorporated in the TUDP model that was used to simulate a full-scale MUCT process optimised for denitrifying EBPR WWTP Hardenberg by Meijer *et al.* (2001) (Chapter 5). On the basis of all these practical experiments, the updated and validated metabolic TUDP model showed that its stoichiometry is fully reliable and can be used and extrapolated without calibration. The combined ASM2 – TUDP model is shown in details in Annex 1.1 (Meijer, 2004). To simulate full-scale EBPR, the metabolic model was combined with the heterotrophic, hydrolytic and autotrophic reactions from ASM2d (Henze *et al.*, 1999). Figure 1.14 shows how the different model structures interact.

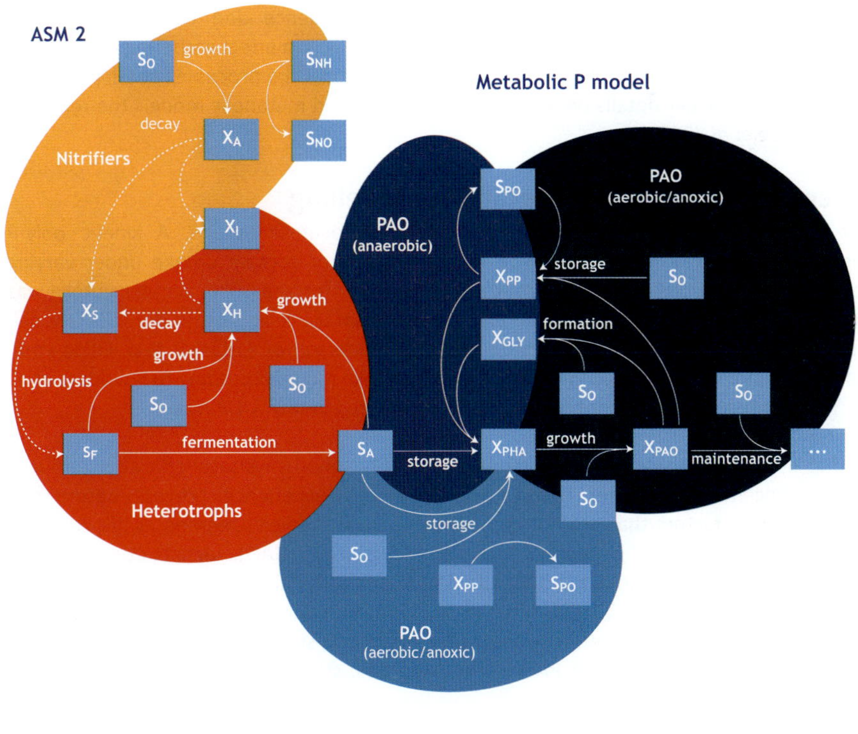

Figure 1.14 Interactions in the integrated TUDP model (Meijer, 2004).

Despite that it would be possible to reformulate the auto- and heterotrophic processes of ASM2d in a metabolic form, such a model would have the same number of yields of the original model. Therefore, it would not be smaller and, moreover, it would not improve the model performance. Therefore, in the TUDP model, the ASM2d processes were maintained in their original form and the integration of the two models was relative simple. Nevertheless, where the two models are merged a new form of substrate competition develops (e.g. between OHO and PAO). Moreover, with the EBPR also the fermentation and hydrolysis processes in the model become more sensitive and two concepts of endogenous respiration/maintenance are used simultaneously.

In the here presented metabolic EBPR model, the proposed kinetic structure results in a set of atypical model reactions (Eq. 1.22 to 1.25). These reactions are the mathematical result of the kinetic formulation (Eq. 1.14 and 1.15), and cannot be seen independently. For those not aware, this could easily lead to misinterpretations of the model matrix, as the individual stoichiometric reactions are not exemplarily for the actual EBPR process. This should be realised when the model is used for educational purposes. However, in the modelling practice, working with the metabolic concept has important advantages over other model approaches. The main advantage is the solid stoichiometric base of the metabolic model. This solid stoichiometric base is largely owed to the inclusion of glycogen and the simultaneous modelling of the counteracting dynamics of glycogen and PHA.

It is clear that when using metabolic information the degrees of freedom in the model can be reduced. Better understanding of the metabolic processes of the organism will close the gap to a fully glass box situation. The increased complexity of processes is consequently reflected in the models. However, improved understanding of the complex interactions within the cell and the introduction of the metabolic approach gives more confidence and consistency in the application of models to describe activated sludge processes. It is in effect gathering information from a lower level of organization to help understand and model the processes at a higher level of organization. For further details on ASM2, ASM2d, ASM3 and metabolic models the reader is referred to Henze *et al.* (2000).

1.6 Other developments on metabolic modelling

Filipe *et al.* (2001) improved the model for anaerobic acetate uptake. A kinetic poly-P dependency was included, which improved the description of acetate uptake under varying initial poly-P concentrations. Also a different pH dependency for anaerobic acetate uptake was suggested that becomes critical when anaerobic substrate uptake is limiting. In the TUDP model, anaerobic acetate uptake was modelled according to Smolders *et al.* (1994a). Also, Filipe *et al.* (1999) proposed improvements for the anoxic acetate uptake model according to Smolders *et al.* (1994a). These improvements were however not incorporated in the TUDP model.

Despite that the EBPR process can reach relatively high phosphorus removal efficiency (effluent phosphorus concentrations lower than 1 mg/L), it may experience process upsets and deterioration due to factors that are not completely understood yet (Oehmen *et al.*, 2007). In this regard, the appearance of glycogen accumulating organisms (GAO), such as *Competibacter* and *Defluviicoccus*, has been linked to the suboptimal operation and even failure of the EBPR process performance (Cech *et al.*, 1993; Satoh *et al.*, 1994; Saunders *et al.*, 2003). Thus, GAO are seen as undesirable microorganisms in wastewater treatment since they do not contribute to the EBPR process but compete with PAO in the anaerobic stage for the same carbon source (VFA).

In 2009, Lopez-Vazquez *et al.* incorporated the influence of carbon source (such acetate and propionate), temperature (from 10 to 30°C) and pH dependencies of PAO and GAO (from pH 6.0 to 7.5) in the metabolic model amended by Murnleitner *et al.* (1997). Thus, using a mechanistic model, Lopez-Vazquez *et al.* (2009) was able to evaluate the carbon source, pH and temperature influence on the PAO and GAO interaction and their effects on EBPR stability aiming at facilitating improved process efficiency and robustness. They concluded that PAO are favoured by temperatures lower than 20°C and pH levels higher than 7.0. Building up on the research carried out by Lopez-Vazquez *et al.* (2009), Oehmen *et al.* (2010) expanded the competition between PAO-GAO to sequential anaerobic-anoxic-aerobic conditions which are typically found in most of the biological nutrient removal systems. This implied to incorporate up to six different biomass groups consisting on *Accumulibacter Types I* and *II* and denitrifying and non-denitrifying *Competibacter* and *Defluvicoccus* in accordance to their observed denitrifying capabilities. Their model also included a multistep denitrifying process (from nitrate to di-nitrogen gas). Overall, the model of Oehmen et al. (2010) with minimum adjustments was able to successfully describe the EBPR biomass activities observed in lab-scale anaerobic-anoxic-aerobic SBR. Since the application to full-scale conditions of the metabolic model developed by Oehmen *et al.* (2010) is not straight-forward (e.g. due to the absence of organic matter and nitrogen oxidation processes as well as practical limitations concerning the elemental balances), recent attempts have been made towards the development of a more friendly-user ASM type model (Ramirez-Higareda *et al.*, 2012). Such an integrated model may prove useful to describe the relevant EBPR microbial populations of interest with the objective of exploring the environmental and operational conditions beneficial for the EBPR process.

1.7 Activated sludge model development history

In this section, the most frequently used activated sludge models are considered to support the modeller in the model selection phase. The focus is on the recent developments of activated sludge models, mainly the family of activated sludge models developed by the International Water Association (IWA) and the metabolic model developed at the Delft University of Technology (the TUDP model). Table 1.10 summarizes the essential features of these and several other activated sludge models.

Table 1.10 Overview of selected activated sludge models (based on Gernaey *et al.*, 2004).

Model	Nitrification	Denitrification	Heterotrophic / autotrophic decay	Hydrolysis	EBPR	Denitrifying PAO	Lysis of PAO / PHA	Fermentation	Chemical P removal	Reactions	State variables	Reference
UCTOLD	●	●	DR, Cst	EA						8	13	Dold *et al.*, 1980, 1991
ASM1	●	●	DR, Cst	EA						8	13	Henze *et al.*, 1987
ASM3	●	●	ER, EA	Cst						12	13	Gujer *et al.*, 1999
UCTPHO	●	●	DR, Cst	EA	●		Cst	●		19	19	Wentzel, 1988, 1989a,b
ASM2	●	●	DR, Cst	EA	●		Cst	●	●	19	19	Henze *et al.*, 1995
ASM2d	●	●	DR, Cst	EA	●	●	Cst	●	●	21	19	Henze *et al.*, 1999
B&D	●	●	DR, Cst	EA	●	●	EA	●		36	19	Barker and Dold, 1997
TUDP	●	●	DR, Cst	EA	●	●	EA	●		21	17	Meijer, 2004
ASM3-bioP	●	●	ER, EA	Cst	●	●	EA			23	17	Rieger *et al.*, 2001

Den. PAO: Denitrifying PAO activity included in the model; DR: death regeneration concept; EA: electron acceptor depending; ER: endogenous respiration concept; Cst: not electron acceptor depending

The ASM1 can be considered as the reference model, since this model triggered the general acceptance of wastewater treatment modeling, first in the research community and later on also in practice. This evolution was undoubtedly supported by the availability of more powerful computers. ASM1 is in essence a consensus model-compromising result of discussions at the time between different modeling groups, most prominently from South Africa, U.S.A., Switzerland, Japan and Denmark. Many of the basic concepts of ASM1 were adapted from the activated sludge model defined by Dold *et al.*, 1980. A summary of the research developments that resulted in ASM1 was given by Jeppsson (1996). Even today, the ASM1 model is still in many cases the state of the art for modeling activated sludge systems (Roeleveld and van Loosdrecht, 2001). ASM1 has become a reference for many scientific and practical projects, and has been implemented (in some cases with modifications) in most of the commercial software available for modeling and simulation of plants for N removal. Copp (2002) reports on experiences with ASM1 implementations on different software platforms. ASM1 was primarily developed for municipal activated sludge plants, to describe the removal of organic carbon compounds and nitrogen compounds, with the corresponding consumption of oxygen and nitrate as electron acceptors. The model furthermore aims at yielding a good description of the sludge production. Chemical oxygen demand (COD) was adopted as the measure of the concentration of organic matter. In the model, the wide variety of organic carbon compounds and nitrogenous compounds are subdivided into a limited number of fractions based on biodegradability and solubility considerations.

The ASM3 model was also developed for biological N removal plants, with basically the same goals as ASM1. The ASM3 model is intended to become the new standard model, correcting a number of defects that have appeared during the usage of the ASM1 model (Gujer *et al.*, 1999). The major difference between the ASM1 and ASM3 model is that the latter recognizes the importance of storage polymers in the heterotrophic activated sludge conversions. Biomass growth directly on external substrate as described in ASM1 is not considered in ASM3. A second difference between ASM1 and ASM3 is that the ASM3 model should be easier to calibrate than the ASM1 model. This is mainly achieved by converting the circular growth-decay-growth (death-regeneration) model by a growth-endogenous respiration model (Figure 1.15).

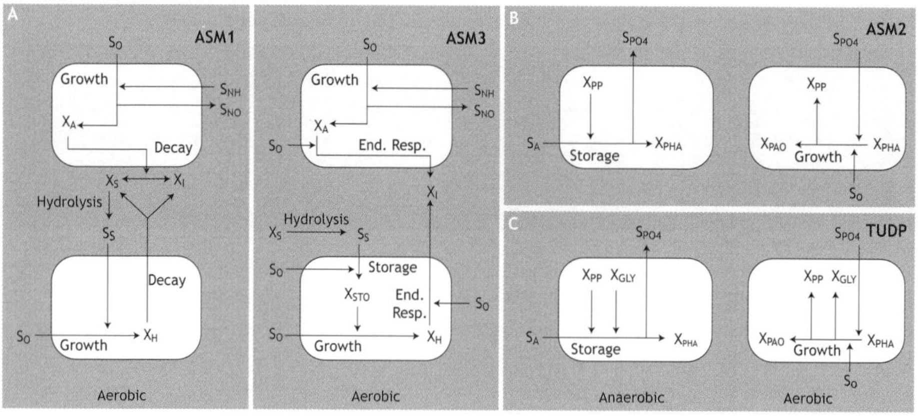

Figure 1.15 Simplified schemes of substrate flows for (A) autotrophic and heterotrophic biomass in the ASM1 and ASM3 models (modified from Gujer *et al.*, 1999), (B) storage and growth of PAO in the ASM2 model (Henze *et al.* 1995), and (C) storage and aerobic growth of PAO in the TUDP model (van Veldhuizen *et al.*, 1999; Brdjanovic *et al.*, 2000). Adapted from Gernaey *et al.*, 2004.

Whereas in ASM1 effectively all state variables are directly influenced by a change in a parameter value, in ASM3 the direct influence is considerably lower allowing a better identification. Koch *et al*., (2000) concluded that ASM1 and ASM3 are both capable of describing the dynamic behaviour in common municipal plants, whereas ASM3 performs better in situations where the storage of readily biodegradable substrate is significant (industrial wastewater) or for plants with substantial non-aerated zones. The ASM3 model can be extended with a EBPR removal module similar to ASM2 (Ky *et al*., 2001; Rieger *et al*., 2001).

The overview of models including EBPR may start with the ASM2 model, which extends the capabilities of ASM1 to the description of EBPR. Chemical P removal (CPR) via precipitation was also included. The ASM2 publication mentions explicitly that this model allows description of EBPR processes, but does not yet include all observed phenomena; most importantly it is based exclusively on aerobic P uptake EBPR behaviour. The ASM2d model builds on the ASM2 model, adding the denitrifying activity of PAO which should allow a better description of the dynamics of phosphate and nitrate. However, it merely allows P uptake to commence in the anoxic reactor with the same kinetics as in the aerobic reactor – it does not take into account the observed reduction in P removal when significant P uptake takes place in the anoxic reactor (Ekama and Wentzel, 1999; Hu *et al*., 2002). Later EBPR models have sought to address this (Hu *et al*., 2007a,b). EBPR modeling in ASM2 is illustrated in Figure 1.15.

The PAO are modelled with internal cellular structure, where all organic storage products are lumped into one model component (X_{PHA}). PAO can only grow on cell internal organic storage material; storage is not dependent on the electron acceptor conditions, but is only possible when fermentation products such as acetate are available. In practice, it means that storage will usually only be observed in the anaerobic activated sludge tanks. The TUDP model combines the metabolic model for denitrifying and non-denitrifying EBPR of Murnleitner *et al*. (1997) with the ASM1 model (autotrophic and heterotrophic reactions). Contrary to ASM2/ASM2d, the TUDP model fully considers the metabolism of PAO, modeling all organic storage components explicitly (X_{PHA} and X_{GLY}), as shown in Figure 1.15 and Annex 1.1.

In some cases, such as high pH (>7.5) and high Ca^{++} concentrations, it can be necessary to add biologically induced P precipitation to the EBPR model (Maurer *et al*., 1999; Maurer and Boller, 1999). Indeed, under certain conditions the EBPR reactions coincide with a natural precipitation that can account for an important P removal effect that is not related to the EBPR reactions included in the models described thus far. The formation of these precipitates, mostly consisting of calcium phosphates, is promoted by the high P concentration and increased ionic strength during the anaerobic P release by the PAO. Model equations and components necessary to describe this precipitation process were given by Maurer and Boller (1999). In general it can be said that the introduction of the ASM model family by the IWA task group was of great importance in this field, providing researchers and practitioners with a standardized set of basic models mainly applicable to municipal wastewater systems, but rather easily adaptable to specific situations such as the presence of industrial wastewater (e.g. Pinzón *et al*., 2007).

1.8 Simulator environments

A wastewater treatment simulator can be described as software that allows the modeller to simulate a wastewater treatment plant configuration. A detailed overview of simulators for wastewater treatment models can be found in Olsson and Newell (1999) and Copp (2002). The Consortium Project Leader has recently co-written the chapter on the choice of models and simulators for WWTP modeling in an educational book which is currently one of the bestselling books of IWA Publishing – the publishing agency of the world's largest International Water Association.

General purpose simulators can be distinguished from specific wastewater treatment simulators. General purpose simulators normally have a high flexibility in their application, but the modeller has to supply (by programming) the models that are to be used to model a specific plant configuration. The latter task can be very time consuming although necessary, as it is needed to debug the implemented model, to avoid running lots of simulations with a model that afterwards is shown to be erroneous for the specific task. As a consequence, general purpose simulators require a skilled user (with programming skills) who fully understands the implications of each line of code in the models. A popular example of a general purpose simulator is MATLAB™/SIMULINK™ (developed by www.mathworks.com). In contrast to general purpose simulators, specific wastewater treatment simulators usually contain an extended library of predefined process unit models, for example a perfectly mixed ASM1 or ASM2d bioreactor, and a 1-dimensional 10-layer settler model. The process configuration to be simulated can easily be constructed by connecting process unit blocks. Pop-up windows allow the modification of the model parameters. Examples of specific commercial wastewater treatment simulators are (in alphabetic order, and some of which are presented in Figure 1.16) AQUASIM (www.aquasim.eawag.ch), BioWin (www.envirosim.com), EFOR (www.dhisoftware.com/efor), GPS-X (www.hydromantis.com), SIMBA (www.ifak-system.com), STOAT (www.wrcplc.co.uk/software) and WEST (www.hemmis.com).

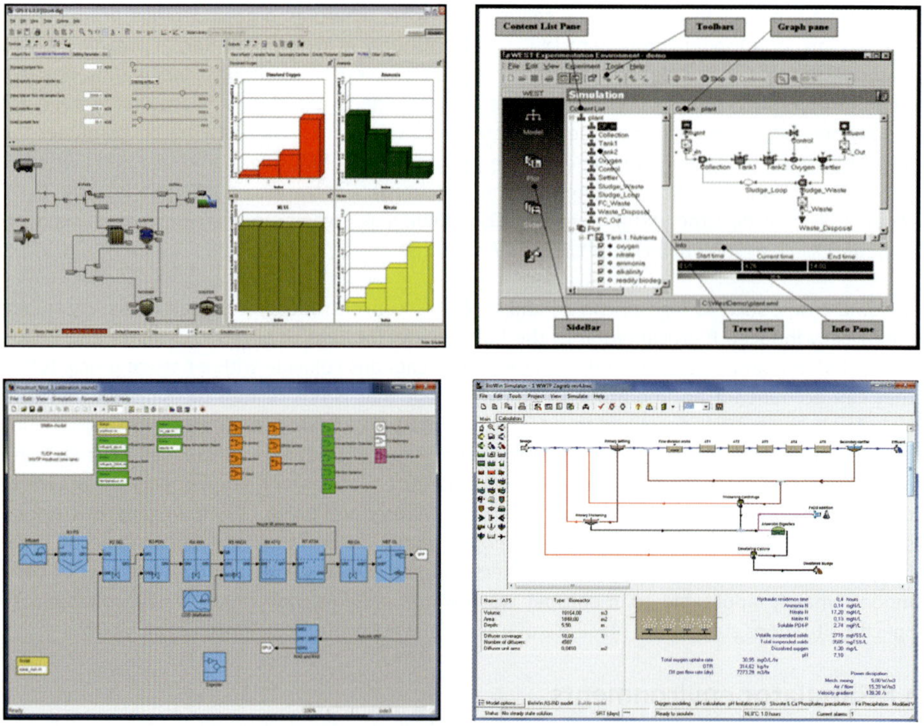

Figure 1.16 WWTP simulators (clockwise): GPS-X, WEST, SIMBA and BioWin (Meijer and Brdjanovic, 2012).

More information about these specific WWTP simulators can be found on their respective websites. On these websites it is often possible to download a demo version of the simulators for evaluation purposes. Specific wastewater treatment simulators allow the modeller to easily produce the desired plant configuration by connecting predefined model blocks (the process flow diagram or PFD). As such, this involves the danger that the user is simulating process configurations without fully understanding the model structure, with the implication that model assumptions and limitations can also easily be overlooked. In the early examples of modeling applications presented in this book SIMBA simulator (combined with Matlab and Simulink) was used as a result of the national agreement in the Netherlands, however SIMBA was gradually replaced by other more user-friendly models such as BioWin (see next section).

1.9 Introduction to general modeling protocols

The key factor for the successful introduction of model-based design of wastewater treatment in the Netherlands was standardization of the methods, models and the simulation software. At an early stage it was decided to use one type of commercial software based on the ASM family of activated sludge models. Thus, protocols and guidelines were developed for the practical use. This approach has proven to be effective for the introduction of this technology in the Netherlands where nowadays the application of activated sludge modeling is a standard practice. Different protocols were developed by the Dutch water boards, including the protocol for influent characterization and the protocol for model-based design. The usefulness of the guidelines is determined by the simplicity and practical applicability of the proposed methods. Therefore, simple methods have been developed allowing small inaccuracies in the calculations rather than complex methods that may make performing slightly more accurate but less understandable. The choice was made to provide individual engineers with tools they would also use in practice. Therefore, self-help groups were encouraged and also organized by the Dutch water boards during the first years of introduction of the new model-based design technology. Finally, the support from the senior management for the introduction and use of the technology and the extra time given for training and education was crucial for the successful introduction in the Netherlands. By doing so, the Dutch water boards have been the main investors in the introduction and standardization of this new design technology in the Netherlands. Currently, most model-based design studies are performed by engineering firms; the application of models has become standard practice and is often required in the terms of reference for WWTP upgrading and design projects. One of the factors determining the successful introduction was the choice of the standard simulation software. At that time, SIMBA/MATLAB/SIMULINK was chosen as the preferred simulator, being one of the most flexible and best developed commercial software packages. The criteria on which SIMBA was selected were price and independency of the developer (Ifak, Germany). The last was seen as very important, since the activated sludge models were not fully developed at that time. Currently, activated sludge models are at the end of the developing stage and widely applied in practice. Therefore this argument no longer holds; most commercial software packages nowadays are based on very similar models (IAWQ ASM type). The protocol used by the Delft modelling group originates from the STOWA protocol for model design in activated sludge modeling (generally referred to as *Model Based Design* or MBD). This protocol was based on published work by Hulsbeek *et al.* (2002) and Meijer *et al.* (2001). After its first publication, this work was further applied and studied in practice by Meijer *et al.* (2002, 2004), and based on extensive practical experience from full-scale and plant-wide MBD studies, adjustments to the protocol were proposed. While the STOWA protocol has its focus on activated sludge modeling only, for plant-wide modeling all the relevant WWTP process units need to be included in the model (such as sludge treatment, eg. ADM: Anaerobic Digestion Model, Batstone *et al.,* 2000). This requires a different (and more general) approach. One of the pioneering examples of plant-wide modeling is presented in Chapter 8 of this book. A comprehensive guidance to modelling activated sludge systems is given elsewhere (Meijer and Brdjanovic, 2012). The general aim of the model protocol is to reduce the complexity of a model-based design project. Herewith the protocol used by the Delft modelling group is presented in which the design tasks are divided into several sub-tasks. Seven different phases in wastewater treatment plant modeling are distinguished according to the general scheme displayed in Figure 1.17.

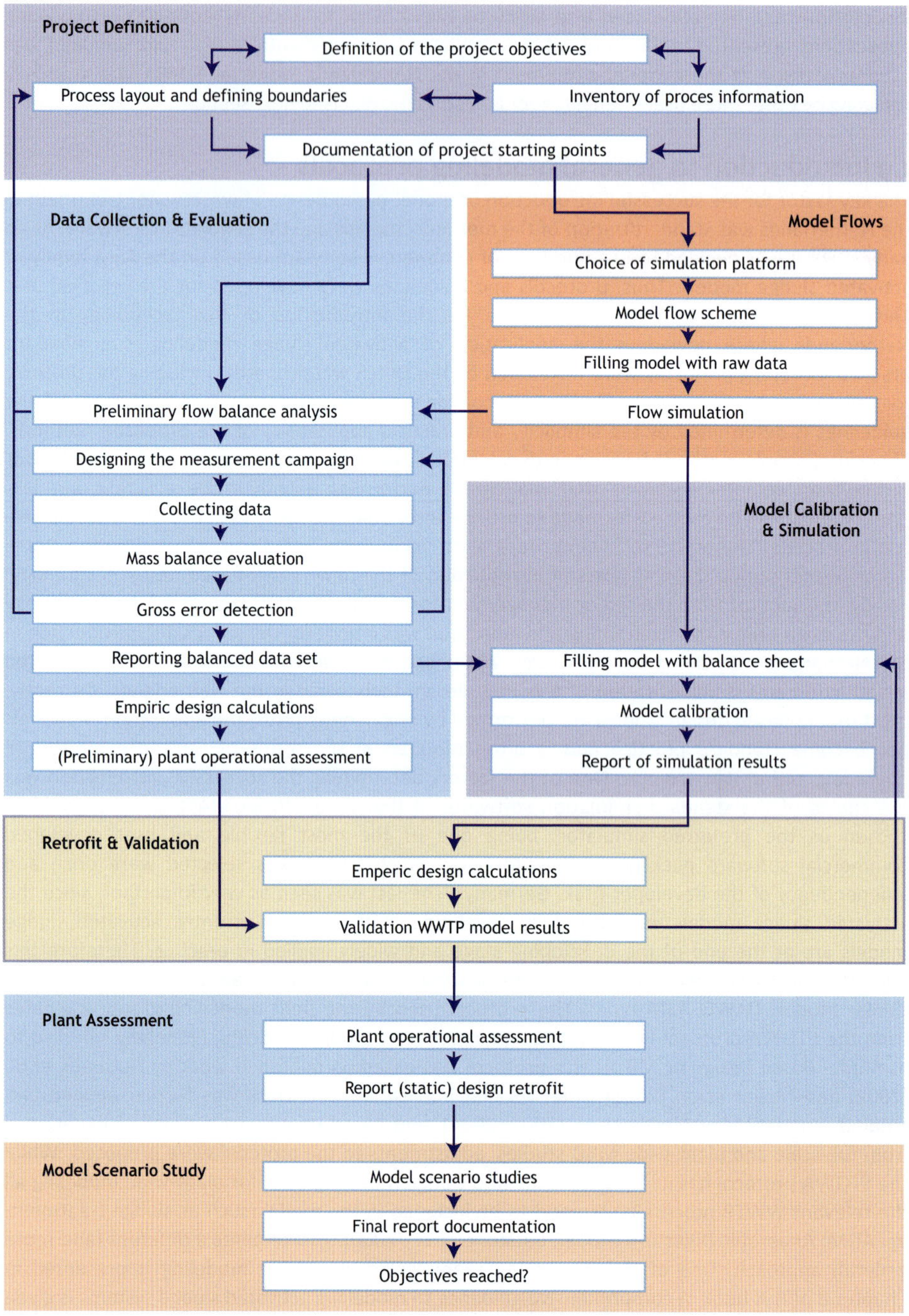

Figure 1.17 General structure of the protocol showing the 7 phases of activated sludge plant modeling (Meijer and Brdjanovic, 2012).

In the protocol each step is addressed in a standard, practical, feasible and (time) realistic manner. In the protocol the practical and applied methodology is presented in a generic and explanatory way. The aim is that the protocol is used as a practical manual and checklist for new design projects, thereby providing a quality standard for modeling design studies. A (model-based) study generally consists of several successive design calculation steps. The order of these calculations is typically determined by the type of information that is needed to perform these calculations; e.g. first the influent quality and quantity is assessed, then the sludge loading is determined and only after that, the oxygen requirement for aeration can be calculated. In the protocol, the design tasks are all ordered in logical successive steps that lead towards the final results of the design study: capacity and volumes of the wastewater treatment plant units. Firstly, in the design protocol, the more traditional performance indicators are calculated e.g. the sludge loading and SRT (sludge retention time or sludge age). Following on from this step, (static) empirical design models, typically in MS Excel, are calculated. Finally, and only if this is required by the nature of the study and allowed by the accuracy of the information, a detailed (and also more complex) process model can be calculated. Therefore, the successive steps in the protocol provide all the necessary design information needed to finally perform, calibrate (correct) and validate (check) the model simulations. Models are typically first simulated in steady state (static simulation) and, but only if needed, dynamic (time variant) simulations are also performed. However, for most model design studies the static approach will be sufficient. Working towards a more detailed plant design (typically in the final stage), a more detailed process model can be developed giving more detailed results on the plant performance. Constructing a full plant model requires more effort and time for analysis and data processing and also a larger financial investment, mainly for additional analytical measurements. Generally, a full plant model (plant-wide model) provides much more valuable information for decision-making. The plant-wide model approach allows to achieving a more precise (and more efficient) design that reduces the risks from design shortcomings, possibly resulting in costly operational failures. In the MBD protocol empirical design rules (e.g. the German ATV design method and STOWA design methods) and BioWin simulations are used together and are fully complementary to assure the quality of the plant design under all conditions.

Brief description of the modelling steps as applied by the Delft modelling group is provided below.

1.9.1 The inception phase

In this first phase, the main goal is to get an accurate and complete project overview. The main goal of this phase is the definition of realistic objectives and to make a realistic project plan for the given time and project budget. Because the model-based design project relies on actual measured data, it needs to be established whether this data is readily available from historical sources and what the quality of such data is. Often additional measurements will be required, and this will affect the planning and project budget. At this stage the problem and the research objectives are defined. This directly determines the physical boundaries of the process that will be studied. To do this efficiently, a process flow diagram (PFD) of the WWTP is made. Also an inventory is carried out of all the available plant information. Therefore, a plant visit is always recommended. Probably not all the required data can be gathered at this stage; however, indicative samples of the data can be acquired to get a first impression on what can one expect of the data collection phase. All the information that is needed to be collected during the inception phase is described in the guidelines (templates/checklists). The result of the inception phase is a complete documentation of the plant inventory including the most important design information, a list of the historical operational data and indicative (average) operational plant information, a process flow diagram and all other collected information that could be of relevance for the rest of the project (e.g. contact data, photographs, guidelines, design reports etc.). Therefore, the inception phase should be seen as a smaller project on its own within the larger modeling project. The time invested here is approximately 4 to 5 days including (at least) one day for visiting the plant. In short, the inception phase covers the following activities:

- Definition of the project objectives;
- Definition of the process lay-out in the process boundaries;
- Collection of initial (estimated) process information;
- Documentation of the project starting points.

1.9.2 The initial model construction

The information collected on the plant in question can be used to construct a preliminary plant model. Based on the definition of the objectives, it will be clear what the specific goals of the project are and, thereby, also the preferred simulation platform and activated sludge and/or anaerobic digestion models. From the process diagram, a (typically simplified) model flow diagram is constructed in the simulation platform. At this stage no calibration is necessary. After 'feeding' the model with the plant design data and preliminary flow data, it will be possible to do an initial simulation of the plant. It is very unlikely that the first simulation will lead to reliable results, however the goal of this step in the protocol is to get an early indication of whether the assumed flow scheme is correct, and also whether the initially provided operational data (at this stage mainly the flow balances) can be fitted in the model. If this is not possible, this can indicate that the preliminary data is incorrect and/or unreliable. In that case, action should be undertaken to acquire better data and information. Testing the data in an early project phase provides the possibility for early feedback to the plant operators and the possibility to resolve this potentially critical problem at an early stage. The model construction phase is a desk study that can be performed within 1 to 2 days. In short, the model construction phase covers the following activities:

- Choosing the simulation platform;
- Constructing the model flow schemes;
- Constructing a non-calibrated model based on non-verified data;
- Checking information on major errors by running a preliminary (quick scan) simulation.

1.9.3 Data acquisition and evaluation

The main goal of this phase is to acquire reliable plant information. Therefore, the historical operational data is collected and checked using mass balance calculations. Also additional measurements are planned with the goal of checking the main operational parameters, mainly flows and the sludge production. Moreover, during this phase, plant data is acquired to be used for the design calculations and in the simulation study as model input, and to validate the model simulation results. Therefore, a comprehensive measurement plan should be made which includes the minimum required information on sampling and analytical measurements to be able to proceed to the next project activities. This plan is based on an initial mass balance analysis; all the data is measured within a preselected (sub) system of mass balances, determined by the initial definition of the process boundaries, as stated in the report objectives of the inception phase. In the data acquisition phase, the actual plant sampling will be executed. It should be taken into account that this is a critical part of the study and also that it will take approximately 2 to 3 weeks from the start of sampling to the moment the final results can be presented for further processing. Again, the additional measured information will need to be checked for major errors and also, mass balance calculations will be used for evaluating the data. The results of the final mass balance calculations will be reported and will serve as input for the already constructed model. An initial plant assessment is also possible, based on the checked information. Therefore, multiple empirical design calculations, as well as calculations of traditional key design parameters such as the SRT, sludge loading, oxygen consumption and also more complicated static empirical design methods (e.g. published by STOWA, WEF or ATV) are applied here. The retrofit design phase can also be performed only using the historical data, e.g., without performing a model-based simulation. Depending on the amount of data that needs to be collected, which largely depends on the definition of the project objectives, the data acquisition and evaluation phase will take 2 to 4 weeks including all the practical work related to the sampling and laboratory analysis. In short, this phase covers the following activities:

- Flow balance analysis of the plant;
- Designing a sampling campaign;
- Executing the sampling and lab analysis;
- Collecting detailed historical operational plant data;
- Performing mass balance calculations;
- Detecting major errors in the measured and historical plant data;
- Reporting a checked and balanced data set;
- Calculating an empirical design using the measured data;
- Conducting an initial plant assessment.

1.9.4 The model simulation and calibration phase

At the beginning of this phase a non-calibrated model is already available. A data report is also available which contains the checked and balanced input data that is ready to be used for the model input and calibration. When the model is calibrated and preformed as expected (according to the checked and balanced data), a model calculation report is made in the form of overview/summary tables. This phase of the modeling project is described by the STOWA protocol as presented in these guidelines. The STOWA protocol handles in detail all the steps needed for the construction and calibration of the activated sludge system. From a mathematical modeling perspective, the activated sludge model is the most complicated part of the total (plant-wide) model, and therefore the additional attention when calibrating the activated sludge system is justified. The time needed to invest in this phase heavily depends on the experience of the modeller. An experienced modeller will generally be able to produce a calibrated model and report the results within 5 days. If an inexperienced modeller is working on the project, 2 to 3 additional weeks should be reserved for the modeller to get acquainted with the use of the software, the background of ASM models and model calibration, and probably follow a training course on this topic. In summary, this phase covers the following activities:

- Loading the model with checked and balanced input data;
- Performing steady-state simulations;
- Calibrating the model;
- Reporting the calibrated model results.

1.9.5 The model retrofit and validation

In this phase of the project, the calibrated simulation results are subject to multiple empirical design calculations (as was the measured data in the previous stage), and calculations of traditional key design parameters such as the SRT, sludge loading, oxygen consumption and also more complicated empirical design methods (e.g. published by STOWA, WEF or ATV) are also applied. This serves three main goals, namely: (i) a plant operational assessment, (ii) checking the information for accuracy and detecting inconsistencies, and (iii) determining whether the model simulation results are within acceptable design boundaries. For the model validation, the checked and balanced data, including key operational and design parameters, is compared to the results of the calibrated model. The design data is typically presented in a double column (side-by-side) table for easy comparison. In principle, the retrofit design can be performed only on historical data without performing model-based simulations. For the majority of (simple) studies the plant assessment can be based on the traditional retrofit design calculations which are sufficient for the defined project objectives. Performing a full model-based study is only advisable in cases where the traditional approach is not sufficient. On the other hand, it should be taken into consideration that when an extensive measurement campaign is performed, the execution of the model simulations is only a little extra work that may provide a lot more detailed and accurate information on the plant operation. In conclusion, it is advisable to decide at an early project phase (before commencing with plant measurements) whether modeling is necessary (preferably at the first phase of this protocol). Depending on the experience of the designer and availability of design tools (e.g. a template MS Excel spreadsheet with all the required calculations), the retrofit design, which is a desk study, will take 2 to 5 days. In short, this phase covers the following activities:

- Preliminary retrofit design calculations based on historical plant data;
- Preliminary plant operational assessment based on historical plant data;
- Empirical design calculations based on the results of the calibrated model;
- Presentation of the historical data and model results together;
- Validation of the model results.

1.9.6 The operational plant assessment

During this stage, the major errors are filtered out of the measurements. However, smaller model calculation errors and measurement errors will still be present, probably resulting in small differences between the checked and balanced data and the data coming from the calibrated model. These (smaller or sometimes larger) differences are a useful indication of the accuracy of the design study; in the final design proposal the appropriate safety factors will be determined based on these results. A safety factor is an introduced margin of uncertainty on e.g. proposed volumes and plant capacities, ensuring that the design takes into account the model uncertainties. Therefore it is advisable to indicate precisely where the model does not fit the data as this information is valuable for the final phase of the design. In the process of comparing the historical data and the model results, firstly the model is validated (or not) and after that the final operational plant assessment is made. The results of all the design calculations are carefully evaluated and conclusions are drawn on the plant operation. Interpreting the design data requires expert judgment. Assessing the plant performance will take 1 to 2 days depending on the experience of the designer. In short, this phase covers the following activities:

- Making an operational plant assessment;
- Reporting the plant assessment.

1.9.7 The model scenarios

In the preceding steps, only the current operation of the wastewater treatment plant was evaluated. In the scenario study, the validated model can be used to predict the plant operation under different conditions (e.g. increased load, additional volume, different temperature, etc). This phase is fully dependent on the definition of the objectives and will be different for each new study. The general advice is to strictly define the amount and type of simulation runs (scenarios) that will be studied; typically 5 to 8 scenarios is the maximum. Above this number, the amount of presented data will make it difficult to distinguish between the different scenario calculations and the whole study becomes meaningless. Depending on the type of scenario study, a typical study presenting 5-8 scenarios can take up to 5 to 8 days.

1.10 The STOWA protocol

The original STOWA protocol for modeling activated sludge systems (STOWA 1996, 1999) was published by Hulsbeek *et al.* (2002) and is based on the work on influent characterization carried out by Roeleveld and van Loosdrecht (2001) and Meijer *et al.* (2001). The protocol is generally applied in practice as depicted by Figure 1.18. The protocol gives a logical sequence of activities needed to model a wastewater treatment plant, such as formulation of modeling objective(s), plant description, data collection, construction of the model, flow characterization (using specific protocols also presented in this report), model fine-tuning (calibration), model verification using other sets of data (validation), and reporting. This approach has been applied and verified in many modeling studies and adopted as good modeling practice, as also shown in this book.

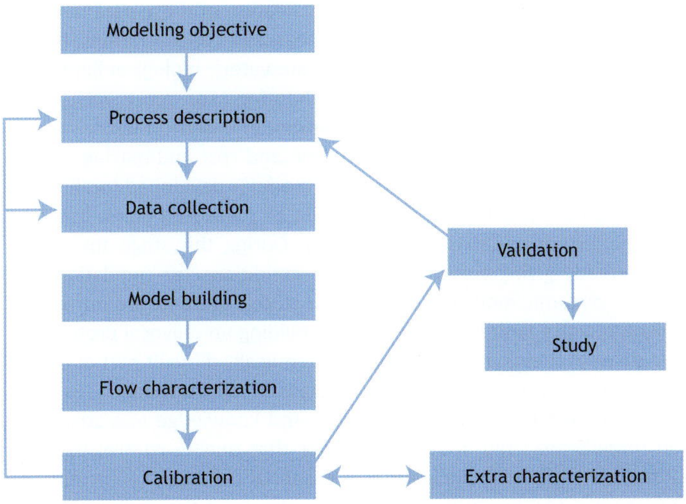

Figure 1.18 The STOWA modeling approach for constructing a calibrated model describing the activated sludge system. In the plant-wide model, this scheme can be applied to model the water line only.

At first glance the STOWA protocol looks different from the approach shown in the section 1.9. However, it should be taken in account that the plant-wide modelling approach covers a broader scope than only activated sludge modeling. The STOWA protocol is mainly focused on modeling the water line (activated sludge system). The general modelling guidelines which were used in most of cases presented in this book also cover organizational and logistical matters on data collecting, planning time and resources, presenting information, methods for data evaluation, etc. Also from the technical perspective, the used guidelines have a broader scope than only describing the water line; they focus on describing the full plant model (WWTP model or plant-wide model) which also often includes anaerobic digestion, primary settling, thickeners, secondary settlers, internal process water flows, etc. Thereby the STOWA protocol supports the general guidelines with practical, detailed, structured and stepwise applicable advice on modeling the activated sludge plant (water line). Practical use of activated sludge models for wastewater treatment plant design and upgrade was encouraged in the Netherlands in three ways; first of all, in 1995, the simulation software SIMBA® was chosen by the Dutch Foundation of Applied Water Research (STOWA) as the preferred software. This made it possible for a country-wide community of users to form and the opportunity to build national experience with the use of (dynamic) simulation software. Following this success, in 1998 STOWA recommended the use of a standard methodology for characterizing the influent (published internationally by Roeleveld and van Loosdrecht in 2001). In 2002, STOWA introduced a protocol for activated sludge modeling. Since 1995 almost all wastewater treatment plants in the Netherlands have been modelled by staff from different water boards, with or without the support of consultants. With wider use of the technology, the need for a better standardization and some form of quality control also became apparent. This led to the development of a standard methodology for characterizing the influent and for developing a calibrated model. In both cases, easy-to-use practical methods were preferred to scientific exactness. Originally, the simulation studies started with a long (1 to 4 week) intensive measurement campaign collecting and analyzing hundreds of samples, and even potentially checking many of the model parameters by respirometry or other tests. This is an expensive and lengthy procedure. It can be shown that many parameters in full-scale systems are not very sensitive (although they might be so in e.g. laboratory batch experiments) and therefore an experimental evaluation of these parameters in practice is often not useful. The same is true for some of the measured concentrations. Here it should be noted that under some conditions (e.g. industrial wastewater), biological limitations can occur and these can be useful specifically for troubleshooting. For most wastewater treatment plants; however, it is fully justifiable to develop the model based on historical data and correctly executed measurements, thereby saving much time and cost and making the use of modeling in everyday practice more feasible. When an initial model is developed (preliminary simulations), it can be used to evaluate which parameters should be measured more, based on the sensitivity to model parameters or variation over the day. During this stage the readily available flow variations are already giving sufficient dynamic conditions for simulation. The STOWA guidelines (protocol) for dynamic modeling of activated sludge systems are introduced to Dutch wastewater modelling practice with the objective of providing an univocal protocol for modeling of activated sludge systems (being an important and technically difficult part of the full modeling protocol), that will give a standardized and easier transfer of knowledge. The main purpose of the protocol was to reflect the present experience and knowledge in a structured way so that the simulation models are used in such a manner that quality control becomes possible. A more standardized use of dynamic simulations is essential because it is increasingly being used for optimization studies and for studying different aspects in the design phase of WWTP. Not only the quality control but also the planning should therefore be improved. The amount of sampling and testing should be kept at a satisfactory but minimal level to allow for a cost-effective use of models. Design and execution of sampling programs was demonstrated in several chapters of this book.

1.11 Influent characterization guidelines

Influent characterization is an important step in the effective application of the ASM models. The characteristics were commonly determined according to the STOWA protocol (STOWA 1996, 1999) that are also presented and explained in these guidelines (Roeleveld and van Loosdrecht, 2001; Hulsbeek et al. 2002). The influent characterization is mainly based on applying filtration methods to raw measured influent, thereby determining the slowly degradable COD. A BOD measurement of the influent is carried out to determine the inert (non-biodegradable) COD. In practice, the influent characterization is not a difficult analysis to make, and therefore is suitable to be taken up in the daily (routine) practice of operational data acquisition of wastewater treatment plants. Although one can argue whether the filtration methods proposed by the STOWA protocol are accurate, in practice the full-scale simulation models are not too sensitive for the division between slowly and readily degradable COD. Only the estimation of the inert COD from a BOD test involves certain uncertainty. Nervetheless, by measuring the $BOD_{1,2,3,4,5,7,10}$ (Roeleveld and van Loosdrecht, 2001) and essentially using the model to describe the BOD as a function of the incubation time, the method becomes more reliable. Nevertheless it also becomes laborious and is still vulnerable to analytical errors. It also seems from detailed calibrations that the fraction of inert COD in the influent depends on the sludge age of the WWTP. Because of all the uncertainties, we propose using the fraction of inert particulate COD in the influent as a main calibration factor. It is concluded that the wastewater characterization methods are not difficult to apply. Valuable information from these methods can be obtained not only on the influent but also on the expected plant performance. Therefore, it is recommended to incorporate such influent characterization measurements into the standard practice of operational data acquisition on a regular basis, e.g. every two weeks.

1.12 Model calibration

In the modeling protocol all the successive steps lead to the calibration procedure. The detailed procedure to calibrate the activated sludge system is described in the calibration protocol included in the guidelines. To fit the model results to the reality, the protocol indicates specific model parameters to be adjusted within an acceptable range. With the protocol at hand, the often complex task of model calibration becomes a controlled and reproducible exercise. In general, it is unlikely that the first simulation will match the observed values; usually a model will need to be calibrated. Based on practical experience it is concluded that nowadays the ASM models are advanced to the level that any 'normal', well described, domestic wastewater treatment plant can be successfully modelled by changing (calibrating) only a few preselected model parameters. If the required changes are very different from the default parameter settings, it is highly likely that the model description or input is not correct. In that case, the input data has to be fixed first before proceeding with calibration. The calibration protocol itself provides a base for problem solving. When the model cannot be fitted to the measured data within a reasonable parameter range, this indicates the actual plant is working differently than the collected information is suggesting. This will require looking back and evaluating the processed data and finding the error. Here, the plant model (that cannot be calibrated in a reasonable range) gives the direction for where to find the problem in the measurement data (e.g. incorrect analytical measurement) or in the actual process (e.g. incorrect process description). This needs to be addressed before the model is used for further calculations. Thereby, the calibration protocol provides a useful tool that can be applied to increase the accuracy of the model study and to obtain reliable results within a controlled time frame. When the model is calibrated, the operation of the WWTP can be simulated thereby matching the measured data to the level of a satisfactory description of the existing conditions. However, these are not the final results of the model study. From this point onwards the model can prove its value by being used for all kinds of extrapolation; e.g. simulating not (yet) existing conditions. These types of 'what if' studies are typically referred to as scenario studies (or scenario analyses). The model is used to check how the plant will perform under full future capacity, in other words whether the plant is indeed capable of reaching the designed effluent standards. If the design is already fully loaded, the model can be used to alter the design and extend the plant to test the effect on the effluent quality. Simulations may also reveal a possibility for operational improvements, savings (energy, chemicals), and an improvement in efficiency in terms of effluent quality, biogas production, sludge minimization, and aeration requirements. The power to carry out realistic (dynamic and static) scenario studies is unique for modeling; there is no other technique by which these type of predictions can be made within the same margin of accuracy. On the other hand, these results can only be achieved if all the recommended steps are precisely followed according to the guidelines including the STOWA protocol. According to the standard protocol, full-scale practice does not include (laboratory) experimental work, as it is considered too complicated to be performed in wastewater treatment practice. However, when building a knowledge base on the use and application of activated sludge modeling this should also be done in cooperation with training and educational institutions (universities and institutes), and also by stimulating national research that is adopting this kind of experimental work. Modern measurement equipment is available nowadays to considerably speed up laboratory batch reactor experiments (batch tests) and greatly facilitate their use in practice. This was (and is) also done in the Netherlands by universities (e.g. TU Delft) and research institutes (e.g. UNESCO-IHE), often stimulated by the Dutch water boards, and has been one of the success factors for the broader introduction of this technology and putting these state-of-the-art modeling approaches into practical use. The following lab-scale batch tests are typically performed during research into full-scale activated sludge systems (Brdjanovic *et al.* 2000; see also Chapter 3):

- Respirometry test;
- Nitrification test;
- Denitrification test;
- Anaerobic P-release;
- Aerobic P-uptake;
- Anoxic P-uptake;
- Endogenous P-release;
- PAO denitrifying activity.

1.13 The stepwise data approach to data acquisition

In the model design protocol that is presented here, a methodology is proposed for structured data acquisition. Information for plant design is acquired in several steps, each with increasing detail and accuracy of the collected data. The first step is the inventory phase where all the information possible is collected ranging from photos, plant visits, lay-outs, operational data, construction data, analytical measurements, aerial photographs, etc. The inventory report (in this study referred to as the Inception Phase*)* typically presents all the collected information in tabular overviews. For this purpose, a generic plant inventory checklist was developed (as presented in the appendix). The second step of the data acquisition involves the collection of more detailed (historical operational) information typically obtained from operational databases (e.g. data stored from SCADA systems). Typical historical data collected (time ranges) is flow measurements, energy consumptions, temperature measurements, control set-points, concentration measurements, waste sludge production etc. In the third step of the data acquisition, additional analytical measurements can be performed (typically grab samples) specifically for the assessment of the plant operation and also to be used in the model study. This data is mainly: the composition of the influent (influent characterization), effluent fractions and the activated sludge composition (activated sludge characterization). In the protocol there is a description of how these additional analytical measurements should be collected. The results are summarized in a practical measurement plan that is ready to be applied.

1.14 Measurements planning

In general, design projects that involve measuring actual plant data are known to be difficult to plan. Besides all the practical and logistical problems of gathering information on site, weather conditions should also be taken in account. In general dry weather conditions are preferred for sampling and therefore at least 3 days of dry weather should precede a sampling campaign. On top of that, it will take 1 to 2 weeks before the analytical data has been measured. Therefore in general it will take 3 to 4 weeks from the start of the sampling campaign before the raw measured data is processed and presentable, and before it is known if the sampling and measurements have all been carried out with the required accuracy, and if the campaign is considered useful for the model design study. The main risk in the design project is getting all the required (sometimes not routinely measured) information on time and with satisfactory accuracy to be used for the plant assessment. For the model-based study, extra attention should be given to the project planning and specifically to the planning of the acquisition via sampling.

Because measuring and processing plant data is time consuming and also costly, it is worthwhile investing time in thorough preparation thereby also reducing the risk of exceeding the budget and the available time. One of the difficulties of executing sampling campaigns on site is effective communication with the staff (laboratory and operators) on what, where, when and how should samples be taken and analysed. To facilitate this communication, the protocol includes a sampling program plan together with information on measurement types, handling, sampling location in reactors and flows, sampling frequency and the analytical method and preparation. For budget management, a cost calculation is also included for the analytical work.

An efficient plan for the data acquisition only should include the minimal required amount of information needed for the design study, and at the same time this data should be available with the highest possible accuracy. In practice, the accuracy of the data will determine how easy it is to construct a model and carry out the study. Typically, the model designer is confronted with conflicting (operational and plant design) information, which usually causes the less experienced designer to struggle. This was the main reason for the development of the modeling protocol: to familiarize inexperienced designers with a practical stepwise protocol simplifying the design tasks. In the protocol it is emphasized how project information can be collected, also focusing on data accuracy and thereby providing not only a reliable model study, but also making the study easier to perform (which is generally the case when the data quality is better). The modeling protocol presented here provides all the necessary information to plan the project regarding the time and resources, including standards for the MS project planning schemes, table layouts and checklists. Thereby the protocol supports the completion of a model design project in time and within the budget and also provides a basis for a qualitatively well performed design study.

1.15 Standards for project presentations

A practical and time-saving feature of the design protocol presented here are standards for processing and presenting the often large amounts of data. After gathering the information in several steps, the data is processed and presented in templates - standard tabular overviews. A standard method for presenting the information is proposed for several reasons as the tables serve as a checklist for the required plant information. Not having to come up with a new table format every time saves time. Standardized presentation of data also makes it easier to compare simulation studies, and to use and judge the results. Several sample tables are presented in this project, such as:

- Initial plant inventory data: contact information;
- Initial plant inventory data;
- Reactor volumes and dimensions;
- Hydraulic design (flows);
- Influent concentrations and flow;
- Influent loads;
- Evaluation of the primary settler operation;
- Influent characterization (concentrations of calculated fractions);
- Activated sludge characterization (concentrations of sludge fractions);
- Measurement data overview (concentrations and loads of raw and corrected data);
- Mass balance tables;
- Sludge retention time (design calculation);
- Retrofit presentation table (including the retrofit and model results);
- Secondary clarifier design (inputs and design results of 5 calculation methods).

1.16 Errors and inconsistent information

Inconsistencies in actual plant data will regularly be found and in order to establish realistic safety factors it is essential to keep note of all the (mostly expert-based) data corrections. It should be realized that the introduction of safety factors, however necessary, is also the main factor determining the cost of a plant design or upgrade; there is a direct relation between the implemented safety factor and the volumes and capacity of the process units and technical equipment, and thereby the investment and running cost of a project. For this reason it is proposed in the protocol, after checking the data for consistency, to present the corrected data set in a tabular overview indicating all the adapted measurement errors, estimations and corrected data (in the tables e.g. printed bold or underlined). This final and corrected data set is used to validate and calibrate the model and directly leads to the design results (e.g. reactor volumes or effluent quality). It is important to keep track of all the data adjustments because this information is used to determine the required safety factor used for the model-based design. An example of error detection and data reconciliation is given in Chapter 4.

1.17 Model accuracy

Within the framework of each design project, information is gathered according to the available possibilities within a given timeframe and budget. A range of information sources can be used such as interviews with plant operators and plant management, historical measured operational plant data, additional analytical measurements, original design documents and reports on plant operation. Most of the time, some of the data will be unavailable or will be available but inaccurate. These gaps in information will need to be filled in based on expert judgment and sound reasoning, supported by literature references. While doing so, it is possible that new errors are introduced. In general, plant information is likely to contain errors and therefore model design calculations will not always (fully) represent the reality. On the other hand, the model design method presented here in itself serves as a check on the plant information; the model relies on general rules of mass conservation (mass balances) and therefore, all the input and output data needs to closely correspond. Therefore, when the model cannot be validated by the plant measurements, this indicates model or measurement errors. When this occurs, it should be realized that the model results do not necessarily represent reality and the appropriate safety factors need to be applied when interpreting a (model) design. When evaluating a model design study, the possibility of errors should always be taken in account (the designer must use common sense), and the results of a (model or classical) design study also need to be seen in that context. Therefore it is good practice to present the results of a model design study in a way that it is clear to the reader where data adjustments and corrections were necessary to fit the model to the measurements.

To decide on how accurate in practice a model-based design should be depends on the design questions to be answered (the objectives of the study). This can be explained based on the following example where two totally different design questions are compared in relation to the required model accuracy:

- The first question could be: how accurate does a model need to be if one wants to use it to design a new (non-existent) WWTP if there is only influent information available about the general demographic growth of a community over 15 years? Here, probably a simple (and less accurate) model would be acceptable; the error in this study would be largely determined by the error in the estimation of the demographic forecast. To cope with this large uncertainty of the input data (the influent), large safety factors would also be required in the plant design.

- A very different question could be: how accurate does a model need to be to predict ammonium effluent peaks in stormy weather conditions at an existing plant? Here the model error probably needs to be less than 5%. This would require extensive plant measurements of a range of operational conditions and model calibration and validation and maybe even dynamic simulation.

For both the above design questions, model-based design is useful but, however, the required model accuracy will be totally different. To answer the first question, constructing a simple model based on readily available plant and operational information would probably be sufficient. However, to answer the second question, it is probable that detailed (e.g. time dependent) operational information and process measurements are needed with all the costs and time involved. From this example it becomes evident that before making a model, the question for what purpose the model will be used should be answered.

1.18 Modeling and modern wastewater management

The most important advantages of the use of models in wastewater treatment are:

- Getting insight into plant performance;
- Evaluating possible scenarios for upgrading;
- Evaluating new plant design;
- Supporting management decisions;
- Developing new control schemes;
- Providing operator training.

Modeling forces the modeller to make his or her work explicit. Qualitative comparisons are often found in the literature such as 'better', 'larger', 'smaller', 'higher', etc. Such comparisons are not very useful and are of a subjective nature, as, for instance, the perception of 'large' or 'small' by a researcher in the laboratory or by a person operating a wastewater treatment plant is not necessarily the same. When it comes to modeling it is not possible to use descriptive elements, but it is necessary to use quantitative inputs for sizes, rates, and conversions as the model requires numbers as input. This also forces modellers to become quantitative and objective in their approach and thus the process knowledge gets a better definition. Of course, one can do without modeling, but very often by making a model one makes a framework that takes into account everything which is considered relevant. It furthermore forces structured and more extensive data collection, and encourages the modeller to be organized. It often exposes knowledge and data limitations and/or incorrect data (such as SRT or flows), supports efforts to improve quality of data, and enhances good plant monitoring practices. Therefore it is not surprising that getting insight into plant performance (quantification of information, mass balances and data reconciliation) and learning about the wastewater treatment plant in question can be even more important than the modeling itself.

The second main reason for using models is the possibility to save time and money whilst selecting the technology and process. A quantitative rather than qualitative comparison of the system performances allows in many cases for easier decision-making and a rapid comparison of options. In comparison with the qualitative description such as 'one system is more efficient than the other', model results showing that 'one system is 2% (or 20%) more efficient than the other' is much more informative and useful. If important information or selection criteria are quantified (such as purification efficiency, effluent quality, sludge production, oxygen requirements, etc.), the application of modeling for evaluating possible scenarios for upgrading will make the comparison more effective and faster than a discussion on such issues that are usually empirical, intuitive, long and often cumbersome. For the purpose of evaluating upgrade scenarios, it is not necessary to make a very accurate model by performing an extensive calibration procedure, as the real uncertainty is associated with the model inputs and not with the model parameters. It is considered much more useful to use trends for comparison as small differences are not relevant in the context of the usual design horizon used in wastewater treatment engineering. In the case of evaluating new plant design, again it is not necessary to have a fine-tuned model due to uncertainty in process conditions in the coming 10 or 20 years. For primary plant design usually static (steady state) models are used while dynamic modeling is applied for sensitivity analysis and optimizing the design. An additional challenge is the fact that wastewater has an extremely complex and uncertain composition. Wastewater flow rate and concentrations are of a highly dynamic nature and are very difficult to control, despite certain limited possibilities to influence the wastewater composition (see Chapter 9). Many processes take place within the wastewater treatment plant, some of them are relevant to the treatment and many of them are not. However, many of them occur simultaneously, even in a single process unit. In order to handle such a complex situation, there is a need for a model to support the understanding of these relevant processes. Despite the fact that from the design perspective, modeling as such is usually not strictly necessary, it is becoming increasingly used as a part of the design process. By applying statistical methods to the occurrence of worst-case scenarios, significant savings can be made and the plant can still achieve its effluent quality standards for about 95% of the time. Often in traditional design all the worst case scenarios are assumed to occur simultaneously leading to a highly unlikely scenario.

Another good reason for using models is the possibility of decreasing or minimizing risks. By using models, 'what if' scenarios can be examined in a quantitative way in respect of what the effects of potential risks are. Such a glass-box type (as opposite to black-box type) quantification is invaluable in the evaluation and selection of acceptable risk, rejecting risks that cannot be taken, and in identification of upfront measures that can be taken to mitigate or control such risks. For example, questions such as 'what will happen if the flow rate doubles?' and 'what is the effect of such an increase on effluent quality?' can be properly addressed by using models. Furthermore, models allow for minimization of risks that are related to scaling-up of the systems (lab-scale vs. pilot-scale vs. large-scale). Related risks originate from the fact that, for example, mixing conditions, load variations etc., are different for full-scale and lab-scale installations. From the perspective of process control, pilot-scale gives a much faster response in comparison with full-scale plants with greater inertia.

Furthermore, the application of models improves knowledge transfer and decision-making. Wastewater treatment engineering and environmental engineering in general are multidisciplinary fields requiring knowledge of different disciplines, such as microbiology, biochemistry, and physical, biological and mechanical engineering. In addition, each expert group involved, be they operators, engineers or scientists, usually has its own perspective of the same subject. By phrasing the subject into a mathematical context the same communication tool (language) is used. Such a multidisciplinary approach allows for a better description of the reality, each discipline delivering its own input for a better understanding of the reality that can be incorporated into the model in a structured, organized and quantitative way. Model-related communication was greatly improved after the introduction of ASM1 in 1987. Prior to the introduction of ASM1, at least five or six different ways of modeling wastewater treatment plants were described. Each model had its different approaches in writing-up, in notation, and in implementation of equations, which made it extremely difficult to understand the models and their results. The uniform context and standardization introduced by ASM1, in terms of notations, symbols and structure, made comparison of results and knowledge transfer much easier and further encouraged modeling applications.

Models are nowadays invaluable tools for training. For example, the plant operator can safely investigate by means of modeling what can happen if one takes certain action at a treatment plant without risking eventually upsetting the operation of the plant. Moreover, models can be used to transfer knowledge from design engineers to operators and, of course, in academia, where modeling is increasingly becoming a part of the curricula for engineers and scientists worldwide. From the perspective of process control, in practice there are no direct model-based controllers functional yet, as their application still remains only of scientific interest. In practice simple controllers are tuned based on the model, which allows for much quicker optimization of the control strategy at the full-scale installations. In the framework of integrated urban water system modeling, wastewater treatment modeling is an important component and it is necessary to link up wastewater treatment with the sewer system (to take into account effects of, for example, combined sewer overflows or processes taking place in the sewerage system) on the one side, or the receiving water's quality and quantity, on the other. Integrated modeling is becoming an increasingly popular tool to support decision-making at the level of urban water system management as it brings objectivity and gives quantitative insight into relevant differences between options.

In fact, the application of models has a place during the entire lifecycle of a wastewater treatment plant including the design, construction, and operation and evaluation stages (Figure 1.19). The spectrum of possible use of models during the WWTP lifecycle is shown in Figure 1.20. The practice revealed that the most cost saving is possible when models are used at the early stage of the plant lifecycle (Figure 1.21). As is elaborated earlier, the modeling goal determines the type and complexity of the model to be applied. Figure 1.22 depicts how model complexity increases as the plant lifecycle matures. Figures 1.19 to 1.22 are adopted from Meijer and Brdjanovic, 2012.

It can be concluded that the largest savings are possible at the plant design phase because:

- Modeling helps to quantify scenarios at the early design stage;
- Quantification helps to speed up the decision making process;
- There is considerable uncertainty in the early design phase (e.g. influent composition);
- Large safety margins are needed (usually from 150 to 200%);
- Models can be simple (no calibration needed!);
- During the design phase investment in models and modeling work is usually paid back.

Furthermore, practice shows that the highest operational risks are at the operational stage because:

- Modeling helps to reduce operational risks;
- Operational problems are often complex;
- More accurate models are required (e.g. ASM models);
- Possible financial savings are often less;
- Not only reduction in costs will show if modeling is useful.

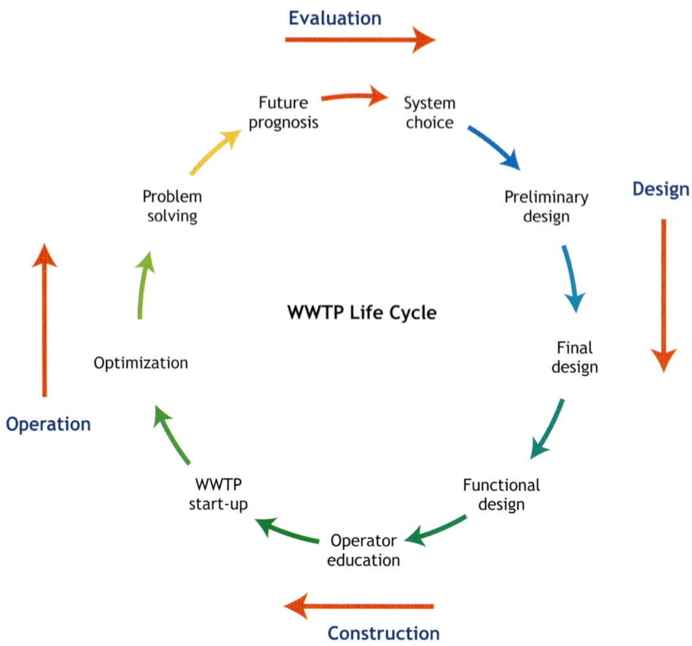

Figure 1.19 Wastewater treatment plant lifecycle (Meijer and Brdjanovic, 2012).

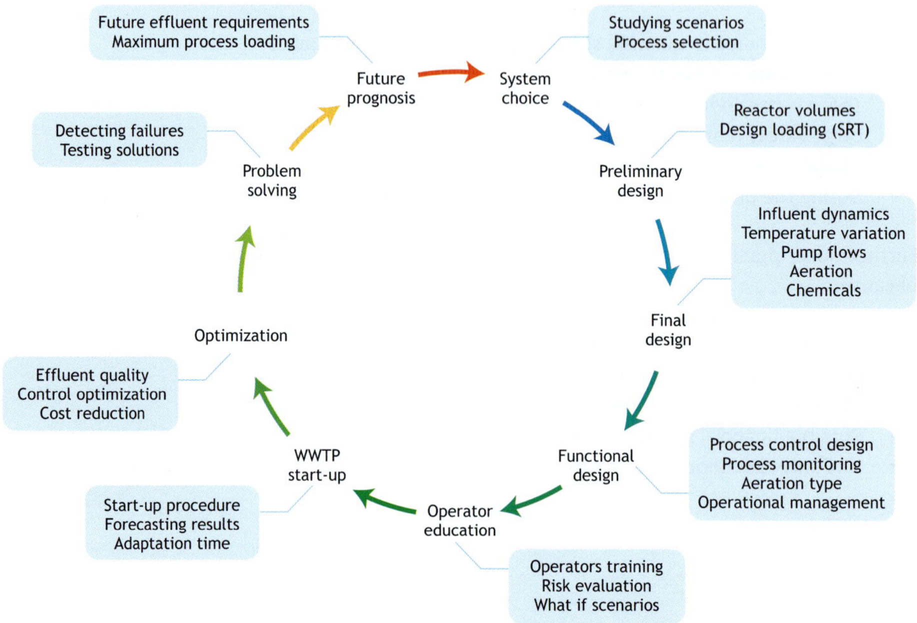

Figure 1.20 Modeling applications at different stages of the plant lifecycle (Meijer and Brdjanovic, 2012).

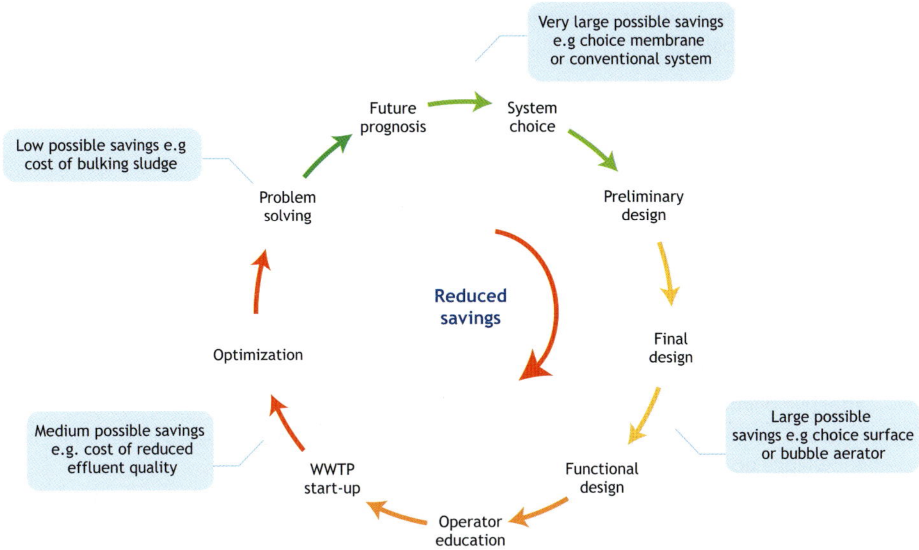

Figure 1.21 Possibility for costs saving during the WWTP lifecycle (Meijer and Brdjanovic, 2012).

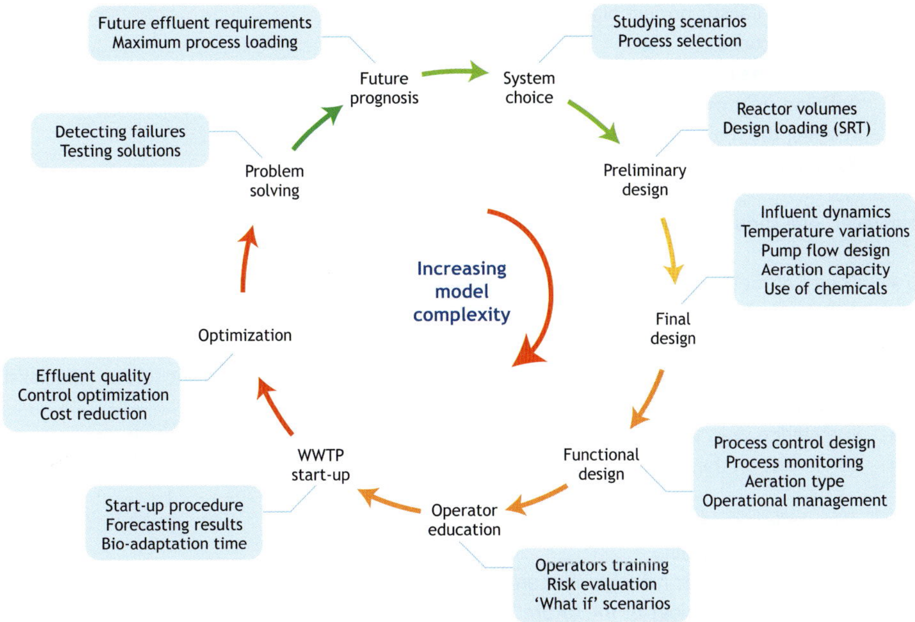

Figure 1.22 Model complexity and increasing operational risks during the plant lifecycle (Meijer and Brdjanovic, 2012).

Reasons to introduce standards and models in wastewater management at institutional level are (the top-down approach):

- To standardize WWTP operation and management;
- For quality assurance and control;
- To improve efficiency and reduce costs;
- As a knowledge base (organize process documentation);
- To improve internal and external communication (standardization of information);
- For planning and decision-making.

Reasons for engineers to use standards and models can be to (the bottom-up approach):

- Do the work better by having better insight in the process and to make better and more effective plant designs;
- Improve personal skills by working with 'state of the art' technology and keeping their knowledge up to date;
- Make the work dynamic by having a low-cost and safe platform to test ideas for better WWTP operation;
- Make the work more easy by creating a reusable model (knowledge base);
- Increase reliability of their work by supporting decisions in communication with higher management;
- Have the most effective tools for process optimization and process control.

Success factors for using modeling in design can be summarized as:

- Working according to a protocol;
- Making a realistic project plan;
- Being practical;
- Getting a 'helicopter' view of the modeling project;
- Defining clear modeling goals;
- Keeping the model as simple as possible;
- Using a standard calibration method.

However, one should be aware that by doing modeling several bottlenecks may be identified:

- Choice of methods and software is important - one standard is required;
- Different approach towards WWTP information is needed;
- There is a continuous need to invest in education (life-long learning);
- Sharing of knowledge through a modeling platform, meetings, internet forum, specialist groups etc., is crucial;
- Knowledge sinks when not used, thus investing in successive demo projects is necessary;
- Consensus on these statements should be reached: "the use of models....
    - saves money,
    - improves quality,
    - is effective for management and decision making,
    - is the technology of the future".

Finally, the use of models in wastewater treatment has the following main advantages:

- Cost reduction, especially at the design phase;
- Improved management and quality control;
- The optimal technological/process design is obtained;
- Works with modern 'state of the art' technology;
- Application of innovative approaches;
- Development of designs at low cost, fast and with confidence.

1.19 Conclusions

Models can serve as extremely useful tools in the design and operation of wastewater treatment plants, and in research into the behaviour of these plants. For design, models can provide guidance in identifying the key design parameters and can quantify system parameters to ensure optimal performance. For operation, models can provide quantitative predictions as to the effluent quality to be expected from a design or existing system, and allow the effect of system or operational modifications to be assessed theoretically. For research, models allow testing of hypotheses in a consistent and integrated fashion, to direct the attention to issues that are not obvious from the physical system. This will lead to a deeper understanding of the fundamental behavioural patterns controlling the system response. In this manner, models provide a defined framework that can give a direction for further investigations. However, this framework does have disadvantages; it can restrict innovative new developments that do not fall within the boundaries of the framework. Also, in modeling and using models, it should be remembered that models are only our rationalization of behavioural patterns of parameters that we conceive to be of importance when describing a particular system. Due to this rationalization, the models need to be rigorously verified by conforming to internal mass balances and adequately validated against appropriate experimental tests, and the conditions within which the model is expected to operate successfully need to be clearly defined. A model can be deemed successful if it fulfils the expectations that people have of it.

References

Alexander W.V., Ekama G.A. and Marais G.v.R. (1980) The activated sludge process Part 2 - Application of the general kinetic model to the contact stabilization process. *Wat. Res.* 14, 1737-1747.

Barker P.S. and Dold P.L. (1997) General model for biological nutrient removal activated sludge systems: Model presentation. Wat. Environ. Res., 69, 969-984.

Batstone D. J., Keller J., Newell R. B. and Newland M. (2000) Modeling anaerobic degradation of complex wastewater. I: model development. Biores. Tech. 75(1), 67-74.

Brdjanovic D., Logemann S., van Loosdrecht M. C. M., Hooijmans C. M., Alaerts G. J. and Heijnen J. J. (1998) Influence of temperature on biological phosphorus removal: process and ecological studies. Wat. Res. 32(4), 1035-1048.

Brdjanovic D., van Loosdrecht M.C.M., Versteeg P., Hooijmans C.M., Alaerts G.J. and Heijnen J.J. (2000) Modeling COD, N and P removal in a full-scale WWTP Haarlem Waarderpolder. Wat. Res. 34, 846-858.

Brink I.C., Wentzel M.C. and Ekama G.A. (2007) New developments in modeling wastewater treatment plants – Using stoichiometry to build a plant-wide mass balance based steady state WWTP model. Proc. 10th IWA Conference on Design, Operation and Economics of WWTP, Vienna, 9-13 Sept.

Casey T.G., Ekama G.A., Wentzel M.C. and Marais G.vR (1994) An hypothesis for the causes and control of F/M filamentous organism bulking in nitrogen (N) and nutrient (N & P) removal activated sludge systems. Wat. Sci. Tech. 29(7), 203-212.

Cech, J.S., Hartman, P. (1993) Competition between phosphate and polysaccharide accumulating bacteria in enhanced biological phosphorus removal systems. Water Res 27, 1219-1225.

Coen F., Vanderhaeghen B., Boonen I., Vanrolleghem P.A., Van Eyck L. and Van Meenen P. (1996) Nitrogen removal upgrade of a WWTP within existing reactor volumes: A simulation supported scenario analysis. Wat. Sci. Tech. 34(3/4), 339-346.

Copp J.B. (2002) The COST Simulation Benchmark: Description and Simulator Manual. Office for Official Publications of the European Community, Luxembourg. pp. 154. ISBN 92-894-1658-0.

de Kreuk M.K., Picioreanu C., Hosseinin M., Xavier J.B., and van Loosdrecht M.C.M. (2007) Kinetic model of a granular sludge SBR: Influences on nutrient removal. Biotech. Bioeng. 97(4), 801-815.

Dold P.L., Ekama G.A. and Marais G.v.R. (1980) A general model for the activated sludge process. Prog. Wat. Tech., 12(6), 47-77.

Dold P.L., Wentzel M.C., Billing A.E., Ekama G.A. and Marais G.v.R. (1991) Activated sludge simulation programs: Nitrification and nitrification/denitrification systems (version 1.0). Water Research Commission, PO Box 824, Pretoria 0001, South Africa. TT 52/91, ISBN 0 947.

Ekama G.A. and Marais G.v.R. (1978) The dynamic behaviour of the activated sludge process. Research Report W27, Dept. of Civil Eng., University of Cape Town, Rondebosch, 7701, RSA.

Filipe C. D. M. and Daigger G. T. (1999) Evaluation of the capacity of phosphorus-accumulating organisms to use nitrate and oxygen as final electron acceptors: A theoretical study on population dynamics. Wat. Environ. Res. 71(6), 1140-1150.

Filipe C. D. M., Daigger G. T. and Grady C. P. L. (2001) Stoichiometry and kinetics of acetate uptake under anaerobic conditions by an enriched culture of phosphorus-accumulating organisms at different pHs. Biotechnol. Bioeng. 76(1), 32-43.

Gabb G.M.D., Still D.A., Ekama G.A., Jenkins D. and Marais G.v.R. (1991) The selector effect on filamentous bulking in long sludge age activated sludge system. Wat. Sci. Tech. 23, 867-877.

Gujer W. and Larsen T.A. (1995) The implementation of biokinetics and conservation principles in ASIM. Wat. Sci. Tech. 31(2), 257–266.

Gernaey K.V., Van Loosdrecht M.C.M., Mogens Henze M., and Jørgensen S.B. (2004) Activated sludge wastewater treatment plant modeling and simulation: state of the art. Env. Modeling & Software 19, 763-783

Gujer W. and Henze M. (1991) Activated sludge modeling and simulation. Wat. Sci. Tech. 23, 1011-1023.

Gujer W., Henze M., Mino T. and van Loosdrecht M.C.M. (1999) Activated Sludge Model No. 3. Wat. Sci. Tech 39(1), 183-193.

Henze M., Grady C.P.L. Jr., Gujer W., Marais G.v.R. and Matsuo T. (1987) Activated Sludge Model No. 1. IAWQ Scientific and Technical Report No. 1, London, UK.

Henze M., Gujer W., Mino T., Matsuo T., Wentzel M.C.M. and Marais G.v.R. (1995) Activated Sludge Model No. 2. IWA Scientific and Technical Report No. 3, London, UK.

Henze M., Gujer W., Mino T., Matsuo T., Wentzel M.C., Marais G.v.R. and van Loosdrecht M.C.M. (1999) Activated Sludge Model No.2d. ASM2D. Wat. Sci. Technol., 39(1), 165-182.

Henze M., Gujer W., Mino T., van Loosdrecht M.C.M. (2000) Activated Sludge Models ASM1, ASM2, ASM2d and ASM3. IWA Scientific and Technical Report No. 9, London, UK.

Henze M., van Loosdrecht M.C.M., Ekama G.A., Brdjanovic D. (2009) Biological Wastewater Treatment: Principles, Modeling and Design. ISBN: 9781843391883. IWA Publishing, London, UK.

Hu Z., Wentzel M.C. and Ekama G.A. (2002) The significance of denitrifying polyphosphate accumulating organisms in biological nutrient removal activated sludge systems. Wat. Sci. Tech. 46(1/2), 129-138.

Hu Z., Wentzel M.C. and Ekama G.A. (2007a) A general model for biological nutrient removal activated sludge systems - Model development. Biotech. Bioeng. 98(6) 1242-1258.

Hu Z., Wentzel M.C. and Ekama G.A. (2007b) A general model for biological nutrient removal activated sludge systems - Model evaluation. Biotech. Bioeng. 98(6) 1259-1275.

Hulsbeek J.J.W., Kruit J., Roeleveld P.J. and van Loosdrecht M.C.M. (2002) A practical protocol for dynamic modeling of activated sludge systems. Wat. Sci. Tech. 45(6), 127-136.

Jeppsson U. (1996) Modeling aspects of wastewater treatment processes. PhD thesis, Lund Institute of Technology, Sweden. (available online: http://www.iea.lth.se/publications)

Koch G., Kuhni M., Gujer W. and Siegrist H. (2000) Calibration and validation of activated sludge model no. 3 for Swiss municipal wastewater. Wat. Res. 34, 3580-3590.

Krebs P. (1995) Success and shortcomings of clarifier modeling. Wat. Sci. Tech. 32 (2), 181-191.

Kuba T., Murnleitner E., van Loosdrecht M. C. M. and Heijnen J. J. (1996) A Metabolic Model for Biological Phosphorus Removal by Denitrifying Organisms. Biotechnol. Bioeng. 52(6), 685-695.

Ky R.C., Comeau Y., Perrier M. and Takacs I. (2001) Modeling biological phosphorus removal from a cheese factory effluent by an SBR. Wat. Sci. Tech. 43(3), 257-264.

Lopez-Vazquez C.M. (2009) The competition between polyphosphate-accumulating organisms and glycogen-accumulating organisms: temperature effects and modelling. PhD dissertation. Delft University of Technology. CRC Press/Balkema. Leiden, The Netherlands. ISBN: 9780415558969.

Maurer M., Abramovich D., Siegrist H. and Gujer W. (1999) Kinetics of biologically induced phosphorus precipitation in waste-water treatment. Wat. Res. 33, 484-493.

Maurer M. and Boller M. (1999) Modeling of phosphorus precipitation in wastewater treatment plants with enhanced biological phosphorus removal. Wat. Sci. Tech. 39(1), 147-163.

Meijer S.C.F., van Loosdrecht M.C.M. and Heijnen J.J. (2001) Metabolic modeling of full-scale biological nitrogen and phosphorus removing WWTP's. Wat. Res. 35, 2711-2723.

Meijer S.C.F., van der Spoel H., Susanti S., Heijnen J.J. and van Loosdrecht M.C.M. (2002) Error diagnostics and data reconciliation for activated sludge modeling using mass balances. Wat. Sci. Tech. 45(6), 145-156.

Meijer S.C.F. and Brdjanovic D. (2012) A Practical Guide to Activated Sludge Modelling, ISBN: 9789073445260, UNESCO-IHE.

Meijer S.C.F. (2004) Theoretical and practical aspects of modeling activated sludge processes. PhD Thesis Delft University of Technology, ISBN 90-9018027-3.

Monod J. (1949) The growth of bacterial culture. Pasteur Institute. Paris.

Murnleitner E., Kuba T., van Loosdrecht M.C.M. and Heijnen J.J. (1997) An integrated metabolic model for the aerobic and denitrifying biological phosphorus removal. Biotech. Bioeng. 54, 434-450.

Oehmen A., Lopez-Vazquez C.M., Carvalho G., Reis M.A.M., van Loosdrecht M.C.M. (2010) Modelling the population dynamics and metabolic diversity of organisms relevant in anaerobic/anoxic/aerobic enhanced biological phosphorus removal processes. Water Res 44(15): 4473-4486.

Oehmen, A., Lemos, P.C., Carvalho, G., Yuan, Z., Keller, J., Blackall, L.L., Reis, M.A.M. (2007) Advances in enhanced biological phosphorus removal: from micro to macro scale. Water Res 41(11), 2271-2300.

Olsson G. and Newell B. (1999) Wastewater treatment systems. Modeling, diagnosis and control. IWA Publishing, London, UK.

Petersen B., Gernaey K., Henze M. and Vanrolleghem P.A. (2002) Evaluation of an ASM1 model calibration procedure on a municipal-industrial wastewater treatment plant. J. Hydroinfor. 4, 15-38 .

Pinzón-Pardo A.L., Brdjanovic D., Moussa M.S., López-Vázquez C.M., Meijer S.C.F., Van Straten H.H.A., Janssen A.J.H., Amy G. and van Loosdrecht M.C.M. (2007) Modeling of an oil refinery wastewater treatment plant. Env. Tech. 28, 1273-1284.

Ramirez-Higareda B.L., Lopez-Vazquez C.M., Hooijmans C.M., Brdjanovic D. (2012), Modelling the effects of process configurations on the occurrence of microbial populations and their impact on the performance of enhanced biological phosphorus removal (EBPR) systems. IWA Nutrient Removal and Recovery 2012: Trends in Nutrient Removal and Recovery. September 23-25th, Harbin, China.

Rieger L., Koch G., Kühni M., Gujer W. and Siegrist H. (2001) The EAWAG bio-P module for activated sludge model No. 3. Wat. Res. 35, 3887-3903.

Roeleveld P.J. and van Loosdrecht M.C.M. (2001). Experience with guidelines for wastewater characterisation in The Netherlands. Wat. Sci. Tech., 45(6), 77-87.

Satoh, H., Mino, T., Matsuo, T. (1994) Deterioration of enhanced biological phosphorus removal by the domination of microorganisms without polyphosphate accumulation. Water Sci Technol 30(6), 203-211.

Saunders, A.M., Oehmen, A., Blackall, L.L., Yuan, Z., Keller, J. (2003) The effect of GAO (glycogen accumulating organisms) on anaerobic carbon requirements in full-scale Australian EBPR (enhanced biological phosphorus removal) plants. Water Sci Technol 47(11), 37-43.

Smolders G. L. F., van der Meij J., van Loosdrecht M. C. M. and Heijnen J. J. (1994a) Model of the anaerobic metabolism of the biological phosphorous removal process: stoichiometry and pH influence. Biotechnol.Bioeng. 43(6), 461-470.

Smolders G. L. F., van der Meij J., van Loosdrecht M.C.M. and Heijnen J.J. (1994b) Stoichiometric model of the aerobic metabolism of the biological phosphorus removal process. Biotechnol.Bioeng. 44(7), 837-848.

Smolders G. L. F. (1995) A metabolic model of the biological phosphorus removal; stoichiometry, kinetics and dynamic behaviour. Ph.D. Thesis Delft University of Technology, Delft, the Netherlands, 208 p.

Smolders G. L. F., van der Meij J., van Loosdrecht M. C. M. and Heijnen J. J. (1995a) Structured Metabolic model for anaerobic and aerobic stoichiometry and kinetics of the biological phosphorus removal process. Biotechnol. Bioeng. 47, 277-287.

Smolders G. L. F., Klop J. M., van Loosdrecht M. C. M. and Heijnen J. J. (1995b) A metabolic model for the biological phosphorus removal process: Effect of the sludge retention time. Biotechnol. Bioeng. 48, 222-233.

STOWA (1996) Method for influent characterisation. Report No. 96-08 (in Dutch). Ammersfoort, The Netherlands.

STOWA (1999) Influent characterisation of raw and pre-treated wastewater. Report No. 99-13 (in Dutch). Ammersfoort, The Netherlands.

Smolders G. L. F., Bulstra D. J., Jacobs R., van Loosdrecht M. C. M. and Heijnen J. J. (1995c) A metabolic model for the biological phosphorus removal process II: Validation during start-up conditions. Biotechnol. Bioeng. 48, 234-245.

van Haandel A.C., Ekama G.A. and Marais G.v.R. (1981) The activated sludge process Part 3 - Single sludge denitrification. Wat. Res. 15(10) 1135-1152.

van Veldhuizen H.M., van Loosdrecht M.C.M. and Heijnen J.J. (1999) Modeling biological phosphorus and nitrogen removal in a full scale activated sludge process. Wat. Res. 33, 3459-3468.

Warner A.P.C., Ekama G.A. and Marais G.v.R. (1986) The activated sludge process Part 4 - Application of the general kinetic model to anoxic-aerobic digestion of waste activated sludge. Wat. Res. 20(8) 943-958.

Wentzel M.C. Loewenthal R.E., Ekama G.A. and Marais G.v.R. (1988) Enhanced poly-phosphate organism cultures in activated sludge systems - Part 1: Enhanced culture development. Water SA, 14(2) 81-92.

Wentzel M.C., Ekama G.A., Loewenthal R.E., Dold P.L. and Marais G.v.R. (1989a) Enhanced polyphosphate organism cultures in activated sludge systems Part II - Experimental behaviour. Water SA, 15(2) 71-88.

Wentzel M.C., Dold P.L., Ekama G.A. and Marais G.v.R.(1989b) Enhanced polyphosphate organism cultures in activated sludge systems Part III - Kinetic model. Water SA, 15(2) 89-102.

Wentzel M.C. and Ekama G.A. (1997) Principles in the modeling of biological wastewater treatment plants, in Microbial community analysis: The key to the design of biological wastewater treatment systems. IWA Scientific and Technical Report No. 5.

Wild D., Schulthess R.V. and Gujer W. (1994) Synthesis of denitrification enzymes in activated sludge; modeling with structured biomass. Wat. Sci. Tech. 30(6), 113-122.

NOMENCLATURE

Symbol	Description	Unit
b_A	Specific rate of endogenous mass loss of nitrifying organisms	1/d
b_H	Specific rate of endogenous mass loss of ordinary heterotrophic organisms (OHO)	1/d
F/M	Food to microorganism ratio or load factor (LF)	gCOD/gVSS.d
f_H	Unbiodegradable fraction of the OHO	mgCOD/mgCOD
f_N	Nitrogen content of VSS	mgN/mgVSS
FSA	Free and saline ammonia	mgN/L
K	Half saturation constant	
k_H	Maximum specific hydrolysis rate of SBCOD by OHO under aerobic conditions	mgCOD/mgCOD.d
K_I	Half saturation constant for inhibition compound	mg/L
K_l	External transfer coefficient	m/h
K_N	Half saturation constant for growth of organisms with nitrogen (FSA)	mgN/L
$K_{N,A}$	Half saturation constant for growth of nitrifiers with nitrogen (FSA)	mgN/L
$K_{N,H}$	Half saturation constant for growth of OHOs with nitrogen (FSA)	mgN/L
K_O	Half saturation constant for dissolved oxygen	mgO_2/L
$K_{O,A}$	Half saturation constant for nitrifiers for dissolved oxygen	mgO_2/L
$K_{O,H}$	Half saturation constant for OHOs for dissolved oxygen	mgO_2/L
K_S	Half saturation concentration for soluble organics utilization	mgCOD/L
K_x	Half saturation concentration for utilization SBCOD by OHO	mgCOD/mgCOD.d
q	Specific conversion rate	L/h
Q_{in}	Influent flow rate	m^3/h
Q_{out}	Effluent flow rate	m^3/h
r_i	Observed transformation rate for process i	$ML^{-3}T^{-1}$
S	Soluble concentration in bulk liquid	mgCOD/L
S_{HCO}	Bicarbonate concentration	mg/L
S_I	Soluble unbiodegradable COD concentration	mgCOD/L
S_{in}	Influent substrate concentration	mgCOD/L
S_{KI}	Inhibition compound concentration	mg/L
S_{max}	Saturation concentration	$gCOD/m^3$
S_N	Nitrogen concentration (ammonia or nitrate)	mgN/L
S_{NH}	Free and saline ammonia concentration	mgFSA-N/L
S_{NO}	Nitrate concentration	$mgNO_3$-N/L
S_O	Dissolved oxygen concentration	mgO_2/L
S_{out}	Effluent substrate concentration	mgCOD/L
S_S	Soluble readily biodegradable (RB)COD concentration	mgCOD/L
t	Time	h
V	Reactor volume	m^3
$v_{j,i}$	General stoichiometry term in model matrix for component i in process j	
X	Biomass concentration	$gCOD/m^3$
X_A	Nitrifier biomass concentration	mgCOD/L
X_H	Ordinary heterotrophic (OHO) biomass concentration	mgCOD/L
X_I	Unbiodegradable particulate organics from influent wastewater	mgCOD/L
X_S/X_H	SBCOD/OHO concentration ratio	mgCOD/mgCOD
X_S	Slowly biodegradable (SB)COD concentration	mgCOD/L
$X_{STO,S}$	Intra-cellularly stored organic concentration	mgCOD/L
X_{TSS}	TSS concentration in reactor	mgTSS/L
Y_H	Yield of OHOs	mgCOD/mgCOD

Abbreviation	Description
ADM	Anaerobic digestion model
ASM	Activated sludge model
BOD	Biological oxygen demand
COD	Chemical oxygen demand
CSTR	Complete stirred tank reactor
DO	Dissolved oxygen
DR	Death regeneration
EA	Electron acceptor
EBPR	Enhanced biological phosphorus removal
ER	Endogenous respiration
GAO	Glycogen accumulating organisms
IWA	International Water Association
OUR	Oxygen utilization rate
OHO	Ordinary heterotrophic organisms
PAO	Phosphorus accumulating organisms
PHA	Polyhydroxyalkanoates
RBCOD	Readily biodegradable COD
SBCOD	Slowly biodegradable COD
SRT	Sludge retention time
TKN	Total Kjeldahl nitrogen
TSS	Total suspended solids
TUDP	Delft University of Technology EBPR model
VFA	Volatile fatty acid

Greek symbol	Description	
A	Symbol representing a stoichiometric formula	
H	Reduction factor for utilization of SBCOD under anoxic conditions	
M	Specific growth rate of organisms	1/d
μ_A^{max}	Maximum specific growth rate of nitrifiers	1/d
μ_H	Specific growth rate of OHOs	1/d
μ_H^{max}	Maximum specific growth rate of OHOs	1/d
μ^{max}	Maximum specific growth rate of organisms	1/d
ρ_j	Kinetic rate of process j	$ML^{-3}T^{-1}$

Figure 1.23 The International Short Course on Modeling of Activated Sludge Wastewater Treatment has been offered in Delft jointly by UNESCO-IHE and Delft University of Technology (TUD) for more than 15 years. From 1993-1996 the course was devised and carried out by the activated sludge modeling pioneer, the late Professor G.v.R. Marais of the University of Cape Town, South Africa. The photo features Prof. Mark C.M. van Loosdrecht of TUD delivering a lecture on modeling protocols as a part of the Masters Programme at UNESCO-IHE to the Sanitary Engineering class 2007/09 (photo: V. Becker).

Annex 1.1 Combined ASM2 and TUDP model

Stoichiometric matrix and component composition matrix

			Component →	1	2	3	4	5	6
				S_O	S_F	S_A	S_{NH}	S_{NO}	S_{N2}
↓ Process				gO_2/m^3	$gCOD/m^3$	$gCOD/m^3$	gN/m^3	gN/m^3	gN/m^3
1	r_h^O	Aerobic Hydrolysis	$gCOD_{XS}/d$		$1-f_{SI}$		$c_{N,1}$		
2	r_h^{NO}	Anoxic Hydrolysis	$gCOD_{XS}/d$		$1-f_{SI}$		$c_{N,1}$		
3	r_h^{AO}	Anaerobic Hydrolysis	$gCOD_{XS}/d$		$1-f_{SI}$		$c_{N,1}$		
Regular Heterotrophic Organisms X_H									
4	r_{SF}^O	Aerobic Growth on S_F	$gCOD_{XH}/d$	$-(1/Y_H-1)$	$-1/Y_H$		$c_{N,4}$		
5	r_{SA}^O	Aerobic Growth on S_A	$gCOD_{XH}/d$	$-(1/Y_H-1)$		$-1/Y_H$	$c_{N,5}$		
6	r_{SF}^{NO}	Anoxic Growth on S_F	$gCOD_{XH}/d$		$-1/Y_H$		$c_{N,6}$	$-\frac{(1/Y_H-1)}{2.86}$	$\frac{(1/Y_H-1)}{2.86}$
7	r_{SA}^{NO}	Anoxic Growth on S_A	$gCOD_{XH}/d$			$-1/Y_H$	$c_{N,7}$	$-\frac{(1/Y_H-1)}{2.86}$	$\frac{(1/Y_H-1)}{2.86}$
8	r_{fe}^{AN}	Fermentation	$gCOD_{SF}/d$		-1	1	$c_{N,8}$		
9	r_{HL}	Heterotrophic Lysis	$gCOD_{XH}/d$				$c_{N,9}$		
Phosphorus Accumulating Organisms X_{PAO}									
10	r_{SA}^{AN}	Anaerobic Storage of S_A	$gCOD_{SA}/d$			-1			
11	r_M^{AN}	Anaerobic Maintenance	gP/d						
12	r_{SA}^{NO}	Anoxic Storage of S_A	$gCOD_{SA}/d$			-1		$-\frac{(1-Y_{SA}^{NO})}{2.86}$	$\frac{(1-Y_{SA}^{NO})}{2.86}$
13	r_{PHA}^{NO}	Anoxic PHA Consumption	$gCOD_{PHA}/d$				$c_{N,13}$	$-\frac{(1-1/Y_{PHA}^{NO})}{2.86}$	$\frac{(1-1/Y_{PHA}^{NO})}{2.86}$
14	r_{PP}^{NO}	Anoxic Storage of poly-P	gP/d				$c_{N,14}$	$-\frac{(1/Y_{PP}^{NO})}{2.86}$	$\frac{(1/Y_{PP}^{NO})}{2.86}$
15	r_{GLY}^{NO}	Anoxic Glycogen Formation	$gCOD_{GLY}/d$				$c_{N,15}$	$-\frac{(1/Y_{GLY}^{NO}-1)}{2.86}$	$\frac{(1/Y_{GLY}^{NO}-1)}{2.86}$
16	r_M^{NO}	Anoxic Maintenance	$gCOD_{PAO}/d$				$c_{N,16}$	$-1/2.86$	$1/2.86$
17	r_{PHA}^O	Aerobic PHA Consumption	$gCOD_{PHA}/d$	$1/Y_{PHA}^O-1$			$c_{N,17}$		
18	r_{PP}^O	Aerobic Storage of poly-P	gP/d	$-1/Y_{PP}^O$			$c_{N,18}$		
19	r_{GLY}^O	Aerobic Glycogen Formation	$gCOD_{GLY}/d$	$1-1/Y_{GLY}^O$			$c_{N,19}$		
20	r_M^O	Aerobic Maintenance	$gCOD_{PAO}/d$	-1			$c_{N,20}$		
Autotrophic Nitrifying Organisms X_A									
21	r_A^O	Autotrophic Growth	$gCOD_{XA}/d$	$1-4.57/Y_A$			$c_{N,21}$	$1/Y_A$	
22	r_{AL}	Autotrophic Lysis	$gCOD_{XA}/d$				$c_{N,22}$		

		Component →	1	2	3	4	5	6
			S_O	S_F	S_A	S_{NH}	S_{NO}	S_{N2}
↓ Composition			gO_2	gCOD	gCOD	gN	gN	gN
1	COD	gCOD	-1	1	1		-2.86	...
2	TOC/COD	gC/gCOD		...	0.4			
3	Nitrogen	gN		$i_{N,SF}$	$i_{N,SA}$	1	1	1
4	Phosphorus	gP		$i_{P,SF}$	$i_{P,SA}$			
5	Ionic charge	mole			-1/64	+1/14	-1/14	
6	TSS	g						

7	8	9	10	11	12	13	14	15	16	17	18
S_{PO}	S_I	S_{HCO}	X_I	X_S	X_H	X_{PAO}	X_{PP}	X_{PHA}	X_{GLY}	X_A	X_{TSS}
gP/m^3	$gCOD/m^3$	$mole/m^3$	$gCOD/m^3$	$gCOD/m^3$	$gCOD/m^3$	$gCOD/m^3$	gP/m^3	$gCOD/m^3$	$gCOD/m^3$	$gCOD/m^3$	g/m^3
$c_{P,1}$	f_{SI}	$c_{e,1}$		-1							$c_{TSS,1}$
$c_{P,1}$	f_{SI}	$c_{e,1}$		-1							$c_{TSS,1}$
$c_{P,1}$	f_{SI}	$c_{e,1}$		-1							$c_{TSS,1}$
$c_{P,4}$		$c_{e,4}$			1						$c_{TSS,4}$
$c_{P,5}$		$c_{e,5}$			1						$c_{TSS,5}$
$c_{P,6}$		$c_{e,6}$			1						$c_{TSS,6}$
$c_{P,7}$		$c_{e,7}$			1						$c_{TSS,7}$
$c_{P,8}$		$c_{e,8}$									$c_{TSS,8}$
$c_{P,9}$		$c_{e,9}$	$f_{XI,H}$	$1- f_{XI,H}$	-1						$c_{TSS,9}$
Y_{PO}^{AN}		$c_{e,10}$					$-Y_{PO}^{AN}$	Y_{SA}^{AN}	$1-Y_{SA}^{AN}$		$c_{TSS,10}$
1		$c_{e,11}$					-1				$c_{TSS,11}$
Y_{PO}^{NO}		$c_{e,12}$					$-Y_{PO}^{NO}$	Y_{SA}^{NO}			$c_{TSS,12}$
$c_{P,13}$		$c_{e,13}$				$1/Y_{PHA}^{NO}$		-1			$c_{TSS,13}$
$c_{P,14}$		$c_{e,14}$				$-1/Y_{PP}^{NO}$	1				$c_{TSS,14}$
$c_{P,15}$		$c_{e,15}$				$-1/Y_{GLY}^{NO}$			1		$c_{TSS,15}$
$c_{P,16}$		$c_{e,16}$				-1					$c_{TSS,16}$
$c_{P,17}$		$c_{e,17}$				$1/Y_{PHA}^{O}$		-1			$c_{TSS,17}$
$c_{P,18}$		$c_{e,18}$				$-1/Y_{PP}^{O}$	1				$c_{TSS,18}$
$c_{P,19}$		$c_{e,19}$				$-1/Y_{GLY}^{O}$			1		$c_{TSS,19}$
$c_{P,20}$		$c_{e,20}$				-1					$c_{TSS,20}$
$c_{P,21}$		$c_{e,21}$								1	$c_{TSS,21}$
$c_{P,22}$		$c_{e,22}$	$f_{XI,A}$	$1- f_{XI,A}$						-1	$c_{TSS,22}$

7	8	9	10	11	12	13	14	15	16	17	18
S_{PO}	S_I	S_{HCO}	X_I	X_S	X_H	X_{PAO}	X_{PP}	X_{PHA}	X_{GLY}	X_A	X_{TSS}
gP	gCOD	mole	gCOD	gCOD	gCOD	gCOD	gP	gCOD	gCOD	gCOD	g
	1		1	1	1	1		1	1	1	
	…		…	…	…	0.334 (α)		0.334	0.375	…	
	$i_{N,SI}$		$i_{N,XI}$	$i_{N,XS}$	$i_{N,XH}$	$i_{N,BM}$				$i_{N,BM}$	
1	$i_{P,SI}$		$i_{P,XI}$	$i_{P,XS}$	$i_{P,XH}$	$i_{P,BM}$	1			$i_{P,BM}$	
-1.5/31		-1					-1/31				
			$i_{TSS,XI}$	$i_{TSS,XS}$	$i_{TSS,BM}$	$i_{TSS,BM}$	$i_{TSS,PP}$	$i_{TSS,PHA}$	$i_{TSS,GLY}$	$i_{TSS,BM}$	1

Component composition factors i_N, i_P and i_{TSS}					
1	Nitrogen content of inert soluble COD, S_I	$i_{N,SI}$	0.01	$gN/gCOD_{SI}$	Henze *et al.*, 1999; ASM2d
2	Nitrogen content of soluble substrate, S_F	$i_{N,SF}$	0.03	$gN/gCOD_{SF}$	"
3	Nitrogen content of inert particulate COD, X_I	$i_{N,XI}$	0.03	$gN/gCOD_{XI}$	Meijer (2004)
4	Nitrogen content of particulate substrate, X_S	$i_{N,XS}$	0.03	$gN/gCOD_{XS}$	Henze *et al.*, 1999; ASM2d
5	Nitrogen content of biomass, X_H, X_{PAO}, X_{AUT}	$i_{N,BM}$	0.07	$gN/gCOD_{BM}$	"
6	Phosphorus content of inert soluble COD, S_I	$i_{P,SI}$	0	$gP/gCOD_{SI}$	"
7	Phosphorus content of soluble substrate, S_F	$i_{P,SF}$	0.01	$gP/gCOD_{SF}$	"
8	Phosphorus content of inert particulate COD, X_I	$i_{P,XI}$	0.01	$gP/gCOD_{XI}$	"
9	Phosphorus content of particulate substrate, X_S	$i_{P,XS}$	0.01	$gP/gCOD_{XS}$	"
10	Phosphorus content of biomass, X_H, X_{PAO}, X_{AUT}	$i_{P,BM}$	0.02	$gP/gCOD_{BM}$	"
11	Ratio Total Suspended Solids to X_I	$i_{TSS,XI}$	0.75	$gTSS/gCOD_{XI}$	"
12	Ratio Total Suspended Solids to X_S	$i_{TSS,XS}$	0.75	$gTSS/gCOD_{XS}$	"
13	Ratio Total Suspended Solids to biomass X_H, X_{PAO}, X_{AUT} $(CH_{2.09}O_{0.54}N_{0.2}P_{0.015})_n$	$i_{TSS,BM}$	0.90	$gTSS/gCOD_{BM}$	Smolders *et al.*, 1994b
14	Ratio Total Suspended Solids to X_{PP} $(Mg^{2+}_{1/3}K^{+}_{1/3}PO_3)_n$	$i_{TSS,PP}$	3.23	$gTSS/gP_{PP}$	Smolders *et al.*, 1994a
15	Ratio Total Suspended Solids to X_{PHA} $(C_4H_6O_2)_{1/4}$	$i_{TSS,PHA}$	0.6	$gTSS/gCOD_{PHA}$	Doi, 1990
16	Ratio Total Suspended Solids to X_{GLY} $(C_6H_{10}O_5)_{1/6}$	$i_{TSS,GLY}$	0.84	$gTSS/gCOD_{GLY}$	Stryer, 1975
17	Ratio COD to Oxygen (S_O)	$i_{COD,O}$	-1	$gCOD/gO_2$	
18	Ratio COD to Nitrate (S_{NO})	$i_{COD,NO}$	-2.86	$gCOD/gN_{SNO}$	

Stoichiometric parameters

Hydrolysis of particulate substrate					
1	Fraction of inert COD generated in hydrolysis	f_{SI}	0	$gCOD_{SI}/gCOD_{(XH+XPAO)}$	Henze *et al.*, 1999; ASM2d
Regular heterotrophic organisms, X_H					
1	Heterotrophic yield for growth on substrate	Y_H	0.63	$gCOD_{XH}/gCOD$	"
2	Fraction of inert COD generated in biomass lysis	$f_{XI,H}$	0.10	$gCOD_{XI}/gCOD_{XH}$	"
Autotrophic nitrifying organisms, X_A					
1	Autotrophic yield for growth	Y_A	0.24	$gCOD_{XA}/gN_{SNH}$	"
2	Fraction of inert COD generated in biomass lysis	$f_{XI,A}$	0.10	$gCOD_{XI}/gCOD_{XA}$	"
Phosphorus accumulating organisms, X_{PAO}					
1	ATP produced per NADH or P/O ratio	δ	1.85	moleATP/moleNADH	Smolders *et al.*, 1994b
2	Observed biomass ratio TOC over COD	α	0.334	$gC_{PAO}/gCOD_{PAO}$	"
4	Anaerobic yield for phosphate release	Y_{PO}^{AN}	$0.184 \times pH - 0.94 \approx 0.35$	$gP_{SPO}/gCOD_{SA}$	Smolders *et al.*, 1994a
5	Yield for anaerobic formation of PHA from S_A	Y_{SA}^{AN}	1.50	$gCOD_{PHB}/gCOD_{SA}$	"
6	Observed yield for anoxic phosphate release	Y_{PO}^{NO}	0.23	$gP_{SPO}/gCOD_{SA}$	Kuba *et al.*, 1994
7	Yield for anoxic formation of PHA from S_A	Y_{SA}^{NO}	$\frac{(8.3 \cdot Y_{PO}^{NO} - 4.9 + 8 \cdot \delta)}{9 \cdot \delta} \approx 0.71$	$gCOD_{PHB}/gCOD_{SA}$	Meijer (2004)
8	Anoxic yield for degradation of X_{PHB}	Y_{PHA}^{NO}	$\frac{19 + 3 \cdot \delta/\alpha}{4 + 9 \cdot \delta} \approx 1.72$	$gCOD_{PHB}/gCOD_{PAO}$	+*Murnleitner *et al.*, 1997
9	Anoxic yield for formation of X_{GLY}	Y_{GLY}^{NO}	$\frac{12.6 + 2 \cdot \delta/\alpha}{9 + 6 \cdot \delta} \approx 1.18$	$gCOD_{GLY}/gCOD_{PAO}$	"
10	Anoxic yield for formation of X_{PP}	Y_{PP}^{NO}	$\frac{57 + 9 \cdot \delta/\alpha}{28 + 4 \cdot \delta} \approx 3.02$	$gP_{PP}/gCOD_{PAO}$	"
11	Aerobic yield for degradation of X_{PHB}	Y_{PHA}^{O}	$\frac{9.4 + 3 \cdot \delta/\alpha}{2 + 9 \cdot \delta} \approx 1.39$	$gCOD_{PHB}/gCOD_{PAO}$	+Murnleitner *et al.*, 1997
12	Aerobic yield for formation of X_{GLY}	Y_{GLY}^{O}	$\frac{12.6 + 4 \cdot \delta/\alpha}{9 + 12 \cdot \delta} \approx 1.11$	$gCOD_{GLY}/gCOD_{PAO}$	"
13	Aerobic yield for formation of X_{PP}	Y_{PP}^{O}	$\frac{57 + 18 \cdot \delta/\alpha}{28 + 4 \cdot \delta} \approx 4.42$	$gP_{PP}/gCOD_{PAO}$	+*Murnleitner *et al.*, 1997

* Parameters containing typesetting errors in *Henze et al.*, 1999 (ASM2d).
\+ Parameters containing rounding errors in Murnleitner *et al.* (1996) and van Veldhuizen *et al.* (1999).

Kinetic parameters for hydrolysis, X_H and X_A

Hydrolysis of particulate substrate					
1	Hydrolysis rate (T)	k_h	$3.0 \times e^{(0.0406 \cdot (T-20))}$	$gCOD_{XS}/gCOD_{(XH+XPAO)}.d$	Henze *et al.*, 1999; ASM2d
2	Anoxic hydrolysis reduction factor	η_{NO}	0.8	-	Meijer, 2004
3	Anaerobic hydrolysis reduction factor	η_{fe}	0.2	-	Henze *et al.*, 1999; ASM2d
4	Saturation / inhibition coefficient for oxygen	K_O	0.2	gO_2/m^3	"
5	Saturation / inhibition coefficient for nitrate	K_{NO}	0.5	gN_{SNO}/m^3	"
6	Saturation coefficient for particulate COD (T)	K_X	$0.1 \times e^{(-0.110 \cdot (T-20))}$	$gCOD_{XS}/gCOD_{(XH+XPAO)}$	"
Heterotrophic microorganisms, X_H					
1	Maximum heterotrophic growth rate (T)	μ_H	$6.0 \times e^{(0.069 \cdot (T-20))}$	$gCOD_{XH}/gCOD_{XH}.d$	"
2	Maximum fermentation rate (T)	q_{fe}	$3.0 \times e^{(0.069 \cdot (T-20))}$	$gCOD_{SF}/gCOD_{XH}.d$	"
3	Heterotrophic decay rate (T)	b_H	$0.4 \times e^{(0.069 \cdot (T-20))}$	$gCOD_{XH}/gCOD_{XH}.d$	"
4	Reduction factor for denitrification	η_{NO}	0.8	-	"
5	Saturation / inhibition coefficient for oxygen	K_O	0.2	gO_2/m^3	"
6	Saturation coefficient for growth on S_F	K_F	4.0	$gCOD_{SF}/m^3$	"
7	Saturation coefficient for fermentation of S_F	K_{fe}	20.0	$gCOD_{SF}/m^3$	* Gujer *et al.*, 1995; ASM2
8	Saturation coefficient for growth on Acetate	K_{Ac}	4.0	$gCOD_{SA}/m^3$	Henze *et al.*, 1999; ASM2d
9	Saturation / inhibition coefficient for nitrate	K_{NO}	0.5	gN_{SNO}/m^3	"
10	Saturation coef. for Ammonium (nutrient)	K_N	0.05	gN_{SNH}/m^3	"
11	Saturation coefficient for Phosphate (nutrient)	K_P	0.01	gP_{SPO}/m^3	"
12	Saturation coefficient for alkalinity (HCO_3^-)	K_{HCO}	0.1	$moleHCO_3^-/m^3$	"
Autotrophic microorganisms, X_A					
1	Autotrophic growth rate (T)	μ_A	$1.0 \times e^{(0.105 \cdot (T-20))}$	$gCOD_{XA}/gCOD_{XA}.d$	"
2	Autotrophic decay rate (T)	b_A	$0.15 \times e^{(0.110 \cdot (T-20))}$	$gCOD_{XA}/gCOD_{XA}.d$	"
3	Saturation coefficient for oxygen	$K_{A,O}$	0.5	gO_2/m^3	"
4	Saturation coefficient for Ammonium	K_{NH}	1.0	gN_{SNH}/m^3	"
5	Saturation coefficient for Phosphate (nutrient)	K_P	0.01	gP_{SPO}/m^3	"
6	Saturation coefficient for alkalinity (HCO_3^-)	K_{HCO}	0.5	$moleHCO_3^-/m^3$	"

(T) Temperature dependant parameters and their reference.
* Parameters containing typesetting errors in *Henze et al.*, 1999 (ASM2d).

Kinetic parameters for X_{PAO}

Phosphorus accumulating organisms, X_{PAO}					
1	Maximum anaerobic acetate uptake rate (T)	q_{Ac}	$8.0 \times e^{(0.090\cdot(T-20))}$	$gCOD_{SA}/gCOD_{PAO}.d$	This thesis; (T) Brdjanovic et al., 1998
2	Anaerobic maintenance rate (T)	m_{AN}	$0.05 \times e^{(0.069\cdot(T-20))}$	$gP_{PP}/gCOD_{PAO}.d$	Smolders et al., 1995; (T) Murnleitner et al.. 1997
3	Maximum anoxic acetate uptake rate (T)	q_{Ac}^{NO}	$1.5 \cdot q_{Ac} \times e^{(0.090\cdot(T-20))}$	$gCOD_{SA}/gCOD_{PAO}.d$	Kuba et al., 1994; (T) Brdjanovic et al., 1998
4	PHA degradation rate (T)	k_{PHA}	$5.51 \times e^{(0.121\cdot(T-20))}$	$gCOD_{PHA}/gCOD_{PAO}.d$	This thesis; (T) Brdjanovic et al., 1998
5	Glycogen formation rate (T)	k_{GLY}	$0.93 \times e^{(0.118\cdot(T-20))}$	$gCOD_{GLY}/gCOD_{PAO}.d$	Meijer, 2004
6	Poly-phosphate formation rate (T)	k_{PP}	$0.10 \times e^{(0.031\cdot(T-20))}$	$gP_{PP}/gCOD_{PAO}.d$	Murnleitner et al., 1997; (T) Brdianovic et al.. 1998
7	Observed oxygen consumption for	m_{OC}	0.096	$gO_2/gCOD_{PAO}.d$	Smolders *et al.*, 1994b
8	Aerobic maintenance rate (T)	m_O	$\frac{3\cdot\delta\cdot m_{OC}}{3.2+\delta/\alpha} \approx 0.06 \times e^{(0.069\cdot(T-20))}$	$gCOD_{PAO}/gCOD_{PAO}.d$	+° Murnleitner et al., 1997
9	Anoxic maintenance rate (T)	m_{NO}	$\frac{6\cdot\delta\cdot m_{OC}}{6.3+\delta/\alpha} \approx 0.09 \times e^{(0.069\cdot(T-20))}$	$gCOD_{PAO}/gCOD_{PAO}.d$	°
10	Saturation reduction factor for PP formation	g_{PP}	0.22	-	Murnleitner *et al.*, 1997
11	Reduction factor for denitrifying P removal	η_{NO}	0.8	-	Meijer, 2004
12	Saturation coefficient for poly-P formation	K_{PO}	1.0	gP_{SPO}/m^3	°
13	Saturation coefficient for growth on acetate	K_{Ac}	4.0	$gCOD_{SA}/m^3$	Henze *et al.*, 1999; ASM2d
14	Saturation / inhibition coefficient for nitrate	K_{NO}	0.5	gN_{SNO}/m^3	°
15	Saturation / inhibition coefficient for	K_O	0.2	gO_2/m^3	°
16	Saturation coefficient for f_{PHA}	K_{fPHA}	0.2	$gCOD_{PHA}/gCOD_{PAO}$	Meijer, 2004
17	Saturation coeff. for Phosphate (nutrient)	K_P	0.02	gP_{SPO}/m^3	°
18	Saturation coefficient for NH_4 (nutrient)	K_N	0.05	gN_{SNH}/m^3	°
19	Maximum poly-phosphate fraction of PAO	f_{PP}^{max}	0.35	$gP_{PP}/gCOD_{PAO}$	Wentzel *et al.*, 1989
20	Maximum glycogen fraction of PAO	f_{GLY}^{max}	0.5	$gCOD_{GLY}/gCOD_{PAO}$	Brdjanovic *et al.*, 1998
21	Saturation coefficient for poly-P	K_{PP}	0.01	gP_{PP}/m^3	Switch
22	Saturation coefficient for glycogen	K_{GLY}	0.01	$gCOD_{GLY}/m^3$	°
23	Saturation coefficient for PHA	K_{PHA}	0.01	$gCOD_{PHA}/m^3$	°
24	Saturation coefficient for f_{GLY}	K_{fGLY}	0.01	$gCOD_{GLY}/gCOD_{PAO}$	°
25	Saturation coefficient for f_{PP}	K_{fPP}	0.01	$gCOD_{PP}/gCOD_{PAO}$	°
26	Saturation coefficient for alkalinity (HCO_3^-)	K_{HCO}	0.01	$moleHCO_3^-/m^3$	°

+ Parameters containing rounding errors in Murnleitner *et al.* (1996) and van Veldhuizen *et al.* (1999).
° Parameters containing typesetting errors in van Veldhuizen *et al.* (1999).
(T) Temperature dependant parameters including their reference.

Kinetic parameters used for the simulation of lab-scale A2 and A/O SBR systems

Phosphorus accumulating organisms, X_{PAO}					
1^a	Maximum anaerobic acetate uptake rate (T) A/O SBR system	q_{Ac}	$9.67 \times e^{(0.090 \cdot (T-20))}$	$gCOD_{SA}/gCOD_{PAO}.d$	Smolders *et al.*, 1995b; (T) Brdjanovic *et al.*, 1998
1^b	Maximum anaerobic acetate uptake rate (T) A2 SBR system	q_{Ac}	$4.3 \times e^{(0.090 \cdot (T-20))}$	$gCOD_{SA}/gCOD_{PAO}.d$	Kuba *et al.*, 1996; (T) Brdjanovic *et al.*, 1998
1^c	Maximum anaerobic acetate uptake rate (T) best fit A2 and A/O SBR system	q_{Ac}	$6.4 \times e^{(0.090 \cdot (T-20))}$	$gCOD_{SA}/gCOD_{PAO}.d$	Murnleitner *et al.*, 1997; (T)Brdjanovic *et al.*, 1998
1^d	Maximum anaerobic acetate uptake rate (T) best fit lab - and full-scale conditions	q_{Ac}	$8.0 \times e^{(0.090 \cdot (T-20))}$	$gCOD_{SA}/gCOD_{PAO}.d$	This thesis; (T) Brdjanovic et al., 1998
2	Glycogen formation rate (T) best fit A2 and A/O SBR system	k_{GLY}	$5.83 \times e^{(0.118 \cdot (T-20))}$	$gCOD_{GLY}/gCOD_{PAO}.d$	Meijer, 2004
3	Saturation coefficient for poly-P formation A2 and A/O SBR system	K_{PO}	3.1	gP_{SPO}/m^3	Murnleitner *et al.*, 1997
4	Saturation coefficient for growth on acetate A/O SBR system	K_{Ac}	32	$gCOD_{SA}/m^3$	Smolders et al., 1995a
5	Saturation / inhibition coefficient for nitrate A2 and A/O SBR system	K_{NO}	1.4	gN_{SNO}/m^3	Murnleitner *et al.*, 1997

Kinetic rate equations (X_S and X_H)

	Process	Kinetic rate equation (r)	Switch function (on/off)
Hydrolysis of particulate substrate, X_S			
1	Aerobic Hydrolysis ($gCOD_{XS}/d$)	$r_h^O = k_h \cdot \frac{X_S/(X_H + X_{PAO})}{K_X + X_S/(X_H + X_{PAO})} \cdot (X_H + X_{PAO})$	
2	Anoxic Hydrolysis ($gCOD_{XS}/d$)	$r_h^{NO} = \eta_{NO} \cdot k_h \cdot \frac{X_S/(X_H + X_{PAO})}{K_X + X_S/(X_H + X_{PAO})} \cdot \frac{S_{NO}}{K_{NO} + S_{NO}} \cdot (X_H + X_{PAO})$	$\cdot \frac{K_O}{K_O + S_O}$
3	Anaerobic Hydrolysis ($gCOD_{XS}/d$)	$r_h^{AN} = \eta_{fe} \cdot k_h \cdot \frac{X_S/(X_H + X_{PAO})}{K_X + X_S/(X_H + X_{PAO})} \cdot (X_H + X_{PAO})$	$\cdot \frac{K_O}{K_O + S_O} \cdot \frac{K_{NO}}{K_{NO} + S_{NO}}$
Heterotrophic microorganisms, X_H			
4	Aerobic Growth on S_F ($gCOD_{XH}/d$)	$r_{SF}^O = \mu_H \cdot \frac{S_F}{S_A + S_F} \cdot \frac{S_F}{K_F + S_F} \cdot \frac{S_O}{K_O + S_O} \cdot X_H$	$\cdot \frac{S_{NH}}{K_N + S_{NH}} \cdot \frac{S_{PO}}{K_P + S_{PO}} \cdot \frac{S_{HCO}}{K_{HCO} + S_{HCO}}$
5	Aerobic Growth on S_A ($gCOD_{XH}/d$)	$r_{SA}^O = \mu_H \cdot \frac{S_A}{S_A + S_F} \cdot \frac{S_A}{K_A + S_A} \cdot \frac{S_O}{K_O + S_O} \cdot X_H$	$\cdot \frac{S_{NH}}{K_N + S_{NH}} \cdot \frac{S_{PO}}{K_P + S_{PO}} \cdot \frac{S_{HCO}}{K_{HCO} + S_{HCO}}$
6	Anoxic Growth on S_F ($gCOD_{XH}/d$)	$r_{SF}^{NO} = \eta_{NO} \cdot \mu_H \cdot \frac{S_F}{S_A + S_F} \cdot \frac{S_F}{K_F + S_F} \cdot \frac{S_{NO}}{K_{NO} + S_{NO}} \cdot X_H$	$\cdot \frac{K_O}{K_O + S_O} \cdot \frac{S_{NH}}{K_N + S_{NH}} \cdot \frac{S_{PO}}{K_P + S_{PO}} \cdot \frac{S_{HCO}}{K_{HCO} + S_{HCO}}$
7	Anoxic Growth on S_A ($gCOD_{XH}/d$)	$r_{SA}^{NO} = \eta_{NO} \cdot \mu_H \cdot \frac{S_A}{S_A + S_F} \cdot \frac{S_A}{K_A + S_A} \cdot \frac{S_{NO}}{K_{NO} + S_{NO}} \cdot X_H$	$\cdot \frac{K_O}{K_O + S_O} \cdot \frac{S_{NH}}{K_N + S_{NH}} \cdot \frac{S_{PO}}{K_P + S_{PO}} \cdot \frac{S_{HCO}}{K_{HCO} + S_{HCO}}$
8	Fermentation of S_F ($gCOD_{SF}/d$)	$r_{fe}^{AN} = q_{fe} \cdot \frac{S_F}{K_{fe} + S_F} \cdot X_H$	$\cdot \frac{K_O}{K_O + S_O} \cdot \frac{K_{NO}}{K_{NO} + S_{NO}} \cdot \frac{S_{HCO}}{K_{HCO} + S_{HCO}}$
9	Heterotrophic Lysis ($gCOD_{XH}/d$)	$r_{HL} = b_H \cdot X_H$	

Kinetic rate equations (X_{PAO} and X_A)

	Process	Kinetic rate equation (r)	Switch function (on/off)
	Phosphorus accumulating organisms, X_{PAO}		
10	Anaerobic storage of S_A ($gCOD_{SA}/d$)	$r_{SA}^{AN} = q_{Ac} \cdot \frac{S_A}{K_A + S_A} \cdot X_{PAO}$	$\cdot \frac{K_O}{K_O+S_O} \cdot \frac{K_{NO}}{K_{NO}+S_{NO}} \cdot \frac{X_{GLY}}{K_{GLY}+X_{GLY}} \cdot \frac{X_{PP}}{K_{PP}+X_{PP}}$
11	Anaerobic Maintenance (gP/d)	$r_M^{AN} = m_{AN} \cdot X_{PAO}$	$\cdot \frac{K_O}{K_O+S_O} \cdot \frac{K_{NO}}{K_{NO}+S_{NO}} \cdot \frac{X_{PP}}{K_{PP}+X_{PP}}$
12	Anoxic storage of S_A ($gCOD_{SA}/d$)	$r_{SA}^{NO} = q_{Ac}^{NO} \cdot \frac{S_A}{K_A + S_A} \cdot \frac{S_{NO}}{K_{NO} + S_{NO}} \cdot X_{PAO}$	$\cdot \frac{K_O}{K_O+S_O} \cdot \frac{X_{PP}}{K_{PP}+X_{PP}}$
13	Anoxic PHA consumption ($gCOD_{XPHA}/d$)	$r_{PHA}^{NO} = \eta_{NO} \cdot k_{PHA} \cdot \frac{X_{PHA}/X_{PAO}}{K_{fPHA} + X_{PHA}/X_{PAO}} \cdot \frac{S_{NO}}{K_{NO}+S_{NO}} \cdot X_{PAO}$	$\cdot \frac{K_O}{K_O+S_O} \cdot \frac{S_{NH}}{K_N+S_{NH}} \cdot \frac{S_{PO}}{K_P+S_{PO}} \cdot \frac{S_{HCO}}{K_{HCO}+S_{HCO}}$
14	Anoxic storage of PP (gP/d)	$r_{PP}^{NO} = \eta_{NO} \cdot k_{PP} \cdot \frac{X_{PAO}}{X_{PP}} \cdot \frac{S_{PO}}{K_{PO}+S_{PO}} \cdot \frac{S_{NO}}{g_{PP} \cdot K_{NO} + S_{NO}} \cdot X_{PAO}$	$\cdot \frac{K_O}{K_O+S_O} \cdot \frac{X_{PHA}}{K_{PHA}+X_{PHA}} \cdot \frac{f_{PP}^{max} - X_{PP}/X_{PAO}}{K_{fPP} + (f_{PP}^{max} - X_{PP}/X_{PAO})}$
15	Anoxic Glycogen formation ($gCOD_{XGLY}/d$)	$r_{GLY}^{NO} = \eta_{NO} \cdot k_{GLY} \cdot \frac{X_{PHA}}{X_{GLY}} \cdot \frac{S_{NO}}{K_{NO}+S_{NO}} \cdot X_{PAO}$	$\cdot \frac{K_O}{K_O+S_O} \cdot \frac{X_{PHA}}{K_{PHA}+X_{PHA}} \cdot \frac{f_{GLY}^{max} - X_{GLY}/X_{PAO}}{K_{fGLY} + (f_{GLY}^{max} - X_{GLY}/X_{PAO})}$
16	Anoxic Maintenance ($gCOD_{XPAO}/d$)	$r_M^{NO} = m_{NO} \cdot \frac{S_{NO}}{K_{NO}+S_{NO}} \cdot X_{PAO}$	$\cdot \frac{K_O}{K_O+S_O}$
17	Aerobic PHA consumption ($gCOD_{XPHA}/d$)	$r_{PHA}^{O} = k_{PHA} \cdot \frac{X_{PHA}/X_{PAO}}{K_{fPHA} + X_{PHA}/X_{PAO}} \cdot \frac{S_O}{K_O + S_O} \cdot X_{PAO}$	$\cdot \frac{S_{NH}}{K_N+S_{NH}} \cdot \frac{S_{PO}}{K_P+S_{PO}} \cdot \frac{S_{HCO}}{K_{HCO}+S_{HCO}}$
18	Aerobic storage of PP (gP/d)	$r_{PP}^{O} = k_{PP} \cdot \frac{X_{PAO}}{X_{PP}} \cdot \frac{S_{PO}}{K_{PO}+S_{PO}} \cdot \frac{S_O}{g_{PP} \cdot K_O + S_O} \cdot X_{PAO}$	$\cdot \frac{X_{PHA}}{K_{PHA}+X_{PHA}} \cdot \frac{f_{PP}^{max} - X_{PP}/X_{PAO}}{K_{fPP} + (f_{PP}^{max} - X_{PP}/X_{PAO})}$
19	Aerobic Glycogen formation ($gCOD_{XGLY}/d$)	$r_{GLY}^{O} = k_{GLY} \cdot \frac{X_{PHA}}{X_{GLY}} \cdot \frac{S_O}{K_O+S_O} \cdot X_{PAO}$	$\cdot \frac{X_{PHA}}{K_{PHA}+X_{PHA}} \cdot \frac{f_{GLY}^{max} - X_{GLY}/X_{PAO}}{K_{fGLY} + (f_{GLY}^{max} - X_{GLY}/X_{PAO})}$
20	Aerobic Maintenance ($gCOD_{XPAO}/d$)	$r_M^{O} = m_O \cdot \frac{S_O}{K_O+S_O} \cdot X_{PAO}$	
	Autotrophic nitrifying organisms, X_A		
21	Autotrophic growth ($gCOD_{XA}/d$)	$r_A^{O} = \mu_A \cdot \frac{S_{NH}}{K_{NH}+S_{NH}} \cdot \frac{S_O}{K_O+S_O} \cdot X_A$	$\cdot \frac{K_{PO}}{K_P+S_{PO}}$
22	Autotrophic Lysis ($gCOD_{XA}/d$)	$r_{AL} = b_A \cdot X_A$	

Chapter 2: WWTP Holten, the Netherlands

Model development and full-scale testing

Van Veldhuizen H.M., van Loosdrecht M.C.M. and Heijnen J.J.

This chapter is based on the paper 'Modelling biological phosphorus and nitrogen removal in a full scale activated sludge process' van Veldhuizen H.M., van Loosdrecht M.C.M. and Heijnen J.J. (1999). Wat. Res. 33, 3459-3468.

2.1 Introduction

An activated sludge process is a complex system in which a range of bacterial conversion and transport processes occur. Kinetics, stoichiometry and transport processes play an important role in the conversion of contaminants. Models are needed for a quantitative evaluation of the processes. These models can become quite complex due to the wide variety of biological processes. In particular, organisms responsible for biological phosphorus removal processes have a complicated cellular physiology (Mino *et al.,* 1996) which is not easy to understand based on external measurements (observations) only.

An IAWQ task group (nowadays called IWA) has proposed a dynamic activated sludge model for the biological removal of COD, N and P, based on direct measurable soluble compounds (ASM2; Gujer et al., 1995). In the Environmental Biotechnology research group of the TUDelft, a metabolic approach was developed to describe the biological phosphorus removal processes based on internal storage compounds (TUDP model). The parameters of the model have been extensively validated by many independent experiments. (Smolders *et al.,* 1994; Kuba *et al.,* 1996 and Murnleitner *et al.,* 1996). A discussion on both approaches is given by Van Loosdrecht (1996) and summarized in the previous chapter.

The goal of the present study was to check the applicability of the metabolic modelling approach developed under laboratory conditions on a full-scale wastewater treatment plant (WWTP). Therefore an integrated activated sludge model for the removal of COD, N and P was developed. For the standard COD and N conversion processes, ASM2 was used as the basis. The mathematical description of the metabolic model for EBPR (Murnleitner *et al.,* 1996) was incorporated in this model structure. The entire model was implemented by means of the computer software package SIMBA 3.0 (based on MATLAB/SIMULINK). The performance of the model was tested on a full-scale WWTP with EBPR under both aerobic and anoxic conditions. First, the sensitivity of the concentrations of ammonium, nitrate and phosphate and the sludge production for the model parameters, the influent composition and several flows of water, sludge and air on the treatment plant was determined. Hereafter, the model was calibrated and validated. In this way, one could evaluate the use of two different approaches for model calibration: a system engineering approach (purely mathematical based on a sensitivity analysis), and a process engineering approach (based on detailed knowledge of the processes and the model structure).

2.2 Model kinetics and stoichiometry

While modelling the complex activated sludge processes, one has to make choices regarding the abstractness of the different parts of the model. The conversion of organic matter (COD) is a complex process since the substrate in sewage as well as the organisms in activated sludge systems are widely varying and have therefore to be modelled on a rather abstract level. The conversions related to the biological nutrient removal (BNR) are, however, performed by more specific microbial groups and related to a more uniform substrate. These can therefore be modelled in more detail.

The TUDP model does not take two extreme cases into account, since they were not part of the original experiments and do not occur in the full-scale WWTP that is subject to this study. These extremes are (i) full depletion of the poly-hydroxy-alkanoates (PHA) pool in PAO (Temmink *et al.*, 1995; Brdjanovic *et al.*, 1997) and, (ii) simultaneous presence of volatile fatty acids (VFA) and electron acceptors such as oxygen or nitrate (Kuba *et al.*, 1996). In order to take these cases into account a more refined model is needed. The metabolic model is based on acetate as substrate. In practice, however, also other VFA and substrates are present in the sewage. This might influence the yield of PHA on VFA. Nevertheless, acetate is the dominant product of the bacterial fermentation processes at neutral pH. In the case described here, the measured VFA in the influent consisted almost entirely of acetate, therefore the acetate-based stoichiometry was used.

In Annex 1.1 the TUDP metabolic model for the EBPR is presented coupled with ASM2 to describe the activities of heterotrophs, autotrophs, and autotrophs as well as the fermentation process (Gujer *et al.*, 1995). The values of the stoichiometric coefficients $c_{i,n}$, $c_{i,p}$, and $c_{i,t}$ are the result of the mass balances over nitrogen, phosphorus and total suspended solids, respectively. The values of the stoichiometric and kinetic parameters of the heterotrophic and autotrophic organisms and of the hydrolysis process originate from ASM2 (Gujer *et al.*, 1995). The parameters for the PAO are taken from the work of Smolders *et al.* (1994), Kuba *et al.* (1996) and Murnleitner *et al.* (1996) and presented in Annex 1.1.

Some of the kinetic parameters of the model are dependent on temperature. Therefore temperature coefficients (θ) are given based on Eq. 2.1.

$$r_T = r_{293} \cdot exp(\theta \cdot (T - 293)) \qquad (2.1)$$

The temperature coefficients of the processes, originating from ASM2, are extracted from this model (Gujer *et al.*, 1995). The temperature coefficients of PAO were not available at the moment of the simulations. Therefore temperature coefficients for the PAO kinetics are taken the same as for heterotrophic growth, i.e. $\theta = 0.069$. The affinity constants are considered to be independent of temperature.

The main difference between the two parts of the total model is the way how the decay and maintenance processes are dealt with (van Loosdrecht, 1996). ASM2 describes such processes assuming the lysis of the organisms, whereby part of the lysis products are reused as substrate (Gujer *et al.*, 1995). Instead, the TUDP model makes use of the maintenance concept, whereby part of the substrates is used for maintenance. PAO always have internal substrate (PHA) available to satisfy the maintenance requirements. Since the use of the latter concept is simpler and well validated from laboratory tests, we believe that this description is acceptable. Thus, the heterotrophic and nitrifying organisms are described by the decay concept and the maintenance concept is used for PAO.

2.2 Process description of WWTP Holten

The WWTP Holten is a modification of University of Cape Town (UCT) type plants with plug flow characteristics, called biological chemical phosphorus and nitrogen removal (BCFS®) process. A complete description of the WWTP can be found in van Loosdrecht *et al.* (1997a). In the BCFS process, part of the phosphorus is removed in a chemical way: a phosphate rich water stream is withdrawn by means of an integrated "settler" at the end of the anaerobic zone, and fed to the pre-thickener where the dissolved phosphate is precipitated with iron-chloride. This procedure is used because of the high sludge retention time applied (about 50 days) which makes it impossible to remove all phosphorus by biological means. To test the activated sludge model only the activated sludge tank together with the secondary settler is simulated. A schematic presentation of this treatment line is given in Figure 2.1.

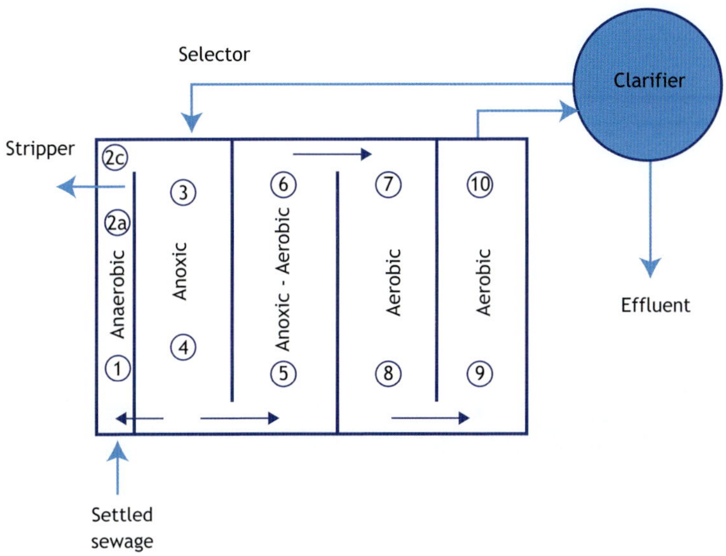

Figure 2.1 Schematic presentation of the aeration basin and the secondary clarifier of the WWTP Holten.

The activated sludge tank is a plug flow reactor, which makes the number of modelled reactors in series rather important. The set-up of the hydraulic model in SIMBA is shown in Figure 2.2 and is based on the physical separation, i.e. that each physically separate compartment is at least 1 tank in the hydraulic model. Compartments 1, 3 and 4 have 2 tanks per compartment because (i) the retention times are relatively short (20 min), (ii) the mixing devices in these tanks only operate during a few minutes per hour to prevent sludge settling and (iii) the ratio between length and width is larger than one. The extent of importance of the set-up of the hydraulic model is tested separately in the sensitivity analysis. Table 2.1 gives the physical compartmentalisation in WWTP Holten and of the model compartmentalisation in SIMBA.

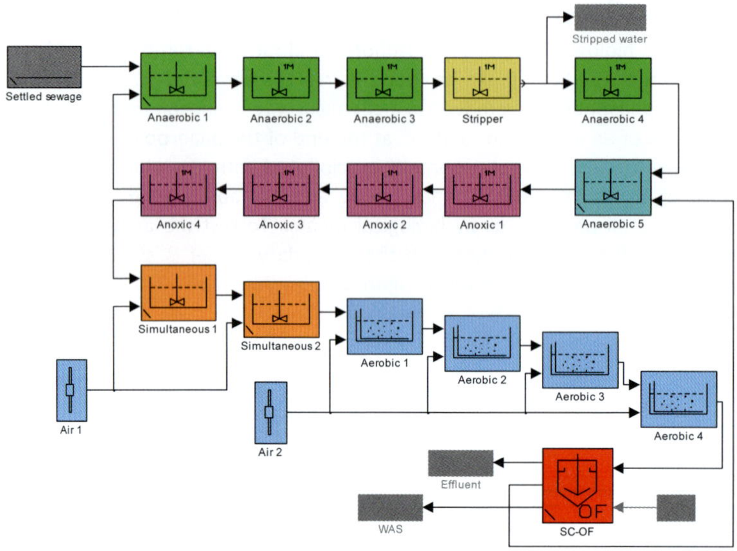

Figure 2.2 SIMBA schematic layout WWTP Holten of the activated sludge tanks combined with secondary settling.

Table 2.1 Division of the physical compartments in WWTP Holten and modelled reactors in SIMBA.

Zone	Physical compartment	Volume (m^3)	HRT (min)	Reactor in SIMBA	Volume (m^3)
Anaerobic	1	175	17	Anaerobic 1	87.5
				Anaerobic 2	87.5
	2a	80.5	58	Anaerobic 3	80.5
	Stripper	55.5	5	Stripper	55.5
	2c	39.0	4	Anaerobic 4	39.0
Anoxic	Selector	79.0	5	Selector	79.0
	3	317.0	18	Anoxic 1	158.5
				Anoxic 2	158.5
	4	395.0	23	Anoxic 3	197.5
				Anoxic 4	197.5
Simultaneous	5	395.0	38	Simultaneous 1	395.0
aerobic/anoxic	6	395.0	38	Simultaneous 2	395.0
Aerobic	7	395.0	38	Aerobic 1	395.0
	8	395.0	38	Aerobic 2	395.0
	9	395.0	38	Aerobic 3	395.0
	10	395.0	38	Aerobic 4	395.0

For the sensitivity analysis and validation of the activated sludge model, it is necessary to have data of the WWTP available. For the sensitivity analysis, available yearly data were used. The average yearly data of the years 1988 through 1992 were taken as a standard. In this period, no significant changes were made to the WWTP. For the validation of the model the WWTP was sampled intensively for two days. Table 2.2 gives the influent data and plant characteristics used in the sensitivity analysis and validation study. Only the flow rate to the treatment plant varied during the day, the concentrations where more or less constant, the latter was probably due to the long retention time in the sewer (0.8-1.5 days). During the sampling period, the pH was also measured at several points in the activated sludge tank and appeared to be about 6.9 everywhere. According to Maurer (1997) no biological induced phosphate precipitation is expected in this case. Also, in previous studies we found no evidence of significant precipitation in the activated sludge tank.

Table 2.2 Influent data and characteristics of WWTP Holten (see also Figure 2.2).

Parameter	Unit	Average of yearly data 1988 - 1992	Measurements November 1995
Influent flow*	m^3/d	5,520	4,757
Total COD in influent	mgCOD/L	354	516
Influent Kj-nitrogen	mgN/L	51	51
Influent phosphorus	mgP/L	8.7	7.3
Supernatant from internal stripper*)	m^3/d	420	450
Return stream from anoxic (4) to anaerobic (1)*	m^3/d	9,600	9,600
Return sludge*	m^3/d	9,930	10,050
Surplus sludge*	m^3/d	70	**

*Measurements are based on registration of the hours of operation multiplied by the capacity of the pumps
**In this case the amount of surplus sludge was not measured, but the sludge concentrations (MLSS) were measured. In the simulation the sludge concentration was kept constant at the measured value of 6.5 g/L with a control loop.

During the measurements at Holten WWTP the filtered and total influent COD, the influent BOD_5 and the fatty acids in the influent were measured. Based on these data, the influent composition for the validation was made according to the method of the Dutch foundation for applied research for water management (STOWA 1996, Roeleveld and Kruit 1998). The composition of COD in the influent for the sensitivity analysis has been estimated, because only information on the total COD and VFA-COD in the influent and the total COD in the effluent was available at that moment. The composition in terms of components of the activated sludge model for both cases is given in Table 2.3.

Table 2.3 Composition of influent COD as used for the simulations of Holten WWTP.

Model component	Values used in sensitivity analysis	Values used in validation and calibration
S_F	116	143
S_A	89	71
S_I	44	24
X_I	0.1	86
X_S	107	192

2.3 Sensitivity analysis

The sensitivity of the concentrations of ammonium, nitrate and phosphate in the effluent and the sludge production for all parameters and process choices was analysed. It should be noticed that the results of this analysis are only valid for WWTP Holten with the current conditions. Since not only the parameters of the activated sludge model can influence the outputs, but also the influent composition and flow scheme, a wide range of analyses was performed. The sensitivity of the following parameters was analysed based on a 10% change of the standard values:

- All stoichiometric (22) and kinetic parameters (42) of the activated sludge model;
- The distribution of COD in the influent over, respectively: S_A/S_F; $(S_A+S_F)/X_S$; X_I/X_S; X_H/X_S;
- The internal recycle flows of sludge, water and mixed liquor (see Table 2.2);
- The air flow to the aerobic zone and the simultaneous aerobic/anoxic zone.

Beside this, the sensitivity of the concentrations of ammonium, nitrate and phosphate at the end of the anaerobic, anoxic and simultaneous zones for the 42 kinetic parameters of the model was analysed.

The sensitivity (*S*) of the above parameters (*p*) with respect to *y* (being sludge production and effluent-ammonium, -nitrate and -phosphate) is a dimensionless number and was calculated as:

$$S = \frac{dy}{dp} \cdot \frac{p}{y} \tag{2.2}$$

where *dp* is the change in the parameter value *p*, and *dy* the change in the output *y*. The influence of the set-up of the hydraulic model of WWTP Holten was analysed based on halving the amount of tanks per zone.

The average temperature was 12°C, therefore the kinetic coefficients were recalculated to this temperature based on the given temperature coefficients. The amount of air supplied to the aerobic zones of WWTP Holten is unknown, but the dissolved oxygen concentrations were measured. The average dissolved oxygen concentrations were 0.2 mg/L for the simultaneous nitrification/denitrification zone (sections 2.5 and 2.6) and 4 mg/L for the aerobic zone (sections 2.7 through 2.10). In the SIMBA model the air supply was adjusted until these values were reached.

2.3.1 Sludge production

The influence on the sludge production of most parameters is negligible. The sludge production was (as expected) only marginally influenced by the distribution of COD over inert particulate organic material and slowly biodegradable substrates (X_I/X_S : $S \approx 0.75$), the heterotrophic yield (Y_H : $S = 0.8$) and the yield of PHA formation on VFA (Y_{PHA} : $S \approx 0.8$). If acetate is the major carbon source for PAO, Y_{PHA} is metabolically fixed and should not be used as a 'calibration' parameter.

2.3.2 Concentrations

The effluent concentrations, especially of ammonium, appeared to have a sensitivity of more than 1 ($S > 1$) towards only 8 of all the stoichiometric and kinetic parameters of the activated sludge model: Y_H, Y_{PHA}, δ, m_{ATP}, μ_{AUT}, b_{AUT}, K^N_{O2}, K^N_{NH4}. The internal concentrations (at the end of the anaerobic, anoxic and simultaneous zones) of nitrate and phosphate were much more sensitive for the kinetic parameters than the effluent concentrations. Fifteen kinetic parameters gave a sensitivity of more than 1 with regard to the internal concentrations. The effluent ammonia and phosphate concentrations are mostly sensitive towards the distribution of COD over the inert particulate material and the slowly biodegradable substrate (X_I/X_S: $S \approx 1.3$). The phosphate effluent concentrations are also sensitive for the distribution of COD over soluble, biodegradable COD and slowly biodegradable COD ($(S_A + S_F)/X_S$: $S \approx 3.6$). The sensitivity of the other distributions (see above) is less than 0.5. The sensitivity of the effluent concentrations for the internal flows of water and sludge is negligible. However, the effluent concentrations are highly sensitive for the air flow, the effluent ammonium concentration has a sensitivity of about 10 for the air flow to the aerobic zone.

2.3.3 Set-up of hydraulic model

Concerning the set-up of the hydraulic model of the WWTP, only the setup of the aerobic zone appeared to have some influence on the effluent concentrations. Decreasing the number of tanks in the aerobic zone of the model from 4 to 2 leads to increase in the effluent ammonium concentration from 0.3 to 0.6 (factor 2).

2.4 Calibration and validation

2.4.1 Performance

One of the major drawbacks of dynamic modelling with a dynamic influent as occurs in practice (daily dynamics) is that it consumes a lot of calculation time. Furthermore, the question is whether dynamic influent modelling will give much more information than that provided by the static influent modelling. Therefore, the calibration was started with a static influent. This means that the complete dynamic model (stoichiometry and kinetics) was used, but that average, constant values for the influent flow and composition were used. This will give an approximate steady-state solution. After this static influent calibration, the model was validated under dynamic conditions, i.e. the daily influent flow profile was introduced. In the calibration a process engineering approach was used; i.e. parameters were changed based on mechanistically reasoning rather than on their sensitivity. After this procedure, it could be compared which values where changed relative to when the parameter sensitivity analyses would have been the basis for the calibration. As already mentioned in the process description, the different concentrations in the influent were obtained during a 2 days sampling program. At the same time, also the concentrations of ammonium, nitrate and phosphate at four different places in the activated sludge tank were measured: the end of the anaerobic, anoxic, simultaneous and aerobic zones. The average values are presented in Figure 2.3.

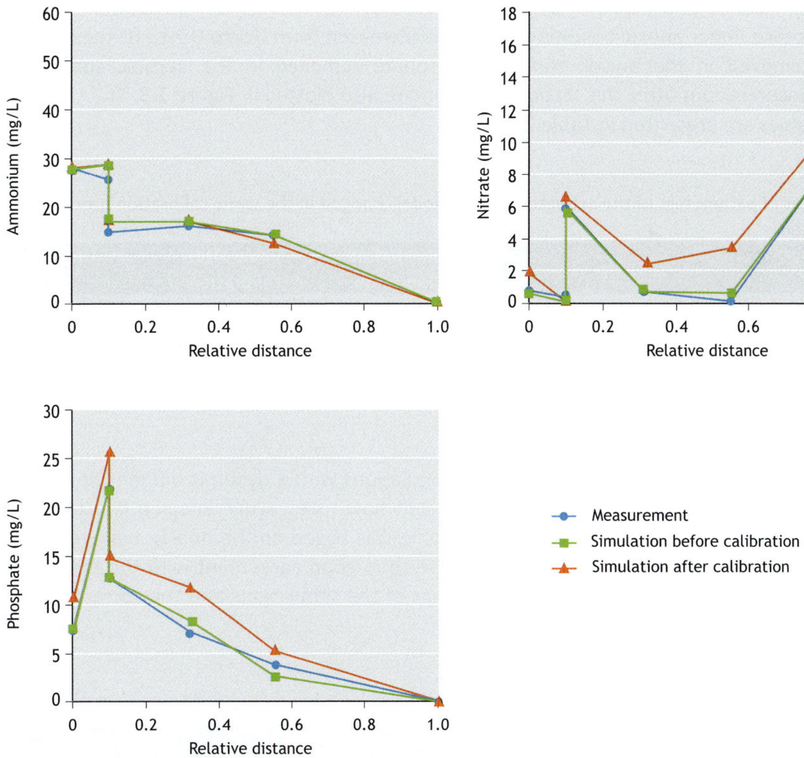

Figure 2.3 Measured and calculated concentrations in WWTP Holten.

It should be noted that the influent concentrations did not change significantly during the day. Before validation and calibration of the model, first the air flows were adjusted to the measured dissolved oxygen concentrations, being around 0.2 mg/L for the simultaneous aerobic/anoxic zone and around 4 mg/L for the aerobic zone. The concentrations calculated with the simulation model using a default set of parameters are presented in Figure 2.3 as well. As can be seen from this figure, the predictions of the effluent concentrations (end aerobic phase) based on the standard values are comparable with the measured concentrations. However, a large deviation is observed between the measured and calculated nitrate and phosphate concentration profiles in the process (at the end of the anaerobic, anoxic and simultaneous zones). Therefore, in the calibration procedure it is logical to change the parameters which have the largest influence (but are not well characterised) on the internal concentrations, measured at the end of the anaerobic, anoxic and simultaneous zones, instead of only fitting the effluent concentrations.

2.4.2 Calibration

To fit the calculated concentration profiles to the measured concentration profiles, three kinetic parameters are changed based on the causality of the parameters on the concentration values, and the relative uncertainty in the original parameter values. Thus, calibration was done heuristically. The high phosphate concentration at the end of the anaerobic zone could be the result of high availability of acetate. This is changed by decreasing the fermentation constant (q_{fe}) by 40%. The phosphate concentrations in the other zones (anoxic, simultaneous) are decreased by increasing the phosphate-uptake rate (k_{PP}) by 40%. The reduction factor for phosphate uptake under anoxic conditions (η_{PNO3}) is increased from 0.5 to 0.7 to fit the ratio of phosphate removed in the anoxic zone to phosphate removed in the aerobic zone. The calculated concentrations after this static calibration are also plotted in Figure 2.3. The changed parameter values are presented in Table 2.4.

Table 2.4 Values of parameters of the activated sludge model changed in the static calibration procedure of WWTP Holten.

Parameter	Value before calibration	Value after calibration	Unit
Fermentation rate (q_{fe})	1.71	1.00	gCOD/gCOD.d
Poly-P uptake rate (k_{pp})	0.05	0.07	gP/gCOD.d
Reduction factor under anoxic conditions (η^{P}_{NO3})	0.50	0.70	

2.4.3 Validation

After the static influent calibration, a number of simulations with a dynamic influent flow were performed. This was executed by simulating different flow rates, while keeping constant the concentrations of the influent parameters. The profile was based on the hourly average flows. The simulated changes in effluent concentrations were in good agreement with the measured concentrations as well as the average concentrations of the simulations and the corresponding measurements (Table 2.5).

Table 2.5 Average calculated concentrations in WWTP Holten with static and dynamic simulations.

Zone	NH_4 (mgN/L)		NO_3 (mgN/L)		PO_4 (mgP/L)	
	Static	Dynamic	Static	Dynamic	Static	Dynamic
Anaerobic	28.80	28.50	0	0	22.00	21.00
Anoxic	16.70	17.00	0.80	1.20	8.40	8.00
Simultaneous aerobic/anoxic	14.50	14.00	0.60	1.10	2.80	2.90
Aerobic	0.16	0.18	14.10	14.50	0.04	0.05

As can be seen from Table 2.4, there is almost no difference between the calculated values of the static and the dynamic influent simulations. The only striking difference is the nitrate concentration at the end of the anoxic and simultaneous zones. Probably this is caused by the fact that under dynamic conditions and during part of the day there is no nitrate left (due to low loading conditions). That means there is no denitrification during part of the day. The rate of denitrification at high nitrate concentrations (high loading conditions) is however almost equal to the rate under average nitrate concentrations. So, under dynamic conditions, the total denitrification capacity in the anoxic zones is lower. Simultaneous denitrification will appear in the aerated zones of WWTP Holten as a result of decreasing oxygen concentrations (which is not controlled) at high loading conditions. Therefore, the total amount of denitrification and the effluent nitrate concentrations are identical under static and dynamic influent conditions.

2.5 Discussion

With the developed dynamic activated sludge model it was possible to obtain a good simulation of the measurements on WWTP Holten without the need for an extensive model calibration. As a start, an available set of default parameters was used. For PAO, the parameters were based on laboratory experiments from cultures enriched on acetate (Smolders *et al.,* 1994; Kuba *et al.,* 1996 and Murnleitner *et al.,* 1996) while the other default parameters are adapted from ASM2 (Gujer *et al.,* 1995). With default parameters the model already provides a good simulation of the WWTP (Figure 2.3). Only three parameters needed to be adjusted to bring the simulated concentration profiles along with the measurements in the different process units. This underlines the extra value of in-process information above effluent measurements. All selected parameters are identified based on process knowledge and no 'black box' mathematical procedures are required. The parameters which are changed in the calibration were not identified as having the highest influence on effluent concentrations, but on the internal concentrations. Thus, for calibration procedures in general, it should be noticed that calibration is not only based on the effluent concentrations, but, maybe even more important, also on internal concentrations. In the calibration, it was preferred to use a process engineering based approach. Had a sensitivity analysis based approach been used, the choice of calibrated parameters would have been different. Moreover, probably also more parameters would have been adapted. Evaluating the sensitivity analysis on the concentrations, which differed from the first simulation with default values, showed that for (i) the P concentration at the end of the anaerobic phase 5, (ii) the P concentration at the end of the anoxic phase 6 and (iii) the relative P-removal in aerobic and anaerobic phases 6 parameters were equal or more sensitive. Some of these parameters are well established values (e.g. yield coefficients). A standard sensitivity analysis based calibration procedure cannot differentiate between more or less defined parameter values. Therefore, a calibration based on process knowledge seems more sensitive. However, this indeed requires more experience. Thus, possibly a combination of a parameter sensitivity analysis and process knowledge is in general the best approach. It would be worthwhile if this could be automated in a simple manner.

Holten WWTP has a pre-settled influent, in which the organic material is mainly composed of soluble COD (46 % of total COD) and VFAs (14% of total COD, 30% of soluble COD). Under these conditions fermentation processes and hydrolysis do not play a major role. Therefore, it was possible to concentrate on the P removal process, which has been extensively studied. However, several WWTPs receive wastewater from aerobic sewers containing lower soluble to particulate organic material (S_S/X_S) ratios. For these cases, fermentation processes and hydrolysis are likely to be more important for the overall performance of the treatment process in general and for P-removal especially. Since these processes are more or less understood, this might cause problems to evaluate the EBPR process.

2.6 Conclusions

A dynamic activated sludge model for biological N and P removal was developed. The model was set-up along the lines of ASM2, but PAO and therefore the EBPR process was modelled according to the previously developed TUDP metabolic model. The ease and practicality to apply the model to describe a full scale WWTP underlines its robustness and reliability. Even before calibration, a reasonable description was already obtained. Only three parameters needed to be adjusted by approximately 40% in order to obtain an accurate description. To execute the calibration, especially internal process measurements proved to be crucial.

Acknowledgements

The authors wish to thank Frank Brandse from the Waterboard "Groot Salland" for his continuous co-operation in our research. Also, the water board "west-Overijssel" is acknowledged for all the data provided from their sewage treatment plant.

References

Barnard J.L (1974) Cut P and N without chemicals. *Water Wastes Eng.* 11(7):33-36.

Barnard J. L. (1975) Biological nutrient removal without the addition of chemicals. *Wat. Res.* 9, 485-490.

Barnard J. L. (1976) A review of biological phosphorus removal in the activated sludge process. *Water SA.* 2(3), 126-144.

Barnard J. L. (1983) Background to Biological Phosphorus Removal. *Wat. Sci. Tech.*, 15(3-4), 1-13.

Brdjanovic D, Slamet A, Van Loosdrecht MCM, Hooijmans CM, Alaerts GJ and Heijnen JJ (1997) Impact of Excessive Aeration on Biological Phosphorus Removal from Wastewater. Accepted for publication in Wat.Res.

Brdjanovic D, van Loosdrecht MCM, Hooijmans CM, Alaerts G. J. and Heijnen J. J. (1998c) Bioassay for glycogen determination in biological phosphorus removal systems. *Wat. Sci. Tech.* 37(4-5), 541-547.

Brdjanovic D., Logemann S., van Loosdrecht M. C. M., Hooijmans C. M., Alaerts G. J. and Heijnen J. J. (1998a) Influence of temperature on biological phosphorus removal: process and ecological studies. *Wat. Res.* 32(4), 1035-1048.

Brdjanovic D., van Loosdrecht M. C. M., Hooijmans C. M., Alaerts G. J. and Heijnen J. J. (1998b) Minimal aerobic sludge retention time in biological phosphorus removal systems. *Biotechnol. Bioeng.* 60(3), 326-332.

Brdjanovic D., van Loosdrecht M. C. M., Hooijmans C. M., Alaerts G. J. and Heijnen J. J. (1999) Innovative methods for sludge characterisation in biological phosphorus removal systems. *Wat. Sci. Tech.* 39(6), 37-43.

Brdjanovic D., van Loosdrecht M. C. M., Versteeg P., Hooijmans C. M., Alaerts G. J. and Heijnen J. J. (2000) Modelling COD, N and P removal in a full-scale WWTP Haarlem Waarderpolder. *Wat. Res.* 34(3), 846-858.

Dold P. L., Wentzel M. C., Billing A. E., Ekama G. A. and Marais G. v. R. (1991) Activated Sludge System Simulation Programs. Water Research Commission, Pretoria, South Africa.

Ekama G. A., Marais G. v. R. and Siebritz I. P. (1984) Biological excess phosphorus removal. *Theory design and operation of nutrient removal activated sludge processes* 7, 1-32. Water Research Commission, Pretoria, South Africa.

Filipe C. D. M. and Daigger G. T. (1999) Evaluation of the capacity of phosphorus-accumulating organisms to use nitrate and oxygen as final electron acceptors: A theoretical study on population dynamics. *Wat. Environ. Res.* 71(6), 1140-1150.

Filipe C. D. M., Daigger G. T. and Grady C. P. L. (2001a) Stoichiometry and kinetics of acetate uptake under anaerobic conditions by an enriched culture of phosphorus-accumulating organisms at different pHs. *Biotechnol. Bioeng.* 76(1), 32-43.

Filipe C. D. M., Daigger G. T. and Grady C. P. L. (2001b) pH as a key factor in the competition between glycogen-accumulating organisms and phosphorus-accumulating organisms. *Wat. Environ. Res.* 73(2), 223-232.

Gujer W, Henze M, Mino T, Matsuo T, Wentzel MC and Marais GvR (1995) The activated sludge model no. 2: Biological Phosphorus Removal. Wat. Sci. Tech. 31 (2), 1-11

Henze M., Gujer W., Mino T. and van Loosdrecht M. C. M. (2000) Activated Sludge Models ASM1, ASM2, ASM2d and ASM3. *IAWQ Scientific and technical report no.9*. London, UK.

Henze M., Gujer W., Mino T., Matsuo T., Wentzel M. C., Marais G. v. R. and van Loosdrecht M. C. M. (1999) Activated Sludge Model No.2d, ASM2d. *Wat. Sci. Tech.* 39(1), 165-182.

Henze M., Gujer W., Mino T., Matsuo T., Wentzel M. C., Marais G. v. R. and van Loosdrecht M. C. M. (1995) Activated Sludge Model No.2, ASM2. *IAWQ Scientific and technical report no.1*. London, UK.

Henze M., Gujer W., Mino T., Matsuo T., Wentzel M. C., Marais G. v. R. and van Loosdrecht M. C. M. (1999) Activated Sludge Model No.2d, ASM2d. *Wat. Sci. Tech.* 39(1), 165-182.

Hong S. -N., Krichten D. J., Kisenbauwer K. S. and Sell R. L. (1982) A biological wastewater treatment system for nutrient removal. EPA Workshop on Biological Phosphorus Removal in municipal wastewater treatement, Annapolis, Md.

Jeppsson U. (1996) Modelling aspects of wastewater treatment processes. Ph.D. thesis, Lund University, Sweden, ISBN 91-88934-00-4.

Koch G., Kühni M., Gujer W. and Siegrist H. (2000) Calibration and validation of Activated Sludge Model No. 3 for Swiss municipal wastewater, *Wat. Res.*, 34(14), 3580-3590.

Kuba T., Murnleitner E., van Loosdrecht M. C. M. and Heijnen J. J. (1996) A Metabolic Model for Biological Phosphorus Removal by Denitrifying Organisms. *Biotechnol. Bioeng.* 52(6), 685-695.

Kuba T., Smolders G.F.S., van Loosdrecht M.C.M., Heijnen J.J. (1993) Biological phosphorus removal from wastewater by anaerobic-anoxic sequencing batch reactor. *Wat. Sci. Tech.* 27 (5-6), 241-252.

Kuba T., Wachtmeister A., van Loosdrecht M. C. M. and Heijnen J. J. (1994) Effect of nitrate on phosphorus release in biological phosphorus removal systems. *Wat. Sci. Tech.* 30(6), 263-269.

Maurer M (1997) Erhöhte Biologische Phosphorelimination - Modellierung der biologischen und chemischen Prozesse. Dissertation der Eidgenössischen Technischen Hochschule Zürich, Nr. 11920 (Enhanced Biological Phosphor Removal - Modelling of the Biological and Chemical Processes. Thesis of the Swiss Federal Institute of Technology, Zürich, Switzerland).

Mino T, Liu WT, Satoh H and Matsuo T (1996) Possible Metabolisms of PAO's and Glycogen Accumulating non-Poly-P Organisms (GAOs) in the Enhanced Biological Phosphate Removal Processes. Med. Fac. Landbouww. Univ. Gent, 61/4a, 1769-1776

Murnleitner E., Kuba T., van Loosdrecht M. C. M. and Heijnen J. J. (1997) An integrated metabolic model for the aerobic and denitrifying biological phosphorous removal. *Biotechnol. Bioeng.* 54, 434-450.

Nicholls H. A. (1975) Full Scale Experimentation of the New Johannesburg Aeration Plants. *Water SA.* 1(3), 121.

Petersen E. E. (1965) Chemical reaction analysis. *Prentice Hall*, Englewood Cliffs, NJ.

Rieger L., Koch G., Kühni M., Gujer W. and Siegrist H. (2001). The EAWAG BioP-Module for Activated Sludge Model No. 3, *Wat. Res.*, 35(16), 3887-3903.

Roeleveld PJ, Kruit J. (1998) Guidelines for wastewater characterisation in the Netherlands. Korrespondenz Abwasser 3:465-468.

Smolders G. L. F. (1995a) A metabolic model of the biological phosphorus removal; stoichiometry, kinetics and dynamic behaviour. Ph.D. Thesis Delft University of Technology, Delft, the Netherlands, 208 p.

Smolders G. L. F., Bulstra D. J., Jacobs R., van Loosdrecht M. C. M. and Heijnen J. J. (1995c) A metabolic model for the biological phosphorus removal process II: Validation during start-up conditions. *Biotechnol. Bioeng.* 48, 234-245.

Smolders G. L. F., Klop J. M., van Loosdrecht M. C. M. and Heijnen J. J. (1995b) A metabolic model for the biological phosphorus removal process: Effect of the sludge retention time. *Biotechnol. Bioeng.* 48, 222-233.

Smolders G. L. F., van der Meij J., van Loosdrecht M. C. M. and Heijnen J. J. (1995b) Structured Metabolic model for anaerobic and aerobic stoichiometry and kinetics of the biological phosphorus removal process. *Biotechnol. Bioeng.* 47, 277-287.

Smolders G. L. F., van der Meij J., van Loosdrecht M. C. M. and Heijnen J. J. (1994a) Model of the anaerobic metabolism of the biological phosphorous removal process: stoichiometry and pH influence. *Biotechnol.Bioeng.* 43(6), 461-470.

Smolders G. L. F., van der Meij J., van Loosdrecht M. C. M. and Heijnen J. J. (1994b) Stoichiometric model of the aerobic metabolism of the biological phosphorus removal process. *Biotechnol.Bioeng.* 44(7), 837-848.

Stouthamer A. H. (1975) The search for correlation between theoretical and experimental growth yields. Microbial Biochemistry, Vol 21.

STOWA (1996) Methode voor influentkarakterisering, inverentarisatie en richtlijnen, STOWA-report 96-08

Temmink H, Petersen B, Isaacs S and Henze M (1995) Recovery of Biological Phosphorus Removal after Periods of Low Organic Loading. Proc. Water Quality International '96, IAWQ 18th Biennial Conference, Singapore, 23-28 june 1996

Van Loosdrecht M. C. M., Brandse F. A. and de Vries A. C. (1998) Upgrading of wastewater treatment processes for integrated nutrient removal – The BCFS® process. *Water Sci. Technol.* 37(9), 209-217.

Van Loosdrecht MCM (1996) A Comparison between Models for Biological Phosphorus Removal. Med. Fac. Landbouww. Univ. Gent, 61/4a, 1753-1760

Van Loosdrecht MCM, Kuba T, Van Veldhuizen HM, Brandse FA and Heijnen JJ (1997a) Environmental Impacts of Nutrient Removal Processes: Case Study, J. Env. Eng., 123 (1), 33-40

Van Loosdrecht MCM, Pot MA and Heijnen JJ (1997) Importance of Bacterial Storage Polymers in Bioprocesses. Wat.Sci.Tech., 35, 41-47

Van Veldhuizen H. M., van Loosdrecht M. C. M. and Heijnen J. J. (1999) Modelling biological phosphorus and nitrogen removal in a full scale activated sludge process. *Wat. Res.* 33(16), 3459-3468.

Wentzel M. C., Dold P. L., Ekama G. A. and Marais G. v. R. (1989a) Enhanced polyphosphate organism cultures in activated sludge. Part III: Kinetic Model. *Water SA*, 15, 89-102.

Wentzel M. C., Ekama G. A., Loewenthal R. E., Dold P. L. and Marais G. v. R. (1989b) Enhanced polyphosphate organism cultures in activated sludge. Part II: Experimental behavior. *Water SA*, 15, 71-88.

Wichern M., Obenaus F. and Rosenwinkel K.-H. (1999) Comparison of the models ASM2d and TU-Delft in simulation praxis, 6. Simba-Anwendertreffen, Magdenburg.

Chapter 3: WWTP Haarlem Waarderpolder, the Netherlands

Model Evaluation of a full-scale bio-P side-stream process

Brdjanovic D., van Loosdrecht M.C.M., Versteeg P., Hooijmans C.M., Alaerts G.J., Heijnen J.J.

This chapter is based on "Modelling COD, N and P removal in a full-scale WWTP Haarlem Waarderpolder" by Brdjanovic D., van Loosdrecht M.C.M., Versteeg P., Hooijmans C.M., Alaerts G.J., Heijnen J.J., (2000). Water Research 34(3), 846-859, and on "Modelling Biological Phosphorus Removal in Activated Sludge Systems" by Brdjanovic D. (1998) *Ph.D. Thesis*, A.A. Balkema Publishers, pp.251, ISBN 90 54104155.

3.1 Introduction

Several different mathematical models for the simulation of the activated sludge process in wastewater treatment are available. The most recent models allow for dynamic simulation of complex activated sludge systems and include carbon oxidation, nitrification, denitrification, chemical and biological phosphorus removal, and even attempts to model the formation of filamentous organisms (bulking) have been made. Besides different microbial conversion reactions, these models may also take into account dynamic feeding regime, influence of temperature, pH, dissolved oxygen, various hydraulic patterns, process control, and effects of gas/liquid mass transfer processes.

In order to promote the development and facilitate the application of models for design and operation of biological wastewater treatment systems, the IAWPRC (later called IAWQ and most recently IWA) formed in 1983 the Task Group on Mathematical Modeling for Design and Operation of Biological Wastewater Treatment. Four years later the Task Group proposed a general model for removal of organic matter, nitrification and denitrification, called Activated Sludge Model no. 1 - ASM1 (Henze *et al.,* 1987). Since its appearance, this model has greatly encouraged the use of mathematical models. The increased requirement for nutrient removal during the last decade created a need to extend ASM1 by inclusion of enhanced biological phosphorus removal (EBPR). In 1989, Wentzel *et al.* presented a kinetic model for aerobic EBPR. This model served as a basis for development of the Activated Sludge Model no. 2 - ASM2 (Henze *et al.,* 1994). Soon after, Mino *et al.* (1995) extended ASM2 by including the glycogen metabolism and EBPR under anoxic conditions. Simultaneously, Smolders and co-workers (1994a) developed a structured metabolic model for EBPR (also incorporating the glycogen metabolism) which served also as a basis for a metabolic model for denitrifying EBPR (Kuba *et al.,* 1996). In their integrated metabolic model for aerobic and anoxic EBPR (so-called "TUDP model"), Murnleitner *et al.* (1997), successfully described the two known EBPR processes with the same kinetic equations and parameters. The metabolic model is based on the bioenergetics and stoichiometry of the bacterial metabolism, and describes all relevant metabolic reactions underlying the metabolism of phosphors accumulating organisms (PAO) by six independent reactions: two for the anaerobic metabolism and four for the aerobic/anoxic metabolism. In this model the ATP/$NADH_2$ ratio (called δ value) is the only model parameter that is different for aerobic and anoxic EBPR. The process reaction rates are described by four kinetic relations and two maintenance terms. A comparison between the ASM2 and the TUDP model is discussed elsewhere (Van Loosdrecht, 1996).

The Task Group proposed ASM2d (Henze *et al.* 1998) - a minor extension of ASM2 by inclusion of two additional processes to account for denitrifying EBPR. At the same time the Task Group moreover, proposed a structured model for simulation of oxygen consumption, sludge production, nitrification and denitrification: ASM3 (Gujer *et al.* 1998). ASM3 replaces ASM1 and includes several improvements compared to ASM1, such as addition of storage of organic substrates as a new process, and replacement of lysis process with endogenous respiration process. This approach in ASM3 is more related to the TUDP model than to ASM1.

Presently, it is widely agreed that the models for simulation of activated sludge systems are suitable for application to complex full-scale wastewater treatment plants (WWTP). ASM1 has been used for more than a decade as a tool for modeling the removal of organic matter and for nitrification and denitrification processes; considerable experience with this model has been acquired. However, for ASM2, the situation is different. It has not been validated extensively due to the fact that the model became available only recently, and because its accuracy is being debated.

While the information on the application of ASM2 is scarce, the TUDP model model was already tested and verified over a range of SRT values (Smolders *et al.* 1995a) and oxygen or nitrate as electron acceptor (Murnleitner *et al.*, 1997), as well as during both start-up and steady state conditions (Smolders *et al.* 1995b) - all in lab-scale SBR systems using EBPR cultures enriched under laboratory conditions and at a temperature of 20°C. The stoichiometry and to a less extend the kinetics of EBPR processes are considered to be well defined by this model. In this model the part of ASM2 that is related to phosphate removal was replaced with the TUDP model. This combined model was for the first time applied on the full-scale WWTP Holten (BCFS® process) in the Netherlands (Van Veldhuizen *et al.*, 1999, see Chapter 2).

In this study, the model was used to check the performance of Phostrip®-like process at WWTP Haarlem Waarderpolder, also located in the Netherlands with objectives to evaluate (i) how a complex model can be applied to a complex full-scale plant (conventional anaerobic-aerobic activated sludge system with side-stream bio-chemical P removal), (ii) influent and sludge characterization procedures for EBPR modeling, (iii) the use of batch tests to validate the model, and (iv) different alternative EBPR process schemes.

3.2 Materials and methods

3.2.1 Configuration of WWTP Haarlem Waarderpolder

The WWTP Haarlem Waarderpolder was built in 1969. In 1995 the installation was retrofitted and expanded to its present state (Figure 3.1). The plant was designed for removal of organic matter, nitrogen and phosphorus from domestic and industrial wastewaters of Haarlem and five smaller nearby communities, in total 160,000 P.E. (Table 3.1).

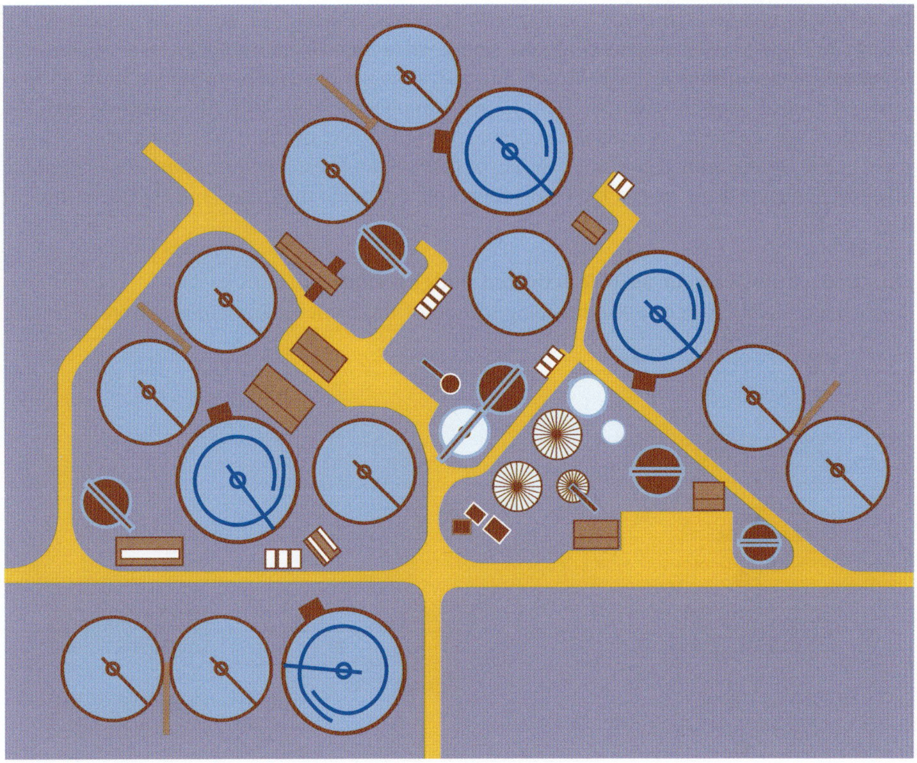

Figure 3.1 Layout of the WWTP Haarlem Waarderpolder.

Table 3.1 Designed and recorded data of the WWTP Haarlem Waarderpolder.

Load	Unit	Designed (1994)	Recorded (1996)	Recorded (1997)
Plant loading	P.E.	160,000	109,000	109,100
Hydraulic loading				
Average flow	m^3/h		1,375	1,333
Dry weather flow	m^3/h	2,175		
Storm weather flow	m^3/h	7,000		
Sewage strength				
BOD_5 (54 $gBOD_5/P.E.d$)	kgBOD/d	8,640 (54)	3,907 (39)	5,082 (47)
Kj-N (12 gKj-N/P.E.d)	kgN/d	1,920 (12)	1,437 (14.2)	1,436 (13.2)
P_{total} (1.8 gP/P.E.d)	kgP/d	288 (1.8)	227 (2.2)	203 (1.9)
Effluent quality				
BOD_5	$mgBOD_5/L$	10	5	7
$COD_{filtered}$	mgCOD/L		29	34
COD_{total}	mgCOD/L		36	42
N_{total}	mgN/L	10	4.2	4.9
NO_3-N	mgN/L		3.5	3.4
NH_4-N	mgN/L		2.5	2.8
P_{total}	mgP/L	1	0.66	0.60
TSS	mgTSS/L	15	9	12

The removal of organic matter and nitrogen is accomplished in a biological system which consists of four parallel activated sludge lines, while phosphate removal takes place in a combined biological-chemical side-stream process where the side-stream sludge from all four activated sludge lines is treated. Due to its complexity from both the process and the operational point of view, this plant can be considered as one of the most complicated sewage works in the Netherlands. The plant consists of three main lines: a conventional anoxic/aerobic activated sludge process line, a bio-P selection and P-precipitation line, and sludge treatment line.

3.2.1.1 Conventional anoxic-aerobic activated sludge treatment

Raw sewage is transported to the pumping station of the plant via combined sewerage (Figure 3.2).

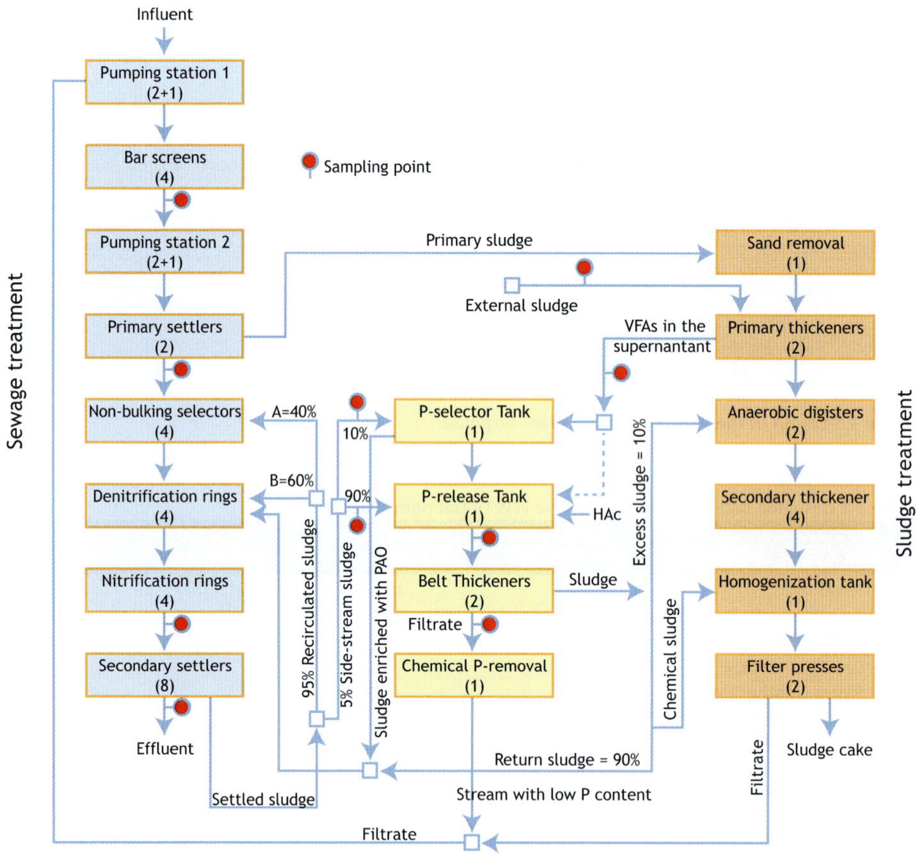

Figure 3.2 Simplified block process scheme of the WWTP Haarlem Waarderpolder (numbers in brackets indicate the number of process units).

After screening, wastewater is elevated for a second time and divided into two primary settling tanks. Settled sewage from each primary settler is split into two lines, prior to being introduced to in total four activated sludge units. Each activated sludge unit comprises of three inter-connected compartments (Figure 3.3 and 3.4), namely, a central compartment (non-bulking selector, V=125 m^3), an inner non-aerated ring (denitrification tank, V=4,330 m^3), and an outer, aerated ring (nitrification tank with a total volume of 4,330 m^3 of which around 85% is aerated). Beside the outer ring, also the non-bulking selector is equipped with an aeration facility. However, the non-bulking selector is not aerated during normal plant operation. The aeration in the nitrification ring is provided by a set of twenty submerged small bubble aerators per unit. Activated sludge is circulated using four propellers located in each of four units. While the mixed liquor flows from the non-bulking selector to the denitrification ring and further to the non-aerated section of nitrification ring via weirs, a high internal recirculation (5,000-8,000 m^3/h) from the non-aerated section of nitrification ring into the denitrification ring is obtained via a submerged tunnel located at the bottom of the structure (under the weir).

Recirculation is forced by the action of a propeller placed at the inlet of the tunnel (Figure 3.3 and 3.4). A combination of diffused aeration, agitation by propellers and a high internal recirculation makes the activated sludge in this unit well mixed (except in the comparatively small non-bulking selector which is designed as a plug-flow system). The activated sludge leaves the nitrification ring via the weir prior to splitting into two secondary settlers. After settling, the purified sewage is discharged into the nearby recipient.

A large fraction (95%) of the settled activated sludge (secondary sludge) is returned to the activated sludge unit in order to maintain the desired biomass content (currently around 6 g MLSS/L). Around 40% of this 95% is pumped to the non-bulking selector (recirculated sludge A) and remaining 60% to the denitrification ring (recirculated sludge B). The remainder (5%) of the settled sludge (side-stream sludge) is pumped to phosphorus treatment line.

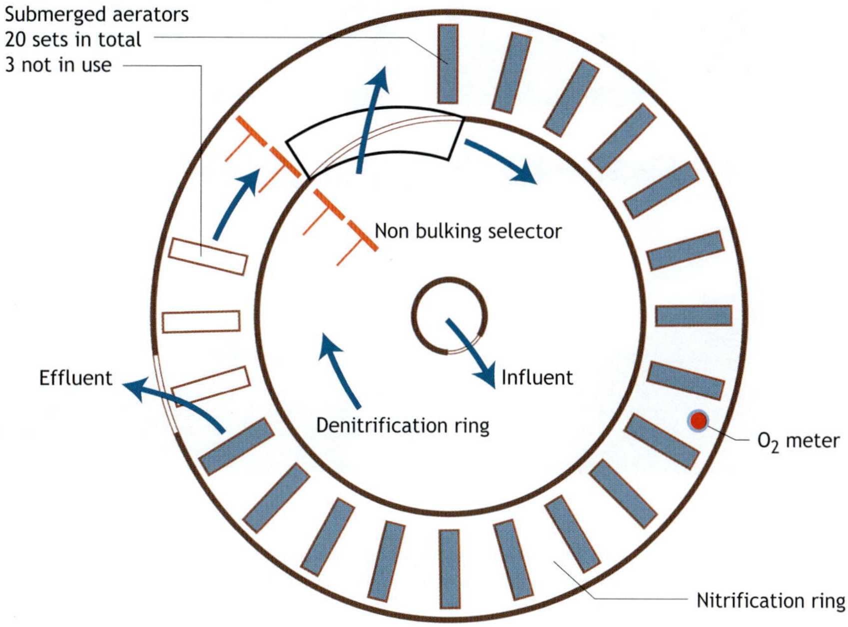

Figure 3.3 Simplified scheme of activated sludge unit at the WWTP Haarlem Waarderpolder.

Figure 3.4 Activated sludge unit: (top left) under full operation, (top right) under construction, (middle left) central compartment (plug-flow non-bulking selector), (middle right) one of four propellers, (bottom left) part of the diffused aeration system and weir to the secondary settlers and (bottom right) connections between the denitrification and the nitrification ring (the weir and the tunnel for internal recirculation) (Photos: Versteeg P.).

3.2.1.2 Phosphorus selection and precipitation line

Side stream sludge is introduced to two plug-flow agitated anaerobic tanks, each of 875 m^3, called the P selector tank (loaded with 10% of the side-stream sludge) and the P release tank (loaded with remaining 90%). P selector and P release tanks serve all four activated sludge units and have an identical structure in which the same processes take place. However, there are two major differences between these two tanks: firstly, the source of volatile fatty acids (VFA) needed for P release in the P selector tank is the supernatant of the primary thickener, while in the P release tank concentrated acetic acid is added to the sludge. About two thirds of the primary thickener's supernatant, (containing substantial amount of VFA as a result of fermentation process that takes place in this unit) is introduced to the P selector where the VFA were used by the PAO in the sludge. Alternatively, there is a provision for sharing this portion of supernatant between the P selector and the P release tanks, which actually occurred at the time when sampling program at the plant was carried out. Due to the dilution effect caused by high thickener supernatant flow fivefold higher biomass content occurs in the P release tank in comparison with the P selector. Secondly, the function of the P selector and P release tanks in this plant is different; since there is no provision of an anaerobic tank in the main stream, the P selector is needed to achieve and maintain enough PAO in the system by means of recirculation of the sludge from the P selector to denitrification ring. The purpose of the P release tank (as being a part of a side-stream process) is (besides selecting for PAO) to achieve a maximal phosphate release at very high phosphate concentration from the biomass into the solution. The phosphate-rich mixed liquor flows into two belt thickeners where partial liquid-solid separation takes place (dry solids content in the thickened sludge is 6%). While 90% of the thickened sludge is returned to the denitrification ring, the remaining part (excess sludge) is wasted via sludge treatment line. The sludge wasting rate results in a total sludge retention time (SRT) in the system of 84 days. Phosphate-rich filtrate from belt thickeners is further treated by addition of calcium-hydroxide to precipitate phosphorus dissolved in the filtrate. Phosphate precipitation takes place in a small settler from which the phosphate-rich chemical sludge is withdrawn for further treatment.

3.2.1.3 Sludge treatment line

Three different types of sludge are produced at this plant: primary sludge, activated excess sludge collected in the secondary settlers, and phosphate-rich chemical sludge. The sludge collected in primary settlers (primary sludge) is, after de-sanding, introduced into the primary gravity thickeners. In addition, the surplus activated sludge (5,360 kg DS/d) from the nearby WWTP Schalkwijk (plant capacity of 80,000 P.E.) is pumped into the primary thickeners, mixed with primary sludge, and thickened. Excess activated sludge is pumped and mixed with the primary sludge to anaerobic digesters, further thickened in a secondary gravity thickener and introduced to the homogenization tank. In this tank the phosphate-rich chemical sludge is also added and mixed with the already mineralized and thickened mixture of primary and excess sludge. The mixture of three sludges is then transported to band filter presses where the final liquid-solid separation occurs resulting in a sludge cake containing 20% DS. This sludge 'cake' is transported from the plant by means of trucks. All water streams resulting from the sludge treatment are via internal sewerage returned to the beginning of the plant.

3.2.2 Influent characterization

To make the application of mathematical models for simulation of activated sludge systems (like ASM1 and ASM2) more uniform, the Dutch water authorities recently selected the simulation program SIMBA® as a modeling standard in the Netherlands. Furthermore, the influent characterization procedure was standardized as well (STOWA, 1996; Roeleveld and Kruit, 1998). The model variables, parameters to be measured and a set of equations needed for the influent characterization according to Dutch guidelines are presented in Table 3.2. A detailed derivation of the equations used in this characterization procedure is given in Annex 3.1.

Table 3.2 Influent characterization according to Dutch guidelines and combined ASM2 and TUDP model

Influenct characteristics required by the combined ASM No. 2 and TUDP model			Wastewater characteristics (T= 12.6°C, pH=7.6)		Equations for determination of the influent characteristics (Roeleveld and Kruit, 1998)	Parameters needed to be measured for the influent characterizatioon (Roeleveld and Kruit, 1998)[a]		
Symbol	Name	Unit	Settled sewage	Thickener supernatant		Symbol	Name	Unit
Soluble components					**Soluble components**			
S_{O2}	Oxygen (negative COD)	$gCOD/m^3$	2.1	0.05	$S_A = COD_{VFAs}$	$COD_{tot,inf}$	Total influent COD	$gCOD/m^3$
S_F	Readily biodegradable organics	$gCOD/m^3$	73	73	$S_I = 0.9COD_{VFAs}$	$COD_{filt,inf}$	Influent COD filtered	$gCOD/m^3$
S_A	Volatile fatty acids	$gCOD/m^3$	32.4	49.9	$S_S = COD_{filt,inf} - S_I$	$COD_{filt,inf}$	Effluent COD filtered	$gCOD/m^3$
S_{NH4}	Ammonium and ammonia nitrogen	gN/m^3	46.5	55	$S_F = S_S - S_A$	COD_{VFAs}	COD from VFAs	$gCOD/m^3$
S_{NO3}	Nitrate and nitrite nitrogen	gN/m^3	0.1	0	$S_{NH4} = KjN-\sum(i^NX + i^NS)$	BOD_{5inf}	Influent COD	$gBOD/m^3$
S_{PO4}	Inorganic soluble phosporus	gP/m^3	6.4	6	$S_{PO4} = P_{total} - \sum(i^NX + i^NS)$	BOD_{5eff}	Effluent COD	$gBOD/m^3$
S_I	Soluble inert organic matter	$gCOD/m^3$	22	22		NH_4	Ammonia	gN/m^3
S_{ALK}	alkalinity	mol/m^3	5	5	**Particulate components**	PO_4	Phosphate	gP/m^3
Particulate components					$X_S = \alpha COD_{susp,inf}$	k_{BOD}	BOD constant	1/d
X_I	Particulate inert organic matter	$gCOD/m^3$	63	50	$X_I = (1-\alpha)COD_{susp,inf}$	Y_{BOD}	Yield factor for BOD	
X_S	Slowly biodegradable substrate	$gCOD/m^3$	141	166	$X_H = 0$			
X_H	Active heterotrophic biomass	$gCOD/m^3$	0.01	0.01	$X_{PHA} = 0$			
X_{PAO}	Phosphate accumulating organisms	$gCOD/m^3$	0.01	0.01	$X_{AUT} = 0.1\text{-}1.0$			
X_{PP}	Poly-phosphate	gP/m^3	0.001	0.001	$X_{PAO} = 0.1\text{-}1.0$			
X_{PHA}	Poly-hydroxy-alkanoates	$gCOD/m^3$	0.001	0.001	$X_{pp} = 0.1\text{-}1.0$			
X_{GLY}	Glycogen	$gCOD/m^3$	0.001	0.001				
X_{AUT}	Active autotrophic biomass	$gCOD/m^3$	0.01	0.01	**Other components**			
X_{TSS}	Mixed liquor suspended solids	$gMLSS/m^3$	153	153	$\alpha = [(BOD_{tot}/(1-Y_{H.BOD}) - S_S]/COD_{susp,inf}$			
X_{MeOH}	Metal-hydroxides	$gFe(OH)_4/m^3$	0	0	$BOD_{tot} = BOD_5/(1-e^{-k \cdot 5})$			
X_{MeP}	Metal-phosphate	$gFe(PO)_4/m^3$	0	0	$COD_{susp,inf} = COD_{tot,inf} - COD_{filt,inf}$			
Q	Flow	m^3	5,760	5,520				

[a]In Dutch wastewater treatment practice the majority of the above paramenters are measured as a part of sampling programs routinely performed at their biological nutrient removal plants.

There are two influent flows to the section of the WWTP Haarlem Waarderpolder which needed to be characterized: (1) the influent of the non-bulking selector (settled sewage) and, (2) the influent of the P selector and the P release tanks (supernatant of the primary thickeners). The term influent is further in the text used for these two flows; not for the raw sewage which enters the plant. The term effluent refers to the mixed supernatant of two secondary settlers which are part of the concerned activated sludge line, not to the total effluent of the plant (from all eight secondary settlers). Influent and sludge characterization was only performed using steady-state data obtained from sampling program and plant records (see Figure 3.6 for sampling locations).

3.2.3 Batch experiments

A range of sludge batch tests were executed at the site in November 1996 and April 1997 as part of the experimental program. For this purpose, lab-scale sequencing batch reactors (SBR) were transported and re-assembled at the on-site laboratory of the WWTP (Figure 3.5).

Figure 3.5 Experimental set-up used for the batch tests at the WWTP Haarlem Waarderpolder (Photo: Brdjanovic D.).

Two double jacketed laboratory fermenters (each 1.5 L) with automated operation, control and monitoring were used in all batch tests. The equipment used in the tests is described elsewhere (Brdjanovic *et al.* 1997a). The activated sludge used in the tests was always the fresh sludge taken at the pumping station of recirculated sludge. Each test was repeated at least three times. The tests were performed at controlled temperature of 20°C and pH of 7.0±0.1. Process rates were determined as initial rates by using linear regression method.

3.2.3.1 Anaerobic phosphorus release

Acetate in excess (100-300 mgHAc-C/L or 266-800 mgCOD/L) was instantly added to the reactor under anaerobic conditions. During 3-4 h phosphate and acetate concentrations were measured to determine the maximal released phosphorus and the release rate. At the end of the test surplus acetate always remained in solution, therefore acetate did not limit phosphorus release.

3.2.3.2 Aerobic phosphorus uptake

The sludge must be exposed first to anaerobic conditions in presence of acetate in order to deplete the internal poly-P pool and to increase the PHA level of the biomass so that PHB does not limit the P-uptake process. Acetate (20-25 mg HAc-C/L) was instantly added to the reactor under anaerobic conditions. During 3h, acetate was fully consumed and phosphorus was released into solution. At this point half of the sludge was transferred to another, identical, reactor. The remaining part of the sludge was exposed to aerobic conditions to measure the maximal specific aerobic phosphorus uptake rate ($q_P^{aerobic}$ in mgP/gVSS.h). In case the surplus acetate remains in solution after the anaerobic phase, the sludge should be washed and phosphate should be manually added to the mixed liquor prior to splitting.

3.2.3.3 Anoxic phosphorus uptake

The second part of the sludge cultivated in the foregoing batch test was exposed to anoxic conditions in order to determine the maximal specific anoxic phosphorus uptake rate (q_P^{anoxic} in mgP/gVSS.h). Anoxic conditions were maintained by addition of surplus amount of nitrate at the beginning of the test (28 mgN/L) and by the continuous flushing of the mixed liquor by N_2 gas.

3.2.3.4 Fraction of denitrifying activity of PAO

A bioassay for determination of the fraction of DPAO in activated sludge has recently been proposed (Wachtmeister *et al.*, 1997). The method is based on the fact that DPAO are active under both aerobic and anoxic conditions, whereas aerobic PAO are inactive under anoxic conditions. From the comparison of the P-uptake rates under aerobic ($q_P^{aerobic}$) and anoxic (q_P^{anoxic}) conditions the relative proportion of denitrifying dephosphation activity ($q_P^{anoxic}/q_P^{aerobic}$) in the phosphorus removing organisms was calculated.

3.2.3.5 Nitrification

Ammonia was added under aerobic condition in excess amount (28 mgN/L) at the beginning of the batch test. During the test the conversion of ammonia to nitrate was monitored and the nitrification rate ($q_{nitr}^{aerobic}$ in mgN/gVSS.h) was determined.

3.2.3.6 Denitrification

Acetate (20 mgHAc-C/L) and nitrate in excessive amount (28 mgN/L) were added to the reactor in order to determine the denitrification capacity of the sludge. Acetate, phosphate, nitrate and nitrite concentrations were measured during the test. Denitrification rate (q_{denitr}^{anoxic} in mgN/gVSS.h) was determined as initial nitrate utilization rate.

3.2.3.7 Endogenous phosphorus release

The phosphate release under anaerobic conditions in absence of external organic substrate was measured during 4h. The phosphorus release rate obtained from such test is considered as endogenous or secondary phosphorus release rate ($q_{P,endog}^{anaerobic}$ in mgP/gVSS.h)

3.2.4 Sampling program and analytical methods

The sampling and experimental program was executed in November 1996 and April 1997. During the course of the program, the weather conditions were favorable (there was no rain, and the plant operated under dry weather flow conditions at similar sewage temperature of around

13°C). The on-site sampling locations are shown in Figure 3.2. For influent characterization the following parameters were determined: COD_{total}, $COD_{filtered}$, $BOD_{5,\ total}$, PO_4, VFA (acetate, propionate, butyrate), NH_4, NO_3, pH, DO and temperature, while PHB, PHV, glycogen, MLSS and MLVSS were determined for sludge characterization (Table 3.3).

Table 3.3 Overview of the parameters sampled at different locations of the WWTP Haarlem Waarderpolder as a part of the sampling program performed in November 1996 and April 1997.

Nr.	Sampling point	Parameter															
		CODtotal	CODfiltered	BOD5	PO4	Acetate	Propionate	Butyrate	NH4	NO3	MLSS	MLVSS	PHA	Glycogen	Temperature	DO	pH
1	Raw sewage: after screens				■												
2	Settled sewage: after PST	□	□	□	□	□	■	■	□	□					■	■	■
3	Activated sludge: after aerobic tank				■				■	■	■	■	■	■		■	
4	Effluent: after the SST				□				□	□							
5	Primary sludge: PST				■												
6	WWTP Schalkwijk sludge: outlet pipe				■	■											
7	Supernatant: primary thickener	■	■	■	■	■			■	■							
8	Side stream sludge: P selector tank				■	■					■	■	■	■		■	■
9	Side stream sludge: P release tank				■	■					■	■	■	■		■	■
10	Filtrate: belt thickener				■												

■ Spot sample
□ 24h composite sample

The sampling points and frequency, and the choice of the parameters were governed by the specific requirements of mathematical models applied on this installation and by the content of the existing records. The analyses were performed according to procedures described in Brdjanovic *et al.* (1997a), and in the Dutch guidelines for sewage and sludge characterization (Roeleveld and Kruit, 1998). The information obtained through the sampling program were combined with the data routinely collected (in weekly intervals) by the staff of the plant. In addition, the influent and the effluent flow records were used together with the internal flow rates obtained from the information on the capacities and operational time of selected pumps at the plant.

3.2.5 Modeling tools

For the simulation of the WWTP Haarlem Waarderpolder the TUDP model replaced the module for P removal of ASM2. The stoichiometry and the kinetics of this combined model is presented in Annex 3.1 (after Van Veldhuizen *et al.* 1999 and Meijer 2001). This combined model was incorporated in the software package SIMBA3.2+® (based on MATLAB 5.1® and SIMULINK 2®) which was used for the simulation of the plant operation. Simulations of plant operation were carried out (using static data) until a steady-state performance was achieved (usual simulation period was 365 days). The scope of modeling included one of the four, arguably identical, activated sludge lines and the biological part of the common section for combined biochemical phosphorus removal (Figure 3.1). A personal computer (processor Pentium II® 233 MHz, 64 Mb) was used for simulation of this plant. Total computing time for present plant configuration as well as for each of three scenarios was up to 2 min from full start-up to steady-state (365 days chosen as simulation time). Computing time was up to 15 sec when the simulations were performed with previously calculated states. For simulation of batch test calculation time was as little as couple of seconds. All plant simulations were performed using the temperature of mixed liquor of 12.6°C as observed in the activated sludge unit, while in the batch sludge tests the temperature was controlled and was 20°C. Although the observed pH value at different locations at the treatment plant varied in the range 6.5 to 7.5, in the present version of SIMBA® it is still not possible to take into account different pH levels in the plant in one simulation.

3.2.6 Modeling strategy

The simulation of the WWTP Haarlem Waarderpolder was performed in four steps of which the first three were interactive. In each step the characteristics of both liquid phase and activated sludge were compared with measured data. The steps are: (a) simulation of the treatment plant operation (both liquid phase and biomass) and model calibration based on measured data (present case), (b) simulation of the batch tests using sludge properties as predicted in the previous step, (c) simulation of the performance of the treatment plant using feedback information from the batch test and their simulation, and (d) simulation of different options for alternative process design and layout.

3.3 Results

3.3.1 Sampling program

The results of sampling program and a selection of routinely collected data by the plant staff are summarized in Annex 3.2. During the entire period of sampling the plant effluent quality was in accordance with Dutch standards.

3.3.2 Influent and sludge characterization

Results of the influent characterization performed by following the procedure described in Annex 3.1 and using data from Annex 3.2 are summarized in Table 3.4. Sludge characteristics were obtained from sampling program (Annex 3.2) and by performing various batch tests as described in section 3.2.3.

3.3.3 Hydraulic set-up of the plant model

Plant design documentation, existing process scheme and current operational mode of the plant were used to create the hydraulic flow scheme of the installation (called Haarlem base case; Figure 3.6) made in SIMBA®. More information about this scheme is given in Table 3.5.

Table 3.5 Major components of the model scheme Haarlem base case.

Component	Nr. of units	Volume (m^3)	Modeled as (SIMBA® blocks)
Non-bulking selector	1 [1)]	125	1 anaerobic reactor
Denitrification ring	1 [1)]	4,380	1 anoxic reactor
Nitrification ring	1 [1)]	4,380	1 aerobic reactor
Secondary settlers	2 [1)]	6,000	1 mixed reactor with a loss of solids
Sludge storage tanks	0 [2)]	4,000 [4)]	1 anoxic (anaerobic) reactor
P selector tank [5)]	1 [3)]	875	1 anaerobic reactor
P release tank [5)]	1 [3)]	875	1 anaerobic reactor
Belt thickeners	2 [3)]	n.a	1 ideal liquid/solid separator

1) Per line; in total there are four identical activated sludge lines
2) Introduced in the model to simulate sludge retention in secondary settlers
3) Serves all four activated sludge lines
4) Represents the volume of the sludge blanket
5) Made of eight identical compartments in serial

Table 3.4 Characterization of the settled sewage at the WWTP "Haarlem Waarderpolder". The characterization was based on the data collected during the period 3rd- 10th April 1998 at DWF and sewage temperature 12.6C and pH 7.6

	No.	Symbol	Name	Unit	Settled Sewage	Thickener's Supernatant	Comment
Soluble Components	1	S_{O2}	Oxygen (negative COD)	$gCOD/m^3$	2.1	0.05	Measured
	2	S_F	Readily biodegradable organics	$gCOD/m^3$	73	73	Calculated using Eq. 3.9 in Annex 3.1
	3	S_A	Volatile fatty acids (mainly acetate)	$gCOD/m^3$	32.4	49.9	
	4	S_{NH4}	Ammonium & ammonia nitrogen	gN/m^3	46.5[1)]	55[2)]	
	5	S_{NO3}	Nitrate & nitrite nitrogen	gN/m^3	0.1	0	Measured
	6	S_{PO4}	Inorganic soluble phosphorus	gP/m^3	6.4[3)]	6[4)]	
	7	S_I	Soluble inert organic matter	$gCOD/m^3$	22	22	From Eq. 3.8, $Y_{BOD} = 0.20$ and $k_{BOD} = 0.23d^{-1}$
	8	S_{ALK}	Alkalinity	mol/m^3	5	5	Default value of the ASM2
Particulate Components	9	X_I	Particulate inert organic matter	$gCOD/m^3$	63	50	Calculated using Eq. 3.14 in Annex 3.1
	10	Xs	Slowly biodegradable substrate	$gCOD/m^3$	141	166	Calculated using Eq. 3.13 in Annex 3.1
	11	X_H	Active heterotrophic biomass	$gCOD/m^3$	0.01	0.01	
	12	X_{PAO}	Phosphate acc. organisms	$gCOD/m^3$	0.01	0.01	
	13	X_{PP}	Poly-phosphate	gP/m^3	0.001	0.001	
	14	X_{PHA}	Poly-hydroxy-alkanoates	$gCOD/m^3$	0.001	0.001	Assumed that there is very little biomass present in the settled sewage, see Annex 3.1
	15	X_{GLY}	Glycogen	$gCOD/ m^3$	0.001	0.001	
	16	X_{AUT}	Active autotrophic biomass	$gCOD/m^3$	0.01	0.01	
	17	X_{TSS}	Mixed liquor suspended solids	$gMLSSL/m^3$	153	162	Calculated using Eq. 3.15 in Annex 3.1
	18	X_{MeOH}	Metal-hydroxides	$gFe(OH)_4/m^3$	0	0	Estimated zero
	19	X_{MeP}	Metal-phosphate	$gFePO_4/m^3$	0	0	
	20	Q	Flow	m^3/day	5,760	5,520	Measured

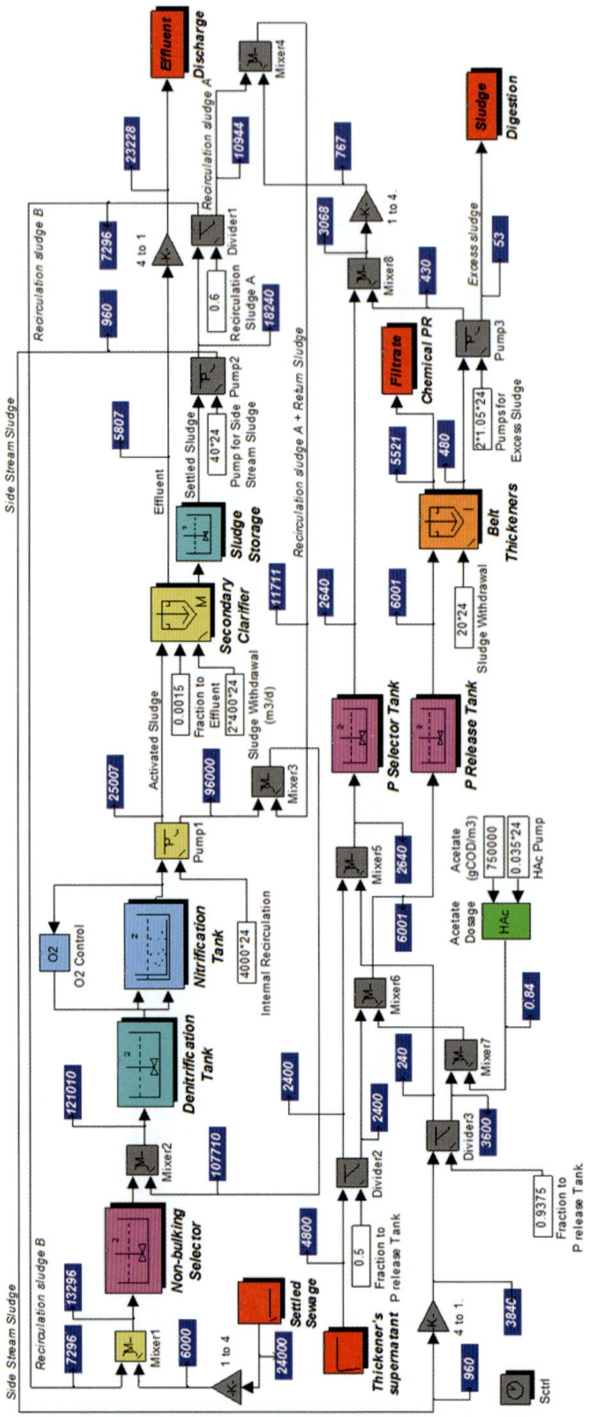

Figure 3.6 Hydraulic scheme of the WWTP Haarlem Waarderpolder model in SIMBA.

Concerning the model hydraulic layout of the plant there are two points that need to be noted: (1) the sludge storage tank was introduced in the return sludge line to account for the conversions in the secondary settlers, allowing a sludge retention time similar to that in settler (sludge blanket only) and, (2) the P release and P selector tanks were each modeled as a one fully mixed anaerobic reactor, while, in fact, they consist of eight interconnected compartments in serial (plug-flow). Both actions were justified; the introduction of a sludge storage tank reduced the nitrate concentration in the sludge to the measured level, and modeling P release (and P selector tank) as one fully mixed tank instead of multiple tanks in series gave the same results for both liquid and solid phase.

The results of simulation of the operation of WWTP Haarlem Waarderpolder (standard case Haarlem 1) are not presented. The composition of liquid phase and the biomass was obtained at various locations throughout the plant (marked as T in Annex 3.2). The model described well the present operation of the plant in terms of both liquid phase and biomass concentrations. Predicted effluent concentrations, like ammonia, nitrate and phosphate, were similar to those observed at the plant. The performance of the P release and P selector tank was also predicted satisfactorily. In case Haarlem 1b, the acetate addition to the P release tank was adjusted to the actual process requirements resulting in a similar, satisfactory performance of the plant. According to the model, the acetate addition could be decreased by up to 30% without the effluent quality being affected. The possible optimization of acetate dosing will be discussed later in the section 3.4.3. Predicted SRT was 84 days.

3.3.4 Model calibration

3.3.4.1 Calibration procedure

Calibration was based on a correct influent and sludge characterization and detailed evaluation of the flow scheme of the treatment plant. Calibration was carried out using static (daily average) data from the treatment plant combined with separate batch experiments with sludge and wastewater. A step-wise calibration procedure was applied in which just a few specific parameters were calibrated on specific plant data according to procedure by Henze *et al.* (1998).

For the initial simulation of the plant performance a default set of parameters was used; the parameters related to COD and N conversions were default parameters of ASM2, and the parameters for P removal originated from lab-scale SBR systems with an enriched culture of PAO grown on acetate.

The first step was to calibrate the SRT and the activated sludge concentration (MLVSS). Since the plant was neither upgraded nor operated completely according to the project documentation, the plant manager indicated that the treatment plant operates at SRT which exceeds the designed value of 28 days at least by twofold. The biomass concentration in the activated sludge unit was adjusted to the observed level of around 4.5 gMLVSS/m^3 by varying the sludge wasting rate and taking into account the amount of biomass leaving the plant via effluent. Proper biomass content of the P release and P selector tank was adjusted by equal distribution of the supernatant of the thickener between them (the amount of side-stream sludge was fixed by the maximal flow rate of four identical submerged pumps). Calculated SRT strongly exceeded the expectations, being 84 days. It is important to note that this first calibration step was based on volatile suspended solids, not on total suspended solids. Total suspended solids (in g/m^3) were calculated from the MLVSS predicted by the model by using measured ash content (~25%) and a conversion factor of 1.37 gCOD/gMLVSS. The SRT was also calculated using a phosphorus balance over the treatment plant which indicated a SRT of 80 days, which is in accordance with the value of 84 days obtained from the calibration procedure.

When the SRT calibration is concerned, an accurate determination of the first order rate constant of the BOD measurement (k_{BOD}) used for influent characterization is of great importance (see further discussion in section 3.4.1). Although the value of k_{BOD} is the characteristic value of a specific plant (may vary from 0.15 to 0.60 1/d), the recommended value of 0.23 1/d (STOWA, 1996) appeared to be the correct one for this plant (there was no measured data on k_{BOD} available). Due to the reasons explained later in the text, it is likely that k_{BOD} needs to be calibrated for each plant.

Once the SRT and the activated sludge concentration were calibrated, the next steps were ammonium and nitrate calibration. By the adjustment of the value of dissolved oxygen concentration at the end of the nitrification ring in the model to the observed level of 1.8 mgO_2/L, very reasonable results were achieved (2.4 $mgNH_4$-N/L and 5.3 mg NO_3-N/L versus 2.5 $mgNH_4$-N/L and 4.6 mg NO_3-N/L in reality as average for April 1997), hence there was no need for further calibration for nitrification and denitrification. In the context of wastewater treatment, it is almost impossible, but also unnecessary to take into account all uncertainties and to fit the model predictions exactly to the measured data. Here, the aim was to predict the observed data within the uncertainty of the measurements themselves.

As far as biological phosphate removal is concerned there was no need for further calibration of the phosphorus concentrations in the system. While P levels were predicted well, the acetate concentration was not fully correct in P selector and P release tanks. There could be at least three possible reasons for it: (a) the actual acetate dosage rate to the P release tank (at the time when the sampling was performed) was higher than the yearly average dosage rate of 22 L/h (this flow rate is estimated by the plant staff based on the records of annual purchase and consumption of 70% acetic acid), (b) default value for the maximal fermentation rate constant (q_{fe}) is not correct in the model, and (c) actual wastewater and sludge flows to P release and P selection tank were different from those obtained from the plant staff (and as such used in the hydraulic scheme of the plant model).

According to our measurements (Annex 3.2) a considerable fraction of acetate was not utilized in both P selector and P release tank, and therefore, was either wasted via excess sludge or returned to the activated sludge unit via return sludge (Figure 3.2). Using the acetate dosing rate of 22 L/h the model predicted overall lower acetate levels in a comparison with the observed, and a complete acetate uptake in the P release tank was forecasted (contradictory to the measurements). To increase the acetate level in the P release tank and to obtain a better fit of the model prediction to the measured data a higher acetate dosage rate of 35 L/h was used. With this increased acetate dosage rate the level of acetate at the beginning of the P release tank was adjusted to the measured value of around 130 mgCOD/L. However, predicted acetate concentration at the end of this tank was still around 20 mgCOD/L lower than measured (data not shown). To further improve prediction of acetate levels in a side stream process, the maximal fermentation rate q_{fe} was tunned on P selector tank. By using a default value of q_{fe} in the model of 3 gCOD/gCOD.d the fermentation process was overestimated resulting in a higher predicted acetate concentration at the end of the P selector tank than at the beginning, which is again in disagreement with the observed situation. This clearly indicated a need to reduce q_{fe} (as it was also the case in a study of van Veldhuizen *et al.* 1999). Reducing the q_{fe} improved a model fit to the measured data in the P selector tank. However, in the P release tank, reduced fermentation rate caused even further lowering of acetate level at the end of the tank. The best model prediction (but still not fully correct) was obtained with q_{fe} of 1.0 gCOD/gCOD.d., which makes the sum of the differences between predicted and measured acetate concentration at the end of the P release and P selection tank minimal. The wastewater and sludge flow rates are considered correct in this study.

The default and the observed phosphate/acetate ratio in this case could not be compared due to the fact that it was practically not possible to obtain adequate influent samples from the P release and selector tank. Therefore, it was decided to keep the default phosphate/acetate ratio (Y_{PO4}=0.36 g P/g COD, Smolders *et al.* 1994a) of the model (see also section 3.3.5.1). The higher PO_4/HAc ratio observed in the P-selector tank in comparison with the P-release tank (1.34 and 0.27 gP/gCOD respectively) may be partially attributed to influence of pH on the anaerobic P release (Smolders *et al*, 1994b). The impact of pH on fermentation process, which can also influence the observed PO_4/HAc ratio, is not investigated yet. Measured pH was 6.5 in P release tank and 7.5 in P selector tank. Lower pH in P-release tank is caused by the addition of acetate and by the presence of CO_2 (produced in glycogen conversion to PHA) in mixed liquor because there is no CO_2 stripping provided. SIMBA 3.2+ does not yet allow for simultaneous simulation of different pH values in different units. The other reasons beside influence of pH may be different fermentation and hydrolysis rates between these two tanks, and/or possible presence of glycogen accumulating non-poly-P bacteria (GAO) in the systems (the latter is discussed in section 3.3.5.1). Due to above complications no further attempt to have a better calibration was made.

Although the batch sludge experiments were used for model validation, some were also used for model calibration. Based on the results of aerobic and anoxic P uptake batch tests, the percentage of denitrifying activity of PAO in this plant was determined to be as high as 80%. Therefore, the reduction factor under anoxic conditions η^{P}_{NO3} was changed from 0.50 to 0.80 to fit the ratio of phosphate removed in the aerobic ring to phosphate removed in the anoxic ring. After the first iteration, the calibration procedure was repeated in a second loop. It is striking that even with the default set of parameters the model predicted the measured parameters of the liquid phase throughout the plant reasonably well (the results not shown). This was to a less extend true for the observed internal biomass concentrations, especially for the glycogen - active biomass ratio (f_{gly}) which was extremely high (2 gCOD/gCOD, results not shown) when the original value of k_{gly} (1.09 gCOD/gCOD.d) was used. Therefore, in the second loop the glycogen formation rate (k_{gly}) was changed to 0.15 gCOD/gCOD.d to fit the predicted data to the normal values of glycogen (~0.4 gCOD/gCOD). Change in k_{gly} influences f_{gly} which further affects f_{pp} and f_{pha} due to back coupling between these parameters in the EBPR model. By decreasing k_{gly}, f_{gly} is reduced, and consequently X_{PAO} increased (this issue is discussed further in section 3.4.2). Based on the simulation results it was judged that no further calibration is needed. The list of parameters adjusted in the calibration procedure is given in Table 3.6.

Table 3.6 List of the model parameters which value was changed in the calibration procedure.

Parameter	Symbol	Unit	Old value	New value
Maximal fermentation rate	q_{fe}	gCOD/gCOD.d	3.00	1.00
Reduction factor under anoxic conditions	η^{P}_{NO3}		0.50	0.80
Glycogen formation rate	k_{gly}	gCOD/gCOD.d	1.09	0.15

3.3.5 Model evaluation

The model evaluation was based on a check of its capability to well describe (1) the liquid phase of the plant flows using static influent data (separate sampling showed that there was no diurnal variation in influent concentration), and (2) the sludge properties determined in batch tests. For this purpose batch experiments were simulated in SIMBA® (Figure 3.7a-f).

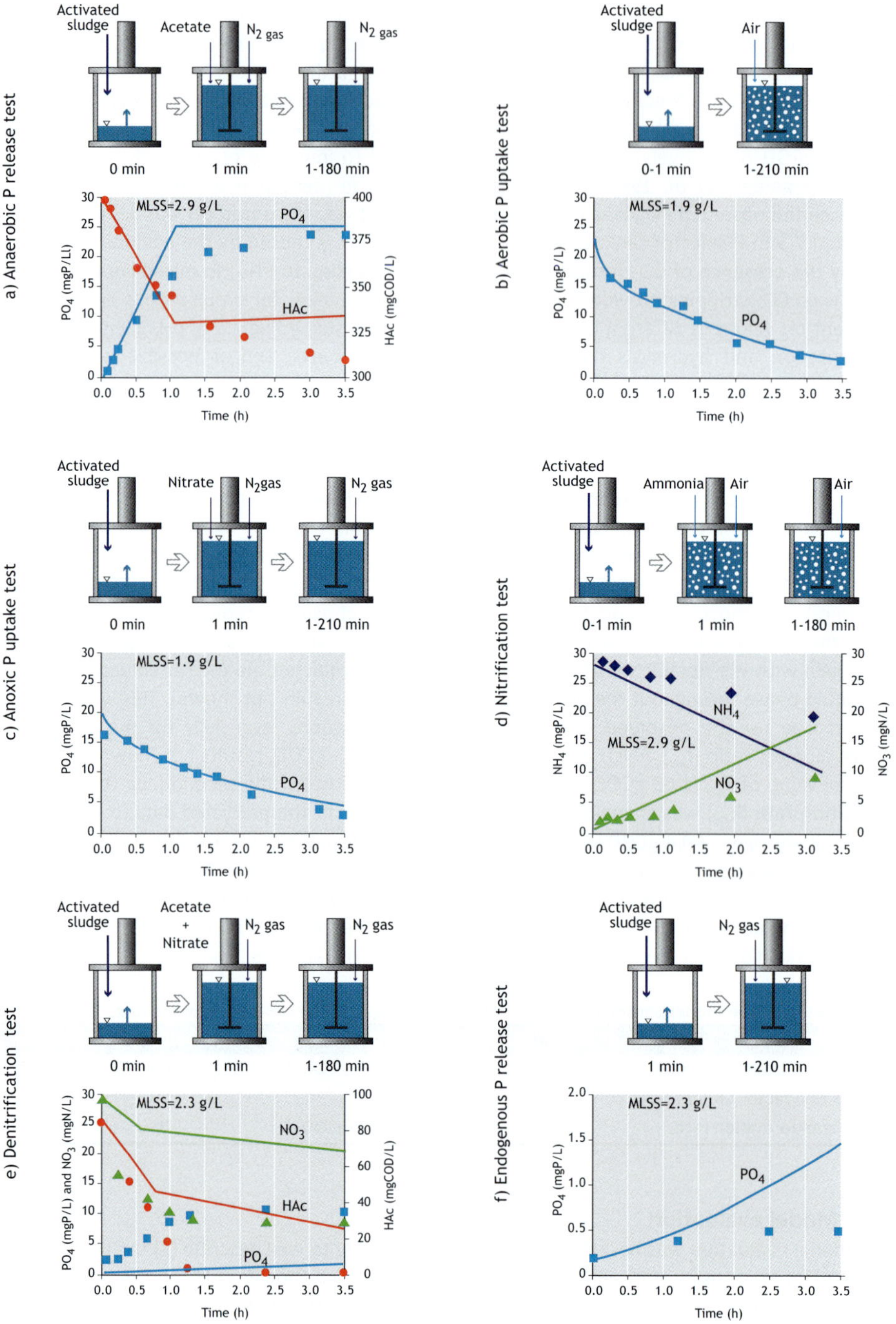

Figure 3.7 Concentration of phosp hate (■) acetate (●), ammonia (◆), and nitrate (▲) in batch tests performed with settled activated sludge at temperature of 20°C and pH 7.0±0.1. The model prediction is represented with solid line. The experimental procedure is given schematically for each test.

3.3.5.1 Maximal anaerobic phosphate release

Observed phosphate and acetate concentrations obtained in sludge batch test were in general well predicted by the model (Figure 3a). In case of activated sludge from Haarlem the maximal P release rate was 6 mgP/gVSS.h. According to model prediction total P release was limited by the poly-P content of the biomass. An average observed phosphate/acetate ratio from five repeated tests was 0.29±0.04 gP/gCOD, similar to the value observed at the WWTP in P release tank (0.27 gP/gCOD), but lower than the default value of the model (0.36 gP/gCOD). The similarity between the PO_4/HAc ratio obtained in batch tests and observed PO_4/HAc ratio in the P release tank is expected due to the fact that the influence of temperature (in tests 20°C and in the plant 12.6°C) should have a marginal influence of anaerobic stoichiometry (Brdjanovic *et al.* 1997a) and that in both cases the same, external, acetate source was used. The default phosphate/acetate ratio was determined in laboratory tests using enriched EBPR culture (Smolders *et al*, 1994b) and was confirmed in tests with activated sludge from full-scale plants. The reason for the comparatively lower PO_4/HAc ratio obtained in batch tests may either be the influence of the extremely long SRT or that in WWTP Haarlem Waarderpolder glycogen accumulating non-poly-P organisms (Cech and Hartman, 1990, 1993; Liu *et al.,* 1994) have accumulated due to oversupply of acetate. GAO are capable of anaerobic utilization of organic substrates which are converted and stored as PHA, while the energy and reduction equivalents are provided by only glycogen degradation without involvement of poly-P. The fact that in all four maximal P release batch tests acetate was still utilized even after full poly-P depletion in the biomass suggests a significant presence of GAO in the plant (Brdjanovic *et al.,* 1997b). Because surplus acetate is available to both PAO and GAO, a competition for substrate, which PAO would win under acetate limited conditions, did not take place here. There are three possible ways to proceed with this issue in this study: (1) to assume there is no GAO in the system and therefore, to lower default PO_4/HAc ratio, (2) to believe there is significant presence of GAO and include their metabolism in the model, or (3) to assume there is significant presence of GAO in the system, do not model their metabolism, retain the default PO_4/HAc ratio, and leave this issue for further research. The last option was adopted in this work. This resulted in a correct description of the PAO population, but the HAc uptake under anaerobic conditions will be underestimated. The remaining HAc will be taken up by the heterotrophs to which group the GAO could be considered to belong.

3.3.5.2 Aerobic and anoxic phosphate uptake

The P uptake tests (Figure 3b and c) showed a very good agreement between predicted and observed data after adjustment of glycogen formation rate (k_{gly}) in the set of model parameters. By comparison of P uptake rate under aerobic and anoxic conditions (2.2 and 1.7 mg P/g VSS.h, respectively) the fraction of denitrifying activity of PAOs was estimated to be 80%.

3.3.5.3 Nitrification

Nitrification process performance in the batch test (Figure 3d) was not predicted very well by the model; observed ammonia conversion to nitrate was approximately 40% lower than predicted. There are two possible reasons for this discrepance: either the amount of nitrifiers (X_{AUT}) predicted by the model in the plant is too high or nitrification rate in the model is too high (or maybe both). To check this assumption two actions were carried out: (1) the decay rate of autotrophs (b_{AUT}) was increased from 0.15 to 0.25 1/d, and (2) maximal growth rate of autotrophic biomass (μ_{AUT}) was reduced from 1 to 0.65 1/d. In both cases, firstly the full-scale plant was simulated, then the batch test with the "new" biomass obtained from a full-scale plant simulation. Each of these changes made very good fit of predicted data to measured ones in a batch test. However, when these new values were fed back into simulation of the treatment plant, nitrification was insufficient, and effluent ammonia and nitrate concentrations were not predicted well (32 mgNH_4-N/L and 1.3 mgNO_3-N/L when the b_{AUT} was increased, and 25 mgNH_4-

N/L and 1.7 mgNO_3-N/L when the μ_{AUT} was decreased). Phosphate effluent concentration changed only very marginally. Consequently, it was not possible to predict the nitrification process in the plant and in the batch test in the same time. The reason for this controversy remains unclear.

3.3.5.4 Denitrification

The model did not predict well observed concentrations in this test (Figure 3e). Especially it is noted that in the denitrification test a very large release of phosphate and a rapid consumption of acetate was observed. The model predicts no P release and much less acetate uptake. This can for the time being directly be explained by the fact that the model does not take simultaneous presence of VFA (acetate) and electron acceptors (in this case nitrate) into account. Under denitrifying conditions (as in this test) acetate is taken up and a part of the uptake results in P release, but now the TCA cycle instead of glycogen is delivering reduction equivalents (Kuba *et al.* 1996). In general, simultaneous presence of acetate and nitrate at WWT plants does not occur often. Although denitrification in sludge batch test was not described satisfactorily, the nitrate concentrations in the treatment plant were predicted well.

3.3.5.5 Endogenous phosphorus release

The results of this test (Figure 3f) suggest that the P release in absence of external substrate is overestimated by the model. This P release can be the result of: (1) maintenance process as used in TUDP model (real endogenous P release) and (2) endogenous processes like decay of heterotrophs by which an extra substrate is supplied (concept used in ASM2) leading to P release. We suggest that this decay of heterotrophs is overestimated. If endogenous respiration, as in ASM3, instead of decay would be used, probably the model and experiments would give the same results. Therefore it was decided not to change the parameters to fit this test.

It can be concluded that these batch tests delivered useful information on how the model performs. Some aspects like the maximal anaerobic P release, anoxic and aerobic P uptake were well described. Other aspects such as P release under denitrifying conditions and the endogenous P release pointed to places where the model could be improved. Finally, the nitrification process needed better parameter estimation for correct fitting of full-scale and batch experiments. Since the default values however already gave good results in the full-scale model we decided to leave the values unchanged.

3.3.6 Alternative EBPR process configurations

The validated model was used to evaluate different thre EBPR process schemes (Table 3.7).

Table 3.7 Comparison between present and alternative process configurations of the WWTP.

Plant configuration		Type of P-removal process	Acetate addition (m^3/d)	Loading[2)]	Comments
Base case	1a	Side stream	0.84[1)]	Total	Normal plant operation
Phostrip	1b	Side stream	0.36	Total	
Alternative	2a	Main stream	0	25%	
A/O	2b	Main stream	0	25%	Selector enlarged
Alternative	3a	Main stream	0	25%	Extra anaerobic tank added
Modified UCT	3b	Main stream	2.50	25%	Extra anaerobic tank added
Alternative *BCFS®*	4	Side stream	0.84	Total	Aerated zone reduced by 25%

[1)] Estimated acetate dosing rate during sampling period based on the measured acetate concentrations in the P release tank. The yearly average acetate dosage rate is 0.53 m^3/d (22 L/h) as 70% acetic acid.

[2)] Total loading to the plant (included all four lines) was applied in cases of Haarlem 1 and 4, while in cases 2 and 3 only loading of one line was applied (see also Figure 3.8).

Thereby it was decided to keep the present physical configuration of the WWTP Harlem Waarderpolder model scheme (volume of tanks, influent flow rate etc.) the same. The model schemes of three alternative EBPR process configurations, namely: (a) A/O, (b) modified UCT and (c) BCFS®, are described below. The calibrated model was used to evaluate BCFS® system, while in A/O and UCT systems k_{gly} was increased from 0.15 to 0.45 gCOD/gCOD d to compensate for comparatively lower SRT (discussed later in detail). A lower SRT has a positive influence on the EBPR. Therefore, in those cases where the removal was not optimal the SRT was reduced to a value which would also under Dutch winter conditions result in a good nitrification (approximately 35 or 15 d aerobic SRT). Simulation results are given in Table 3.8.

Table 3.8 Comparison of selected operational parameters of present and alternative process configurations of WWTP Haarlem Waarderpolder.

Parameter	Unit	Base case Phostrip		Alternative A/O		Alternative Modified UCT		Alternative BCFS
		1a	1b	2a	2b	3a	3b	4
SRT total	day	85	85	34	37	37	37	83
MLVSS aeration tank	mg/L	5,927	5,814	2,595	2,598	2,564	3,043	5,948
Sludge wasting	kg/d	2,979[1)]	2,927[1)]	848[2)]	837[2)]	838[2)]	991[2)]	2,950[1)]
PO_4 effluent	mgP/L	0.3	0.8	7.5	8.0	2.5	0.7	0.3
NH_4 effluent	mgN/L	2.4	2.1	3.5	4.5	2.7	3.5	3.1
NO_3 effluent	mgN/L	5.3	5.9	7.8	2.6	9.3	5.6	5.7

[1)] from the total plant (4 lines)
[2)] from one of four identical lines

3.3.6.1 Alternatives 1a and 1b: A/O system configuration

The elements concerning side stream EBPR were removed. The rest of the plant remained unchanged and was simulated as such (alternative 1a). Since alternative 1a presents, in essence, a non-optimized conventional configuration for COD and N removal, the predicted effluent quality strongly deteriorated (with P being most affected) in comparison with the present case. To achieve better process performance, including biological P removal, an anaerobic tank (larger than the already present non-bulking selector) is needed. The design criteria was that the acetate (from the thickener) entering this anaerobic tank is fully utilized. By increasing the volume of the non-bulking selector from 125 to 1,500 m^3, a main stream (A/O) EBPR was obtained (alternative 1b) resulting in improved process efficiency. However, predicted effluent quality was still less good in comparison with the present case. The plant was simulated with SRT of 36 d; an increased SRT leads to even worse P removal efficiency.

3.3.6.2 Alternatives 2a and 2b: modified UCT system configuration

In alternative 2a an additional anaerobic P release tank (2,000 m^3) was placed before the activated sludge unit in a modified UCT process configuration. Internal sludge recirculation-influent flow ratio between nitrification and denitrification rings was decreased from 1:13 (present case and alternative 1) to 1:5 and between denitrification ring and P release tank was 1:1. Operating sludge age was 37 d. In this alternative, external addition of acetate to P release tank was not applied and the predicted effluent quality was similar to that from alternative 1b. This is likely because the nitrate concentrations in case 1b are already low. A good effluent quality was only achieved after external acetate dosing was introduced (alternative 2b). The amount of acetate needed to be added was sevenfold higher in comparison with optimized acetate addition in alternative 1b. The increased acetate need was predicted by the theoretical evaluation by Smolders *et al.,* 1996 and shows the efficiency of a Phostrip process with respect to HAc use.

3.3.6.3 Alternative 3a and 3b: BCFS system configuration

Recently developed system configuration for chemical and biological P removal with in-line `P stripping' and off-line precipitation (denoted as BCFS process, Van Loosdrecht *et al.,* 1998) was applied. Side stream EBPR was removed, three an-aerobic reactors in series (to simulate a plug-flow tank of 2,000 m^3) were introduced, followed by an in-line stripper (a baffled zone acting as clarifier). Phosphorus that cannot be accumulated by the sludge (and removed via excess sludge) is removed by extracting a part (around 20% of the influent flow) of P-rich supernatant at in-line stripper. The amount of extracted flow equals the flow chemically treated in the original configuration (present case). After chemical treatment, P-free stream is returned to the beginning of the plant. Following in-line stripping, the reminder of the treatment was kept as in the present case; operating SRT was 90 d. The effluent quality was good without need for external acetate dosing.

3.4 Discussion

3.4.1 Influent characterization

As far as the WWTP Haarlem Waarderpolder is concerned, the Dutch standard procedure for wastewater characterization (Roeleveld and Kruit, 1998) proved satisfactory for the model construction. This procedure was primarily developed for COD and N removal modeling; in this case it was used for COD, N and P modeling. The critical part of this procedure was the estimation of the coefficient k_{BOD}. The methodology for k_{BOD} determination is described elsewhere (Roeleveld and Kruit, 1998). This coefficient has a large influence on the determination of particulate inert organic matter (X_I) and slowly biodegradable substrate (X_S). The model is very sensitive towards change in k_{BOD} (and X_I); a relatively small change in k_{BOD} strongly affects the effluent quality and SRT or sludge concentration, especially in the systems with long SRT as it is the case here. The impact of estimated k_{BOD} on the calculated SRT for WWTP Haarlem Waarderpolder is shown in Figure 3.8. Even only a marginal change in this value (for example ± 0.01) changes the calculated SRT by 5 d. It is a very important, but also very difficult task to have k_{BOD} determined at this accuracy. It is therefore expected that the k_{BOD} need to be adjusted in the model calibration for WWTP with long SRT due to the high sensitivity of the model to this parameter and to its relatively rough experimental estimation methodology (inaccuracy certainly higher then 0.01).

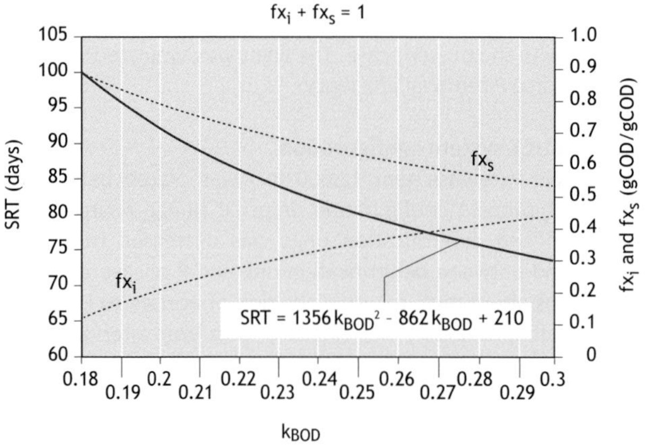

Figure 3.8 Influence of estimated k_{BOD} on calculated SRT and fraction of X_I and X_S of suspended COD.

3.4.2 Model calibration

The concentrations of the soluble components in the liquid phase in the WWTP were predicted well even without any calibration. This was true only to a lesser extend when the polymer (PHB, glycogen and poly-P) concentrations are concerned. This strongly suggests that the model should be calibrated not only on the liquid phase (effluent) concentrations, but also on the polymer concentrations. One of three parameters which were changed in the model was the glycogen formation rate constant (k_{gly}). In the TUDP model k_{gly} directly influences glycogen-active biomass ratio (f_{gly}). If the value of f_{gly} is incorrectly determined by the model, it will lead to inaccurate prediction of the polyphosphate-active biomass ratio (f_{pp}) and poly-hydroxy-alkanoate-active biomass ratio (f_{pha}), which may directly affect the prediction of the observed rates, such as P uptake rate. To check the influence of change in k_{gly} on predicted aerobic P uptake rate (and also anoxic P uptake rate) the WWTP in Haarlem was simulated with three different values of k_{gly} (1.09, 0.45 and 0.15 gCOD/gCOD.d). Clearly, a change in k_{gly} strongly affects the ratio of storage pools to active biomass (Table 3.9), whereby the values for f_{gly} obtained for the high k_{gly} values are not realistic (Smolders *et al.,* 1995a).

Table 3.9 An example of influence of variation in k_{gly} on fraction of storage materials in the biomass (source: simulation standard case Haarlem 1).

		k_{gly} (gCOD/g COD.d)		
Parameter	Unit	0.15 (this study) SRT = 84 days	0.45 (van Veldhuizen *et al.* 1999) SRT = 50 days	1.09 (Murnleitner *et al.* 1997) SRT = 20 days
f_{gly}	gCOD/gCOD	0.49	1.10	1.89
f_{pp}	gP/gCOD	0.18	0.20	0.24
f_{pha}	gCOD/gCOD	0.30	0.24	0.18

The sludge composition from above simulations was used to simulate aerobic and anoxic P uptake batch tests. Each batch test simulation was performed again with three different k_{gly} values. The results showed that this second variation in k_{gly} constant had no impact on predicted P uptake rates. However, a comparison (Figure 3.9) between predictions of batch tests using sludge characteristics obtained from the simulation of the treatment plant performance clearly indicated the need for a lower k_{gly}.

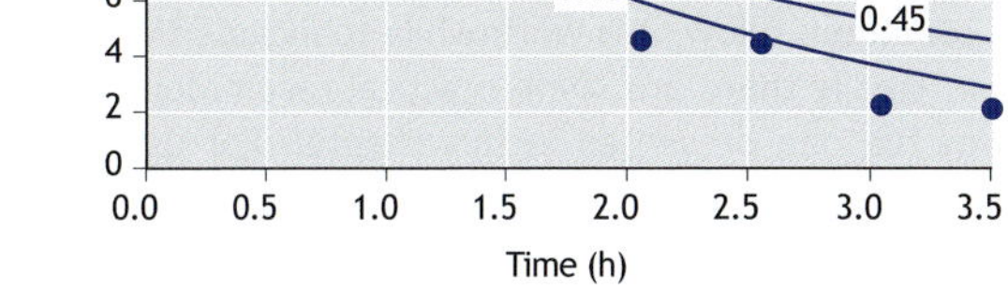

Figure 3.9 Influence of glycogen formation rate constant on aerobic phosphate uptake.

Lower k_{gly} results in a higher formation of the PAO (X_{PAO}), thus comparatively higher P uptake rate is predicted at lower k_{gly}. This indicates that k_{gly} either strongly depends on SRT or the kinetic expression for glycogen in the model is not fully correct. It is interesting to note that in the systems with longer SRT (Table 3.9), the k_{gly} needed to be lowered from 1.09 gCOD/gCOD.d at 20 d SRT (Murnleitner *et al.,* 1997), to 0.45 at 50 d SRT (Van Veldhuizen *et al.,* 1999) and 0.15 at 84 d SRT, to obtain a good prediction of aerobic and anoxic P uptake rate.

3.4.3 Operational aspects

According to project documentation and plant operational record (Table 3.1), since its expansion in 1995, the WWTP Haarlem Waarderpolder has been underloaded. Consequently, this resulted in a long SRT of 85 d. When the plant loading would be increased to its full capacity, according to the model prediction, the SRT would drop to 54 d with biomass concentration in the activated sludge unit of 6 g/L and the plant would still perform well (simulated NO_3, NH_4 and P effluent concentrations were 3 mgN/L, 6.4 mgN/L and 0.4 mgP/L, respectively). The plant simulation results, as well as the acetate measurements, indicate that only a fraction of acetate, fermented in primary thickeners and added to the P selector tank, is utilized in this tank. During normal plant operation the thickener supernatant is only introduced into the P selector. During the period of sampling this supernatant was delivered in equal proportions to both tanks. The additional shift of this flow into P release tank might further reduce the need for external dosing of acetate to P release tank. Additionally, the optimization of acetate dosing could bring extra benefits in case the hypothesis of presence of significant amount of GAO (see above) in this plant is true. If there would be less acetate available, the competition between PAO and GAO for substrate would intensify and, eventually, GAO might be washed out from the system. This will result in even further reduction in requirements for acetate in a side stream process. Since this matter is highly speculative, more detailed research would be needed. Due to low COD/P ratio relative poor P removal was obtained in A/O and UCT process. In those systems insufficient amount of PAO was produced to take up all phosphate. Larger anaerobic tank or shorter SRT would improve such situation. However, latter will certainly negatively affect nitrification process. It was shown that the processes with P stripping are more efficient because part of P is chemically removed. For the strip processes little COD is needed for EBPR (10% of main stream process, Smolders *et al.,* 1996) leading to an efficient process, allowing a high SRT. This improves nitrification and reduces activated sludge production, but also results in chemical sludge production. Furthermore, similarly good results were obtained by BCFS process without need for HAc addition.

3.4.4 Practical aspects

This work is the first case where the validation of a model for combined COD, N and P removal was performed on number of batch tests with fresh activated sludge, in an on-site well equipped laboratory. Although some batch tests were not well predicted by the model, they were very valuable as a source of feedback information used for model calibration. The batch tests also yielded useful information which could not be obtained otherwise for this particular plant. If planned well and possibly already performed elsewhere, the time spend on their execution can be reduced. If the time and the technical conditions allow, it is worthwhile to include them in a modeling process of existing WWT plants. Finally there are some practical recommendations which may help in modeling tasks such as the one presented in this chapter. For users less familiar with SIMBA® simulation software it is recommended to first thoroughly study the software itself and work-out a couple of examples, prior to undertaking any further work on this matter. Good understanding of mathematical models to be used is a must. Once this is done, the next steps should be to study in depth project documentation, investigate the present state of the plant, and to compare the differences between these planned and actual situation (which usually will exist). One should make sure that the most recent and true information is obtained, because it may well happen that due to complexity of the plant some changes occurred which are considered not important to the plant employees, but could well be essential for the modeling. It is crucial that all flow rates and their fate are known, and that the hydraulic scheme is complete and fully correct. In case that a plant consists of two or more parallel lines and only one is chosen to be modeled, one should make sure that there is no difference between them, or if there is, be aware of them or take this into account, if significant. Once a proper layout and hydraulics is obtained influent characterization should be done. Here, it is important to select a good characterization protocol, making certain that the required model inputs and the parameters determined by a protocol match. One must be certain which data are needed to be measured for influent characterization. It is advisable to get a good quality data set routinely measured by the staff of the plant. These data should be studied prior to making selection of the items and frequency of parameters to be included in, very likely always required, an additional sampling program. The design of a sampling program will also be governed by the decision on whether static or dynamic influent data will be used for modeling. The sampling program should be designed in a way to provide sufficient data for confident influent characterization, model calibration and evaluation. Samples should be analyzed as soon as possible, otherwise stored immediately and transported and handled properly. A longer period without rain or snow is desirable for execution of sampling providing easier working conditions and evaluation of data. If data obtained from the plant staff are used, the way how they are analyzed should be known. One needs to be sure to choose a period of very normal plant operation. Obtained data from sampling program should be worked out and evaluated as soon as possible, while there is still possibility to relatively easy repeat some sampling or analyses. Sludge characterization can be performed as a part of the sampling program and/or by executing batch experiments using preferably fresh sludge from the treatment plant. Care should be taken that the sludge is taken from the locations at the treatment plant which best correspond with the purpose of batch test. If possible, batch tests should be performed at the treatment plant itself, if not, sludge could be transported and stored in a dark and cold place for up to a couple of days. It should be checked if there is any difference using fresh and stored sludge. Batch tests should be performed at the same temperatures and pH as observed in the process. Once influent and sludge characterization is done simulation and calibration steps can be performed (for a recommendation concerning this work phase the reader is referred to Henze *et al.* 1998). The time needed for the execution different phases of modeling the plant in Haarlem is given below. This time distribution is subject to many factors such as experience and skills of researcher, availability, quality and quantity of existing information, configuration, size and complexity of the treatment plant, enthusiasm of the local staff, capacity and analytical diversity of the lab, availability of equipment for batch experiments, weather conditions etc. Therefore, Figure 3.10 can be used as indication only.

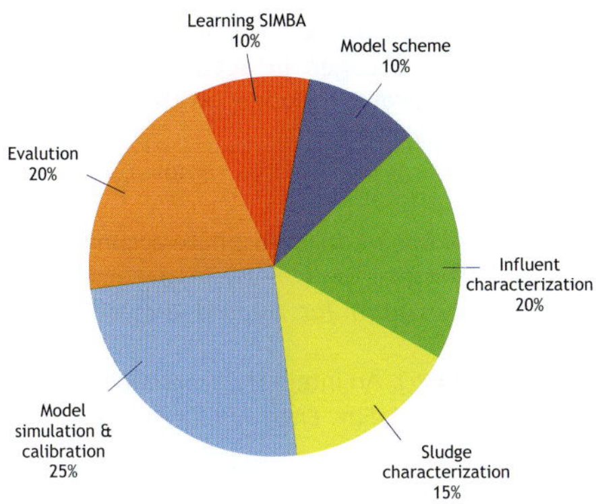

Figure 3.10 Time (in %) spent on various stages of modeling WWTP Haarlem Waarderpolder.

3.5 Conclusions

The combined ASM No. 2 and TUDP model for COD, N and P removal proved well capable of describing the performance of WWTP Haarlem Waarderpolder with adjustment of only three parameters of the model. This modeling exercise was not only carried out to gain more experience from practical application of the model and to find out where the model should be improved, but was also beneficial to better understanding of plant operation and treatment processes and to further optimize the plant performance.

Acknowledgements

Authors would like to thank Dutch Foundation for Water Research (STOWA) and The World Bank for sponsoring this project. We appreciate assistance of process department of Dutch water board Rijnland, Mr. Siem Tromp and Mrs. Astrid Vermaat of WWTP Haarlem Waarderpolder and technical and analytical staff of Klyver Laboratory of TU Delft.

References

Brdjanovic D., Van Loosdrecht M.C.M., Hooijmans C.M.,Alaerts G.J.F.R., and Heijnen J.J. (1997a). Temperature effects on physiology of biological phosphorus removal. *J. Env. Eng.* Vol.123, No.2., pp.144-154.

Brdjanovic D., van Loosdrecht M.C.M., Hooijmans C.M., Mino T., Alaerts G.J. and Heijnen J.J. (1997b). Innovative methods for sludge characterization in activated sludge systems. Conference proceedings. 3rd International Conference on Biological Nutrient Removal from Wastewater. 30 Nov.- 4 Dec., 1997. Brisbane, Australia.

Cech J.S. and Hartman P., (1993). Competition between polyphosphate and polysaccharide accumulating bacteria in enhanced biological phosphorus removal systems", *Wat. Res.*, Vol. 27., No.7, pp.1219-1225.

Gujer W., Henze M., Mino T, Matsuo T, MC Wentzel and GvR Marais (1995). The Activated Sludge Model no 2. 3[trd] *Seminar on Activated Sludge Modeling: Modeling and Microbiology of Activated Sludge Processes.* Kollekole, Denmark.

Gujer W., Henze M., Mino T and Van Loosdrecht M.C.M (1998). Activated Sludge Model no 3. 4th *Seminar on Activated Sludge Modeling: Modeling and Microbiology of Activated Sludge Processes.* Kollekole, Denmark.

Henze M., Grady C.P.L., Gujer W.,Marais G.v.R. and Matsuo T. (1994). Activated sludge model no.1. *IAWQ Scientific and Technical Report No.1.*

Henze M., Gujer W., Mino T., Matsuo T., Wentzel M.C. and Marais G.v.R. (1994). Activated sludge model no.2. *IAWQ Scientific and Technical Report No.3.*

Henze M., Gujer W., Mino T., Matsuo T., Wentzel M.C., Marais G.v.R. and Van Loosdrecht M.C.M. (1998). Activated Sludge Model no. 2d. 4th *Seminar on Activated Sludge Modeling: Modeling and Microbiology of Activated Sludge Processes.* Kollekole, Denmark.

Kuba T., Wachtmeister A., Van Loosdrecht, M.C.M. and Heijnen J.J. (1996). A metabolic model for biological phosphorus removal by denitrifying organisms. *Biotechn. & Bioeng.* Vol. 52, pp.685-695.

Liu W.T., Mino T., Nakamura K., and Matsuo T. (1994). Role of glycogen in acetate uptake and polyhydrohyalkanoate synthesis an anaerobic-aerobic activated sludge with minimized polyphosphate content", *J. Ferment. Bioeng.*, Vol. 77., No. 5, pp. 535-540.

Matsuo T. (1994). Effect of the anaerobic SRT on enhanced biological phosphorus removal, *Wat. Sci. Techn.*, Vol. 30, No. 6, pp. 193-202.

Mino T., Liu W.T., Satoh H., and Matsuo T. (1996). Possible metabolisms of polyphosphate accumulating organisms (PAOs) and glycogen accumulating non-poly-P organisms (GAOs) in the enhanced biological phosphorus removal process. Proceedings 10th *Forum for Applied Biotechnology*, Brugges, Belgium.

Murnleitner E., Kuba T., Loosdrecht van M.C.M. and Heijnen J.J. (1997). An integrated metabolic model for the aerobic and denitrifying biological phosphorus removal. *J. Env. Eng.*, ASCE, Vol. 54, No. 5, pp. 434-450.

Roeleveld P. J. and Kruit J. (1998) Guidelines for wastewater characterisation in The Netherlands. *Korrespondenz Abwasser.* 45(3), 465–468.

Satoh H., Mino T., and Matsuo T. (1994). Deterioration of enhanced biological removal by the domination of microorganisms without polyphosphate accumulation, *Wat. Sci. Techn.* Vol. 30, No. 6., pp. 203-211.

Smolders G.J.F., Van Loosdrecht M.C.M., Heijnen J.J. (1994a) A metabolic model for the biological phosphorus removal process. *Wat.Sci.Techn.* Vol. 31, pp. 79-93.

Smolders, G.J.F., van der Meij J., van Loosdrecht M.C.M. and Heijnen J.J. (1994b). Model of the anaerobic metabolism of the biological phosphorus removal processes; stoichiometry and pH influence". *Biotechnology and Bioengineering*, Vol. 43, pp. 461-470.

Smolders, G.J.F., Krop J.M., Van Loosdrecht M.C.M. and Heijnen J.J., (1995a). A metabolic model for the biological phosphorus removal process I: Effect of the sludge retention time. *Biotech. Bioeng.* Vol. 48, pp. 222-233.

Smolders, G.J.F., Bulstra D.J., Jacobs R.,Van Loosdrecht M.C.M. and Heijnen J.J., (1995b). A metabolic model for the biological phosphorus removal process II: Validation during start-up conditions. *Biotech. Bioeng.* Vol. 48, pp. 234-245.

Smolders G.J.F., Van Loosdrecht M.C.M., Heijnen J.J. (1996) Steady state analysis to evaluate the phosphate removal capacity and acetate requirement of biological phosphorus removing mainstream and sidestream process configurations. *Wat. Res.* Vol. 30, pp. 2748-2760.

STOWA (1996) Methods for influent charactrization (96-08)

Van Loosdrecht M.C.M. (1996) A comparison between models for biological phosphorus removal. *Med. Fac. Landbouww. Univ. Gent*, 61/4a, 1753-1760

Van Loosdrecht M.C.M., Brandse F.A. and de Vries A.C. (1998) Upgrading of wastewater treatment processes for integrated nutrient removal: the BCFS process. *Water Sci. Tech.* 37, 209-217.

Van Veldhuizen H. M., van Loosdrecht M. C. M. and Heijnen J. J. (1999) Modelling biological phosphorus and nitrogen removal in a full scale activated sludge process. *Water Res.* 33(16), 3459–3468.

Wachtmeister A., Kuba T., Van Loosdrecht M.C.M. and Heijnen J.J. (1997). A sludge characterization assay for aerobic and denitrifying phosphorus removing sludge. *Wat. Res.*, Vol. 31, No. 3, pp. 471-478.

Wentzel M.C., Dold P.L., Ekama G.A., Marais G.v.R. (1989). Enhanced poly-phosphate organism cultures in activated sludge systems - part 3: Kinetic Model, *Water SA*, Vol. 15, No. 2, pp. 89-102.

Figure 3.11 UNESCO-IHE participants undergo training at WWTP Haarlem Waarderpolder (Photo: Brdjanovic D.).

Annex 3.1
Influent characterization procedure according to Dutch guidelines (STOWA, 1996)

Characterization of the organic fractions of the influent

The total COD in the influent is a sum of the COD of soluble ($COD_{filt,inf}$) and particulate fractions ($COD_{part,inf}$):

$$COD_{tot,inf} = COD_{filt,inf} + COD_{part,inf} \tag{3.1}$$

Soluble fraction of the COD consists of following components,

$$COD_{filt,inf} = S_A + S_F + S_I \tag{3.2}$$

while the particulate fraction is represented by:

$$COD_{part,inf} = X_I + X_S + (X_H + X_{PAO} + X_{PHA} + X_{GLY} + X_{AUT}) \tag{3.3}$$

For simplicity the contribution of particulate components of the active biomass present in the settled sewage and the supernatant of the primary thickeners to the total COD is neglected. For the modeling purposes a very low value is assigned to these components (Table 3.4). Therefore eq. 3.3 can be written as:

$$COD_{part,inf} = X_I + X_S \tag{3.4}$$

Volatile fatty acids COD is directly measurable organic fraction of the influent COD and is presented as S_A:

$$S_A = COD_{VFAs} \tag{3.5}$$

Readily biodegradable substrate S_S was introduced in the ASM1. In ASM2 it is replaced by:

$$S_F = S_S - S_A \tag{3.6}$$

Introducing eq. 3.6 to eq. 3.2, the fermentable, readily biodegradable organic substrates S_F can be expressed as:

$$S_F = COD_{filt,inf} - (S_A + S_I) \tag{3.7}$$

The S_I in the influent is not affected by the treatment processes and, therefore, leaves the treatment plant with the effluent. This inert fraction can be calculated using the effluent COD measured after filtration ($COD_{filt,eff}$). The $COD_{filt,eff}$ has to be corrected with a factor 0.9 due to some production of S_I during the treatment process. For presence of BOD in the effluent the S_I is further reduced by the factor 1.5 $BOD_{5,eff}$: The factor 1.5 is a conversion factor of BOD_5 to COD.

$$S_I = 0.9\ A\ COD_{filt,eff} - 1.5\ BOD_{5,eff} \tag{3.8}$$

Finally, by combining eq.3.7 with eq.3.8, S_F can be calculated using directly measurable parameters:

$$S_F = COD_{filt,inf} - 0.9\ A\ COD_{filt,eff} + 1.5\ BOD_{5,eff} - S_A \tag{3.9}$$

In ASM1, the slowly biodegradable particulate substrate X_S is given as the difference between the biodegradable fraction of COD (BCOD) and the readily biodegradable substrate S_S:

$$X_S = BCOD - S_S \tag{3.10}$$

The routine BOD measurements are not recommended to be used for the estimation of the BCOD due to the facts that the BOD_5 measurement underestimates the BCOD, and that the BOD_{20} measurement is not yet reliable enough. The Dutch guidelines recommend the estimation of BCOD via determination of the total BOD (BOD_{tot}) from the BOD measurement by the best fit of the BOD-curve to the measured data. The following equation is recommended:

$$BOD_{tot} = BOD_t /(1 - e^{-k\,t}) \tag{3.11}$$

where k_{BOD} is the constant which varies from plant to plant and is in the range 0.15-0.6 d^{-1} for domestic sewage (Roeleveld and Kruit, submitted). Description of the determination of the k_{BOD} from the BOD test is given elsewhere (Roeleveld and Kruit, submitted). It is now possible to calculate the biodegradable COD from the total BOD taking into account the initial value of RBOD will be somewhat higher than the BOD_{tot} calculated by eq.3.11. This is due to the conversion of the fraction of BCOD into an inert fraction during the course of the BOD test. This is corrected by the factor Y_{BOD} which falls in the range of 0.1-0.2 (Roeleveld and Kruit, 1998). The BCOD can be expressed as follows:

$$BCOD = BOD_{tot} /(1 - Y_{BOD}) \tag{3.12}$$

The X_S can be calculated from eq.3.10 and 3.12:

$$X_S = \{[BOD_t /(1 - e^{-k\,t})]/(1 - Y_{BOD})\} - S_S \tag{3.13}$$

Finally, the inert particulate organic material is calculated from the eq.3.4 and eq.3.13:

$$X_I = COD_{part,inf} - X_S \tag{3.14}$$

The concentration of total suspended solids (X_{TSS}) in the influent is estimated from following equation:

$$X_{TSS} = 0.85 \cdot COD_{part,inf} \tag{3.15}$$

Characterization of the N and P fractions in the influent

In ASM2 the nitrogen compounds are for simplicity represented only with its soluble components which are directly measurable parameters: ammonium plus ammonia nitrogen plus Kjeldahl nitrogen (S_{NH3}) and nitrate and nitrite nitrogen (S_{NO3}).The phosphate compounds are represented also as soluble phosphorus (S_{PO4}). A small correction has to be made for an eventual poly-P fraction (X_{PP}) in the influent.

Annex 3.2
Results of the sampling program and data collected by the plant staff (April 1997)

Control point	Medium: sampling point	Parameter																				
			Liquid phase										Biomass							Others		
			COD_{total}	$COD_{filtered}$	$COD_{particulate}$	BOD_5	PO_4	Acetate	Propionate	Butyrate	NH_4	NO_3	MLSS	MLVSS	VSS/TSS	PHB	PHV	PHA	Glycogen	Temperature	PH	DO
		Unit/Date	mgCOD /L	mgCOD/L	mgCOD/L	mgO_2/L	mgP/L	mg COD /L	mg COD /L	mg COD /L	mg N/L	mg N/L	g /L	g /L		mg COD /L	mg COD /L	mg COD /L	mg COD /L	° C		mg O_2/L
	Raw sewage: after screens	8 April '97					6.7	45	9.1	2.5												
T1	**Settled sewage:** after the primary settler	7-8 April '97	329*	117*	212*	135*	6.7*	33*	4.7	0	42*	0.1*								**12.6**	6.7	**2.1**
		8-9 April '97	335*	138*	197*	134*	6.1*	32*			51*	0.1*										
		Average	**332**	**128**	**205**	**135**	**6.4**	**33**			**46**	**0.1**										
T8	**Activated sludge:** the end the aerobic ring	8-9 April '97					0.5						6	4.7	0.76	55	10	65	190			1.6
T10	**Effluent:** after the secondary settler	3-4 April '97		31*		4*	0.9*				1.1*	2.3*										
		9-10 April '97					0.8				1.9	4										
	Primary sludge: outlet of primary settler	10 April '97					6.5															
	Schalkwijk sludge: outlet pipe	9 April '97					6.5	11.7														
T12	**Supernatant:** primary thickener	9 April '97	**360***	**144***	**216***	**157***	**6.0***	**50***			**55***	**0***										**0.3**
		10 April '97					6	37.4	9	3.7												
	Side stream sludge: P selector tank:	8 April '97																				
T17	section 1 (start)						5.7	33.2														
	section 2						7.1	32.1														
	section 3 (middle)						9.6	30.4					5	3.8	0.77	17	3	20	33		7.5	0.1
	section 4						10.5	29.5														
T18	section 5 (end)						11.1	29.2														
	Side stream sludge: P release tank	8 April '97																				
T15	section 1 (start)						22.9	129														
	section 2						25.2	109														
	section 3 (middle)						26.6	107					1.1	0.9	0.78	77	17	94	156		6.5	0.1
	section 4						31.3	96.1														
T16	section 5 (end)						33.2	92.5														
T19	**Filtrate:** belt thickener	8 April '97					41.1															

Numbers printed in bold are used for influent characterization (Table 3.5)
Numbers with * refer to 24h composite samples, without *, to grab samples
Control point refers to the scheme of base case Haarlem 1 given in Annex 3.2

Summary of the data collected at the WWTP Haarlem Waarderpolder by the plant staff on 12th April 1997 (concentration values are obtained from spot samples taken around 10 p.m.).

Sampling point	Parameter	Value	Unit
RAW SEWAGE	Q	27,580	m^3/d
Plant influent	pH	7.75	
	COD_{tot}	495	$gCOD/m^3$
	BOD_5	190	$gBOD_5/m^3$
	Kj-N	56	gN/m^3
	(NO_3+NO_2)-N	0.05	gN/m^3
	P_{tot}	7.5	gP/m^3
	TSS	220	$gTSS/m^3$
	Load	109,002	P.E.
SETTLED SEWAGE	COD	265	$gCOD/m^3$
Biological unit influent	BOD_5	100	$gBOD_5/m^3$
	Kj-N	48	gN/m^3
	P_{otrho}	4.7	gP/m^3
	P_{tot}	6.7	gP/m^3
	TSS	100	$gTSS/m^3$
ACTIVATED SLUDGE	TSS	5,200	$mgTSS/m^3$
Biological unit effluent	SVI	60	mL/g
	Inorganic matter	24	% TSS
P-SELECTION TANK	TSS	860	$mgTSS/m^3$
Settled sludge	P_{otrho}	13.3	$g\ P/m^3$
	pH	7.5	
P-RELEASE TANK	TSS	5,540	$mgTSS/m^3$
Settled sludge	P_{otrho}	41.2	gP/m^3
	pH	6.8	
EFFLUENT	pH	8.05	
Secondary settlers effluent	COD_{tot}	39	$gCOD/m^3$
	COD_{filtr}	31	$gCOD/m^3$
	BOD_5	4	$gBOD_5/m^3$
	Kj-N	5.2	gN/m^3
	NH_4-N	3.2	gN/m^3
	(NO_3+NO_2)-N	8.3	gN/m^3
	N_{tot}	13.5	gN/m^3
	P_{tot}	2.3	gP/m^3
	TSS	7	$gTSS/m^3$
	Load	9,111	P.E.
OVERALL PLANT EFFICIENCY	COD_{tot}	92	%
	BOD_5	98	%
	Kj-N	91	%
	N_{tot}	76	%
	P_{tot}	69	%
	SS	97	%
	Load	92	%
PLANT LOADING	COD_{tot}	13,652	kgCOD/d
	BOD_5	5.240	$kgBOD_5/d$
	Kj-N	1.544	kgN/d
	P_{tot}	207	kgP/d
	TSS	5,516	kgTSS/d

Annex 3.3
Process configurations schemes of the WWTP Haarlem Waarderpolder

Base case: Phostrip®

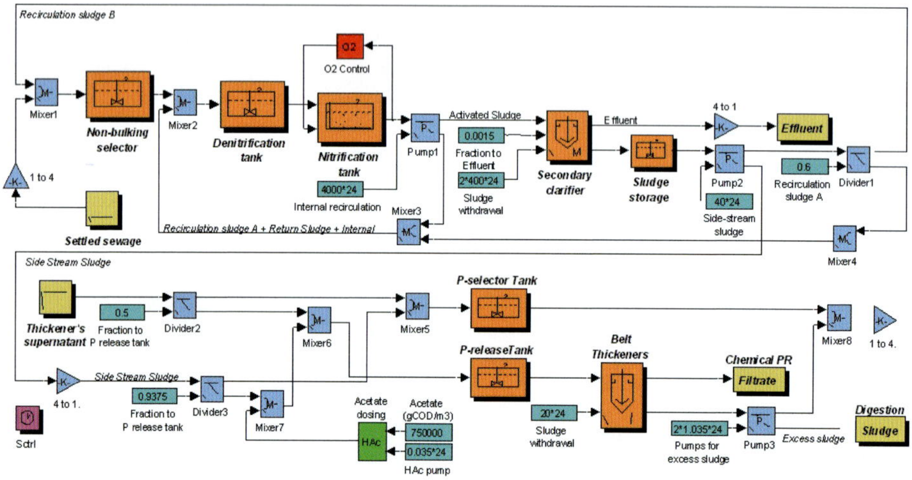

Alternative 1: A/O

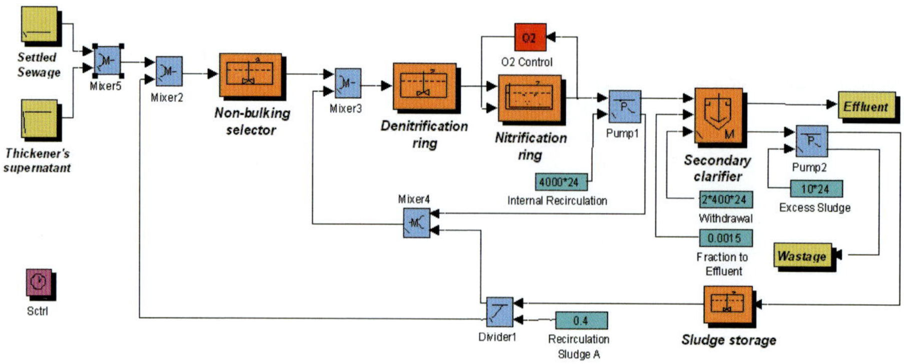

Alternative 2: Modified UCT

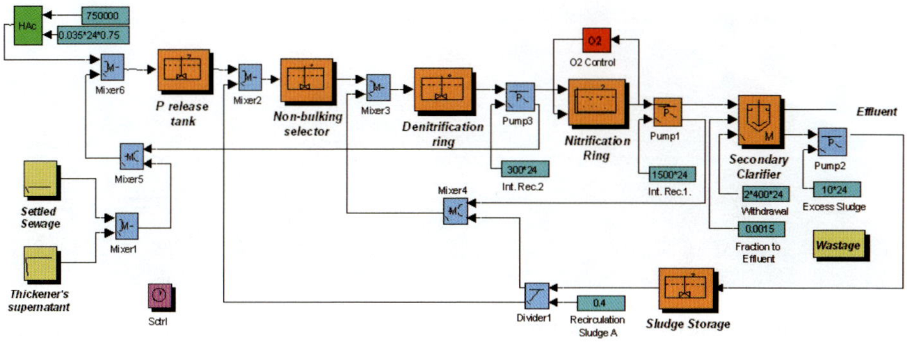

Alternative 3: BSFC

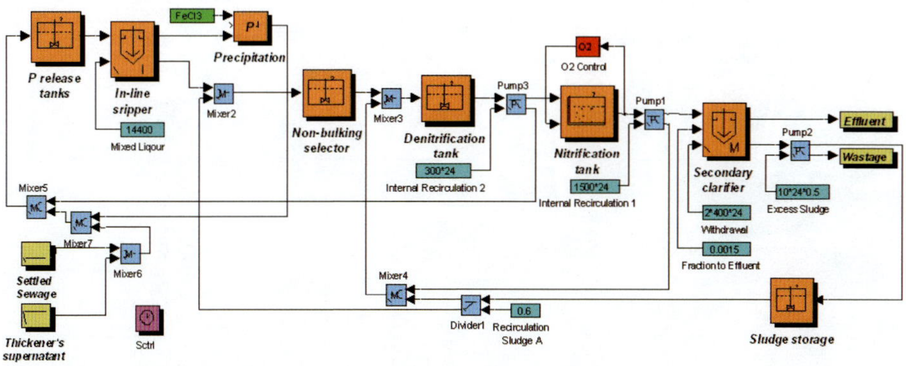

Figure 3.12 Activated sludge tank at WWTP Haarlem Waarderpolder with anoxic inner ring and outer aerobic ring and overflow to secondary clarifier (Photo: Brdjanovic D.).

Figure 3.13 Sanitary Engineering specialization studens of UNESCO-IHE exploring the WWTP Haarlem Waarderpolder (Photo: Brdjanovic D.).

Chapter 4: WWTP Katwoude, the Netherlands

Development of wastewater treatment data verification techniques

Meijer S.C.F., van der Spoel H., Susanti S., Heijnen J. J. and van Loosdrecht M.C.M.

This chapter is based on the paper Meijer S.C.F., van der Spoel H., Susanti S., Heijnen J.J. and van Loosdrecht M.C.M. 'Error diagnostics and data reconciliation for activated sludge modelling using mass balances', (2002). Wat. Sci. Technol. 25(6), 145-156.

4.1 Introduction

Activated sludge models strongly rely on the accuracy of collected data and measurements of the modelled process. In most wastewater treatment plant (WWTP) modelling studies, the accuracy of these data is fully relied upon. Instead, much emphasis is put on calibration of kinetic and stoichiometric parameters of the activated sludge model. Other studies demonstrated that WWTP models are more sensitive towards operational data than to most model parameters. Because measurements are never fully accurate, mass balances never can be closed perfectly. In models however, all flows are balanced. In general, fitting models on erroneous or unbalanced data, will lead to laborious and unjustified model calibration. Therefore, it is essential to check prime data on errors and consistency.

Gross error detection and data reconciliation techniques for biochemical processes have been developed by van der Heijden *et al.* (1994a,b,c). The method was implemented in the free-domain software called Macrobal made by Hellinga *et al.* (1992). All chemical processes can be described by a combination of linear relations based on mass balances (e.g. COD, N, and P). If more measurements are available than required to solve the system of equations, the system is over determined. For such systems gross errors can be detected by evaluating the balance residuals (ε). This is done on the basis of a statistical test. When there is no statistical proof for gross errors, the data can be balanced in a data reconciliation procedure. Finally, this results in a balanced data set without errors and an improved overall accuracy.

Macrobal was originally designed for balancing elements and conversion rates on a molecular level. In this study it was applied on a process level, balancing the mass flows of the full-scale WWTP Katwoude, situated in the Netherlands and managed by the Dutch water board 'Uitwaterende Sluizen in Hollands Noorderkwartier'. In this chapter it is demonstrated how Macrobal can be used for gross error detection and data reconciliation. It is shown how this simplifies the model calibration and WWTP simulation.

4.2 WWTP Katwoude

4.2.1 Process description

The actual capacity of WWTP Katwoude was 86,300 PE (designed capacity 90,000 PE). The plant has a completely mixed non-aerated selector (R1; 270 m^3), a completely mixed anoxic reactor (R2; 1,760 m^3) and a 5 meter deep aerated Carrousel® type reactor (R3; 15,840 m^3). Originally the plant was designed for COD and N removal (Figure 4.1).

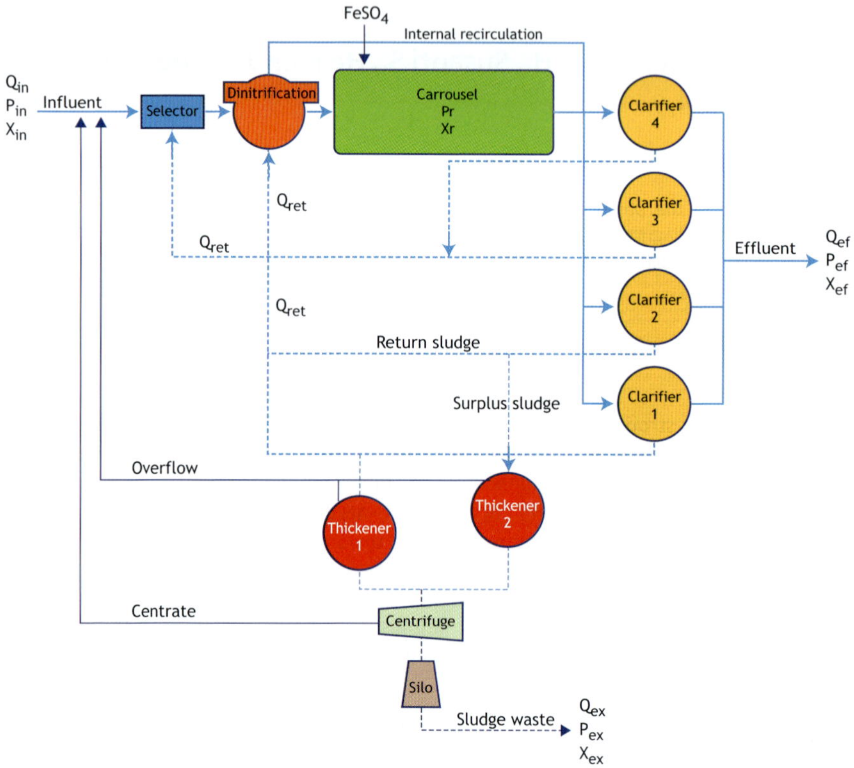

Figure 4.1 Process scheme of WWTP Katwoude.

However, some enhanced biological phosphorus removal (EBPR) activity was observed. Most likely, in summer conditions R1 acted as an anaerobic selection reactor for phosphorus accumulating organisms (PAO). Two surface aerators in R3 controlled oxygen on 1.5 and 0.8 gO_2/m^3 measured on respectively $^3/_5$ and $^4/_5$ length of the tank. Dissolved oxygen (DO) concentration gradients were measured in the longitude, width and depth of the aerobic reactor (data not shown). During the entire year, chemical P precipitation was applied in R3 to assure effluent phosphate (PO_{EF}) below 0.5 gP/m^3. A natural external recycle from R3 (Q_{R3EX}) supplied R2 with nitrate. Q_{R3EX} was estimated from the superficial flow velocity in R3 (0.37 m/s) and the cross-section of the passage connecting R2 and R3 (5×1.3 m^2). Four clarifiers are operated in pairs (CL_{12} and CL_{34}). According to the plant design, flow distribution works FD_1 divided Q_{R3} in favour of CL_{34} with a distribution factor of 0.57. R1 was supplied with RAS by Q_{RS34}, R2 by both Q_{RS12} and Q_{RS34}. FD_2 divided Q_{RS34} in favour of R1 with a distribution factor of 0.543. The waste activated sludge (WAS) from the underflow of CL_{12} (Q_W) is concentrated in two WAS thickeners

(TH). The thickened WAS (Q_{TW}) is dewatered by centrifugation (CE) and transported from the WWTP (Q_{DW}). The overflow from the thickener (Q_{OF}) and centrate (Q_{CE}) are recycled to R1. In Table 4.1 the hydraulic and operational data are presented. The flow scheme of the WWTP is presented in Figure 4.2.

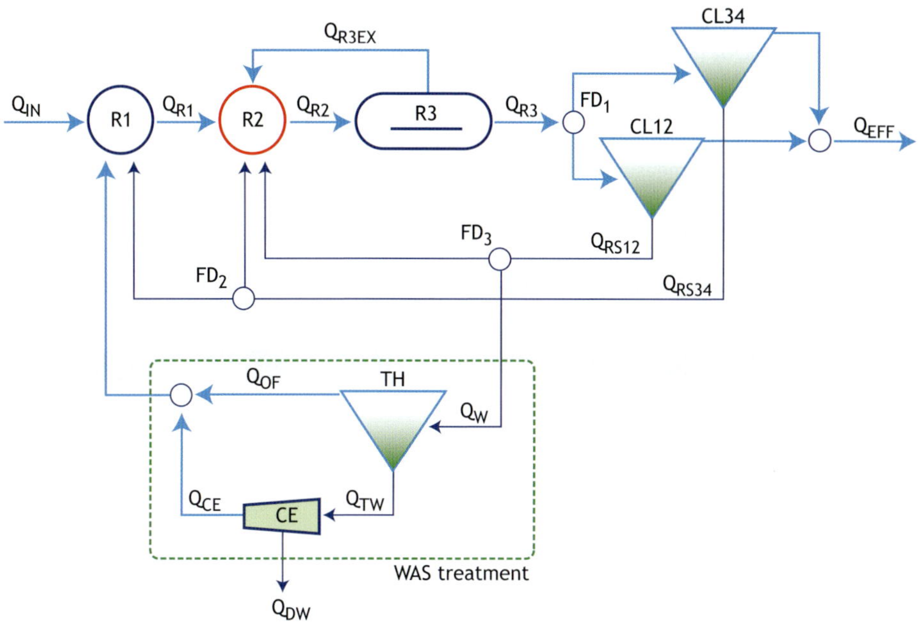

Figure 4.2 Flow scheme of WWTP Katwoude. Dark blue lines are activated sludge flows. Light blue lines are RAS and WAS flows. R1: selector, R2: anoxic reactor, R3: aerated carousel, CL: clarifiers, TH: WAS thickening, CE: dewatering centrifuge and FD: hydraulic flow distribution works.

Table 4.1 Measured operational data and volumes of WWTP Katwoude.

Flow	Average ± s_D m^3/d	Process unit	Volume m^3	Depth m	HRT[(3)] h
Avg. influent yr. 2000, Q_{24}	13,252 ± 5,960	Selector, R1	270	3.5	0.2
8-day avg. influent, Q_{IN8}	12,380 ± 1,270	Denitrification reactor, R2	1,760	5.5	0.4
External recycle R3, Q_{R3EX}	207,792[(1)]	Aerated carousel, R3	15,840	5.0	7.5
Internal recycle R3, Q_{R3IN}	2,077,920[(2)]	Total reactor volume	17,870	-	33.5
RAS from CL_{12}, Q_{RS12}	12,144 ± 1,214	Clarifiers, CL_{12} (2 × 1,395 m^3)	2,790	-	3.1
RAS from CL_{34}, Q_{RS34}	24,960 ± 2,496	Clarifiers, CL_{34} (2 × 1,399 m^3)	2,798	-	2.3
WAS, Q_W	1,227 ± 791	Thickeners, TH (2 × 308 m^3)	615	-	12.8
Thickened WAS, Q_{TW}	154 ± 92	Flow distribution factor, FD_1	0.570	-	-
Dewatered WAS, Q_{DW}	14 ± 4	Flow distribution factor, FD_2	0.543	-	-

(1) Estimated from the superficial flow velocity in R3 (0.37 m/s) and the cross-section of the passage connecting R3 and R2 (1.3×5 m^2).

(2) Estimated from the superficial flow velocity in R3 (0.37 m/s) and the cross-section of reactor R3 (13×5 m^2).

(3) The HRT was calculated from the balanced flows (Q_{IN8}, Q_{R1}, Q_{R2}, Q_{R3}).

4.2.2 Measurements

The pseudo steady state operation of WWTP Katwoude was recorded during an 8-day sampling program (the 23rd of February to the 2nd of March 2001). In this period, samples (among others) were collected daily from the influent (Q_{IN8}), Q_{R1}, Q_{R2}, Q_{R3}, Q_{RS12} and Q_{RS34} Q_{OF}, Q_{CE} and the effluent (Q_{EF}). The influent, centrate and effluent were 24-hour composite samples, whereas the process internal measurements were grab samples during influent peak flow. The samples were analysed on TCOD (total COD), COD_{MF} (COD of the micro-filtrated fraction, 0.45 μ pore diameter, no flocculation), BOD_5 (5 day biological oxygen demand), MLSS (mixed liquor suspended solids), ash fraction of the MLSS, VFA (volatile fatty acids), NH_4^+ (ammonium), NO_3^- (nitrate), TKN (total Kjeldahl nitrogen), TP (total phosphorus) and PO_4^{2-} (ortho-phosphate). Energy use, pump flow-rates (Table 4.1), mixer and aerator operation time and process control measurements were recorded daily. Average measurements of the 8-day sampling program are presented in Table 4.3. Also the average measurements over the year 2000 were collected. These measurements are presented in Table 4.2. The average sewage temperature in the year 2000 was 15 ^{0}C.

Table 4.2 Measurement and simulation results over the year 2000. Simulated results and influent data are printed in blue. The plant was simulated with average influent loads and a temperature of 15°C. TP_{R3} was controlled, by regulating Q_W, resulting in a SRT of 19.8 days.

Sampled flow / reactor	Q_{24} (g/m^3)	R3 (g/m^3)	Q_{EF} (g/m^3)	Q_W (kg/d)
Total COD, TCOD	631 ± 172	-	37.5 ± 7.0	-
X_T+S_T	631	4,662	37.5	4,119
COD particulate, COD_X	-	4,628 ± 514	-	4,132 ± 910
X_T	316	4,628	4.0	4,118
Total phosphorus, TP	9.2 ± 3.8	112.7 ± 13.0	0.6 ± 0.8	100.6 ± 22
S_{PO} + X_{MeP} + X_{PP} + iP	9.2	113	0.6	101.2
Ortho-phosphate	-	-	0.5 ± 0.6	-
S_{PO}	3.6	0.7	0.5	0
Precipitated $FePO_4$	-	-	-	-
X_{MeP}	0	20.3	0.0	18
Tot. Kjeldahl N, TKN	56 ± 15	274 ± 30	2.6 ± 1.5	244.8 ± 54
S_{NH} + iN	56	274	0.9	247
Ammonium	-	-	1.0 ± 1.6	-
S_{NH}	37	0.7	0.7	0
Nitrate	-	-	3.2 ± 3.0	-
S_{NO}	0	3.4	3.3	0

4.2.3 The process model

For all simulations the simulation platform SIMBA© 3.3$^+$ (Alex *et al.*, 1997) was used, operating under Matlab/Simulink© 5.2. In SIMBA© WWTP process models are constructed by connecting CSTR in a desired configuration (Figure 4.2). Biological conversions were calculated with the TUDP model (Annex 1.1). The WWTP was modelled (Figure 4.3) according to the flow scheme given in Figure 4.1 and the data in Table 4.1. R1 and R2 were each modelled as one CSTR. R3 was modelled as five CSTR in series, with an internal recycle Q_{R3IN}. Aeration of R3 (OC_{R3}) was modelled with PI control, which regulated DO_{R3} in the first and third CSTR on respectively 1.5 and 0.8 gO_2/m^3. With this setup, the horizontal DO gradients were approximated. The modelled clarifiers separated solids and water ideally. A small fraction X_{EF} was modelled by a percentile loss of the RAS (0.0005%). Denitrification in the RAS volume was not observed and therefore not modelled. The solids concentration in the AT (X_{R3}) was PI controlled by regulating Q_W. Chemical precipitation of $FePO_4$ was modelled in Q_{R2}. PO_{EF} was PI controlled at 0.5 gP/m^3 by regulating the dosage of $FeCl_3$. The model influent was characterised according to Roeleveld and van Loosdrecht (2001). All simulations were performed with a constant influent flow.

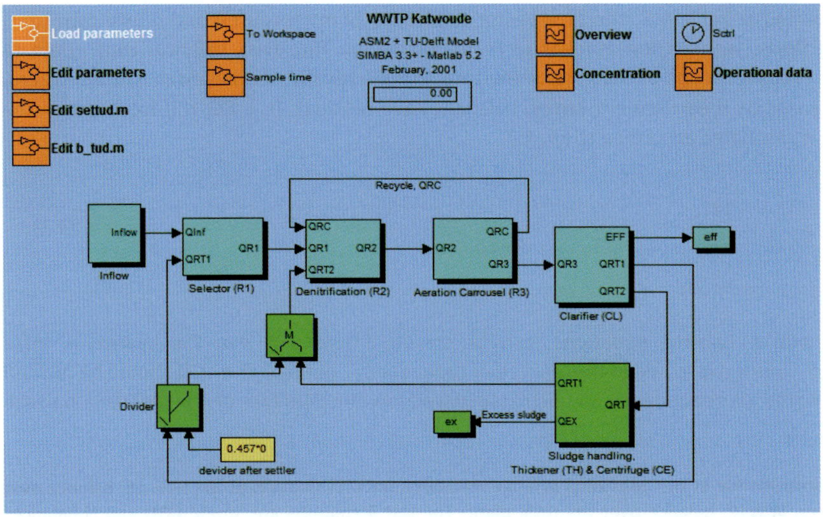

Figure 4.3 SIMBA model of WWTP Katwoude.

Table 4.3 Measurements and simulations of the 8-day sampling program. The flows were sampled daily, at dry weather influent peak flow (8-day average 12,380 m^3/d). Simulated results and influent data are printed blue.

Sampled flow/reactor	Q_{IN8} (g/m^3)	R1 (g/m^3)	R2 (g/m^3)	R3 (g/m^3)	Q_{EF} (g/m^3)
TCOD	773 ± 53	-	-	4,840 ± 1263	43.7 ± 8.3
$X_T + S_T$	773	4,180	4,869	4,845	43
COD_X	379 ± 93	3,999 ± 222	-	4,795 ± 1263	0.0 ± 8.9
X_T	380	4,020	4,806	4,805	4.1
COD_{MF}	393 ± 41	-	-	44.3 ± 0.6	44.0 ± 0.6
$S_A + S_F + S_I$	393	160	63	40	40
TP	12.9 ± 4.5	97.1 ± 5.4	-	116.7 ± 6.5	1.8 ± 0.3
$S_{PO} + X_{PP} + iP$	12.9	98.5	116.5	116.4	1.8
Phosphate	6.4 ± 0.9	5.0 ± 1.4	3.0 ± 0.3	1.8 ± 0.4	1.4 ± 0.2
S_{PO}	6.1	4.6	3.2	1.7	1.7
TKN	61.7 ± 7.6	-	-	285 ± 32	1.9 ± 0.3
$S_{NH} + iN$	61.5	248	286	282	2.0
Ammonium	39.9 ± 4.9	14.0 ± 2.2	5.3 ± 1.5	1.0 ± 0.6	0.7 ± 0.6
S_{NH}	38.3	14.4	5.1	1.2	1.2
Nitrate	0 ± 0	0.7 ± 0.4	0.8 ± 0.6	3.3 ± 1.0	3.2 ± 0.7
S_{NO}	0	0.54	0.4	3.3	3.5

4.2.4 Introducing Macrobal

The free-domain software Macrobal was used for evaluation of operational data. Macrobal was originally designed for balancing chemical elements and conversion rates on a molecular level. In this chapter, the program is used to balance compounds and flows in a full scale WWTP. Operation of Macrobal is explained according to Table 4.4 and 4.5. These tables are drawn up in the Macrobal in and output format. In Table 4.4, *n* columns represent the mass balances, with in each balance the concentrations $C_{i,n}$. All empty spaces in the balances represent zeroes. Negative concentrations are outflows, positive concentrations are inflows. The balances correspond with the first column of Table 4.4, which is indicated as the flow vector Q. This is further specified in Table 4.5. The flow vector Q consists of *m* measured and *u* unknown (thus calculated) flows, with a total number of *m+u=i* elements. The mass balances may relate to individual reactors, groups of reactors or the total WWTP. Ideally, the balances add up to zero, however caused by measurement errors, balances have residuals (ε_n). For the total system the residuals are calculated according to Eq. 4.1.

$$\sum_{i=1}^{i} \overline{Q_i} \times \overline{C_{i,n}} = \overline{\varepsilon}_n \tag{4.1}$$

A system with *n* balance equations, can be solved if *n=u*. For *n>u* the system is 'over determined'. This is expressed in the 'degree of redundancy', defined as *n-u*. It should be noticed that the degree of redundancy is calculated based on the number of *independent* balance equations.

Table 4.4 Reconciliation of the year 2000 average measurements. Evaluation of the SRT. CE: balance over the centrifuge, WWTP: balance over the total plant. The last 4 columns are 'open' COD and N balances (OC_{COD}: oxidised COD, ^{L}NIT: nitrified load, ^{L}DEN: denitrified load, OC_{NET}: net oxygen consumption). Empty spaces in the matrix represent zero values.

			n mass balances with elements $C_{i,n}$							
Defined balance			WWTP	CE	WWTP	CE	WWTP	WWTP	WWTP	WWTP
Balance compound			Flow balance		TP	TP	COD	TKN	NO_3^-	DO
Unit			-	-	gP/m^3	gP/m^3	$gCOD/m^3$	gN/m^3	gN/m^3	gO_2/m^3
Flow vector Q with *i* elements (*m* measured, *u* unknown)	Q_{24}	m^3/d	1		9.2±3.8		631±172	56±15		
	Q_{EF}	m^3/d	- 1		-0.6±0.8		-37.5±7.2	-2.6±1.5	-3.2±3	
	Q_{TW}	m^3/d		1		669±99	-27,468±4,077	-1,630±242		
	Q_{CE}	m^3/d		- 1		-58±46	2,383±1,876	141±111		
	Q_{DW}	m^3/d	- 1	- 1						
	TP_{DW}	kgP/d			- 1	- 1				
	OC_{COD}	kgO_2/d					- 1			- 1
	^{L}NIT	kgN/d						- 1	1	- 4.57
	^{L}DEN	kgN/d					- 2.87		- 1	
	OC_{NET}	kgO_2/d								1

Residuals are calculated according to Eq 4.1.

Balancing data is only useful when no gross errors are apparent. In Macrobal these gross errors are therefore first detected by evaluating the residual (ε) of each balance on the basis of a statistical test (the χ-square distribution). An elaborate description this procedure is found in van der Heijden *et al.* (1994a,b,c). Gross errors are introduced by (i), an incorrect process description (ii), erroneous operational data or (iii), erroneous process measurements. Gross errors in the flow vector can be detected by systematically rejecting measured flows (thus a measured value becomes an unknown value), until the χ-square test is passed. An incorrectly defined balance or measurement error, can be detected by systematically rejecting balances, until the χ-square test is passed. If $n\text{-}u \geq 1$ and no gross errors are apparent, the system can be balanced. Hereby the residuals (ε_n) are minimised and the standard deviation (s_D) of the balanced operational data are reduced. A high degree of redundancy, results in a high accuracy, i.e. a small s_D.

4.3 Error detection and data reconciliation

4.3.1 Estimation of the SRT

For a reliable simulation, the SRT should be known within 95 % accuracy. This was shown by Brdjanovic *et al.* (2000). Therefore the SRT was evaluated on the basis of the TP balance according to procedure described in Chapter 1 and Nowak *et al.* (1999). Primary data indicated the SRT in WWTP Katwoude was 20 days. This required an evaluation over approximately 60 days. For the evaluation, average data over the year 2000 was used, which was readily available from the regular WWTP sampling program. The TP balance was calculated over the sludge dewatering (CE) and over the total WWTP. The system of balances is presented in Table 4.4. Three out of five flows were measured (Q_{24}, Q_{TW} and Q_{DW}), leaving two unknown (Q_{CE} and Q_{EF}). TP was measured in Q_{24}, Q_{TW}, Q_{CE} and Q_{EF}. The TP load in the dewatered flow ($^{L}TP_{DW}$) was calculated from the measured sludge production (^{L}WAS). The system of four balance equations (n=4) and two unknown flows (u=2) had a degree of redundancy of two (n-u=2), which allowed error detection and data reconciliation. From the yearly average data, also the COD and N balance were calculated. For these 'open' balances the degree of redundancy was zero (n-u=0), meaning all loads were calculated and no balancing was performed. The net oxygen consumption (OC_{NET} 3,607 kgO_2/d, Table 4.5), calculated from the COD and N balance, was used as a final check on the calibrated model.

Table 4.5 Balancing results of the year 2000 average measurements. Balanced operational data are according to Table 4.4. No gross errors were detected. The results of the data reconciliation are presented in the last column. The result of the COD and N balance is presented in the last four rows.

			Degree of redundancy = 2 No proof for measurement errors based on the χ-square distribution.		
				Measured $\pm$ s_D	Balanced $\pm$ s_D
Flow vector Q with *i* elements (*m* measured, *u* unknown)	Q_{24}	m^3/d	MB	13,252 $\pm$ 5,960	11,833 $\pm$ 2,210
	Q_{EF}	m^3/d	C	-	11,819 $\pm$ 2,210
	Q_{TW}	m^3/d	MB	153.7 $\pm$ 92.3	165.2 $\pm$ 31.1
	Q_{CE}	m^3/d	C	-	151.2 $\pm$ 31.4
	Q_{DW}	m^3/d	MB	14 $\pm$ 3.8	14 $\pm$ 3.8
	$^{L}TP_{DW}$	kgP/d	MB	100.6 $\pm$ 22	101.5 $\pm$ 10
	OC_{COD}	kgO_2/d	C	-	1,852 $\pm$ 10
	^{L}NIT	kgN/d	C	-	383.9 $\pm$ 10
	^{L}DEN	kgN/d	C	-	346.1 $\pm$ 10
	OC_{NET}	kgO_2/d	C	-	3,607 $\pm$ 10

M: measured, B: balanced, C: calculated

Table 4.5 shows the results of the Macrobal calculations. No gross errors were detected in the operational data. Macrobal was able to balance the operational data caused by the large standard deviation (s_D) of Q_{24}. Hereby, the balanced influent flow surprisingly corresponded to the dry weather influent flow (11,833 $\pm$ 2,210 m^3/d). Based on this observation, we verified that TP_{IN} was sampled at an average flow rate of 11,500 m^3/d, which is in correspondence with the balanced Q_{24}. In the reconciled system, all balances were in accordance. From the raw data the SRT was calculated to be 20.0 $\pm$ 6.7 days. In the balanced system this was 19.8 $\pm$ 4.8 days (17,870 $m^3 \times 112.7 \pm 13$ gP/m^3 divided by 101.5 $\pm$ 13 kgP/d).

4.3.2 Balancing operational data

On the basis of the 8-day sampling program, a system of balances was formulated with the intention to check seven process flows (Q_{IN8}, Q_{RS12}, Q_{RS34}, Q_{R3EX}, Q_W, Q_{TW} and Q_{DW}) and the flow division by FD_2 on possible errors (Table 4.6). According to the flow scheme of the WWTP in Figure 4.1 a system of 12 mass balances was formulated. Flows were balanced over R1, R2, R3, CL, TH and CE (Figure 4.4). Hereby, ammonium was balanced over R1 and R2, and TP was balanced over R1, R2, R3 and CL. On the basis of the relative short hydraulic retention time (HRT), all reactors were assumed to be in a (local) steady state (Table 4.1). Q_{IN8} was slightly higher than Q_{24}, caused by minor rain events during the recorded period. Consequently, the sludge production and the net oxygen consumption were not representative and therefore not used. Instead we relied on the SRT evaluation in the previous section.

Table 4.6 Data reconciliation of the 8-day measurements: evaluation of operational data. Table headings are explained in Figure 4.1. The first balances R1, R2, R3, CL, TH and CE are volumetric flow balances. In the following columns TP and ammonium are balanced.

Flow m^3/d	R1 Flow balances	R2	R3	CL	TH	CE	CL TP gP/m^3	R1 TP gP/m^3	R1 NH_4^+ gN/m^3	R2 TP gP/m^3	R2 NH_4^+ gN/m^3	R3 TP gP/m^3
Q_{IN8}	1							12.9±4.5	39.9±4.9			
Q_{R1}	-1	1						-97.14±5.4	-14.02±2.2	97.14±5.4	14.02±2.2	
Q_{R3EX}		1	-1							116.7±6.5	0.99±0.6	-116.7±6.5
Q_{R2}		-1	1							-116.7±6.5	-5.25±1.5	116.7±6.5
Q_{R3}			-1	1			116.7± 6.5					116.7±6.5
Q_{RS12}		1		-1			-182.4±4.9			182.4±4.9	0.67±0.6	
Q_{EF}				-1			-1.83 ±0.5					
Q_{RS34}	1			-1			-141.5±3.8	141.5±3.8	0.67±0.6			
Q_W				-1	1		-182.4±4.9	182.4±4.9				
Q_{OF}	1				-1				27.8±0.6			
Q_{TW}					-1	1		-860.5±112.7				
Q_{CE}	1					-1		183±5.8	80.5±5.3			
Q_{DW}						-1						

The system contained 12 balance equations (*n*=12) and 6 unknown flows (*u*=6), resulting in a degree of redundancy of *n*-*u*=6. On the basis of the original data, we were not able to balance the system within the 75 % confidence level of the χ-square test (Table 4.7). Therefore, three measured flows (Q_{RS12}, Q_{RS34} and Q_{R3EX}) were rejected. Recalculating the system with Q_{RS12}, Q_{RS34} and Q_{R3EX} as unknown (*n*-*u*=3), the calculated return flow rates (Q_{RS12} and Q_{RS34}) indicated that flow divider FD_2 was fully directed towards R1. After an inspection at sight this was confirmed. Initially, Q_{R3EX} was estimated from the superficial flow velocity in the carousel reactor (0.37 m/s). From the balance calculations a lower average flow was expected. This was however not verified in practice. Therefore, in the final calculation Q_{R3EX} was set as unknown. The balanced value for Q_{R3EX} (66,750 ± 4,500 m^3/d) suggests that the average velocity over the depth of the carousel reactor was 0.12 m/s. The balanced flows are presented in Table 4.7.

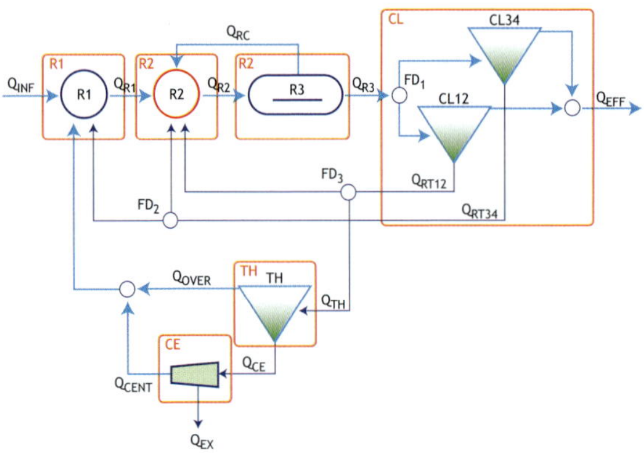

Figure 4.4 Scheme of WWTP Katwoude. Red rectangles are mass balance boundaries for data verification.

Table 4.7 Balancing results of the 8-day sampling program. Two calculations were performed. In the first gross errors were detected in Q_{R3EX}, Q_{RS12} and Q_{RS34}. In the second calculation Q_{R3EX} was rejected, and the system could be balanced.

First calculation Degree of redundancy = 6 99% confidence of a measurement error based on the χ-square distribution		Measured $\pm s_D$ m^3/d	Second calculation Degree of redundancy = 5 No proof for measurement error based on the χ-square distribution	Measured $\pm s_D$ m^3/d	Balanced $\pm s_D$ m^3/d
Q_{IN8}	M	12,380 ± 1,270	MB	12,380 ± 1,270	12,800 ± 737
Q_{R1}	C	-	C	-	38,640 ± 2,390
Q_{R3EX}	MB	207,792 ± 20,779^(1)	MC	rejected	66,750 ± 4,500
Q_{R2}	C	-	C	-	117,300 ± 7,270
Q_{R3}	C	-	C	-	50,520 ± 2,940
Q_{RS12}	MB	23,551 ± 2,355^(2)	MB	12,144 ± 1,214^(3)	11,880 ± 853
Q_{EF}	C	-	C	-	12,790 ± 737
Q_{RS34}	MB	13,553 ± 1,355^(2)	MB	24,960 ± 2,496^(2)	24,710 ± 1,540
Q_W	MB	1,227 ± 791	MB	1,227 ± 791	1,149 ± 291
Q_{OF}	C	-	C	-	999 ± 217
Q_{TW}	MB	153.7 ± 92.3	MB	153.7 ± 92.3	150 ± 75
Q_{CE}	C	-	C	-	136 ± 75
Q_{DW}	MB	14.0 ± 3.8	MB	14.0 ± 3.8	14 ± 3.8

(1) Estimated from superficial flow velocity (0.37 m/s).
(2) Calculated with FD_1 0.57 in favour of CL_{34} FD_2 0.543 in favour of R1.
(3) FD_2 set to 1.

M: measured, B: balanced, C: calculated

4.4 Model calibration and simulation

After error detection and data reconciliation the model of WWTP Katwoude was calibrated. Herweith, the calibration method was used as described in Chapter 1. First the solids were fitted (TP_X, COD_X and TKN_X) on the basis of average data for the year 2000 (Table 4.2 and 4.5). Next, nitrification, denitrification and EBPR were calibrated on the basis of the 8-day sampling program (Tables 4.3 and 4.7). The calculated net oxygen consumption (OC_{NET}) was a final check on the calibrated model (Table 4.5).

4.4.1 Fitting the sludge production

Step 1. Fitting the model on the TP balance

The model was fitted to the year 2000 balanced loads ${}^{L}TP_{IN}$, ${}^{L}TP_{EF}$ and ${}^{L}TP_{DW}$ (Table 4.4 and 4.5). PO_{EF} was controlled at 0.5 gP/m^3 by chemical precipitation. TP_{EF} was calibrated with a percentile loss of solids from the modelled clarifier. For modelling convenience, Q_{DW} was fixed on 14 m^3/d (Table 4.5). Hereby, TP_{R3} was controlled on 112.7 gP/m^3 (Table 4.2) by regulating Q_W, which resulted in ${}^{L}TP_{DW}$ of 101.5 kgP/d. This was in accordance with the balance in Table 4.5. Because TP is strongly related to the solid fraction, fitting the model to TP fixes the SRT.

Step 2. Fitting the solids COD_X balance

Because the COD_X balance is a non-conserved balance, an incorrect load ${}^{L}COD_{IN}$ will generally be compensated by OC_{NET}. In the previous section the SRT was fixed according to the TP balance. Hence, the total amount of MLSS in the WWTP was mainly determined by the influent X_I/X ratio, as inert COD (X_I) accumulates in the WWTP. By adjusting the influent X_I/X ratio to 0.798, we fitted the model to the measured TP_X/COD_X ratio of 0.032. Hereby, all model uncertainties related to the production of X_I and the influent characterisation are lumped up in the influent X_I/X ratio.

Step 3. Fitting COD_S in the effluent

COD_S in the effluent was fitted by adjusting the influent ratio S_I/S_F from 0.146, calculated according to Roeleveld and van Loosdrecht (2001) to 0.134. Increasing S_I in the influent directly effects the effluent concentration, as S_I is not converted. Hereby, S_A in the influent was not changed, as this model component was directly measured from VFA_{IN}.

Step 4. Fitting the TKN balance

Likewise COD, TKN is a non-conserved compound. An incorrect ^{L}TKN in e.g. Q_{EF} or Q_W, generally will be compensated by OC_{NET} via nitrification, and N_2 production via denitrification. The measured fraction TKN_X/COD_X was 0.059. TKN_X was fitted by increasing the activated sludge fractions iN_{XI} and iN_{XS} to 0.06 gN/gCOD. This resulted in a modelled $^{L}TKN_{DW}$ of 247 kgN/d (Table 4.2).

4.4.2 Calibrating nitrification, denitrification and EBPR

In the previous section X_I/X, S_I/S_F, iN_{XI} and iN_{XS} were calibrated. Nitrification, denitrification and EBPR were calibrated on the basis of the 8-day sampling program (Table 4.3). Therefore the influent composition (Table 4.3), balanced flows (Table 4.7) and X_{R3} from the 8-day sampling program were used here. The simulated sewage temperature was 9^0C.

Step 5. Calibrating the net nitrified load

Strong DO gradients were observed in the depth and width of R3. However, only the gradient in the length of the carousel was measured. DO_{R3} was therefore assumed to be less reliable. To simulate NH_{EF}, the DO set-points in R3 were increased to 2 gO_2/m^3. Hereby, the original aerated volume was not changed (9,504 m^3).

Step 6. Calibrating the net denitrified load

The net denitrification (^{L}DEN) was fitted by increasing K_O from 0.2 to 0.3 gO_2/m^3. This caused effluent nitrate (NO_{EF}) to decrease from 3.7 to 3.2 gN/m^3.

Step 7. Calibrating anaerobic phosphate release

From the phosphate balance over R1, the phosphate release in the anaerobic reactor ($^{L}PR_{R1}$) was calculated to be 34 $\pm$ 14 kgP/d. Without calibration, the model predicted a release of 24 kgP/d. Hereby PAO was approximately 1 to 2 % of the total activated sludge. Because the simulated release was within the accuracy of measurement, it was chosen not to calibrate the phosphate release in R1.

Step 8. Evaluating the net oxygen consumption

A final check on the calibrated model was the simulated net oxygen consumption (OC_{NET} 3,594 kgO_2/d). This value approximately matched the calculated value (3,607 kgO_2/d, Table 4.5). On the basis the stoichiometric oxygen consumption and the measured energy use for aeration (6,569 $\pm$ 1,319 kWh/d or 27.8 kWh/PE.), theoretical efficiency of 0.55 kgO_2/kWh is calculated. If alpha factor is assumed of 0.7 and a deficiency factor of 1.4 (average DO at the aerator 3 gO_2/m^3 and an average sewage temperature of 15°C), a net efficiency of 1.12 kgO_2/kWh is expected, which is a realistic value.

4.5. Discussion

4.5.1 Balancing conserved compounds

A compound only can be balanced when it is totally recoverable. Most 'conserved' compounds (COD, TN, TOC), have a component in the gaseous phase (respectively O_2, N_2 and CO_2), which is in general not measured. Theoretically this also accounts for volumetric flows, however in practice evaporation of water is negligible. A compound that under normal conditions exclusively is associated with the solid and liquid fractions is phosphorus. Phosphorus can be recovered from all in and outgoing flows, and therefore is very suitable for balancing purposes. Moreover, TP can be related to the solid fraction (MLSS). This is justified as the activated sludge ratio $TP_X/COD_X/TKN_X$ only changes over a relative long period (3×SRT). Practical advantages of using TP as a 'balancing compound' are: (i) measuring TP is relative simple, (ii) in WWTP's, TP is generally well above the detection limit and (iii) historical TP measurements are often readily available. In this study, ammonium and phosphate balances were used to evaluate the process. Ammonium and phosphate are not conserved and therefore cannot be fully recovered. It was however assumed that (i), ammonium was not oxidised in R1 or R2, (ii) ammonification was negligible and (iii), growth of biomass in R1 and or R2 was negligible. These assumptions indicate that balancing ammonium is only valid when the HTR in R1 and R2 is low, and therefore the contribution of these reactions can be neglected. The phosphate balance is less sensitive to biomass growth. This is the direct result of the biomass composition which contains little P ($CH_{2.09}O_{0.54}N_{0.20}P_{0.015}$, Smolders *et al.*, 1994). For the same reason, also hydrolysis and fermentation have reduced effect on the phosphate balance over R1. When phosphate and ammonium are balanced, it must however be realised that the accuracy of the calculations are negatively influenced by these processes.

4.5.2 Calibrating EBPR

No PAO activity was simulated with a sewage temperature of 9°C. However, if the temperature profile of the preceding year was taken in account, full EBPR was simulated in summer and autumn, whereas some EBPR activity was simulated under winter conditions. This showed that the simulated PAO concentration was not only sensitive to the actual temperature, but also to the preceding temperature profile. Simulation of the WWTP at 15°C, showed that at least 360 days were needed before activity of PAO and ordinary heterotrophic organisms (OHO) was in steady state. Because both species competed for the same substrate (VFA), their growth was limited. Therefore it took a considerable period before steady state was reached. This is in contrary to ammonia oxidizing organisms (AOO), which if allowed by process conditions, recover in approximately 10 days. It can therefore be concluded that PAO activity is generally not in steady state at this plant. To avoid laborious and relative inaccurate simulations with variable temperature, it is suggested to experimentally determine the PAO concentration from an anaerobic phosphate release batch-tests. As initial estimation, it is proposed to calculate the anaerobic phosphate release according to Eq. 4.2.

$$Q_{IN} \times PO_{IN} + Q_{RS34} \times PO_{RS34} + Q_{CE} \times PO_{CE} + Q_{OF} \times PO_{OF} - Q_{R1} \times PO_{R1} + {}^{L}PR_{R1} = 0 \qquad (4.2)$$

From the phosphate balance over R1, the anaerobic phosphate release ${}^{L}PR_{R1}$ was calculated to be 34 ± 14 kgP/d. On the basis of Q_{R1} = 38,640 m^3/d, this corresponds with an increase of 0.9 gP/m^3 in PO_{R1}. Simulating the model with a yearly temperature profile, ${}^{L}PR_{R1}$ was calculated to be 24 kgP/d. This corresponded to an increase of PO_{R1} with 0.6 gP/m^3. Both estimated and simulated releases were smaller than the measurement accuracy (PO_{R1}=5.0 ± 1.4 gP/m^3, Table 4.3). Moreover, in the simulation PAO only made up 2% of the TCOD. On the basis thereof, further calibration EBPR was not justified.

4.5.3 Calibrating N fractions

The TKN fraction in waste sludge was highly sensitive in the model. Incorrect simulation of TKN in the waste sludge (${}^{L}TKN_W$), is compensated by the net oxygen consumption (OC_{NET}) via nitrification and the N_2 production via denitrification. In ASM2d (Henze *et al.*, 1999), it is proposed to use three distinguishable N fractions for the modelled solids (iN_{XI}, iN_{XS} and iN_{BM}) and two for the dissolved compounds (iN_{SF} and iN_{SI}). It is adviced that these values are merely a reference and should experimentally be determined for each situation. iN_{BM} is more or less constant in WWTP (0.07 gN/gCOD), and therefore should not be calibrated. To fit TKN_X, iN_{XI} is the most sensitive parameter, as X_I usually contributes the most to the activated sludge. In a properly functioning WWTP, X_S will be in the range of 1 to 2% of the activated sludge. Therefore, only a small error is made when iN_{XI} and iN_{XS} are calibrated simultaneously. Independent calibration of iN_{XI} and iN_{XS}, as proposed by Koch *et al.* (2000), is not justified as iN_{SF} (iN_{SS}, ASM3) and iN_{SI} cannot be determined with the required accuracy.

In this study ammonium was balanced cover the anaerobic reactor. Therefore S_{NH} in the influent should match the measured influent ammonium concentration. Roeleveld and van Loosdrecht (2001) suggest calculating S_{NH} according to Eq. 4.3.

$$S_{NH} = TKN_{IN} - (iN_{XI} \times X_I + iN_{XS} \times X_S + iN_{BM} \times X_{BM} + iN_{SF} \times S_F + iN_{SI} \times S_I) \qquad (4.3)$$

Instead, it is proposed to fit Eq. 4.3 to the measured NH_{IN} by adjusting iN_{SF} or iN_{XS} in the model.

4.6 Conclusions

Data evaluation based on linear conservation relations gives insight in basic plant performance. For this information, in practice often complete simulation studies are performed. The use of Macrobal instigates a systematic organisation of raw operational data, whereby information is acquired in the benefit of a proper WWTP evaluation. The extra time invested in data and process evaluation will be saved during calibration, as it takes considerably more effort to fit a model on faulty data.

Macrobal software showed suitable for error detection and reconciliation of operational data and process measurements. Therefore ammonium, TP and flow balances were formulated over several process units. Major errors were detected in three process flows. With the help of balancing methods these errors were corrected.

Model calibration mainly relied on fitting the mass balances that were established in the Macrobal data evaluation. Hereby no stoichiometric or kinetic parameters required calibration.

The balanced data set simplified the model calibration. For the calibration the stepwise method described in Chapter 1 was used. Hereby, the SRT was fitted according to the overall TP balance, TP_X/COD_X was fitted with the influent X_I/X ratio, TKN_X/COD_X was fitted with iN_{XI} and iN_{XS}, nitrification was fitted by adjusting the DO set-points and denitrification was fitted by increasing K_O.

References

Alex J., Tschepetzki R. and Bitter U. (1997) SIMBA Simulation of biological waste water treatment. User's manual, Institute of Automation and Communication e. V., Magdeburg.

Brdjanovic D., Logemann S., van Loosdrecht M. C. M., Hooijmans C. M., Alaerts G. J. and Heijnen J. J. (1998) Influence of temperature on biological phosphorus removal: process and ecological studies. *Water Res.* 32(4), 1035-1048.

Brdjanovic D., van Loosdrecht M.C.M., Versteeg P., Hooijmans C.M., Alaerts G.J. and Heijnen J.J. (2000) Modelling COD, N and P removal in a full-scale WWTP Haarlem Waarderpolder. *Water Res.* 34(3), 846-858.

Henze M., Gujer W., Mino T., Matsuo T., Wentzel M. C. and Marais GvR. and van Loosdrecht M. C. M. (1999) Activated Sludge Model No. 2d, ASM2d. *Wat. Sci. Technol.* 39(1), 165-182.

Hellinga C. (1992) Macrobal 2.02. Delft University of Technology, the Netherlands. M.C.M.vanLoosdrecht@TNW.TUDelft.NL. http://www.bt.tudelft.nl/miscell

Koch G., Kuhni M., Gujer W. and Siegrist H. (2000) Calibration and validation of Activated Sludge Model No. 3 for Swiss municipal wastewater. *Water Res.* 34(14), 3580-3590.

Meijer S. C. F., van Loosdrecht M. C. M. and Heijnen J. J. (2001) Metabolic modelling of full-scale biological nitrogen and phosphorus removing WWTP's. *Water Res.* 35(11), 2711-2723.

Nowak O., Franz A., Svardal K., Muller V. and Kuhn V. (1999) Parameter estimation for activated sludge models with the help of mass balances. *Water Sci. Technol.* 39(4), 113-120.

Roeleveld P.J and van Loosdrecht M.C.M. (2001) Experiences with guidelines for wastewater characterization in The Netherlands. *Water Sci. Technol.* 45(6), 145-156.

Van der Heijden R.T.J.M., Heijnen J.J., Hellinga C., Romein B., Luyben K.Ch.A.M. (1994a), Linear constraint relations in biochemical reaction systems: I. Classification of the calculability and the balanceability of conversion rates. *Biotechnol. Bioeng.* 43, 3-10.

Van der Heijden R.T.J.M., Romein B., Heijnen J.J., Hellinga C., Luyben K.Ch.A.M. (1994b) Linear constraint relations in biochemical reaction systems: II. Diagnosis and estimation of gross errors. *Biotechnol. Bioeng.* 43, 11-20.

Van der Heijden R.T.J.M., Heijnen J.J., Hellinga C., Romein B., Luyben K.Ch.A.M. (1994c) Linear constraint relations in biochemical reaction systems: III. Sequencial application of data reconciliation for sensitive detection of systematic errors. *Biotechnol. Bioeng.* 43, 781-791.

Van Veldhuizen H. M., van Loosdrecht M. C.M. and Heijnen J. J. (1999) Modelling biological phosphorus and nitrogen removal in a full scale activated sludge process. *Water Res.* 33(16), 3459-3468.

Figure 4.5 Sampling at WWTP Katwoude (Photo: Meijer S.C.F.).

Figure 4.6 Images of WWTP Katwoude (Photo: Meijer S.C.F.).

Chapter 5: WWTP Hardenberg, the Netherlands

Modelling full-scale start-up of the BCFS® process

Meijer S.C.F., van Loosdrecht M.C.M. and Heijnen J.J.

This chapter is based on Meijer S.C.F., van Loosdrecht M.C.M. and Heijnen J.J. (2001) Metabolic modelling of full-scale enhanced biological phosphorus removing WWTP's. *Water Res.* 35(11), 2711-2723., Meijer S.C.F., van Loosdrecht M.C.M. and Heijnen J.J. (2002) Modelling the start-up of a full-scale biological nitrogen and phosphorus removing WWTP's. Water Res. 36, 4667-4682., and Meijer S.C.F. (2004) Theoretical and practical aspects of modelling biological nutrient removal in activated sludge systems (ISDN 90-9018027-3) Ph.D. Thesis, Delft University of technology, The Netherlands (205p.).

PART 1: Modelling regular operation of WWTP Hardenberg

5.1 Introduction

Modelling WWTP for COD and N removal, has become a standard practice and valuable instrument for design and operation of WWTPs. However, with respect to modelling EBPR, practical use of activated sludge models is still limiting. Partly this is due to the complexity of the EBPR process, partly to limited modelling experience. Essentially, there are two models proposed, ASM2d (Henze *et al.*, 1999) from the IAWQ task group, and an ASM2d integrated metabolic model proposed by Smolders *et al.* (1994a,b), Murnleitner *et al.* (1997) and van Veldhuizen *et al.* (1999). This model is further referred to as the TUDP (Technical University Delft bio-P) model. ASM2d uses a grey box description towards EBPR, as only one organic storage compound is modelled. In the TUDP model, full account is given to the metabolism of phosphorus accumulating organisms (PAO) and therefore all storage compounds are modelled explicitly. Eight cell-internal processes are described involving three storage polymers: poly-hydroxy-alkanoates (PHA), glycogen and poly-phosphate (poly-P). In two previous studies, the TUDP model was used to simulate full-scale EBPR; a side stream P-removal process (WWTP Haarlem Waarderpolder, Brdjanovic *et al.*, 2000, Chapter 3) and a mainstream P-removal process (WWTP Holten, van Veldhuizen *et al.*, 1999, Chapter 2). This chapter contains, besides the simulation of a full-scale BNR process at WWTP Hardenberg (regular steady operation), also simulation of the WWTP start-up, and model-based optimization of the control strategies at this plant. This MUCT-type WWTP (Figure 5.1, Wentzel *et al.*, 1990) was optimised by introduction of denitrifying EBPR according to the novel BCFS® design. During a three day sampling period, the pseudo-steady state (PSS) conditions of the WWTP was monitored. Pseudo-steady state conditions imply that the process is determined by recurring 24-hour influent variations. By definition, no significant changes in the activated sludge composition will occur at such conditions.

In previous research, problems were reported with the metabolic model kinetics (Brdjanovic *et al.*, 2000). In this study, the model is re-evaluated. Hereby focus was given to the verification of operational data, validation of the model stoichiometry and kinetics, the calibration method and the practical applicability of the TUDP model at full-scale conditions.

5.2 Materials and methods

5.2.1 WWTP Hardenberg

WWTP Hardenberg is one of seven WWTPs managed by the Dutch water board "Groot Salland". In 1998 it was upgraded to a BCFS® process (Figure 5.1), which is a MUCT-type design. Typical process characteristics are SVI below 100 mL/g without addition of chemicals, high activity of denitrifying EBPR and $PO_{4,EF}$ below 0.5 gP/m^3. The chemical phosphate stripper shown in Figure 5.1 was not built. An elaborate description of BCFS® and the upgrading philosophy is given in van Loosdrecht *et al.* (1998).

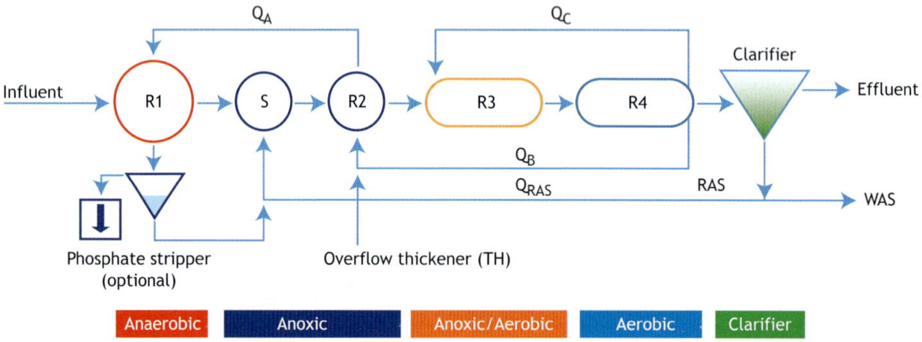

Figure 5.1 Schematic layout of upgraded WWTP Hardenberg. For WWTP Hardenberg, R1, S, R2 and the thickener (not shown) were constructed as one circular reactor. The overflow from the thickener (Q_{OF}) was received by R2.

Table 5.1 Operational data and hydraulic design parameters of the WWTP Hardenberg: Average measurements ± standard deviation. The hydraulic retention time (HRT) was calculated from the listed flows and volumes.

Flow	Average ± SD m^3/d	Process unit	Volume m^3	Depth m	HRT h
Influent, Q_{IN}	6,855 ± 2,387	Anaerobic reactor, R1	1,48	5.0	1.4
Overflow thickener to R2, Q_{OF}	126$^{(1)}$	Selector reactor, S	740	5.0	0.5
Effluent, Q_{EF}	6,738 ± 2,516	Anoxic reactor, R2	2.29	5.0	0.8
Waste activated sludge, Q_W	243 ± 299	Alternate aerated tank, R3	4.19	2.5	0.9
Pump-flow A, Q_A	18,708 ± 263	Aerated carousel, R4	4.19	2.5	0.9
Pump-flow B, Q_B	35,236 ± 334	Clarifier, CL (2 × 2,625 m^3)	5.25	-	6.6
Pump-flow C, Q_C	59,174 ± 3,021	RAS volume	150$^{(2)}$	-	0.3
Recycle activated sludge, Q_{RAS}	12,126 ± 335				
Internal recycles, $Q_{R4,IN}$ and $Q_{R3,IN}$	635,040$^{(3)}$	Total WWTP	18,140	6,981	62.4

$^{(1)}$ Estimated
$^{(2)}$ Estimated from denitrification in the RAS
$^{(3)}$ Estimated from superficial flow velocity measurements

The WWTP was constructed from an anaerobic reactor (R1) with plug-flow characteristics, a selector reactor (S) and an anoxic reactor (R2), both supposed to be completely mixed, followed by two carrousel reactors (R3 and R4) in line. R3 aerated alternately. R4 was fully aerated. The oxygen concentration in R3 and R4 (DO_{R3} and DO_{R4}) was controlled on respectively 0.6 and 2.8 gO_2/m^3 (Table 5.3). In both R3 and R4, significant oxygen gradients were measured (not shown). In the clarifiers (CL), activated sludge was separated from the liquid phase. Cyclic process conditions were provided by the return activated sludge (RAS) from the clarifier underflow (Q_{RAS})

and three recycle pump-flows Q_A, Q_B and Q_C. These flows were ORP controlled on the measurements ORP_{R1}, ORP_{R2} and ORP_{R3} respectively. During the recorded period, dry weather influent caused Q_A, Q_B, Q_C and Q_{RAS} to be more or less constant. In Table 5.1 operational data and reactor volumes are listed. Q_A, Q_B and Q_C were measured on-line. Q_W was calculated from the capacity and operation time of the WAS pump. The overflow from the thickener (Q_{OF}) was estimated. The internal carrousel flows $Q_{R3,IN}$ and $Q_{R4,IN}$ were estimated from the superficial flow velocities in R3 and R4.

5.2.2 Measurements

The pseudo-steady state operation of the WWTP was recorded during a 60 hour measurement campaign from the 23rd to 25th of June 1998. Samples were taken every two hours from the influent (Figure 5.2), R1 (Figure 5.4), R2 (Figure 5.5), R3 and R4, WAS, overflow of the thickener and effluent.

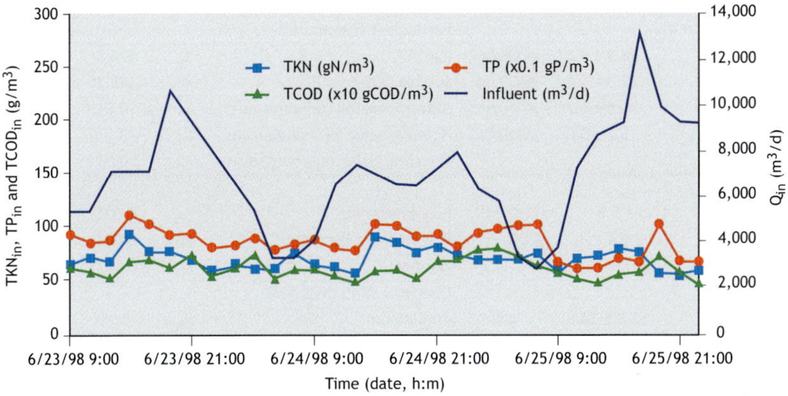

Figure 5.2 Influent measurements at WWTP Hardenberg.

The samples were analysed on COD, COD_{MF} (COD of the micro-filtrated fraction, 0.45 μ pore diameter, no flocculation), BOD_5 (biological oxygen demand), VFA (volatile fatty acids), NH_4 (ammonium), NO_3 (nitrate), TKN (total Kjeldahl nitrogen), TP (total phosphorus), PO_4 (ortho-phosphate) and MLSS (mixed liquor suspended solids). Pump flow rates, control measurements and energy consumption and operation time of mixers and aerators and were recorded every two hours. Average measurement results are presented in Tables 5.2 and 5.3.

5.2.3 The WWTP Hardenberg model

All simulations were performed with the TUDP model (appendix III). SIMBA© 3.3+ was used as simulation platform (Alex *et al.*, 1997), which operated under Matlab/Simulink© 5.2. In SIMBA© models are constructed by connecting CSTR's in a desired process configuration. The WWTP was modelled according to the flow scheme in Figure 5.1 and data from Table 5.1. R1 was modelled as three CSTR's in series. S and R2 were modelled as single CSTR's. R3 and R4 were modelled as six CSTR's with internal recycle $Q_{R3,IN}$ and $Q_{R4,IN}$. With this set-up, longitudinal oxygen gradients in R3 and R4 were simulated sufficiently. CL separated particulate matter from the liquid phase ideally. A residual X_{EF} was modelled as a percentile loss of the RAS. Biological conversions in the RAS volume were simulated by a non-aerated CSTR placed in the RAS flow. DO_{R3} and DO_{R4} were controlled on respectively 0.6 and 2.8 gO_2/m^3 (Table 5.3). X_{R4} was controlled on 4,693 $gCOD/m^3$

by regulating Q_W (Table 5.3). This control simulated manual operation by the plant operators. In the model, Q_A, Q_B, Q_C and Q_{RAS} were constant (Table 5.1). Plant hydraulic scheme in SIMBA is shown in Figure 5.23 and 5.24.

Table 5.2 Influent and effluent data and the characterised model influent. The model influent was calculated according to Roeleveld and van Loosdrecht (2001).

Measurement	Average ± SD	Unit	Simulation		value	Unit
Influent			**Soluble compounds**			
Total COD (TCOD)	603.9 ± 95.5	$gCOD/m^3$	Dissolved oxygen (DO)	S_O	0	gO_2/m^3
Micro-filtrated COD (COD_{MF})	240.5 ± 37.9	$gCOD/m^3$	Fermentable COD (COD_F)	S_F	109	$gCOD/m^3$
5-day BOD (BOD_5)	245.5 ± 48.7	$gCOD/m^3$	Fatty acids (VFA)	S_A	91	$gCOD/m^3$
Total Kjeldahl N (TKN)	68.8 ± 10.1	gN/m^3	Ammonium (NH_4)	S_{NH4}	52	gN/m^3
Ammonium (NH_4)	53.4 ± 8.7	gN/m^3	Nitrate (NO_3)	S_{NO3}	0	gN/m^3
Nitrate (NO_3)	0.1 ± 0.2	gN/m^{-3}	Ortho-phosphate (PO_4)	S_{PO4}	6.6	gP/m^3
Total phosphorus (TP)	8.4 ± 1.4	gP/m^3	Inert COD (COD_I)	S_I	40	$gCOD/m^3$
Ortho-phosphate (PO_4)	5.2 ± 0.8	gP/m^3	Alkalinity (HCO_3)	S_{HCO}	8	mole/L
Total suspended solids (TSS)	308.6 ± 18.2	$gMLSS/m^3$				
Influent volatile fatty acids (VFA)			**Particulate compounds**			
Acetic acid (CH_3COOH)	68.3 ± 9.2	$gCOD/m^3$	Inert COD	X_I	174.8	$gCOD/m^3$
Propionic acid (C_2H_5COOH)	12.3 ± 1.4	$gCOD/m^3$	Soluble COD	X_S	188.7	$gCOD/m^3$
Butyric acid (C_3H_7COOH)	5.4 ± 0.7	$gCOD/m^3$	Autotrophic microorganisms	X_A	0.01	$gCOD/m^3$
Valeric acid (C_4H_9COOH)	5.2 ± 0.6	$gCOD/m^3$	Heterotrophic microorganisms	X_H	0.01	$gCOD/m^3$
			P-accumulating microorganisms	X_{PAO}	0.01	$gCOD/m^3$
Effluent						
Micro-filtrated COD (COD_{MF})	41.5 ± 5.8	$gCOD/m^3$	Poly-phosphate	X_{PP}	0.001	gP/m^3
5-day BOD (BOD_5)	1.9 ± 0.4	$gCOD/m^3$	Poly-hydroxy-butyrate	X_{PHA}	0.001	$gCOD/m^3$
			Glycogen	X_{GLY}	0.001	$gCOD/m^3$
Overflow WAS thickener			**Overflow WAS thickener**			
Total Kjeldahl N (TKN)	55.9 ± 14.6	gN/m^3	Ammonium	S_{NH4}	55.9	gN/m^3
Total phosphorus (TP)	47.2 ± 8.5	gP/m^3	Ortho phosphate	S_{PO4}	47.2	gP/m^3
Ortho-phosphate (PO_4)	39.8 ± 8.9	gP/m^3	Alkalinity, HCO_3	S_{HCO}	8	mole/L

Table 5.3 Measurements and simulation results (in blue).

Paremeter	Symbol	R1	R2	R3	R4	Q_{EF}	Q_W
Total COD	TCOD	3,896±493	4,698±522	4,610±140	4,735±321	42±8	8,615±1,762
($gCOD/m^3$)	X_T+S_T	4	4,8	4,7	4,7	44	8,5
Particulate COD	COD_X	3,839±493	4,654	4,570	4,693±320	0.8±6.4	8,574±1,762
($gCOD/m^3$)	X_T	3,954	4,7	4,7	4,7	5	8,4
Micro-filtrated COD	COD_{MF}	57±	44±5	40±2	42±8	42±6	42±8
($gCOD/m^3$)	S_A+S_F+S_I	56	42	40	40	40	40
Volatile fatty acids	COD_{VFA}	0±1	0	-	-	-	-
($gCOD/m^3$)	S_A	0	0	0	0	0	0
Fermentable COD	COD_F	17	4	-	-	-	-
($gCOD/m^3$)	S_F	16	2	1	0	0	0
Total phosphorus	TP	-	-	-	155±6	0.2	268±17
(gP/m^3)	S_{PO4}+X_{PP}+i_P	124	152	152	152	0.4	274
Ortho-phosphate	PO_4	24±5	7.4±3.9	1.6±2.5	-	0.4±0.8	-
(gP/m^3)	S_{PO4}	24	7.1	1.2	0.3	0.3	0
Total Kjeldahl	TKN	-	-	-	-	2±1	-
nitrogen (gN/m^3)	S_{NH4} + i_N	158	178	173	171	1	307
Ammonium	NH_4	-	6.4±2.6	-	0.6±1.8	0.3	-
(gN/m^3)	S_{NH4}	19	6.7	2	0.4	0.4	0
Nitrate	NO_3	0	0.2±0.1	1±1	2.8±1.6	3.2±0.5	3.0
(gN/m^3)	S_{NO3}	0	0.2	1.9	3.0	3.0	3.0
Dissolved oxygen	DO	-	-	0.6±0.2	2.8±0.9	-	-
(g/m^3)	S_O	0	0	0.4	3.0	3.0	0.1

5.2.4 Model adjustments

In the original TUDP model, hydrolysis (r_h) was a function of ordinary heterotrophic micro-organisms (OHO). In EBPR processes, PAO (phosphorus accumulating organisms) grow at the expense of OHO as both compete for the same substrate. This study showed a X_{PAO}/X_H ratio of 0.51, whereas in simulations of A_2N processes even higher ratios were observed (Hao *et al.*, 2001). This model set-up therefore caused a reduction of the hydrolysis rate. However, according to Goel *et al.* (1998 and 1999), hydrolysis is associated with the total heterotrophic biomass. Therefore in the TUDP model r_h was made a function of X_{PAO} + X_H. The fraction particulate biodegradable matter was hereby redefined as $X_S/(X_{PAO}+X_H)$. Except for $PO_{4,R1}$, the model sensitivity towards the maximum hydrolysis rate (k_h) was relative low (Table 5.6).

5.2.5 Influent characterisation

Influent was measured with a flow proportional sampler. During the recorded period no rain events occurred. The operation of the WWTP was mainly determined by 24-hour recurring dry weather influent dynamics. Hereby sewage pumping stations caused the typical flow dynamics observed in Figure 5.2. Concentration variations in the measured influent were relative small. Variations in the influent loads (concentration × flow), therefore were primarily caused by flow dynamics. To decrease simulation time, the influent was modelled with average concentrations and actual flow data. The actual influent measurements and model influent composition are presented in Table 5.2. The influent characterisation was performed according to Roeleveld and van Loosdrecht (2001). In this method the influent X_I/X ratio is determined from BOD_5 measurements. If however, a COD_X balance is available from which X_I/X can be estimated, it is suggested not to use this method. Firstly, BOD_5 measurements are laborious and secondly, it is questioned if BOD_5 can be determined with the required accuracy for precise determination of X_I/X. The latter argument is of importance, as the simulated SRT has been found highly sensitive towards X_I/X (Brdjanovic *et al.*, 2000; van Veldhuizen *et al.*, 1999). Instead, it is suggested to use the influent X_I/X ratio to fit the COD_X balance in the model (i.e. calibrate the sludge production).

5.3 Data evaluation

5.3.1 Initial simulation

On the basis of the original operational data it was not possible to simulate the WWTP performance. It was observed that (i) the overall TP and solids (COD_X) balance could not be fitted simultaneously, (ii) the COD_X balance over the anaerobic reactor R1 and the clarifier (CL) could not be closed, and (iii) internal ammonium, nitrate and phosphate concentration profiles could not be simulated. It was therefore concluded that operational data and or measurements contained gross errors. These data therefore were checked by evaluating the mass balances over the WWTP.

5.3.2 Evaluation of the SRT

The RAS concentration is a crucial operational parameter. In the model, via Q_W, RAS determines solids in the aeration tank (X_{AT}) and eventually the SRT. Brdjanovic *et al.* (2000) showed that the model is highly sensitive towards the SRT. For a proper simulation, the SRT should be known within 95 % accuracy. Often much lower accuracy is obtained, especially when the SRT is calculated from the RAS concentration (initially in this study 88 %, Table 5.4). RAS measurements are generally inaccurate, especially when grab-samples are used. This is caused by the dynamic operation of the clarifier. Moreover, RAS samples taken from distribution works and transport-pipes with valves and curves are known to be inaccurate caused by non-homogeneous distribution of MLSS. In this study the SRT was estimated from mass balance calculations (Figure 5.3).

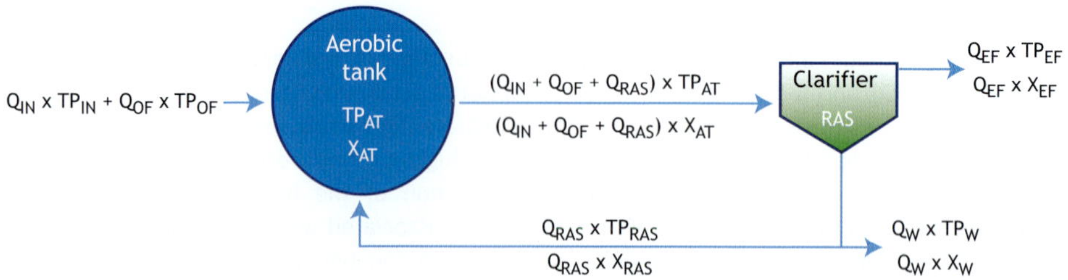

Figure 5.3 Mass balances for SRT estimation. TP and COD_X balances over the WWTP and clarifier (kg/d). The mass balances are presented in Eq. 5.1 to 5.4. Concentrations and flows are used from Tables 5.1, 5.2 and 5.3. Table 5.4 presents the results of the balance calculations.

According to Figure 5.3, four balances were formulated:

(i) the overall TP balance,

$$Q_{IN} \times TP_{IN} + Q_{OF} \times TP_{OF} = Q_{EF} \times TP_{EF} + Q_W \times TP_W \quad (5.1)$$

(ii) the overall flow balance

$$Q_{IN} + Q_{OF} = Q_{EF} + Q_W \quad (5.2)$$

(iii) the clarifier TP balance

$$(Q_{IN} + Q_{OF} + Q_{RAS}) \times TP_{AT} = Q_{EF} \times TP_{EF} + (Q_{RAS} + Q_W) \times TP_W \quad (5.3)$$

(iv) the clarifier solids (COD_X) balance

$$(Q_{IN} + Q_{OF} + Q_{RAS}) \times X_{AT} = Q_{EF} \times X_{EF} + (Q_{RAS} + Q_W) \times X_W \quad (5.4)$$

Table 5.4 Results of SRT and flow estimations. Average measurements ± standard deviation and balanced operational data according to Eq. 5.1 to 5.7.

Mass balance calculations (eq. 5.1 to 5.7)	Average ± SD	Balanced	Unit
Estimation of the SRT			
Return activated sludge flow, Q_{RAS} (Q_{RAS}=1.2xQ_{IN})	12,126 ± 335	8,226	m^3/d
Overflow thickener, Q_{OF}	126[(1)]	102	m^3/d
Waste activated sludge flow, Q_W	244 ± 299	219	m^3/d
Total phosphorus in WAS, TP_W	268 ± 17	278	gP/m^3
Total particulate in WAS, X_W	8,574 ± 1,762	8,411	$gCOD/m^3$
Sludge retention time, SRT	28.9	32.8	d
Estimation of Q_A			
Pump-flow A, Q_A	18,708 ± 263	22,415	m^3/d
MLSS in flow A, X_{QA}	4,654 ± 522	4,895	$gCOD/m^3$
Estimation of Q_B			
Pump-flow B, Q_B	35,236 ± 334	44,536	m^3/d

[(1)] Obtained measurement from grab sample

These mass balances only hold for steady state conditions. Concentrations and flows are obtained from Tables 5.1 and 5.3. The system of four eq. (n=4), had one unknown u (Q_{OF}) and thereby n-u=3 degrees of freedom. Therefore three operational parameters could be estimated. The least reliable parameters were chosen, being Q_W, TP_W and X_W (Table 5.3). For the calculations, the equation solver "Maple V" was used (Char *et al.*, 1991). Initially no satisfactory solutions were found. It was however possible to solve Eq. 5.1 and 5.2, hereby finding solutions for Q_W and TP_W. We relied on accurate measurement of solids (X_{AT}) and TP in R4. Therefore, a gross error in Q_{RAS} was assumed. By adjusting Q_{RAS} to 1.2 times Q_{IN}, as specified in the original plant design (Q_{RAS} = 8,226 m^3/d), it was possible to balance the system within the error of measurements (Table 5.4).

5.3.3 Evaluation of recycle flow A

Because it was not possible to simulate the solids concentration in the anaerobic tank R1 (X_{R1}), the solids balance over R1 (Eq. 5.5) was formulated.

$$Q_{IN} \times X_{IN} + Q_A \times X_{QA} = (Q_{IN} + Q_A) \times X_{R1} \tag{5.5}$$

In this balance, formation and degradation of PAO storage polymers and anaerobic hydrolysis are neglected. Solving Eq. 5.5 for Q_A resulted in 29,012 m^3/d, being well above the maximum pump capacity of 18,000 m^3/d. Therefore, an additional balance was formulated.

$$Q_{IN} \times PO_{4,IN} + Q_A \times PO_{4,R2} + Q_{IN} \times VFA_{IN} \times Y_{PO4} = (Q_{IN} + Q_A) \times PO_{4,R1} \tag{5.6}$$

Eq. 5.6 is the phosphate balance over R1. In this balance the anaerobic PO_4 release is calculated based on VFA_{IN} and the yield according to Smolders *et al.* (1994a, Y_{PO4}=0.36 gP/gCOD$_{Ac}$). Hereby formation of VFA by hydrolysis was neglected. Solving eq. 5.5 and 5.6 simultaneously, resulted in a minimal flow Q_A of 22,415 m^3/d and a solids concentration in flow Q_A (X_{QA}) of 4,895 gCOD/m^3. The high calculated value of X_{QA} could be caused by settling of particulate matter in R2. Therefore, retention of solids in R2 was modelled. Applying the estimated values for X_{QA} and Q_A, the solid and phosphate concentration in the anaerobic reactor were simulated satisfactory (Figure 5.4).

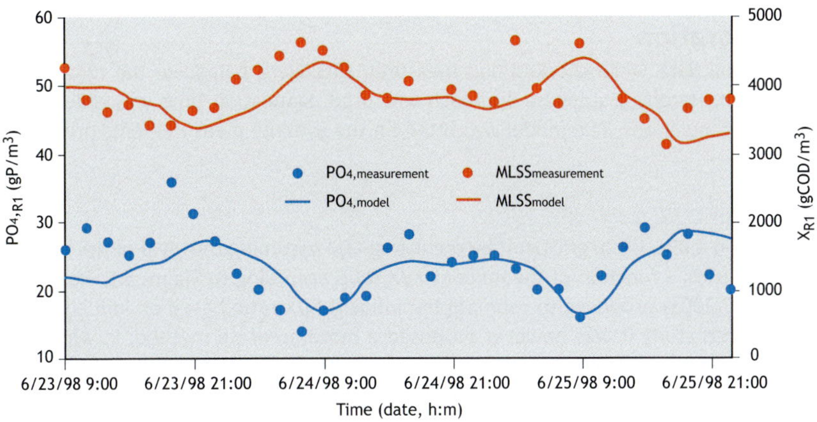

Figure 5.4 Measurement and simulation of X_{R1} and $PO_{4,R1}$.

5.3.4 Evaluation of recycle flow B

Because ammonium and nitrate in reactor R2 could not be simulated, an ammonium balance over R2 was formulated (Eq. 5.7).

$$Q_{IN}\times NH_{4,IN}+Q_{OF}\times NH_{4,OF}+Q_{B}\times NH_{4,R4}+Q_{RAS}\times NH_{4,EF}=(Q_{IN}+Q_{OF}+Q_{B}+Q_{RAS})\times NH_{4,R2} \quad (5.7)$$

In this balance, nitrification, ammonification and biomass growth in R2 were neglected. From eq. 5.7, flow Q_B was estimated under the assumptions that (i), ammonium in R2 was accurate (6.4 ± 2.6 gN/m^3) (ii), ammonium in the effluent and R4 were negligible and (iii), $Q_{RAS}=1.2 \times Q_{IN}$. From Eq. 5.7 it was calculated that Q_B was 44,536 m^3/d, which is 1.27 times higher than the original value (35,000 m^3/d). Because the measurement of Q_B showed unreliable, in the simulation Q_B was used to calibrate ammonium and nitrate in R2 (Table 5.3, Figure 5.5). This resulted in a calibrated flow of Q_B=44,300 m^3/d.

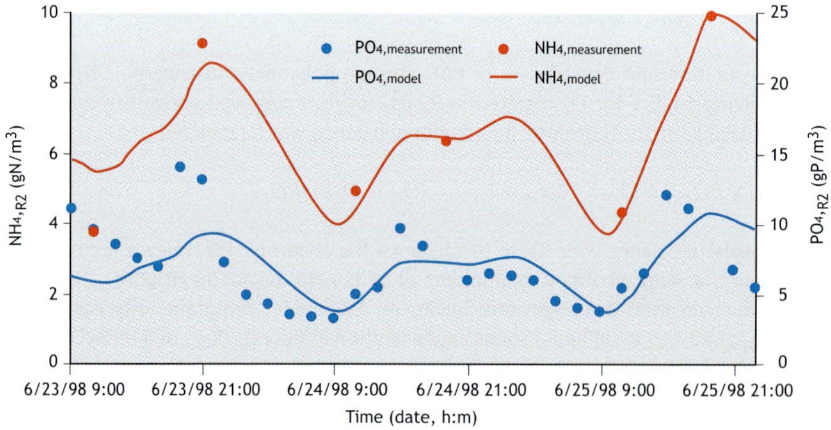

Figure 5.5 Measurement and simulation of $NH_{4,R2}$ and $PO_{4,R2}$.

5.4 Model calibration

After the operational data were checked and measurements were balanced, the model was ready for calibration. Hereby a stepwise approach was used. Simulation time was reduced by starting with static simulations. The model was fitted on the average measurements presented in Table 5.3.

Step 1. Calibrating the COD_X balance

X_{R4} was controlled on $4,693 \pm 320$ gCOD/m^3 by regulating Q_W. Assuming the yield of the process was constant, the SRT is a function of the influent X_I/X ratio according to Figure 5.6. Roeleveld and van Loosdrecht (2001) proposed to estimate the influent X_I/X ratio based on influent BOD_5 measurements. In this study it was however proposed a more accurate method, in which the WAS production is estimated from the overall TP balance (Eq. 5.1). On the basis of the WAS production, the COD_X balance was closed by calibrating the influent X_I/X ratio. The calibrated ratio was found to be 0.507.

Step 2. Calibrating nitrification

The overall nitrified load (^{L}NIT kgN/d) could be simulated without calibration. However, during the calibration procedure unrealistic low phosphate in R4 caused growth limitation of ordinary autotrophic micro-organisms (OAO). Therefore, temporarily the phosphate affinity for growth of OAO ($K_{N,PO4}$) was reduced to 0.001 gP/m^3.

Step 3a. Calibrating denitrification in the RAS volume

Some denitrification occurred in the RAS volume (0.3 gN/m^3, Table 5.3). From the calculated denitrified load ($^{L}NO_{3,RAS}$), the RAS volume was estimated to be 150 m^3.

5.4.1 Simultaneous nitrification and denitrification

Simultaneous nitrification and denitrification (SND) often occurs in full-scale WWTP's. Non-ideal mixing and (partial) settling of activated sludge in the aeration tank causes oxygen gradients and diffusion limitations. Therefore, the actual anoxic volume often is higher than the designed anoxic volume. With the COD and N balance, the overall denitrified load (^{L}DEN, kgN/d) is calculated (Nowak *et al*, 1999a). To fit denitrification and thereby correct for the actual anoxic volume, the calibration parameter K_O was introduced. In the model, K_O replaces the oxygen affinity for PAO ($K_{P,O}$) and OHO ($K_{H,O}$). The sensitivity of K_O towards nitrate in the effluent ($NO_{3,EF}$) is a function of X_I/X. This relation is shown in Figure 5.6 The net effect of increasing K_O is an increase of the anoxic activated sludge fraction (Figure 5.7), as if the oxygen penetration depth in the sludge flocs is decreased, hereby causing an anoxic zone in the centre of the floc (Pochana and Keller, 1999).

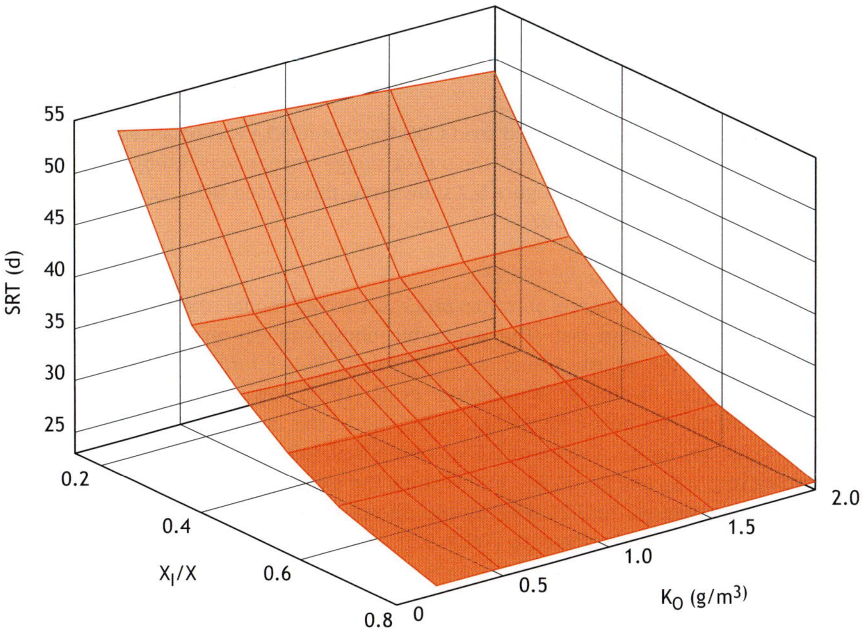

Figure 5.6 SRT as a function of the influent X_I/X ratio and K_O. With X_I/X the COD_X balance is fitted. This process is also effected by K_O.

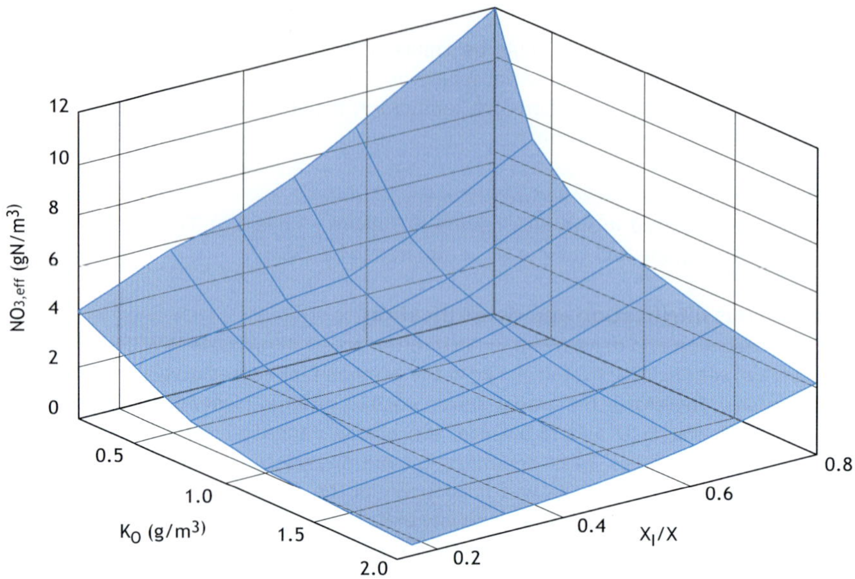

Figure 5.7 $NO_{3,EF}$ as a function of the influent X_I/X ratio and K_O. K_O is used to fit ^{L}DEN. This process is also effected by X_I/X.

Step 3b. Calibrating the anoxic sludge fraction

After the nitrified load (^{L}NIT) was fitted and solid COD fraction (COD_X) was balanced (step 1 and 2), the denitrified load (^{L}DEN) was fitted (i.e. the anoxic sludge fraction was fitted). Therefore K_O was increased from 0.2 to 0.7 gO_2/m^3. Figure 5.7 shows how this caused $NO_{3,EF}$ to decrease from 6.5 to 3.2 gN/m^3 in relation to X_I/X=0.507.

Step 4. Calibrating the anoxic phosphate uptake

Because EBPR has a limited influence on other processes in the model, the phosphate profile over the WWTP was calibrated last. The anaerobic phosphate concentration ($PO_{4,R1}$) was simulated correctly with the adjusted kinetics for hydrolysis. Poly-P formation depends on the electron acceptor (oxygen or nitrate). Murnleitner *et al.* (1997) however, observed a reduced electron acceptor affinity for poly-P formation. In the model this is expressed in the reduction factor g_{PP} (<1). Originally for g_{PP} a value of 0.1 was proposed, hereby increasing the affinity for oxygen according to $g_{PP} \times K_O$ and $g_{PP} \times K_{P,NO3}$. In this study the oxygen affinity for poly-P formation was reduced by increasing g_{PP} to 0.25. Hereby the phosphate profile over the WWTP ($PO_{4,R2}$, $PO_{4,R3}$ and $PO_{4,R4}$) was simulated sufficiently (Table 5.3).

Step 5. Calibrating the net oxygen consumption

The COD_X balance was fitted by adjusting the influent X_I/X ratio. There is however, a second (independent) balance related to COD_X, being the COD and N balance (Nowak *et al.*, 1999a). COD and N balances are linked via denitrification and therefore have to be solved simultaneously. From the COD and N balance it was calculated that the net oxygen consumption OC_{NET} was 2,631 kgO_2/d (Table 5.5). The simulated OC_{NET} was 2,597 kgO_2/d, which indicated that the calibration was sufficient.

Table 5.5 The COD and N mass balance calculation (based on Nowak *et al.*, 1999a).

COD and N mass balance calculation		Balanced	Unit
Oxygen for NH_4 oxidation (nitrification)	OC_{NIT}	1,595	kgO_2/d
Oxygen for COD oxidation	OC_{COD}	1,037	kgO_2/d
Total oxygen consumption	OC_{NET}	2,631	kgO_2/d
COD oxidised with NO_3 (denitrification)	$^{L}COD_{DEN}$	958	kgCOD/d
Total COD oxidised	$^{L}COD_{BD}$	1,995	kgCOD/d
Nitrification rate	^{L}NIT	349	kgN/d
Denitrification rate	^{L}DEN	334	kgN/d

5.5 Discussion

5.5.1 Fitting models on faulty data

The P balance is a closed balance, indicating that all in and outgoing TP loads can be recovered in the liquid phase. In contrast to the P balance, the COD and N balance has compounds in the gaseous phase (O_2 and N_2). Usually this is not measured. When a model is fitted on faulty data, the COD and N balance is 'closed' by over- or under-estimating O_2 and N_2. The P balance however, cannot be fitted on faulty data unnoticed, as all loads are generally measured. Consequently, EBPR models cannot be fitted on faulty measurements (ASM2d, TUDP model), whereas models lacking this balance can (ASM1, ASM3). The P balance is an extra independent balance coupled to the solids balance (COD_X), and therefore useful for error detection and estimation of the SRT. For a proper simulation, the SRT has to be known with accuracy higher than 95 % (Brdjanovic *et al.*, 2000). In modelling practice, this means that the P balance should fit.

5.5.2 Sensitivity analysis

To evaluate the sensitivity of the estimated operational data (Table 5.4) and calibrated parameters, a sensitivity-analysis was performed. Eq. 5.8 shows the sensitivity (*s*) expressed as relative change of the modelled concentration (y) divided by the relative change of the calibrated parameter (p) according to van Veldhuizen *et al.* (1999).

$$s = \frac{dy/y}{dp/p} \qquad (5.8)$$

Results of the analysis are presented in Table 5.6. It is seen that the proposed calibration parameters (g_{PP}, K_O, k_h, X_I/X) are sensitive to specific processes. This simplifies the calibration procedure. The estimated operational data however, show a more general sensitivity (Q_A, Q_B, Q_W and Q_{RAS}). The high sensitivity of operational data is expressed by the inability to calibrate the model while using erroneous data. Performing simulations with erroneous data generally leads to laborious and unjust calibrations of kinetic and/or stoichiometric model parameters. It was also observed that model kinetics generally are less sensitive than stoichiometric parameters and operational data (not shown).

Table 5.6 Sensitivity analysis of model parameters and operational data. The most sensitive parameters are presented. The sensitivity was calculated according to Eq. 5.8. Only sensitivity above 0.1 is shown.

		Calibration parameters				Operational data			
y\p	Reactor	g_{PP}	K_O	k_h	X_I/X	Q_A	Q_B	Q_W	Q_{RAS}
PO_4	1	0.1	0.2	0.1	-0.4	-1.1	-0.3		
	2	0.3	0.5	0.2	-0.8	-0.4	-1.3		-0.3
	3	0.4	0.5		-0.3		-0.4		-0.4
	4	0.5	0.5			0.4			-0.5
NO_3	1		-11.0		2.7				
	2		-7.7		1.7		1.5		-0.2
	3		-2.3		1.1		-0.5		-0.2
	4		-1.6		0.7		-0.5		-0.2
NH_4	1					-0.6	-0.3		
	2					-0.1	-0.9		-0.1
	3					-0.2	-0.4		-0.2
	4		-0.1			-0.4			-0.1
COD_S	1		0.1	0.2					
	2								
COD_X	1				0.5	-1.4		-0.7	0.4
	2				0.5	-0.1		-0.7	0.4
	3				0.5	-0.1		-0.7	0.4
	4				0.5	-0.1		-0.7	0.4
X_{PAO}	1		0.2	0.2	-0.6	-1.9		-0.6	0.2
	2		0.2	0.2	-0.6	-0.5		-0.6	0.2
	3		0.2	0.2	-0.6	-0.5		-0.6	0.2
	4		0.2	0.2	-0.6	-0.5		-0.6	0.2
X_A	1					-1.12		-0.2	0.1
	2					0.2		-0.2	0.1
	3					0.2		-0.2	0.1
	4					0.2		-0.2	0.1

5.5.3 A heuristic calibration approach

The proposed calibration method is rather qualitative than quantitative. Calibration parameters were selected on the basis of process knowledge rather than mathematical sensitivity analysis. This approach resulted in the introduction of the black box calibration parameter K_O that was introduced to correct for hydrodynamic and physical shortcomings of the plant model. Additionally the calibration parameter X_I/X was introduced to balance COD_X for a given SRT.

Several mathematical parameter calibration methods were developed (Weijers *et al.*, 1997; Yuan *et al.*, 1997; von Sperling, 1993; Wanner *et al.*, 1992). In this study however, a largely heuristic approach was used for several reasons. Firstly, it was observed that the model was more sensitive towards operational data than towards model parameters. Therefore, evaluation of input data was preferred over calibration of model parameters. Secondly, mathematical calibration procedures usually neglect the influence of faulty operational data, and instead, focus on model kinetics and stoichiometric parameters. Finally, it was observed that after balancing faulty operational data and measurements, only minor calibration was required to fit the model. It is the opinion of the author that model calibration in first instance should be focused on the evaluation of operational data. From previous experience with the TUDP model (van Veldhuizen *et al.*, 1999; Brdjanovic *et al.*, 2000; this study) it was concluded that only minor calibration was needed after correcting errors in the operational data. Table 5.7 gives an overview of model parameters that were calibrated for simulation of WWTP Haarlem Waarderpolder (Hrlm), WWTP Holten (Hltn) and WWTP Hardenberg (Hdbg). For none of these studies the model stoichiometry required calibration. Caused by shortcomings of the activated sludge model (ASM1, ASM2d, TUDP and ASM3), the parameters K_O and X_I/X will need adjustment for each simulated WWTP. This should be considered when extrapolating the model to new conditions. In this simulation study, model (EBPR) kinetics was found to be less sensitive. This generally will be the case for low loaded WWTP's at pseudo steady state conditions. Under such conditions, only kinetic rates that are limiting will be sensitive for calibration. Therefore, such conditions are not suitable for a proper kinetic model evaluation.

Table 5.7 Full scale calibration parameters. Calibration results of simulations of Hltn (WWTP Holten by van Veldhuizen *et al.*, 1999), Hrlm (WWTP Haarlem Waarderpolder by Brdjanovic *et al.*, 2000), and Hdbg (WWTP Hardenberg).

Calibration parameters	Hltn	Hrlm	Hdbg	Default	Reference
Maximum fermentation rate, q_{fe}	1	1	3[(4)]	3	Henze, 1999
Poly-P storage rate, K_{PP}	0.07	0.05	0.05	0.05	Murnleitner, 1997
Glycogen storage rate, K_{GLY}	1.09	0.15	1.09	1.09	Murnleitner, 1997
Anoxic PAO reduction factor, η_{PNO3}	0.7	0.8	0.8	0.5[(2)]	Murnleitner, 1997
X_I influent fraction, X_I/X	0.57[(1)]	0.67[(1)]	0.49[(1)]	-	This chapter
Affinity reduction factor, g_{PP}	0.1	0.1	0.25	0.1	Murnleitner, 1997
DO affinity, K_O	0.2[(3)]	0.2[(3)]	0.7	0.2	Henze, 1999

[(1)] COD_X balance calibrated based on the influent X_I/X ratio
[(2)] Default value proposed to be set to 0.8
[(3)] Denitrification calibrated by adjusting aeration
[(4)] After re-definition of the hydrolysis kinetics

5.5.4 The calibration procedure

The proposed stepwise calibration resulted in an orderly and relative quick calibration. The most sensitive parameters were calibrated first (X_I/X, K_O, k_h, g_{PP}). The sensitivity analysis in Table 5.6 shows that the selected calibration parameters primarily affected the intended processes, leaving other process barely affected. Therefore the amount of iteration loops could be reduced. Iteration is however inevitable, as processes in the model are linked. Figure 5.6 shows this effect for the COD_X balance, which is effected by calibration of denitrification (K_O). Increasing denitrification results in a decrease of the SRT, caused by different anoxic and aerobic yields. This should then be corrected by readjusting X_I/X, which again affects denitrification etc. However, because this effect was relative small, only one iteration was required.

5.5.5 Balancing solids

Figure 5.6 shows the sensitivity of X_I/X towards the SRT. X_I is not converted and therefore accumulates in the WWTP. When X_{AT} is controlled, increasing X_I/X leads to an increase of ^{L}WAS (i.e. decreasing SRT). The relation between SRT and X_I/X is not linear, caused by increasing contribution of lysis and maintenance at higher SRT. Extrapolation of the model to a higher SRT causes a shift in the COD_X balance as was described by Nowak *et al.* (1999b). In the model, X_I is defined as being inert. In practice however, hydrolysis of the inert fraction will take place very slowly. In the model this effect should be corrected by lowering the influent X_I/X ratio.

5.5.6 Calibrating K_O

In the TUDP model, K_O is used to fit ^{L}DEN and thereby correct for the actual anoxic volume, which is caused by oxygen diffusion limitations in sludge flocs and concentration gradients in the aeration tank caused by non-ideal mixing. In an attempt to model these processes explicitly, Alex *et al.* (1999) proposed a hydrodynamic mixing model. Such a model will be of limited use as long as (i), diffusion limitations in flocs are not properly described and (ii), the proposed mixing model requires new (unknown) parameters for vertical liquid exchange. Instead, the aeration tank was simulated with a relative simple hydrodynamic set-up and approximation of the actual (fuzzy-logic) oxygen control. Hereby, the modelling error that was introduced in the anoxic sludge fraction was corrected by calibrating K_O. According to Figure 5.8, increasing K_O from 0.2 to 0.7 caused an increase of denitrification of approx. 30 % in R3 and 10 % in R4. Hereby aerobic conversions decreased proportionally. This resulted in a decrease of $NO_{3,EF}$ from 6.5 to 3.2 gN/m^3 for X_I/X=0.507 (Figure 5.7).

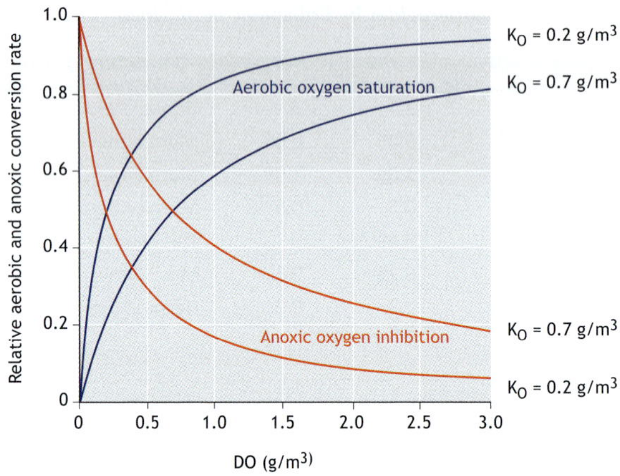

Figure 5.8 The aerobic and anoxic conversion rate as function of K_O. K_O was increased from 0.2 to 0.7 gO_2/m^3 to increase ^{L}DEN.

5.5.7 The COD and N balance

The final calibration step (5), was a check on the calibrated model. The COD and N balance is related to the COD_X balance via ^{L}NIT, ^{L}DEN, OC_{COD} and OC_{NET}. This relation is shown in Figure 5.9. Calculation of the COD and N balance corresponded with the modelled OC_{NET}, providing a check on the calibrated model.

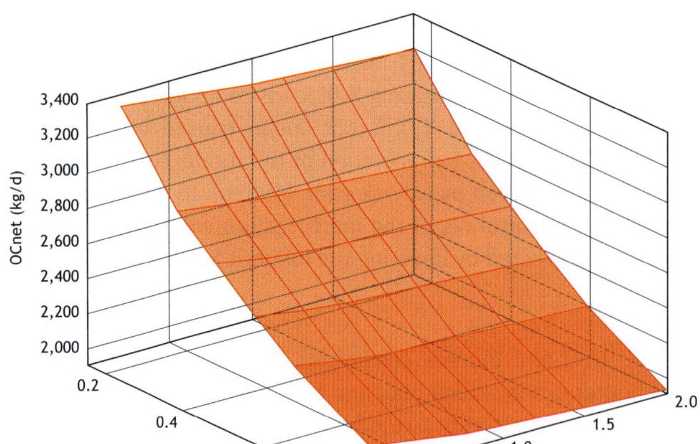

Figure 5.9 OC_{NET} as a function of the influent X_I/X ratio and K_O. Increasing K_O enlarges the anoxic conversion rate while aerobic rates are decreased accordingly, which slightly effects OC_{NET}.

5.6 Conclusions on the modelling of regular plant operation

Without changing the stoichiometric matrix of the TUDP model, the pseudo steady state of WWTP Hardenberg was simulated, hereby predicting effluent as well as process internal concentration profiles correctly. Based on this and previous research, it is concluded that the stoichiometric matrix of the TUDP model, that was developed from lab experiments, can be extrapolated to full-scale conditions without calibration (Annex 1.1). Correct estimation of operational data, in particular the SRT, showed decisive for an accurate simulation. This was caused by high model sensitivity towards operational data. Adjustment of erroneous operational data was therefore preferred over calibration of model parameters. Based on mass balance calculations, five process flows and three measured concentrations were estimated. Model (EBPR) kinetics was found less sensitive and therefore less suitable for model calibration. This will generally be the case for low loaded (i.e. over dimensioned) WWTP at pseudo steady state conditions. Two calibration parameters were introduced, K_O and X_I/X. Both require calibration for each simulated system. Calibration of K_O proved a straightforward black box solution to modelling diffusion limitations in activated sludge processes. The influent ratio X_I/X was calibrated to balance solids in the model for a given SRT. The calibration method that was proposed in this chapter was rather qualitative than quantitative, selecting calibration parameters on the basis of processes knowledge, rather than on sensitivity analysis. The stepwise calibration method provided an orderly and relative quick calibration, minimising iteration loops. Only three model parameters were changed, each parameter calibrating a specific balance in the model. Calculation of the COD and N balance proved as useful final check on the calibrated model.

PART 2: Modelling start-up of WWTP Hardenberg

5.7 Introduction

In Part 1 and studies by van Veldhuizen *et al.*, 1999 and Brdjanovic *et al.*, 2000, the metabolic model was applied to simulate a full-scale WWTP under pseudo steady state conditions. Such conditions imply that the process is determined by recurring 24-hour influent dynamics, whereas the activated sludge composition remains unchanged. Under such conditions the model kinetics are insensitive. Kinetics however, becomes sensitive when a shift of the microbiological population is induced. This would be the case under start-up conditions. To evaluate the model kinetics, the start-up of a full-scale UCT-system was evaluated in a simulation study.

5.8 Materials and methods

5.8.1 The start-up procedure

WWTP Hardenberg, managed by the Dutch water board 'Groot Salland', is one of seven WWTP's upgraded according to the BCFS®-design. BCFS® is a MUCT-type design, optimised for denitrifying EBPR. The process scheme of the original plant, WWTP Hardenberg-Old (H-Old), and the upgraded process are shown in Figure 5.1. The old WWTP was originally designed for COD removal and nitrification, however some denitrification and EBPR were measured. In 1998, WWTP H-Old was upgraded by introducing three additional process units, R1, S and R2, with new process flows Q_A, Q_B and Q_C. During the reconstruction, WWTP H-Old remained in operation. The existing carrousel reactors R3 and R4, and the clarifier units (CL) were integrated in the newly built configuration. The start-up was initiated instantly by redirection of the process flows. A start-up is a transition from an initial to a final pseudo steady state. For the start-up of WWTP Hardenberg, the initial state was the pseudo steady state of WWTP H-Old (t=0, the 4th of February 1998). The final state was measured after 140 days, corresponding with the pseudo steady state of WWTP Hardenberg recorded on the 23rd to the 25th of June 1998. The inoculum for the start-up was activated sludge from WWTP H-Old. At that time, R1, S and R2 were filled with effluent.

5.8.2 Recording the original WWTP

The pseudo steady state of the original WWTP (H-Old) was recorded during 60-hours (the 20th to the 22nd of January 1998). Every two hours, samples were collected from the influent, contact tank (S), aeration tank (R4), WAS, thickener overflow and effluent. The samples were analysed on MLSS, TCOD, COD_{MF}, BOD_5, VFA, TKN, NH_4^+, NO_3^-, TP and PO_4^{3-}. Pump flow rates (Table 5.8), dissolved oxygen (DO) and the oxidation reduction potential (ORP) and energy use and operation time of mixers and aerators were recorded every two hours. The average measurement results are presented in Table 5.9.

Table 5.8 WWTP H-Old operational and hydraulic data. Average flows are presented with standard deviation (s_D). Hydraulic retention time (HRT) was calculated.

Flow	Average ± s_D m^3/d	Process unit	Volume m^3	Depth m	HRT h
Influent, Q_{IN}	5,761 ± 2,788	Selector reactor, S	40	2.5	0.06
Overflow thickener, Q_{OF}	97	Aerated carousel, R3	4,190	2.5	11.7
Effluent, Q_{EF}	5,313 ± 2,817	Aerated carousel, R4	4,190	2.5	11.7
Waste activated sludge, Q_W	452 ± 331	Clarifiers, CL	2,625	-	3.7
Return activated sludge, Q_{RAS}	11,362 ± 3,487	RAS volume	1,000	-	2.1
Internal recycle R4, $Q_{R4,IN}$	635,040				
Internal recycle R3, $Q_{R3,IN}$	635,040	Total WWTP	11,045	-	46

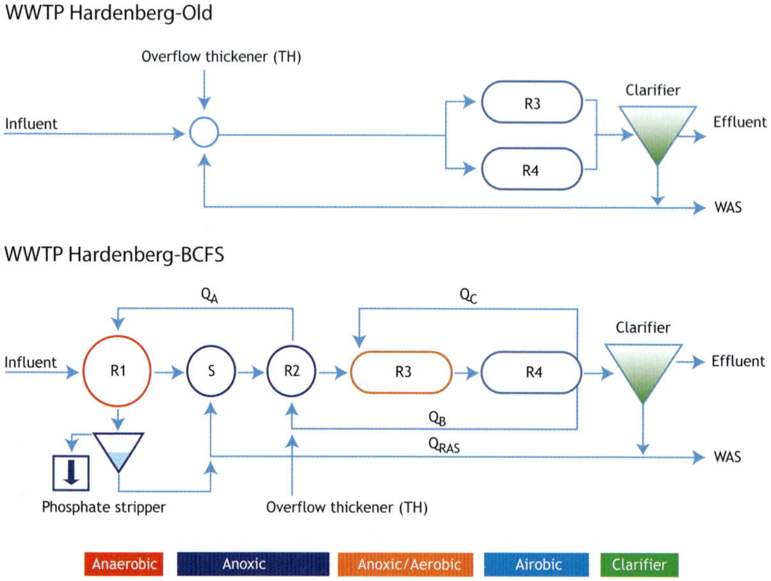

Figure 5.10 Basic plant scheme of the old and new (upgraded) WWTP Hardenberg. Carrousels R3 and R4 were integrated in the new process. R1, S, R2 and the thickener (not shown), were constructed as one circular reactor (van Loosdrecht *et al.*, 1998). R1 was a plug flow reactor. S, R2 and R3 and R4 were completely mixed. The phosphate stripper was not applied. R2 received an internal loading from the overflow of the thickener (Q_{OF}). Q_A, Q_B and Q_C were ORP set-point controlled (van Loosdrecht *et al.*, 1998).

Table 5.9 Measurement results of WWTP H-Old. Average measurements ± standard deviation (s_D) and simulation results (blue). The measurements were recorded on the 20th to the 22nd of January 1998.

Flow		$FePO_4$ X_{MeP} g/m^3	Fe(OH) X_{MeOH} g/m^3	DO S_O gO_2/m^3	NO_3^- S_{NO3} gN/m^3	NH_4^+ S_{NH4} gN/m^3	TKN $S_{NH4}+i_N$ gP/m^3	PO_4^{3-} S_{PO4} gP/m^3	TP $S_{PO4}+X_{PP}+i_P$ $gCOD/m^3$	- S_F $gCOD/m^3$	VFA S_A $gCOD/m^3$	COD_{MF} S_T $gCOD/m^3$	COD_X X_T+S_T $gCOD/m^3$	TCOD X_T+S_T $gCOD/m^3$
Q_{IN}	measured	-	-	0	0.1 ± 0.0	-	67 ± 7	-	7.7 ± 1.0	137 ± 26	69 ± 9	239 ± 30	388 ± 67	627 ± 77
	simulated	0	0	0	0,0	51	67	4	7.8	137	69	240	388	628
Q_{OF}	measured	-	-	0	0.1 ± 0.0	57 ± 5	93 ± 14	-	30 ± 8	295	295	623 ± 52	1126 ± 327	1749 ± 344
	simulated	0	0	0	0	51	94	18	30	294	295	623	1126	1749
R2	measured	-	-	-	17 ± 4	-	-	1 ± 1	-	21 ± 22	12 ± 13	66 ± 27	4416 ±797	4481 ± 777
	simulated	157	119	0	14	19	188	1.6	86	48	24	105	4070	4175
R4	measured	-	-	3.7 ± 0.3	28 ± 3	0.2 ± 0.0	-	0.6 ± 0.0	-	-	-	34 ± 2	4113 ± 244	4147 ± 243
	simulated	160	125	3.7	27	0.7	170	0.6	86	0	0	34	4057	4090
Q_{EF}	measured	-	-	-	28 ± 3	0.01 ± 0.04	1.6 ± 0.3	0.7 ± 0.1	0.8 ± 0.1	-	-	36 ± 5	3 ± 4	39 ± 4
	simulated	0.2	0.1	3.1	28	0.8	1.4	0.7	0.8	0	0	34	6	39
Q_W	measured	-	-	-	25 ± 2	0.3 ± 0.0	-	-	130 ± 15	-	-	34 ± 2	6272 ± 642	6306 ± 642
	simulated	247	191	0	24	1	259	0.7	131	0	0	33	6213	6247

5.8.3 Measuring the start-up

During the 140-day start-up (the 4th of February to the 22nd of June 1998), influent, R1, R2, R3 and R4, effluent, WAS and thickener were sampled. Samples were analysed on MLSS, TP, phosphate, ammonium and nitrate During the first 60 days, phosphate, ammonium and nitrate in R2 ($NH_{4,R2}$, $NO_{3,R2}$ and $PO_{4,R2}$) were measured on-line. Every two hours, process control measurements (DO and ORP) and energy use and operation time of mixers and aeration and were recorded. Growth of PAO was estimated regularly from phosphate release batch-tests (Brdjanovic *et al.*, 1997).

5.8.4 Models

For all simulations SIMBA© 3.3+ was used as simulation platform, operating under Matlab/Simulink© 5.2 (Alex *et al.*, 1997). Biological conversions were calculated with the TUDP model (Annex 1.1). The influent was characterised according to Roeleveld and van Loosdrecht *et al.* (2001). For the simulation of WWTP H-Old, the measurements from Table 5.9 were used. For the start-up simulation, the influent compositions measured in January and June were averaged.

5.8.4.1 Model of the old WWTP

The old WWTP was modelled according to the flow scheme in Figure 5.10. The contact tank was modelled as one CSTR. R3 and R4 were modelled as 6 CSTRs in series with internal recycle flows $Q_{R3,IN}$ and $Q_{R4,IN}$ (Table 5.8). DO_{R3} and DO_{R4} were both controlled on 3.7 gO_2/m^3 in the third CSTR (Table 5.9), hereby approximating the longitudinal DO gradients in the aeration tanks (not shown). The solids concentration in R4 (X_{R4}) was controlled on 4,113 $gCOD/m^3$ by regulating the waste flow (Q_W, Table 5.9).

In the model, particulate material was separated from the effluent ideally. The residual solids concentration in the effluent (X_{EF}), was modelled as a percentile of the RAS concentration. Biological conversions in the RAS were modelled with a non-aerated CSTR placed in the RAS flow. Chemical P precipitation was modelled according to Henze *et al.* (1999). Dosage of $FeCl_3$ was controlled on the average $PO_{4,EF}$ of 0.8 gP/m^3 (Table 5.9). During the start-up, no chemical precipitation was applied. This resulted in a wash-out of initially present $FePO_4$ and $Fe(OH)_3$.

5.8.4.2 Model of the new WWTP

The BCFS®-process was modelled as described in Part 1, according to the flow scheme in Figure 5.10. The actual flow rates Q_A, Q_B and Q_C were modelled, as they largely influenced the first 50 days of the start-up (Figure 5.11). During dry weather conditions, Q_{RAS} was controlled at 8,552 m^3/d. During rainy weather conditions (Q_{IN} > 8,552 m^3/d), Q_{RAS} was controlled proportionally to the influent flow (Q_{IN}). This operation resulted in a more or less constant sludge concentration in the process (Figure 5.12). Measurements of Q_W were not used as these were inaccurate (Part 1). Instead, the solids concentration in R4 (X_{R4}) was controlled on the actual concentration by regulating Q_W (Figure 5.12). R3, R4 and CL were initialised according to the simulation of the old WWTP (Table 5.11). R1, S and R2 were initialised with all concentration set to zero.

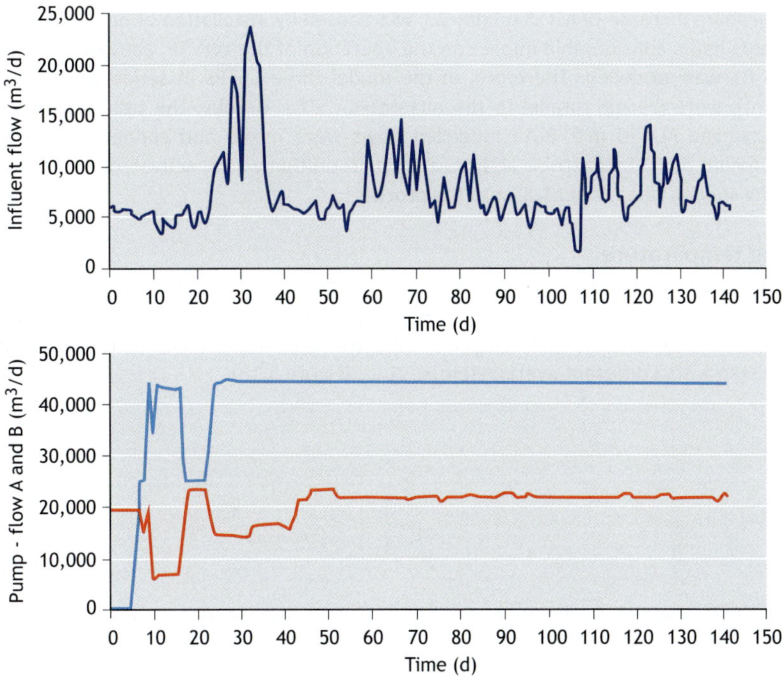

Figure 5.11 Measurement of Q_{IN}, Q_A and Q_B during the start-up. The average dry weather flow was 6,855 m^3/d. During the first 50 days, Q_A (red line) and Q_B (blue line) were operated manually, after which the flows were kept constant. During the start-up, some rain-events occurred.

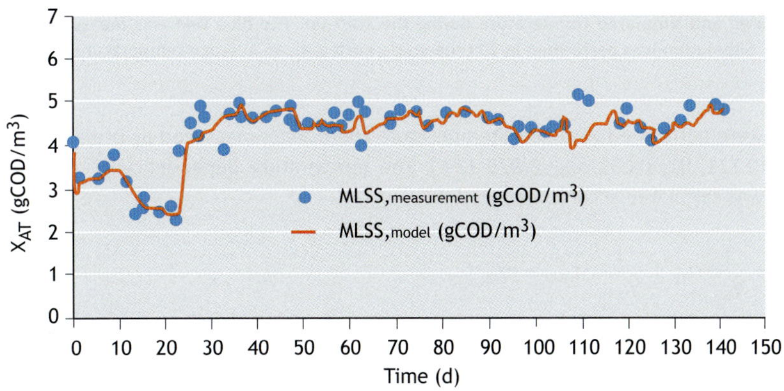

Figure 5.12 Measurement and simulation of X_{R4} during the start-up. Dots were measured, the line represent simulation. In the model MLSS was controlled on the measured concentration (dots). The drop of concentration during days 10 to 25, was caused by anaerobic sludge retention.

5.8.5 Solids retention in the anaerobic reactor

From day 1 to 23, insufficient mixing occurred in R1, which facilitated settling of solids in the 5 meter deep and 2 meter wide plug-flow reactor. During this period, X_{AT} decreased significantly

(Figure 5.12). The sharp increase of MLSS on day 23, was caused by installation of new mixers in R1. As these events had a considerable impact on the operation of the WWTP, activated sludge accumulation in R1 was modelled. Therefore, in the model three CSTRs in series with solids retention (740 m^3), were placed parallel to the anaerobic CSTRs. Hereby, the total anaerobic volume was unchanged (1,470 m^3). Both modelled flows were mixed and connected to the selector. On simulation-day 23, anaerobic activated sludge retention was deactivated. Hence, we could simulate the sudden release of MLSS into the process.

5.8.6 Modelling temperature

During the start-up, temperature increased significantly. The model showed highly sensitive towards temperature, and therefore it was modelled (Figure 5.14). SIMBA© 3.3+ however, did not allow variable temperature simulations. Therefore, the simulation was performed in 10 time-steps, each step with a different average temperature (Figure 5.13).

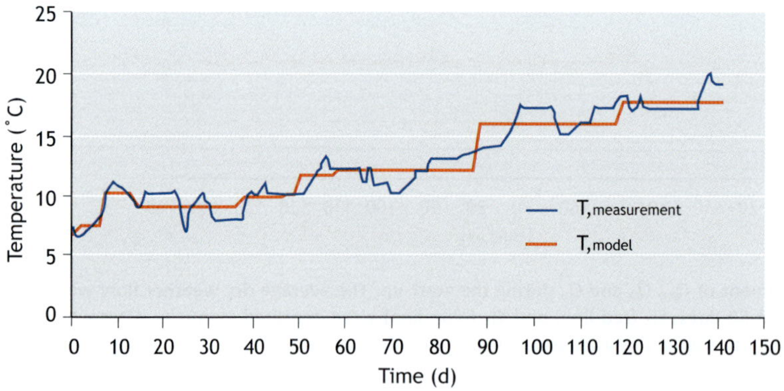

Figure 5.13 Measured and simulated temperature during the start-up. The blue line was measured, the red line simulated. Simulation was performed in 10 time-steps, each with an average temperature.

The simulations were performed with EBPR temperature coefficients measured by Brdjanovic *et al.* (1998) (θ_{PHA}=0.121, θ_{PP}=0.031, θ_{O2}=0.079 1/°C). The temperature dependency of glycogen formation (θ_{GLY}) was calculated according to Eq. 5.9.

$$r_O = \frac{-1}{Y_{PHA/O}} \cdot r_{PHB} + \frac{1}{Y_{PP/O}} \cdot r_{PP} + \frac{1}{Y_{GLY/O}} \cdot r_{GLY} + m_O \cdot C_X \qquad (5.9)$$

Equation 5.9 can be rewritten into Eq. 5.10.

$$r_{GLY}(T) = \left[Y_{GLY/O} \times \left(r_O + \frac{-1}{Y_{PP/O}} \cdot r_{PP} + \frac{1}{Y_{PHA/O}} \cdot r_{PHA} + m_O \cdot C_X \right) \right]_{(T=20)} \times \theta_{GLY}^{(T-20)} \qquad (5.10)$$

In Eq. 5.10, the observed rates r_O, r_{PP} and r_{PHA} were measured as a function of temperature by Brdjanovic *et al.* (1998). θ_{GLY} could be fitted on the basis of Eq. 5.10. Hereby, the contribution of maintenance was neglected. This resulted in a temperature coefficient for glycogen formation of 0.118 1/°C. Figure 5.14 shows the EBPR temperature sensitivity in relation to ASM2d.

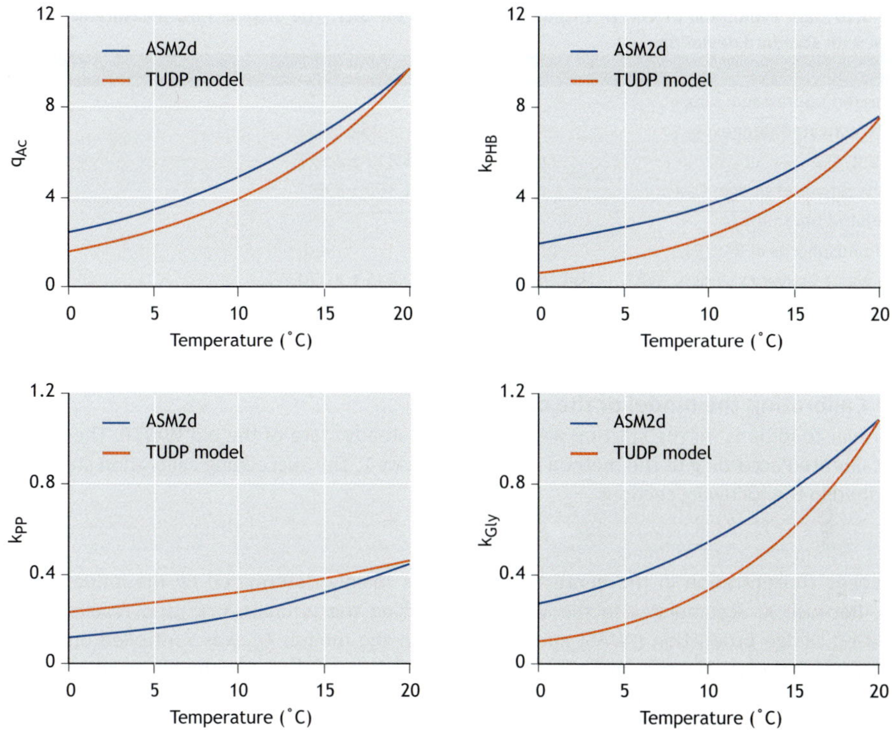

Figure 5.14 EBPR temperature dependencies. The dependency of the TUDP model is shown by blue lines, of ASM2d by red. On the x-axes temperature is plotted, on the y-axes the maximum rate; (a) Acetate uptake (θ_{Ac}=0.090 1/°C), (b) PHA degradation (θ_{PHA}=0.121 1/°C), (c) poly-P formation (θ_{PP}=0.031 1/°C) and (d) glycogen formation (θ_{GLY}=0.118 1/°C). The ASM2d temperature coefficient was 0.069 1/°C (Henze *et al.*, 1999).

5.9 Model calibration and simulation

5.9.1 Data evaluation

Measurements are often inaccurate. Therefore mass balances often do not fit completely. In models however, all flows and compounds are balanced. Fitting a model on erroneous data will lead to laborious and unjustified model calibrations. It is therefore essential to check prime data on errors and consistency, before using it in simulations. Hereby especially the SRT is a sensitive operational parameter. For WWTP H-Old, the SRT was evaluated using the balancing method in Chapter 3. On the basis of data from Tables 5.8 and 5.9, the overall flow and TP balance, and TP and solids balance over the clarifier were calculated (Figure 5.12). In the system of four balance equations (*n*=4), TP_{AT} was unknown (*u*=1). Therefore the degree of redundancy was 3 (*n-u*). We chose to evaluate Q_W, Q_{EF} and Q_{RAS}. From the calculations it was shown that all flows could be estimated within their standard deviation (s_D). It was therefore concluded that no gross errors were present. The result of the mass-balance calculations is presented in Table 5.10. It is shown that the SRT calculated from the measured flows is considerably smaller than calculated after balancing.

Table 5.10 Data evaluation of WWTP H-Old: estimation of the SRT. The original flow measurements are shown with standard deviations (s_D).

	Unit	Avg. ± s_D	Balanced
Estimated operational data			
Waste activated sludge, Q_W	m^3/d	452 ± 331	330
Effluent, Q_{EF}	m^3/d	5,313 ± 2,817	5,530
Return activated sludge, Q_{RAS}	m^3/d	11,362 ± 3,487	10,004
Calculated parameters			
Total phosphorus in R4, TP_{AT}	gP/m^3	n.d.	85.5
Load waste activated sludge, $^L WAS$	$kgCOD_X/d$	2,835 ± 2,370	2,050
Sludge retention time, SRT	d	12.3 ± 11	16.7

5.9.2 Calibrating the model of the old WWTP

The initial conditions for the start-up was the pseudo steady state of the old WWTP. The model was calibrated according to the method described in Part 1. The succeeding calibration steps are described in the following sections.

Step 1. Fitting the solids balance

The solids concentration in the aeration tank (X_{AT}) is mainly determined by the influent X_I/X ratio. Because X_I accumulates in the WWTP, increasing the influent X_I/X ratio results in an increased sludge production ($^L WAS$) and vice versa. In the model, X_{AT} was controlled on 4,113 $gCOD_X/m^3$ (Table 5.11) by regulating Q_W. To fit Q_W and $^L WAS$ according to the balanced values in Table 5.10, the influent X_I/X ratio was adjusted to 0.507. An identical value for X_I/X was found for WWTP Hardenberg (Part 1).

Step 2. Calibrating nitrification

Nitrification was simulated sufficiently without calibration (Table 5.9).

Step 3. Calibrating denitrification

A difference in nitrate measured in the effluent and the RAS flow ($NO_{3,EF}$ and $NO_{3,RAS}$, Table 5.9), indicated that 3 gN/m^3 was denitrified in the RAS volume. This effect was modelled with three CSTRs placed in the RAS flow. By approximation this corresponded with the actual RAS volume (1,050 m^3). Denitrification ($^L DEN$) was calibrated by increasing K_O from 0.2 to 0.7 gO_2/m^3 (Table 5.9), which matched the calibrated K_O for WWTP Hardenberg (Part 1).

Step 4. Calibrating EBPR

The calibrated model of WWTP H-Old calculated absence of PAO. However an anaerobic phosphate release batch-test, showed some PAO activity. From the maximum phosphate release rate (0.036 gP/gCOD.d), q_{Ac} and Y_{PO4}, the initial PAO concentration was calculated to be 33 $gCOD/m^3$. Clearly, this low concentration could be neglected when simulating the old WWTP. However, for the start-up simulation, the initial PAO concentration showed to be sensitive. Therefore, also the initial values for f_{PP}, f_{GLY} and f_{PHA} had to be estimated. From a phosphate release batch-test simulation, the aerobic f_{PP} was estimated to be 0.16 $gP/gCOD_{PAO}$, which matched the value found for WWTP Hardenberg (0.17 $gP/gCOD_{PAO}$, Part 1). In the batch test simulation, the aerobic fractions f_{GLY} and f_{PHA} were insensitive. These fractions were therefore directly obtained from the simulation of the WWTP (0.38 $gCOD_{GLY}/gCOD_{PAO}$ and 0.02 $gCOD_{PHA}/gCOD_{PAO}$). The simulation results of WWTP H-Old are presented in Table 5.11.

Table 5.11 Simulated activated sludge compositions. The pseudo steady state (PSS) of the old WWTP was the initial state of the start-up. The WWTP was simulated without chemical P precipitation and f_{GLYmax}=0.4.

Component	Symbol	Unit	PSS of H-Old	Final state start-up	PSS of Hdbg
Inert particulate	X_I	$gCOD/m^3$	2,828	3,620	3,550
Solid substrate	X_S	$gCOD/m^3$	63	10	7
Heterotrophic microorganisms	X_H	$gCOD/m^3$	1,091	352	294
Autotrophic microorganisms	X_A	$gCOD/m^3$	75	52	43
P accumulating organisms	X_{PAO}	$gCOD/m^3$	32.9[1]	531	572
Poly-hydroxy-alkanoate	X_{PHA}	$gCOD/m^3$	0.7[2]	13	9
Glycogen	X_{GLY}	$gCOD/m^3$	12.5[2]	265	216
Poly-phosphate	X_{PP}	gP/m^3	5.3[1]/5.7[2]	92	98
Iron-hydroxide	X_{MeOH}	g/m^3	125	1	-
Iron-phosphate	X_{MeP}	g/m^3	160	6	-.
Total particulate phosphorus	TP_X	gP/m^3	86	148	152
Particulate COD	COD_X	$gCOD/m^3$	4,090	4,843	4,692
Total Kjeldahl nitrogen	TKN_X	gN/m^3	170	175	171

[1] Estimated from a phosphate release batch-test simulation
[2] Obtained from a simulation of WWTP Hardenberg (Hdbg)

5.9.3 Calibrating the start-up

To simulate the start-up, model calibration was required. Hereby the proposed stepwise method was applied. Whereas this method proved to be quick and simple for pseudo steady state calibrations, for dynamic conditions it was laborious, as for each adjusted parameter a complete dynamic (in our case 140 days) simulation run was required. To decrease simulation time, initial calibration was performed with average flows and controller set-points.

Step 1. Fitting the COD_X balance

Mass balancing dynamic systems can be inaccurate. Therefore the SRT of the dynamic system was evaluated differently. Because P is stored in the activated sludge, there is a direct relation between TP_{AT} and the SRT. The SRT during the start-up was therefore evaluated on the basis of the accumulation of TP_{AT} (Figure 5.15).

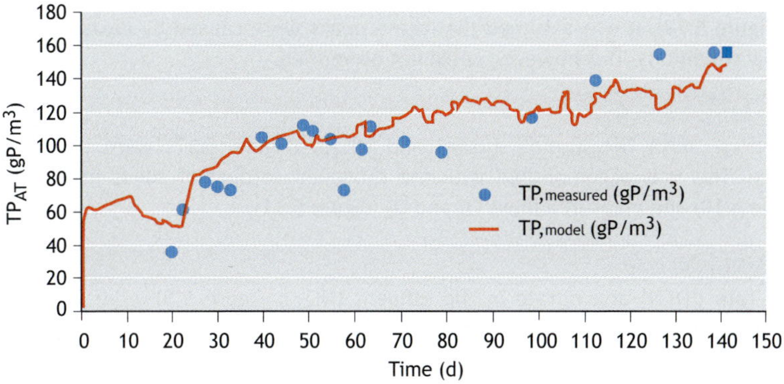

Figure 5.15 Measured and simulated accumulation of TP_{AT}. Dots were measured concentrations, the line is simulation result. The square symbol at 141 days represents the PSS of WWTP Hardenberg.

Hereby, the influent X_I/X ratio was set to 0.507, according to the previous simulations. Accumulation of TP_{AT} was predicted sufficiently as was observed from Figure 5.15. This indicated that the SRT was simulated correctly. With the actual Q_A (Figure 5.11) and modelled settling in R1, also X_{R1} was simulated satisfactory (not shown).

Step 2a. Calibrating effluent ammonium

In the model nitrification was mainly determined by DO_{R3} and DO_{R4}. DO_{R4} was PI-controlled based on the measured DO profile (Figure 5.16).

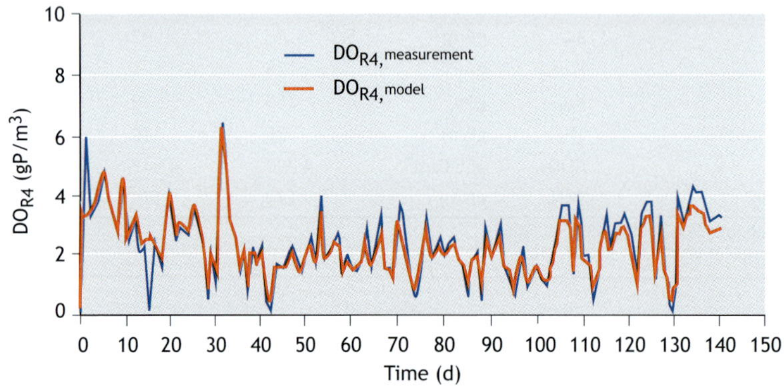

Figure 5.16 Measured and simulated DO_{R4} control. The red line is simulation result, the blue line measured. DO_{R2} was measured, however not plotted, as the concentration continuously was below 0.5 gO_2/m^3.

DO_{R3} was regularly below the measurement detection and therefore less accurate. Moreover, $NH_{4,EF}$ showed sensitive towards DO_{R3}, as it was in the range of the DO saturation constant ($K_{A,O}$=0.5 gO_2/m^3). Therefore we chose to fit $NH_{4,EF}$ by calibrating DO_{R3}. Between days 10 and 23, anaerobic digestion of settled X_{R1}, caused a high ammonium loading of the WWTP. During this period, nitrification was calibrated by increasing DO_{R3} to 2 gO_2/m^3. From day 23 and on, with ending of anaerobic activated sludge retention, $NH_{4,EF}$ was simulated sufficiently with actual DO_{R3}. $NH_{4,EF}$ peaks on days 60, 85, 110 and 135, were not predicted. Because these peaks did not also occur in R2 (Figure 5.17), it was assumed that these peaks were caused by measurement errors or operational conditions. This however, could not be verified.

Step 2b. Calibrating $NH_{4,R1}$ and $NH_{4,R2}$

After calibrating $NH_{4,EF}$, both $NH_{4,R1}$ (not shown) and $NH_{4,R2}$ (Figure 5.17) were predicted sufficiently accurate. This was accomplished with the modelled settling of solids in R1, the balanced Q_{RAS} (Table 5.10) and measured flows Q_A and Q_B (Figure 5.11).

Step 3. Fitting the denitrified load

The denitrification rate (^{L}DEN) and nitrate in the effluent ($NO_{3,EF}$ Figure 5.9) were fitted by increasing K_O according to the method in Chapter 3. The same value was found as in the previous simulations (0.7 gO_2/m^3). Denitrification in the RAS volume could not be verified, as $NO_{3,RAS}$ was not measured regularly.

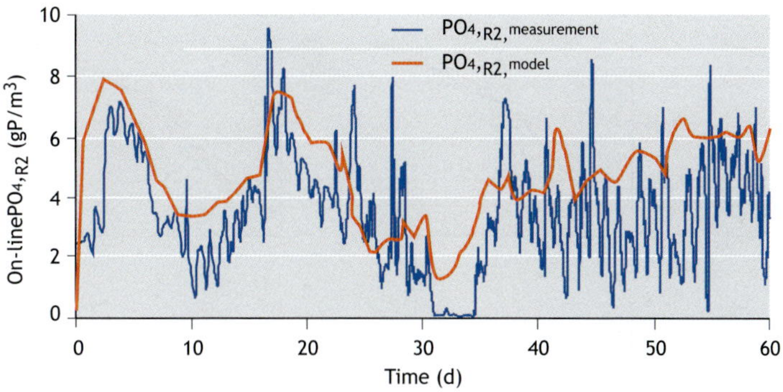

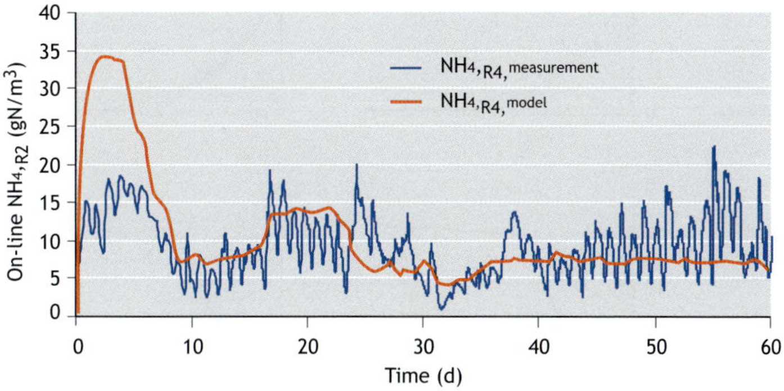

Figure 5.17 Measured and simulated $PO_{4,R2}$ and $NH_{4,R2}$ during the first 60-days. Red line represents simulation results, thin blue on-line measurements. $NO_{3,R2}$ was measured, however not plotted, as the concentration continuously was below 0.5 gN/m^3.

5.10 Evaluation of the TUDP model

5.10.1 Sensitivity analysis

On the basis of a dynamic sensitivity analysis the EBPR kinetics was evaluated. The sensitivity of k_{GLY}, k_{PP} and k_{PHA} was analysed, as these kinetic parameters are directly related to growth of PAO (Figure 5.18 and 5.19). In total seven simulations were performed; the default parameter value (k_{GLY}=0.93, k_{PP}=0.45, k_{PHA}=7.55), a 20% decrease and a 20% increase of the individual rate values. Hereby, all model distortions were eliminated by applying average flows and controller set-points. The applied temperature was 10°C, the total simulation time 100 days. In the simulations, only PAO, phosphate in the effluent ($PO_{4,EF}$), the anaerobic phosphate concentration ($PO_{4,R1}$) and the anaerobic fractions f_{PP} and f_{GLY} and the aerobic fraction f_{PHA} showed sensitive towards the kinetic rates. This is shown in Figure 5.18 (PAO_{AT} a,b,c; $PO_{4,EF}$ d,e,f; $PO_{4,R1}$ g,h,i), and Figure 5.19 (f_{PP} j,k,l; f_{GLY} m,n,o; f_{PHA} p,q,r).

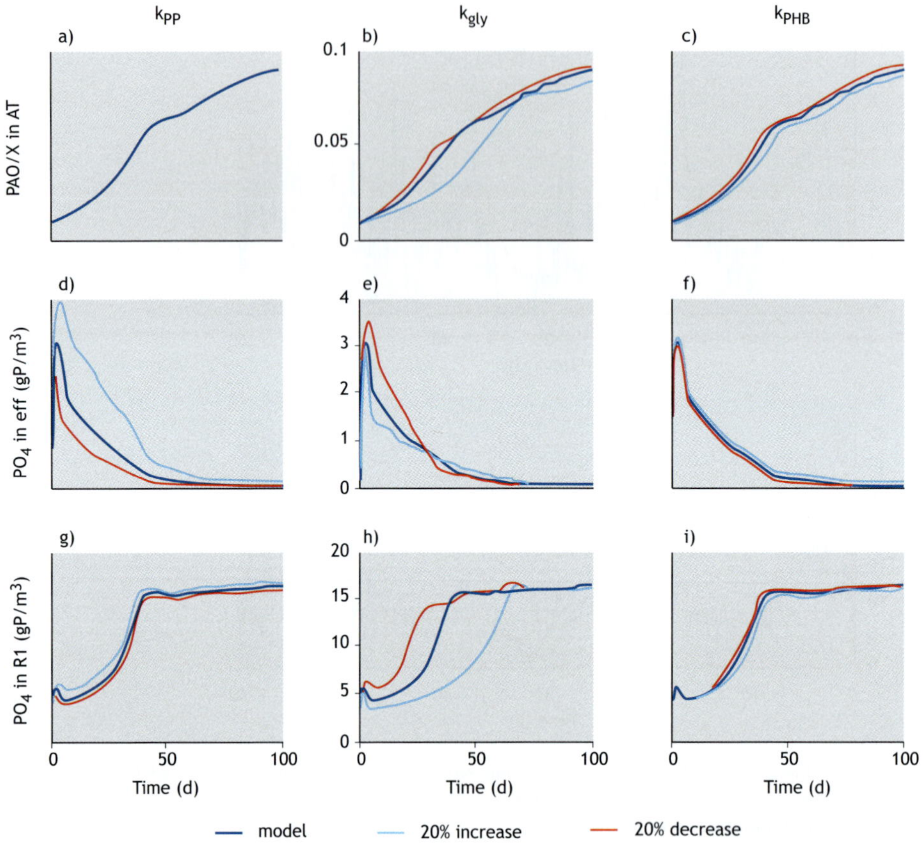

Figure 5.18a-i Sensitivity of EBPR kinetics. Each figure represents the effect of a 20% in- (light blue lines) and decrease (red lines) of kinetic rates k_{GLY}, k_{PP} and k_{PHA} on PAO (a to c), $PO_{4,EF}$ (d to f) and $PO_{4,R1}$ (g to i). Dark blue lines are simulations with default parameter values. The sensitivity analysis was performed with average flows and controller set-points and a temperature of 10°C.

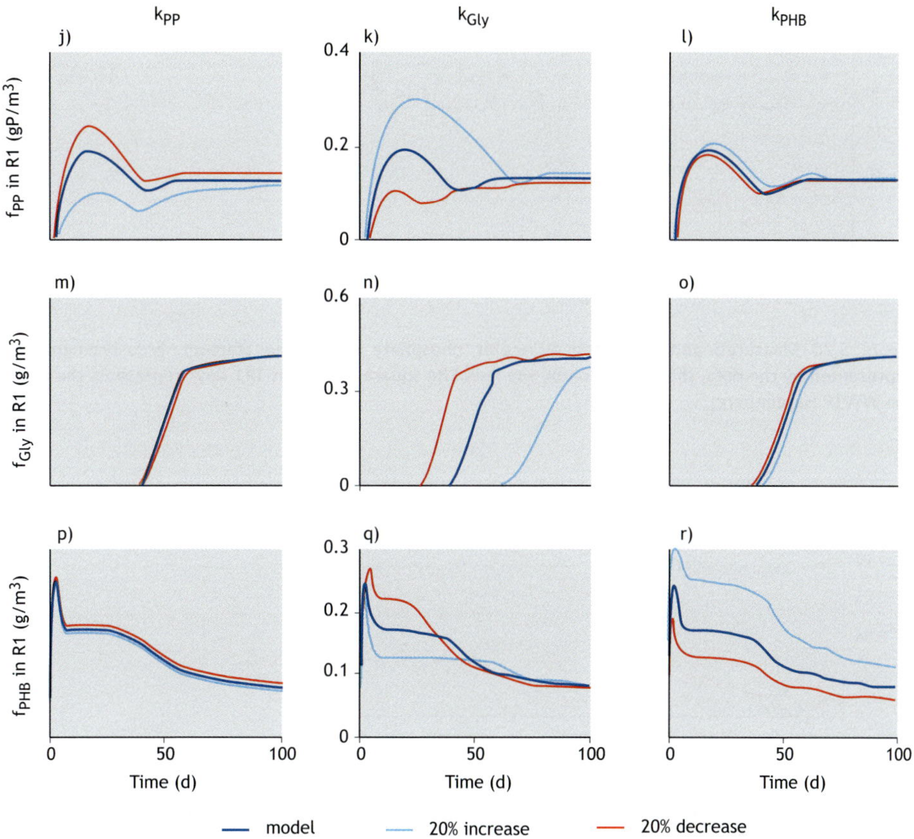

Figure 5.19j-r Sensitivity of EBPR kinetics. Each figure represents the effect of a 20 % in- (light blue lines) and decrease (red lines) of kinetic rates k_{GLY}, k_{PP} and k_{PHA} on f_{PP} (j to l), f_{GLY} (m to o) and f_{PHA} (p to r). Dark blue lines are simulations with default parameter values. The sensitivity analysis was performed with average flows and controller set-points and a temperature of 10°C.

5.10.2 Calibrating EBPR

According to the calibration method presented in Meijer (2004) and in detail in Meijer and Brdjanovic (2012), EBPR is calibrated as last. Anaerobic P release and poly-P formation are modelled as separate processes. Independent calibration of these processes was thus possible.

Step 4a. Calibrating anaerobic phosphate release

$PO_{4,R1}$ (Figure 5.20) is via anaerobic phosphate release, related to growth of PAO (Figure 5.21). In the model, phosphate release is coupled to storage of PHA via Y_{PO4} (0.36 gP/gCOD$_{Ac}$, pH 6.8, Smolders *et al.*, 1994a). PHA is partly utilised for glycogen and poly-P formation and maintenance. The remaining PHA is available for growth of PAO. Theoretically, decreasing k_{PP} and k_{GLY} and increasing q_{Ac} should result in more PAO. In practice is poly-P formation often phosphate limited as usually phosphate in the effluent is low. Growth of PAO is insensitive towards increasing k_{PP} (Figure 5.18a and 5.18g). Decreasing k_{PP} however, immediately results in more phosphate in the effluent ($PO_{4,EF}$, Figure 5.18d). In a well-designed WWTP, PHA formation is limited by the availability of VFA_{IN}. Therefore, growth of PAO will be insensitive for increased q_{Ac}, however highly sensitive for decreased q_{Ac}, as this directly affects the storage of PHA.

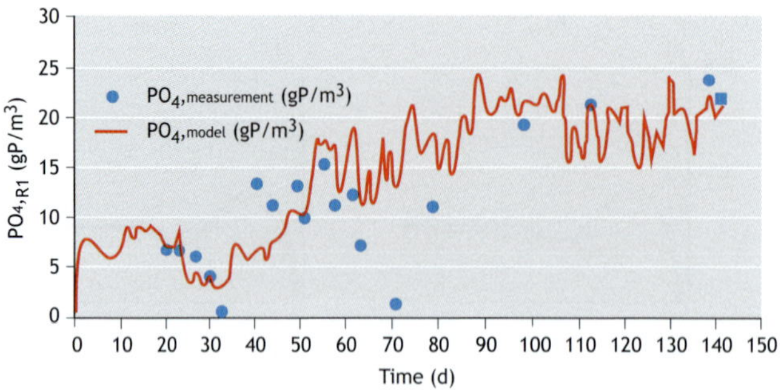

Figure 5.20 Measured and simulated anaerobic phosphate during the start-up. Measurements are represented by the dots, the simulation by the line. The square symbol at 141 days represents the PSS of the WWTP Hardenberg.

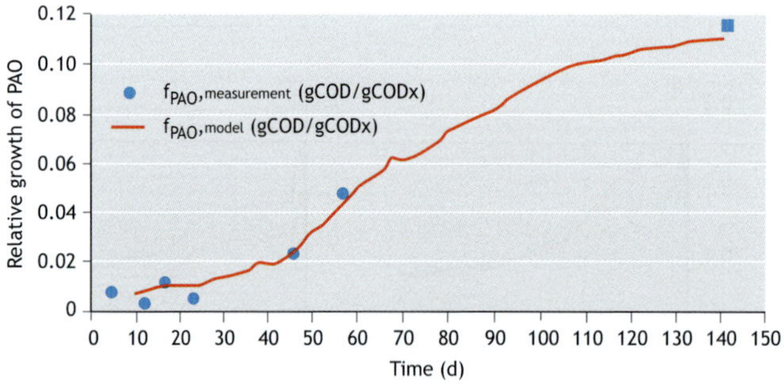

Figure 5.21 Measured and simulated growth of PAO. Data represented by the dots were calculated from anaerobic phosphate release batch-tests, the line was simulated. The square at 141 days represents the PSS of the WWTP.

Where PHA and poly-P formation kinetics is well described (Murnleitner *et al.*, 1997, Smolders *et al.*, 1994a/b), glycogen formation kinetics is less certain (Brdjanovic *et al.*, 2000). The model sensitivity towards k_{GLY} was however relative high compared to k_{PP} and q_{Ac} (Figure 5.18). Therefore, we chose to fit growth of PAO by adjusting k_{GLY}. This was validated on the basis of $PO_{4,R1}$. By decreasing k_{GLY} with 15 % to 0.93 $gCOD_{GLY}/gCOD_{PAO}.d$, acceptable simulation results were obtained (Figure 5.20 and 5.21).

Step 4b. Calibrating anoxic PP formation

To fit anoxic poly-P formation a unique calibration parameter was required. Sensitivity analysis showed that poly-P formation was sensitive towards k_{PP}. This however accounted both anoxic and aerobic poly-P formation. A more specific calibration parameter was the effective saturation constant for nitrate expressed as $g_{PP} \times K_{P,NO3}$ (with $K_{P,NO3}$=0.5 gN/m^3 and reduction factor g_{PP}<1). In the kinetic expression r_{PP}, g_{PP} originates from the observation that PP formation is less sensitive towards DO and nitrate (Murnleitner *et al.*, 1996). During the start-up, $NO_{3,R2}$ was constantly in the range of $K_{P,NO3}$ (0.2 to 0.5 gN/m^3). Hence, g_{PP} showed an increased sensitivity. For simulation of WWTP Hardenberg (Chapter 3), g_{PP} was determined on 0.25 (-). Simulating the start-up with this value resulted in the $PO_{4,R2}$ profile presented in Figure 5.17. This profile was established with actual Q_A and Q_B, balanced Q_{RAS} and calibrated $NO_{3,EF}$ (step 3) and $PO_{4,R1}$ (step 4a).

5.11 Discussion

The primary goal of the start-up, was to obtain a stable process in the shortest time possible. We were therefore limited to record the WWTP as it was operated. This included calamities, as the start-up was a single event. The scientific goal of the start-up simulation was to evaluate the TUDP model kinetics. Hereby, our focus was on long term population dynamics. Therefore sporadic grab sampling was sufficient. For (full-scale) evaluation of short term TUDP model kinetics, we refer to Part 1 and previous studies by van Veldhuizen *et al.* (1999) and Brdjanovic *et al.* (2000), Chapter 2 and 3, respectively.

On the basis of the start-up, the sensitivity of EBPR kinetics was evaluated. A similar start-up evaluation was performed by Smolders *et al.* (1995) at lab-scale conditions. From short-cycle SBR experiments, Murnleitner *et al.* (1996) observed that k_{PHA} and k_{PP} were the most sensitive kinetic parameters for EBPR. This study shows that during a start-up, k_{GLY} is the most sensitive parameter for growth of PAO (Figure 5.18b). The modelled glycogen kinetics showed typical on-off dynamics (Figure 5.19m,n,o). This indicated that glycogen formation is over sensitive (i.e. critical) in the model.

To validate the kinetics, accurate measured data of glycogen, poly-P and PHA under dynamic conditions are required. In this study these measurements were not performed for two reasons. Firstly, it was expected that high recycle factors in the WWTP would result in small glycogen and PHA gradients. Distinction between anaerobic, anoxic and aerobic glycogen and PHA concentrations would therefor become difficult. Secondly, in full-scale WWTP PAO are not the only microbes storing glycogen and PHA. This further affects the measurement accuracy. On the basis thereof, we chose to rely on anaerobic phosphate release batch-tests for the validation of EBPR kinetics.

This study showed that during a start-up, glycogen formation is a key process. More (lab-scale) research however, would be needed for better determination and validation of the glycogen kinetics at start-up conditions. From this chapter, it was concluded that under PSS conditions the sensitivity of EBPR kinetics is strongly reduced. Therefore the current glycogen kinetics is sufficient for simulation of such conditions.

5.11.1 Influent characterisation

From the influent measurements in January and June, it was observed that seasonal influences were negligible and that dynamics in the influent loads primarily were caused by flow dynamics. This is typical for Dutch sewers, which generally are flat, resulting in concentration and flow dispersion. The start-up was therefore simulated with average dry weather influent loads and measured Q_{IN}, including several rain-events (Figure 5.11). In reality, rain events did cause some variation. However, for the purpose of this study, the influent was simulated sufficiently, as we were mainly interested in long term dynamic effects.

5.11.2 Simulation of the old WWTP

The model of WWTP H-Old was considered adequately calibrated, as the simulation was in agreement with the measurements (Tables 5.9 and 5.10). Calibration was done independent of the calibration of WWTP Hardenberg (Part 1). Nevertheless, both models were fitted with identical parameter values for K_O and X_I/X. It should be emphasised that for both models these parameters were selected on the basis of high sensitivity towards the denitrification rate (^{L}DEN) and the sludge production (^{L}WAS).

5.11.3 Modelling chemical P precipitation

During the start-up simulation, chemical precipitation was inactivated. X_{MeP} therefore acted as an inert fraction that was removed from the WWTP via the WAS in approximately 70 days. During this time, loss of $FePO_4$ (X_{MeP}) was compensated by accumulation of PP. The net accumulation of TP_{AT} was well simulated (Figure 5.15). Neglecting chemical precipitation processes during the start-up was therefore justified.

5.11.4 Modelling anaerobic solids retention time

The method that was used to model solids retention in R1 during the first 23 days was critical. The sudden increase of MLSS on day 23 (approximately 2,000 $gCOD/m^3$, Figure 5.12), implicated that 25,780 kg MLSS was accumulated in R1. In the model, we assumed this was a gradual process, taking place in the bottom half of the reactor (740 m^3). In 23 days, X_{R1} accumulated to a maximum of 34.8 $kgCOD/m^3$. This concentration corresponded to the measured MLSS in the thickener (32 $kgCOD/m^3$). During anaerobic sludge retention, X_{R1} was subject to hydrolysis, fermentation, lysis and anaerobic maintenance. This caused an increasingly high phosphate and ammonium loading of the WWTP. Our model of the settling process was supported by $PO_{4,R2}$ and $NH_{4,R2}$ measurements, as these trends were simulated fairly well (Figure 5.17). The simulated effluent however, did not fully correspond with the measurements (Figure 5.22). $PO_{4,EFF}$ and $NH_{4,EF}$ peaks on day 20, were caused by sudden increase of Q_A (Figure 5.11), while $PO_{4,R1}$ and $NH_{4,R1}$ were high. At the same time, nitrification and the phosphate uptake rate were low, caused by the reduced MLSS in the WWTP. In an attempt to simulate this effect (step 2a), DO_{R3} was increased to 2 gO_2/m^3. Because there were no other reasonable options to increase aerobic conversions, we chose not to further calibrate the effluent.

5.11.5 Dynamic evaluation of operational conditions

Unlike PSS conditions, a dynamic WWTP does not give the opportunity for detailed evaluation of operational data. The start-up dynamics were in great extent determined by operational conditions. In previous studies it was shown that calibrating model kinetics without gross error detection is inaccurate. However, most operational problems occurred in the first 25 days, after which a proper kinetic evaluation was possible. It showed possible to accurately evaluate the SRT on the basis of accumulation of TP_{AT} (Figure 5.15). k_{GLY} was calibrated on the basis of $PO_{4,R1}$ and anaerobic phosphate release batch-tests between days 30 and 65 (Figure 5.20 and 5.21). In this period, a maximum sensitivity for k_{GLY} was observed, as growth of PAO was in an 'exponential' phase (Figure 5.18b,h). Decreasing k_{GLY} 15 % to 0.93 ($gCOD_{GLY}/gCOD_{PAO}.d$), resulted in a sufficient simulation of $PO_{4,R1}$ and growth of PAO. The activated sludge composition was relative insensitive towards the calibrated value of k_{GLY} (Table 5.11 and Figure 5.18b,h). This indicated that the WWTP was mainly stoichiometric determined. The start-up was simulated with identical parameter values (X_I/X, K_O and g_{PP}) as for the steady state simulations of the old and new WWTP. This resulted in an adequate simulation of the start-up, as observed in Figures 5.15, 5.17, 5.20, 5.21 and 5.22.

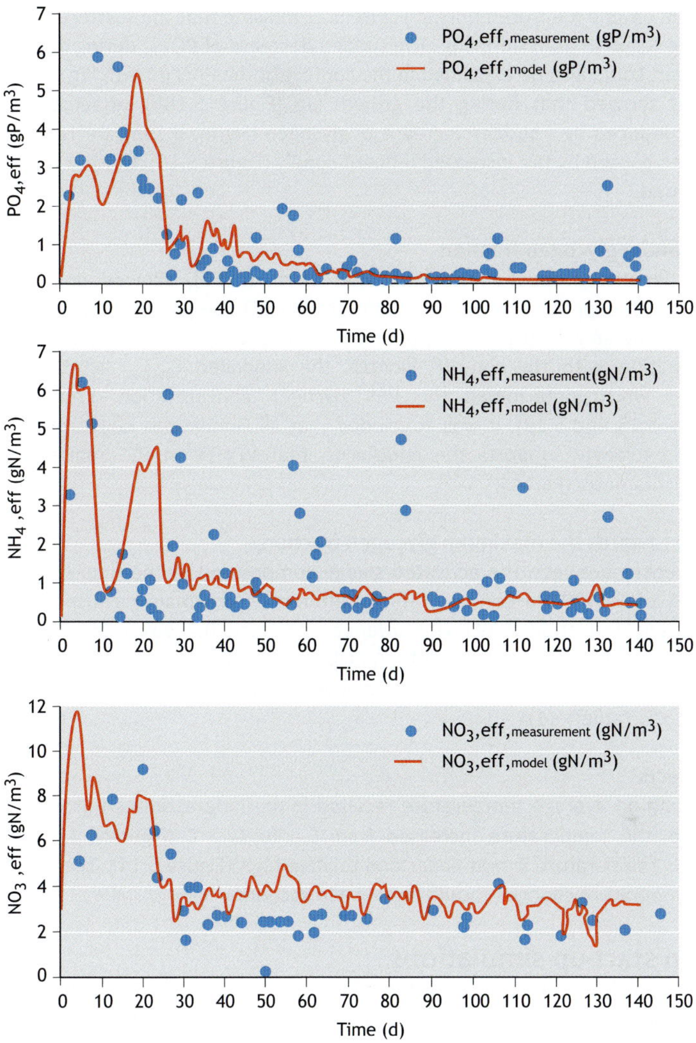

Figure 5.22 Measured and simulated effluent concentrations. $PO_{4,EF}$ (top), $NH_{4,EF}$ (middle) and $NO_{3,EF}$ (bottom) during the start-up. Measurements are presented by the dots, simulation results by the line.

5.11.6 Interpretation of the start-up dynamics

5.11.6.1 Glycogen kinetics

Simulation of PAO showed a distinctive delay on day 60 (Figure 5.22). This also was observed in the sensitivity analysis (Figure 5.18a,b,c). Here the delay coincided with sudden increase of glycogen (Figure 5.19m,n,o). When $PO_{4,EF}$ becomes limiting (Figure 5.18e), the maximum anaerobic phosphate concentration is reached ($PO_{4,R1}$ Figure 5.18h). This causes a sudden increase of f_{GLY} to f_{GLYmax} (Figure 5.19n). It could not be answered if this 'on-off' behaviour is realistic, since accurate glycogen measurements were not available.

The strong dynamics of the on-line measurements in Figure 5.22 indicate that the scattered data in Figure 5.20 could be the result of grab sampling. The sudden decrease of $PO_{4,R1}$ during days 60 to 70 (Figure 5.20) however, could also be explained in the context of the PAO growth lag (Figure 5.21). Sensitivity analysis showed that during the growth lag (Figure 5.18b), anaerobic f_{PP} decreased (Figure 5.19k), induced by a sudden increase of glycogen (Figure 5.19n). Decrease of $PO_{4,R1}$ also could have been caused by the increased influent supply (Figure 5.11). Neither theory however, could be confirmed.

5.11.6.2 Modelling a maximum glycogen fraction

The glycogen formation kinetics proposed by Murnleitner *et al.* (1997), allowed unlimited accumulation of glycogen. This especially caused problems at high SRT (Brdjanovic *et al.*, 2000). Modelling f_{GLYmax} showed to be an effective and simple measure to prevent unrealistic glycogen accumulation. In combination with the 'on-off' kinetics the modelled f_{GLYmax} acted as a stoichiometric parameter which determined the PSS glycogen concentration. From the sensitivity analysis (Figure 5.18 and 5.19), it was concluded that all simulations converge to a similar steady state. This strongly supports the conclusion that WWTP in PSS mainly are determined by model stoichiometry (Part 1).

5.11.6.3 Model sensitivity towards the maximum glycogen fraction

Results show small differences between the activated sludge composition of the final state of the start-up and the PSS of WWTP Hardenberg. These differences are mainly caused by a different value for f_{GLYmax}. For simulation of WWTP Hardenberg the ratio 0.4 was used, whereas the start-up was simulated with a ratio 0.5 ($gCOD_{GLY}/gCOD_{PAO}$, Brdjanovic *et al.*, 1998). The sensitivity of f_{GLYmax} is however low, caused by the relative small glycogen fraction in the activated sludge (4.6 to 5.5%, Table 5.11).

5.11.6.4 Temperature effects

Simulating the start-up with an average temperature resulted in an underestimated growth of PAO. During the start-up, the temperature increased from 6.8 to 17.7°C (Figure 5.13). The sensitivity of PAO towards temperature is non-linear and relative high (Figure 5.14). Therefore, when modelling EBPR, temperature changes should be taken in account.

5.12 Conclusions on start-up simulations

On the basis of a start-up simulation of a full-scale WWTP, it was possible to evaluate the (long term growth) kinetics of the metabolic model. A sensitivity analysis showed that the glycogen formation rate and the temperature were most sensitive in the model. By including a temperature profile in the model, simulation results improved significantly. By carefully fitting the glycogen formation, a proper simulation of the start-up and accumulation of PAO over a 140-day period was possible.

In accordance with previous observations, a high sensitivity towards operational data was observed. This underlines the importance of a proper data evaluation. During the start-up, the sludge retention time could be evaluated accurately on the basis of accumulation of phosphorus in the activated sludge. The start-up could be simulated using identical calibration parameter values for K_O, g_{PP} and X_I/X as for previous simulations.

The glycogen formation rate was found to be sensitive in the model. A maximum glycogen fraction was introduced in the kinetic rate equation for glycogen formation to avoided model instability. With this adjustment, problems relating to unrealistic accumulation of glycogen in the model, as reported in several previous studies, were solved. It was observed that when glycogen formation was not calibrated properly, the value of the maximum glycogen fraction determined the steady state glycogen concentration in the model. The overall model sensitivity of towards this parameter was however low.

This study shows that glycogen formation is a key process during start-up conditions. In the model glycogen formation is sensitive. Therefore a more accurate determination and validation of glycogen formation under start-up conditions is required. For most WWTP simulations however, the current EBPR kinetics are sufficient, as this and previous research showed that the model sensitivity towards EBPR kinetics is low under (pseudo) steady state conditions.

References

Alex J., Tschepetzki R. and Bitter U. (1997) SIMBA Simulation of biological waste water treatment. User's manual, Institute of Automation and Communication e. V., Magdeburg.

Alex J., Tschepetzki R., Jumar U., Obenaus F. and Rosenwinkel K. H. (1999) Analysis and design of suitable model structures for activated sludge tanks with circulating flow. *Water Sci. Technol.* 39(4), 55-60.

Baeza J., Gabriel D. and Lafuente J. (1999) Expert supervisory system for a pilot WWTP. *Environmental Modelling and Software* 14(5), 383-390.

Belanche L., Valdes J. J., Comas J., Roda I. R. and Poch M. (1999), Towards a model of input-output behaviour of wastewater treatment plants using soft computing techniques. *Environmental Modelling and Software* 14(5), 409-419.

Brandse F. A. and de Vries A. C. (1998) Werkwijze voor het zuiveren van afvalwater. Dutch patent, 1003711, 17p.

Brandse F. A. personal communication (2001)

Brdjanovic D., van Loosdrecht M. C. M., Hooijmans C. M., Alaerts G. J. and Heijnen J. J. (1997) Temperature effects on physiology of biological phosphorus removal. ASCE *Journal of Env. Eng.* 123(2), 144-154.

Brdjanovic D., Logemann S., van Loosdrecht M. C. M., Hooijmans C. M., Alaerts G. J. and Heijnen J. J. (1998) Influence of temperature on biological phosphorus removal: process and ecological studies. *Water Res.* 32(4), 1035-1048.

Brdjanovic D., van Loosdrecht M. C. M., Versteeg P., Hooijmans C. M., Alaerts G. J. and Heijnen J. J. (2000) Modeling COD, N and P removal in a full-scale WWTP Haarlem Waarderpolder. *Water Res.* 34(3), 846-858.

Char B.W., Geddes K.O., Gonnet G.H., Leong B.L., Monagan M.B. and Watt S.M. (1991) Maple V Language reference manual. Springer Verlag, ISBN 0-387-94124-X, 267p.

Ekama G. A., Marais G. v. R., Siebritz I. P. (1984) Biological excess phosphorus removal. Theory, design and operation of nutrient removal activated sludge processes, Water Research Commission, Pretoria, South Africa, 7, 1-32.

Garcia-Sanz M., Ostolaza J. X. (2000) QFT-control of a biological reactor for simultaneous ammonia and nitrate removal. *System analysis Modelling Simulation* 38(2), 353-370.

Goel R., Mino T., Satoh H. and Matsuo T. (1998) Comparison of hydrolytic enzyme systems in pure culture and activated sludge under different electron acceptor conditions. *Water Sci. Technol.* 37(4-5), 335-343.

Goel R., Mino T., Satoh H. and Matsuo T. (1999) Modeling hydrolysis processes considering intracellular storage. *Water Sci. Technol.* 39(1), 97-105.

Gujer W., Henze M., Mino T. and van Loosdrecht M. (1999) Activated Sludge Model No. 3. *Water Sci. Technol.* 39(1), 183-193.

Hao X., van Loosdrecht M. C. M., Meijer S. C. F. and Qian Y. (2001) Model Based Evaluation of two BNR processes – UCT and A_2N. *Wat. Res.* 35(12), 2851-2860.

Harnett D. L. (1982) Statistical methods. Addison-Wesley Publishing Company, Reading Massachusetts.

Heduit A. and Thevenot D. R. (1992) Elements in the interpretation of platinum electrode potentials in biological treatment. *Wat. Sci. Technol.* 26(5-6), 1335-1344.

Hellinga C., van der Heijden R. T. J. M., Heijnen J. J., Romein B. and Luyben K. C. A. M. (1992) An improved method for measurement reconciliation and gross error diagnosis for (bio)chemical processes. IFAC-Symp.Ser.10, 235-37.

Henze M., Gujer W., Mino T., Matsuo T., Wentzel M. C. and Marais GvR. and van Loosdrecht M. C. M. (1999) Activated Sludge Model No. 2d, ASM2d. *Wat. Sci. Technol.* 39(1), 165-182.
Janssen A. J. H., Meijer S.C.F., Bontsema J and Lettinga G. (1998) Application of the redox potential for controlling a sulphide oxidising bio-reactor. *Biotechnol. Bioeng.* 60(2), 147-155.
Lie E. and Welander T. (1994) Influence of dissolved oxygen and oxidation-reduction potential on the denitrification rate of activated sludge. *Wat. Sci. Technol.* 30(6), 91-100.
Meijer S. C. F., van Loosdrecht M. C. M. and Heijnen J. J. (2001), Metabolic modelling of full-scale enhanced biological phosphorus removing WWTP's. *Water Res.* 35(11), 2711-2723.
Menzi S. and Steiner M. (1995) Model-based control for nitrogen-eliminating wastewater treatment plants. IEEE Conference on Control Applications Proceedings 842-847.
Murnleitner E., Kuba T., van Loosdrecht M. C. M. and Heijnen J. J. (1997), An integrated metabolic model for the aerobic and denitrifying biological phosphorous removal. *Biotechnol. Bioeng.* 54, 434-450.
Nowak O., Franz A., Svardal K., Muller V. and Kuhn V. (1999a) Parameter estimation for activated sludge models with the help of mass balances. *Water Sci. Technol.* 39(4), 113-120.
Nowak O., Svardal K., Franz A. and Kuhn V. (1999b) Degradation of particulate organic matter - A comparison of different model concepts. *Water Sci. Technol.* 39(1), 119-127.
Pochana K. and Keller J. (1999) Study of factors affecting simultaneous nitrification and denitrification (SND). *Water Sci. Technol.* 39(6), 61-68.
Roeleveld P.J and Van Loosdrecht M.C.M. (2001) Experiences with guidelines for wastewater characterisation in The Netherlands. *Wat. Sci. Technol.* 45(6), 145-156.
Smolders G. L. F., van der Meij J., van Loosdrecht M. C. M. and Heijnen J. J. (1994a) Model of the anaerobic metabolism of the biological phosphorous removal process: stoichiometry and pH influence. *Biotechnol.Bioeng.* 43(6), 461-470.
Smolders G. L. F., van der Meij J., van Loosdrecht M. C. M. and Heijnen J. J. (1994b) Stoichiometric model of the aerobic metabolism of the biological phosphorus removal process. *Biotechnol.Bioeng.* 44(7), 837-848.
Smolders G. L. F., Bulstra D. J., Jacobs R., van Loosdrecht M. C. M. and Heijnen J. J. (1995) A metabolic model of the biological phosphorus removal process: Validation during start-up conditions. *Biotechnol.Bioeng.* 48(3), 234-245.
Stephanopoulos G. (1984) Chemical process control: An introduction to theory and practice. Prentice-Hall, Inc., Englewood Cliffs, New Jersey (696p.)
Van Loosdrecht M.C.M., Brandse F.A. and de Vries A.C. (1998) Upgrading of wastewater treatment processes for integrated nutrient removal – The BCFS®-process. *Wat. Sci. Technol.* 37(9), 209-217.
Van Loosdrecht M. C. M. and Henze M. (1999) Maintenance, endogeneous respiration, lysis, decay and predation. *Wat. Sci. Technol.* 39(1), 107-117.
Van Veldhuizen H. M., van Loosdrecht M. C.M. and Heijnen J. J. (1999) Modelling biological phosphorus and nitrogen removal in a full scale activated sludge process. *Water Res.* 33(16), 3459-3468.
Van Veldhuizen H. M., van Loosdrecht M. C.M. and Heijnen J. J. (1999) Modelling biological phosphorus and nitrogen removal in a full scale activated sludge process. *Water Res.* 33(16), 3459-3468.
Von Sperling M. (1993) Parameter estimation and sensitivity analysis of an activated sludge model using Monte Carlo simulation and the analyst's involvement. *Water Sci. Technol.* 28(11-12), 219-229.
Yuan Z., Vanrolleghem P. A. and Vansteenkiste G. C. (1997) Modelling error identification of activated sludge models. *Water Sci. Technol.* 36(5), 81-88.
Wanner O., Kappeler J. and Gujer W. (1992) Calibration of an activated sludge model based on human expertise and on a mathematical optimisation technique, a comparison. *Water Sci. Technol.* 25(6), 141-148.
Wentzel M. C., Ekama G. A. and Marais G. v. R. (1992) Processes and modelling of nitrification denitrification biological excess phosphorus removal systems - a review. *Water Sci. Technol.* 25(6), 59-82.
Wentzel M. C., Ekama G. A., Dold P. L., Marais G. v. R. (1990) Biological excess phosphorus removal. Steady state process design. *Water SA.* 16(1), 29-48.
Weijers S. R. and Vanrolleghem P. A. (1997) Procedure for selecting best identifiable parameters in calibrating Activated Sludge Model No. 1 to full-scale plant data. *Water Sci. Technol.* 36(5), 69-79.
Wen Ch. H. and Vassiliadis C. A. (1998) Applying hybrid artificial intelligence techniques in wastewater treatment. *Engineering Applications of Artificial Intelligence* 11(6), 685-705.

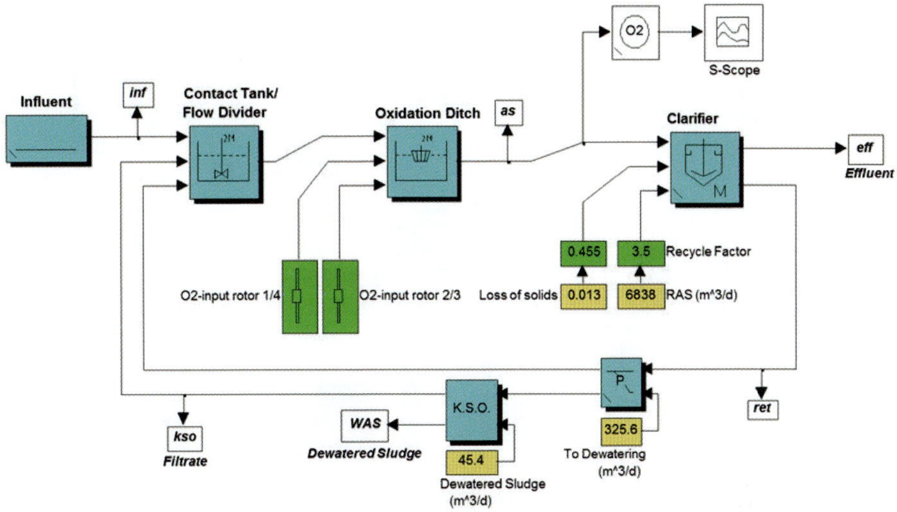

Figure 5.23 SIMBA process scheme of the original WWTP Hardenberg.

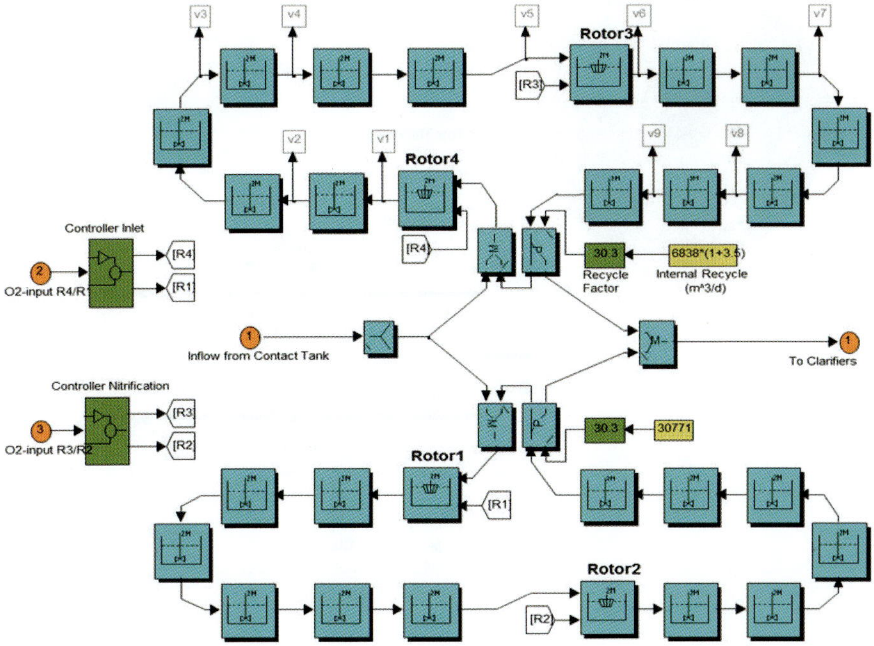

Figure 5.24 SIMBA process scheme of the carousel at WWTP Hardenberg.

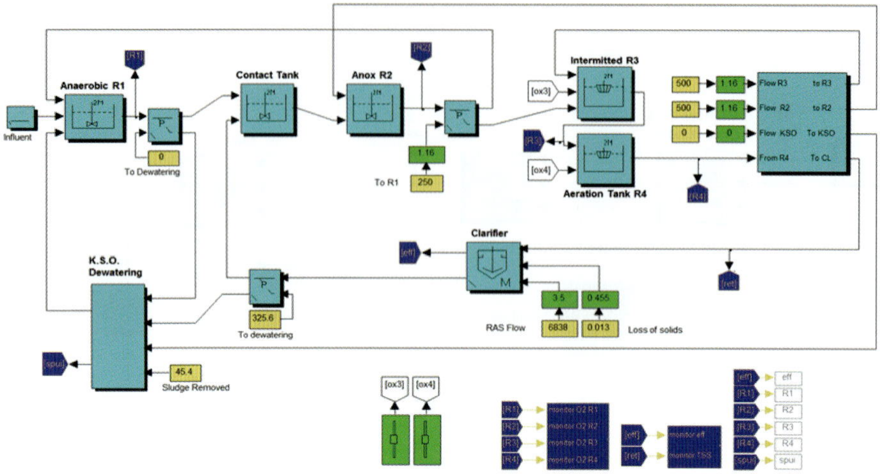

Figure 5.25 SIMBA process scheme of the new EBPR BCFS WWTP Hardenberg.

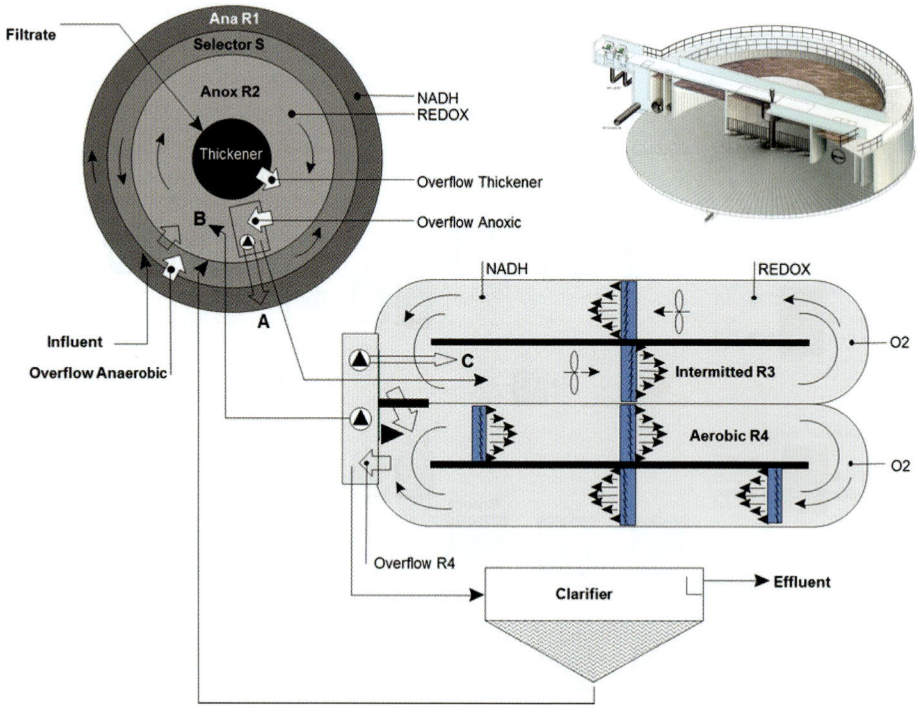

Figure 5.26 Layout of the activated sludge section of the new WWTP Hardenberg with integrated BCFS process unit with anaerobic, anoxic and thickening zones.

Chapter 6: WWTP Shell Godorf, Germany

Optimization of oil refinery wastewater treatment

Pinzon A., Brdjanovic D, Moussa M., Lopez-Vazquez C.M., Meijer S.C.F., Van Straaten H.H.†, Janssen A., Van Loosdrecht M.C.M., and Amy G.

This chapter is based on Pinzon A., Brdjanovic D, Moussa M., Lopez-Vazquez C.M., Meijer S.C.F., Van Straaten H.H.†, Janssen A., Van Loosdrecht M.C.M., and Amy G., (2007) Modelling of an oil refinery wastewater treatment plant. Environ. Technol., 28(11):1273-84

6.1 Introduction

Since the introduction of the Activated Sludge Model No. 1 (ASM1) in late 80's (Henze *et al.*, 1987), a substantial amount of information and experience has been gained from its application to a number of full-scale (mainly municipal) wastewater treatment plants (Coen *et al.*, 1997; van Veldhuizen *et al.*, 1999; Brdjanovic *et al.*, 2000; Koch *et al*,. 2000; Wichern *et al.*, 2001; Printemps *et al.*, 2004). Due to recent developments, improvements and extensions of the models, their application for the simulation of treatment of industrial effluents has become increasingly feasible (Orhon, 1998; Barañao and Hall, 2004; Boursier *et al.*, 2004).

In this study the Activated Sludge Model No. 3 (ASM3) (Gujer *et al.*, 1999), although originally developed for its application to domestic wastewater treatment plants, was selected to simulate the operation of an installation treating effluents from an oil refinery. The fact that ASM3 was introduced to overcome certain deficiencies of ASM1 and its capability to comparatively improve the description of the different biological processes involved in the activated sludge were the main reasons for its selection. After analyzing the most common influent characterization protocols used for activated sludge models, such as WERF, HSG, BIOMATH and STOWA (Sin *et al.*, 2005), the Dutch protocol developed by the Dutch Applied Water Research Foundation (STOWA in Dutch) (Hulsbeek *et al.*, 2002) was selected due to its simplicity and practice-oriented approach. So far, the STOWA characterization protocol has proved successful and has become the standard protocol used in the Netherlands (Roeleveld and van Loosdrecht, 2002).

The goals of this study were to: (i) evaluate the capability of ASM3 to satisfactory describe the performance of a complex full-scale industrial wastewater treatment plant, (ii) assess the applicability of the STOWA protocol for the characterization and calibration of an oil refinery influent, (iii) examine the usefulness of batch tests for the determination of activated sludge kinetics parameters as well as for model calibration, and (iv) further optimize plant performance by modifying the operational parameters and/or altering the existing plant configuration.

6.2 Materials and Methods

6.2.1 Wastewater treatment plant configuration

The oil refinery selected for the execution of this study has the following characteristics: (i) it is operated at an average wastewater temperature of around 34°C, (ii) it applies external methanol addition to support denitrification, and (iii) despite its complexity it is well monitored. The treatment plant consists of an activated sludge system operating in intermittent aeration mode designed for the biological removal of organic matter (COD) and nitrogen (N) from the oil refinery process effluents.

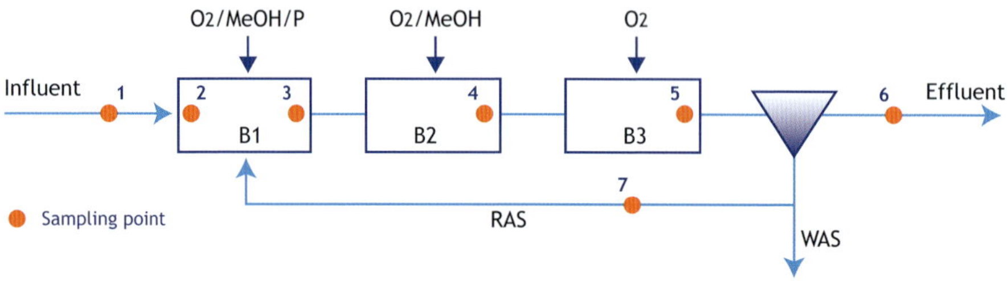

Figure 6.1 Simplified schematic of the oil refinery activated sludge system subject to this study.

Following the preliminary and primary treatment units (oil-water separation, settling, equalization, and flocculation-flotation unit), the pre-treated process water (5,122 m^3/d, 1,639 kgCOD/d, 527 kgN/d) is introduced into the activated sludge system that consists of three aeration basins in series (herein named B1, B2 and B3) and two parallel sedimentation tanks (Figure 6.1). Basins B1 and B2 are intermittently operated and, thus, the activated sludge is exposed to aerobic (40 minutes) and anoxic conditions (20 minutes) in one-hour cycles (24 cycles a day). The aeration in the activated sludge tanks is carried out through the addition of pure oxygen by means of injectors. Due to the relatively low influent COD/N ratio (2.4 gCOD/gN) and the fact that two third of the influent flow is directed to an aerated reactor, an external carbon source (methanol) is needed to supply the required COD for denitrification during the anoxic periods. The last basin (B3) is only operated in an aerobic mode. From the hydraulic perspective, the biological basins are assumed to be completely mixed. pH is controlled at approximately 7.0 through the addition of NaOH to B1. To avoid phosphorous limitation for microbial growth, H_3PO_4 is added to B1. After the secondary settling tank, the final effluent is discharged into receiving water. Recycled activated sludge (RAS) is returned from the settling tanks to the biological basin B1 (flow ratio of 1:1 with respect to influent). The waste activated sludge (WAS) is introduced into a sludge thickener, and the thickened sludge is incinerated off-site. The applied sludge wasting rate results in a total sludge retention time (SRT) in the plant of about 40 days. The overall plant performance in the year 2005 fully complied with the corresponding regulations, as shown in Table 6.1.

Table 6.1 Performance of the oil refinery wastewater treatment plant and current EU effluent regulations.

	Concentrations (mg/L)				
	COD	BOD_5	N	P	TSS
Oil refinery treatment plant effluent	37	6.4	4	<1	10
EU regulations	125	25	15 - 10	2 - 1	35 - 60

6.2.1 Influent characterization

Pre-treated wastewater was characterized through the implementation of the standard Dutch influent characterization protocol STOWA (Roeleveld and van Loosdrecht, 2002). Routinely collected operational data for the year 2005 were provided by the plant staff. For modeling purposes, these data were insufficient. Therefore, an additional tailor-made sampling campaign was carried out for the determination of the parameters of interest (such as COD_{total}, $COD_{filtered}$, NH_4^+, NO_2^-, NO_3^-, TKN, TSS, VSS, DO, pH, and alkalinity) concerning the influent stream to reactor B1 (sampling point No. 1 as shown in Figure 6.1).

6.2.1 Sampling campaign

Taking into account both the initial modeling results using default model parameters and regularly collected data from the plant, a tailor-made sampling program was designed in order to: (i) acquire additional data for modeling purposes, (ii) distinguish the performance of the three biological basins and, (iii) better understand the plant operation. A three-day intensive sampling campaign took place from March 22nd to 24th, 2006, following the plan provided in Table 6.2 at the locations of interest (as indicated in Figure 6.1). The sampling points, frequency, and the choice of parameters were governed by the specific requirements of ASM3 and by the data available in the existing records.

Table 6.2 Sampling schedule applied during the three-day campaign.

Location	COD	COD_{filt}	TKN	TP	NH_4^+	NO_2^-	NO_3^-	TSS	VSS	PO_4^{3-}	T	DO	pH	Alk
1	●	●	●	●	●	●	●	●		●	○	○	○	○
2		●	●	●	●	●	●			●	○	○	○	
3		●			●	●	●				○	○	○	
4		●			●	●	●	●	●		○	○	○	
5		●			●	●	●	●	●		○	○	○	
6	●	●	●	●	●	●	●	●		●	○	○	○	○
7				●		●	●	●		●				

● Daily grab sample ○ On-line measurement

6.2.1 Experimental program

In order to obtain a proper set of plant-specific data regarding biomass activity, nitrification and denitrification, batch tests were repeatedly executed using the same experimental setup at both an on-site plant laboratory and UNESCO-IHE laboratory located in Delft. A double jacketed batch reactor with a maximal operating volume of 1.5 L was used. The experiments were performed under similar environmental conditions as at the full-scale plant (at controlled pH 7.0 ± 0.3 and temperature 34°C). The operation of the reactor was fully automated regarding pH and dissolved oxygen (DO) control. Batch tests were performed to determine: (i) the volumetric (r_N) and specific ammonia uptake rate (k_N), (ii) the volumetric (r_D) and specific nitrate uptake rate (k_D) and (iii) the actual anoxic yield under methanol conditions ($Y_{CODMeOH}$). Process rates were determined based on the initial rates using the linear regression method. The description of the experimental procedures can be found elsewhere (Brdjanovic *et al.*, 1997).

6.3 Modeling tools

All simulations were performed with ASM3 (Gujer *et al.*, 1999). The simulations of the operation of the plant were carried out using the software package SIMBA™ 4.2 combined with MATLAB® 7.0 and SIMULINK® 6.0 using average influent data for the period of study. Simulations were executed until a steady-state performance was achieved. Overall, dynamic simulations were not carried out because of the presence of buffering/equalization tanks in the pre-treatment step that, to a large extent, levels out any potential influent variation. However, due to its importance regarding plant operation and performance, methanol and oxygen additions were modeled taking into account the intermittent dynamic and cyclic operation of the biological tanks B1 and B2 (40 minutes oxygen addition followed by 20 minutes of methanol dosing). The dynamic methanol-DO pattern was included when modeling the tanks B1 and B2 reflecting the actual plant operation. Basin No. 3 (B3) was modeled as an aeration tank with a constant input of oxygen. Methanol was introduced in the model as an inflow stream of rapidly biodegradable COD (S_S). The three basins were each modeled as a Completely-mixed Stirred Tank Reactor (CSTR). Secondary settlers were modeled as ideal settlers (i.e. 100% separation between solids and water). However, in order to account for any potential losses of suspended solids in the effluent, a percentile loss of solids from the settlers was included. The DO level at all three basins was set to 2.2 gO_2/m^3 based on routine records and profiles observed during the sampling campaign. All plant simulations were performed using an average mixed liquor temperature of 34°C as observed at the full-scale system.

6.4 Calibration strategy

The model calibration of the refinery wastewater treatment plant was performed following a stepwise approach in which each step included the cross comparison of predicted data against measured data regarding both the liquid phase and biomass. The five steps applied were namely: (i) simulation of plant operation and subsequent calibration based on collected data for the year 2005, (ii) simulation and calibration of batch tests using sludge properties as predicted using sampling data (iii) simulation and calibration of the performance of the full-scale treatment plant using the outcomes of calibration of batch tests, (iv) validation of the model using nitrogen measurements recorded inside three basins during the sampling campaign (according to Table 6.2), and (v) simulation of different scenarios with the aim of further optimizing the plant operation.

6.5 Results

6.5.1 Influent characterization

Table 6.3 summarizes the results of the influent characterization according to the STOWA protocol. Regarding slowly biodegradable substrates (X_S) and particulate inert organic material (X_I), it was calculated that the particulate COD fraction accounted for approximately 10% of the total COD – considered as a rather low percentage in comparison with domestic wastewaters (40-60%) (Henze, 1992; Gujer *et al.*, 1999). A proper estimation of this parameter is important for the following two reasons: (i) overestimation of X_S can be reflected directly at the denitrification potential of wastewater (resulting in less S_S readily available for denitrification) (Roeleveld and van Loosdrecht, 2002), and (ii) according to Orhon (1998) and Brdjanovic *et al.* (2000), the incorrect determination of X_I may cause prediction inaccuracies since it is very sensitive to the solids retention time (X_I has major influence in SRT estimation at modeling level). In this study, the calculated fraction seems to be reasonable, since the measured fraction with regard to total COD (of 10%) was sufficiently accurate to predict the mixed liquor suspended solids (MLSS) in the reactor as well as to predict the SRT applied at the plant. On the other hand, the estimated readily biodegradable organic substrate (S_S), was found to be higher in comparison to the domestic wastewater fraction which has been estimated at values of about 20-40% (Henze, 1992; Gujer *et al.*, 1999; Koch *et al.*, 2000). Moreover, the recorded effluent concentrations versus the COD influent concentrations provide an indication of the biodegradability of the organic compounds present in the wastewater (Table 6.3).

Table 6.3 Measured average influent and effluent data and subsequent influent characterization according to STOWA.

Measured records					Influent characteristics for ASM3			
Description	Symbol	Influent	Effluent	Unit	Description	Symbol	Influent	Unit
					Soluble components			
Total COD	COD_{tot}	319.0	33.7	$gCOD/m^3$	Oxygen (negative COD)	S_{O2}	0.3	$gCOD/m^3$
COD filtered[1]	COD_{filt}	287.1	30.3	$gCOD/m^3$	Readily biodegradable organics	S_F	216.9	$gCOD/m^3$
BOD_5	BOD_5	121.1		$gBOD/m^3$	Volatile fatty acids	S_A	39.9	$gCOD/m^3$
Ammonium	NH_4	93.0	0.9	gN/m^3	Ammonium and ammonia nitrogen	S_{NH4}	102.8	gN/m^3
Nitrate	NO_3	0.0	2.8	gN/m^3	Nitrate and nitrite nitrogen	S_{NO3}	0	gN/m^3
Total P	P_{tot}	0.8	0	gP/m^3	Inorganic soluble phosphorus	S_{PO4}	0	gP/m^3
TSS	TSS	25.0	10.1	g/m^3	Soluble inert organic matter	S_I	30.3	$gCOD/m^3$
Temperature	T	34.0	34.0	°C	Alkalinity	S_{ALK}	9.1	Eq/m^3
TKN[1]	TKN	205.5		gN/m^3	Particulate compounds			
					Particulate inert organic matter	X_I	13.9	$gCOD/m^3$
					Slowly biodegradable substrate	X_S	18.0	$gCOD/m^3$
					Active heterotrophic biomass	X_H	0	$gCOD/m^3$
					Glycogen	X_{STO}	0	$gCOD/m^3$
					Active autotrophic biomass	X_{AUT}	0	$gCOD/m^3$
					Mixed liquor suspended solids	X_{TSS}	23.9	$gMLSS/m^3$

[1] Extended to historical data based on results from the sampling program

6.5.2 Sampling campaign

The average results from the sampling campaign at different locations of the wastewater treatment plant are presented in Table 6.4. Compared to historical plant measurements (Table 6.3), the concentrations obtained during the three-day sampling campaign executed in March 2006 are approximately 20% lower. Nevertheless, the removal performance was still satisfactory according to regulations and operational plant records. The results of the sampling campaign confirmed that the addition of methanol indeed takes place during periods when the oxygen input is shut off, and at recorded rate of 4.0 to 4.5 L/min in tank B1, and 0.5 L/min in tank B2 (Figure 6.2).

Table 6.4 Average values of concentrations (mg/L) obtained by the sampling campaign.

Location	COD	COD_{filt}	TKN	NH_4-N	NO_3-N	NO_2-N	TSS	VSS
1	262.3	248.3	87.2	74.1	0.3	0	11.0	
2		73.9		25.1	13.2	0.7	5,921	3,423
3		37.4	175.5	1.4	12.3	0.3		
4		33.1		0.6	2.6	0		
5		32.3		0.6	2.1	0		
6	30.7	29.4	1.1	0.1	1.6	0	8.0	
7				0.7	1.7	0	12,237	

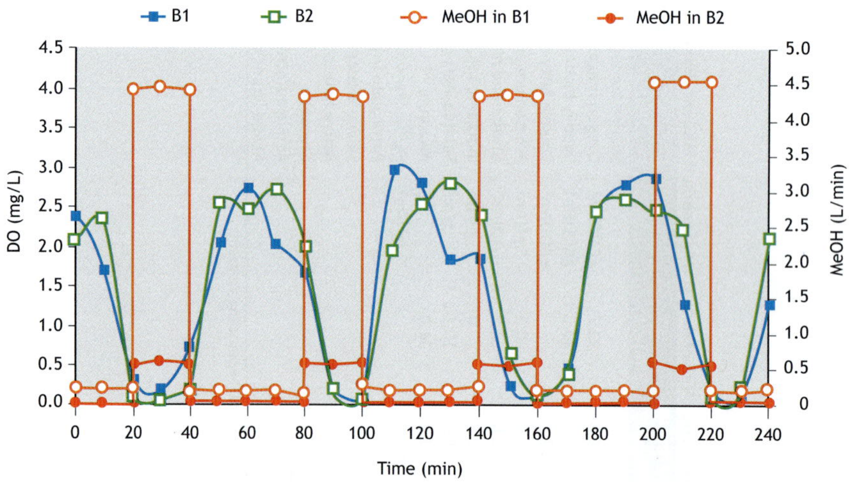

Figure 6.2 Example of dissolved oxygen and methanol concentrations pattern in the biological basins during a selected period of 4 cycles (basin B3 was constantly aerated at a DO of 2.2 mg O_2/L).

6.6 Experimental campaign

6.5.1 Nitrification test

A typical conversion profile of ammonia, nitrate and nitrite during nitrification batch tests with sludge taken from the refinery plant is shown in Figure 6.3. In the nitrification tests, complete ammonia oxidation was achieved in the first four hours and only limited nitrite accumulation was observed. The batch reactor was operated for 24 hours in order to detect any further nitrite accumulation as a result of the faster growth of ammonia oxidizers compared to that of nitrite oxidizers at temperatures above 15 to 20°C (Hellinga and van Loosdrecht, 1999; Moussa *et al.*, 2004). However, at the end of the nitrification test, complete nitrite depletion was observed (data not shown). Using linear regression, the volumetric ammonia uptake rate (r_N) of 9.55 mgN/L.h and specific ammonia uptake rate (k_N) of 3.1 mgN/gVSS.h were calculated.

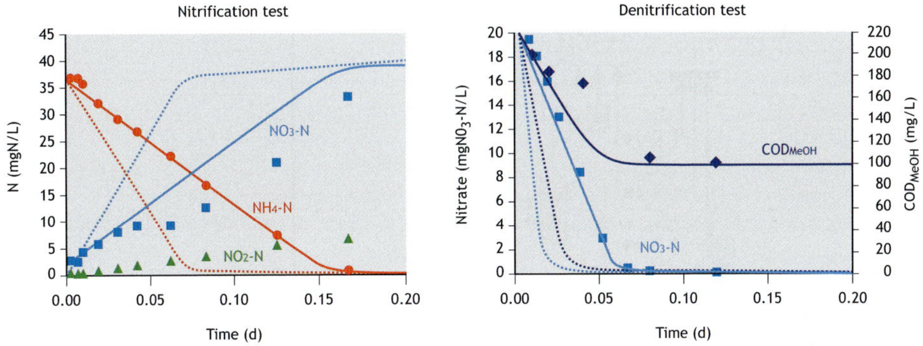

Figure 6.3 Typical results of nitrification and denitrification batch test performed with sludge from the oil refinery plant (NH_4-N (●), NO_3-N (■) NO_2-N (▲), COD_{MeOH} (♦). Dashed and solid lines present results of simulation before and after calibration of the ASM3 model, respectively.

6.5.2 Denitrification test

A typical profile of nitrate and methanol during the denitrification batch tests with sludge from the refinery plant is shown in Figure 6.3. Using linear regression, a volumetric nitrate uptake rate (r_D) of 14.2 mgN/L.h and a specific nitrate uptake rate (k_D) of 4.6 mgN/gVSS.h, were calculated. The methanol to nitrate degradation ratio was found to be 3.4 gMeOH/gNO_3-N. Moreover, since the amount of methanol was exactly known, the corresponding yield coefficient (Y_{ODMEOH}) could be calculated from experimental results using the COD conservation equation:

$$\Delta COD_{MeOH} = \Delta X_B + 2.86 \Delta NO_3 \quad (6.1)$$

Consequently, $Y_{CODMEOH}$ was found to be 0.45 $gCOD_{XB}/gCOD_{MeOH}$. This resulting yield of 0.45 obtained during the denitrification batch experiment was found to be very close to the literature data of 0.44 $gCOD_{XB}/gCOD_{MeOH}$ (Purschert *et al.*, 1996; Purschert *et al.*, 1999) and of 0.36 ± 0.05 $gCOD_{XB}/gCOD_{MeOH}$ for methanol-based denitrification (Chudoba *et al.*, 1989). Likewise, the measured methanol to nitrate ratio of 3.45 gMeOH/gNO_3-N was similar to other reported values (Timmermans and Haute, 1983; Purschert *et al.*, 1996; Purschert *et al.*, 1999).

6.5.3 Hydraulic set-up of the plant model

Plant design documentation, the existing process scheme, and the current operational mode of the plant were used to create the hydraulic flow scheme of the installation made in SIMBA (Figure 6.4). More information about this scheme is provided in Table 6.5.

Table 6.5 Plant hydraulic data.

Stream	Flow rate (m^3/d)	Reactor	Volume (m^3)
Influent	5,222 ± 683	B1	2600
RAS	5,271 ± 685	B2	1080
WAS	55.4 ± 9.7	B3	600
Methanol	2.5	Settlers	710
H_3PO_4	0.3		

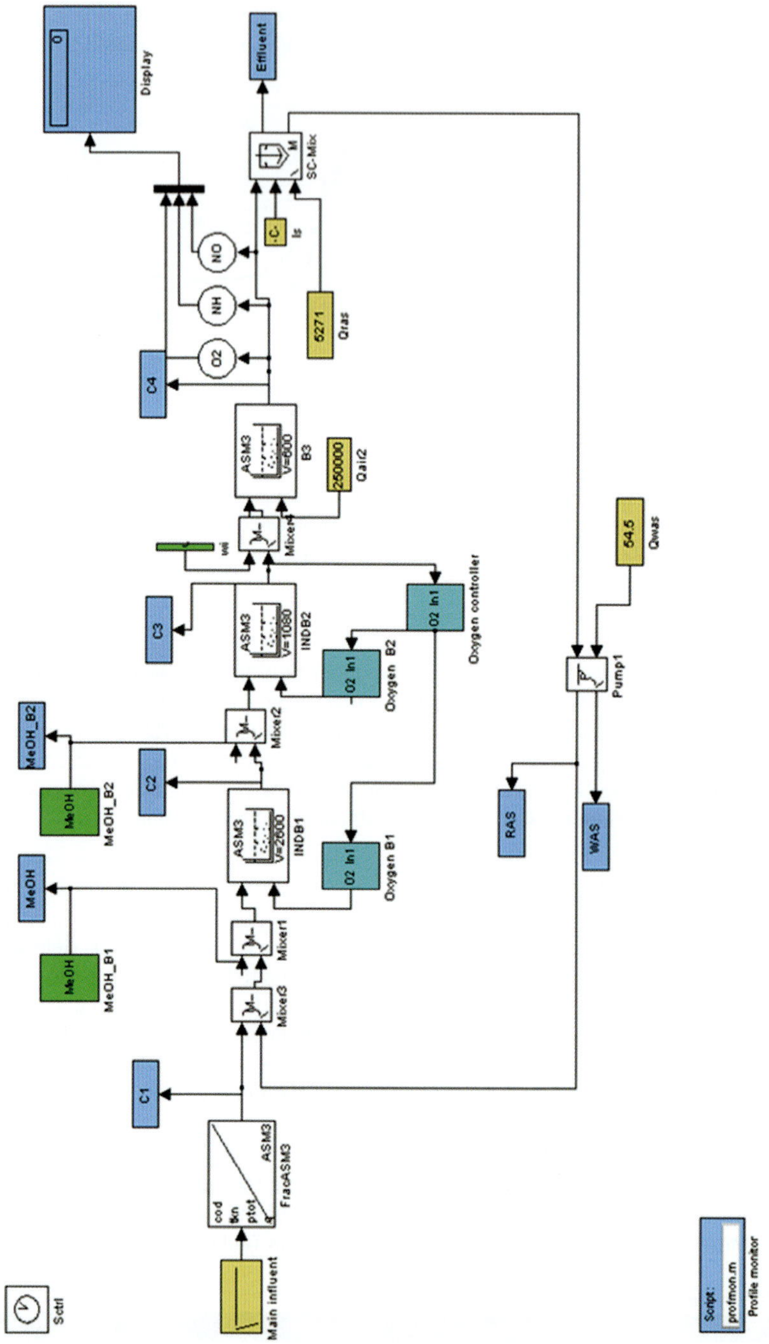

Figure 6.4 Hydraulic scheme of the oil refinery wastewater treatment plant in SIMBA™ environment.

6.5.4 Model calibration and simulation

At this treatment plant, the most extensively recorded parameters are effluent COD_{total}, NH_4 and NO_3 content, and TSS and VSS in the biological basins. Therefore, model calibration was carried out to reasonably predict those parameters. The calibration was performed using steady-state data from the treatment plant operation, combined with results from batch tests. The calibration protocol, as applied elsewhere, was used in this study (van Veldhuizen *et al.*, 1999; Brdjanovic *et al.*, 2000; Meijer *et al.*, 2001; Hulsbeek *et al.*, 2002; Moussa *et al.*, 2004). This implies that only a few selected parameters were fitted to actual data (firstly solids, secondly the effluent COD, and finally the N balance was checked). In addition nitrification and denitrification process parameters were calibrated. For the initial simulation of the performance of the plant, the default ASM3 kinetic and stochiometry set of parameters were used (Gujer *et al.*, 1999).

The total suspended solids (TSS) values were calculated by the predicted biomass COD concentration of the activated sludge based on the measured ash content of approximately 40% of 6,573 mg/L and a conversion factor of 1.42 gCOD/gMLVSS (McCarty, 1965; Warner *et al.*, 1986; Nowak *et al.*, 1999). However, it is reported that COD concentration in the process is strongly determined by the influent X_I/X_S ratio (Meijer *et al.*, 2002). As inert COD (X_I) accumulates in the sludge, increasing X_I leads to increased COD in the system. Hence, the X_I fraction was increased from 6% (as in the initial influent characterization) to 9% to fit the observed MLVSS concentration in the biological tanks. As a result of adjusting the X_I fraction the model satisfactorily predicted the COD concentration in the effluent and MLVSS concentration in the biological tanks (Table 6.6).

Table 6.6 Summary of plant measured and predicted model concentrations in the effluent and biomass (in mg/L).

Data set		Effluent			Biomass	
	COD_{total}	NH_4-N	NO_3-N	TSS	VSS	TKN
Average measured	33.70 ± 9.10	0.80 ± 1.20	2.70 ± 3.40	6,573 ± 458	4,009 ± 438	206
Predicted default ASM3	41.00	0.02	6.60	6,673	4,070	165
After preliminary calibration X_I of 9% and iN_{XI} = 0.031	41.00	0.08	6.70	6,385	3,894	208
After calibration with nitrification batch feedback μ_A= 0.5 1/d	41.00	0.20	7.00	6,408	3,908	212
After calibration with denitrification batch feedback Y_{XSTO} = 0.5 g COD_{XSTO}/gCOD_{Ss}	41.00	0.14	2.50	6,177	3,768	208
After calibration with denitrification batch feedback Y_{XSTO} = 0.5 and h_{NO}=0.3	41.00	0.17	4.00	6,269	3,823	206

Unlike COD, TKN is a non-conservative parameter since an incorrect TKN load will be compensated by the oxygen consumption and N_2 production in the process (Meijer *et al.*, 2001; Roeleveld and van Loosdrecht, 2002). The amount of nitrogen available for oxidation determines the autotrophic oxygen demand and the required denitrification capacity. The first step is therefore to ensure that the N-content of the sludge predicted by the simulations is close or equal to the measured value (which in this case was 0.036 gN/g sludge COD). This was achieved by increasing the fraction of N for inert particulate COD (iN_{XI}) from 0.020 to 0.031 gN/gCOD.

After the adjustment of the N content of the sludge the effluent ammonia concentration was predicted well without any further calibration. However, since the effluent concentrations are, in general, very low the model tended to overestimate the N conversion rates. Therefore N conversions were checked based on the results from batch tests.

The nitrification process in the batch test (Table 6.3) was initially not predicted sufficiently well by the model. As observed, the actual ammonia conversion to nitrate was lower than predicted. There are two possible explanations for this: (i) either the amount of nitrifiers (X_A) predicted by the model was too high or, (ii) the nitrification rate is too high. To check which explanation is more reasonable, the growth rate of autotrophic biomass (μ_{AUT}) was reduced while the autotrophic decay rate (b_A) was increased. When μ_{AUT} decreased from 1.0 to 0.5 1/d a very good fit of predicted data to measured data was obtained (as seen in Figure 6.3). On the other hand, b_A had to be increased as much as five times with respect to the default value in order to closely reach the observed nitrification rate, but still fitting less accurately in comparison with μ_{AUT}. Hence, b_A was not taken into further consideration for calibration of the batch test due to its low sensitivity. When the adjusted μ_{AUT} value of 0.5 1/d was implemented in the full-scale plant model, the nitrogen concentrations in the effluent only slightly increased (0.20 mgNH_4-N/L and 7.0 mgNO_3-N/L) in comparison to the values obtained by the simulations when using the default μ_{AUT} value (0.08 mgNH_4-N/L and 6.7 mgNO_3-N/L, respectively). Since these differences are practically negligible, it was decided that further calibration regarding nitrification was not necessary.

Similar to nitrification, the model initially did not predict sufficiently well the denitrification process in batch tests (Figure 6.3) given the fact that the observed nitrate volumetric depletion rate was slightly lower than the predicted one. Additionally, under laboratory conditions methanol consumption to nitrate conversion was estimated at 5.1 $gCOD_{MeOH}$/gNO_3-N, whereas the model predicted a consumption of twice as much (10.1 $gCOD_{MeOH}$/gNO_3-N). This can be explained by the fact that ASM3 employs storage mechanisms to describe growth and decay processes: in ASM3, soluble substrate S_S is first converted to storage compounds (X_{STO}) and then used by bacteria, whereas as reported elsewhere (Purschert and Gujer, 1999), methanol is assumed to not be converted to storage products but consumed directly by microorganisms. Following the agreed calibration procedure, it was decided to reduce the anoxic yield of stored products per S_S ($Y_{STO,NO}$) from 0.8 to 0.5 $gCOD_{XSTO}/gCOD_{SS}$ in order to fit the amount of methanol taken up per nitrate consumed. The choice was based on: (i) the uncertainty in the formation of storage polymers, (ii) the fact that C_1-carbon compounds in general give a lower growth yield than other carbon compounds (Nyberg *et al.*, 1992; Henze *et al.*, 1997; Purschert and Gujer, 1999), (iii) the fact that the largest contribution to denitrification, in the plant under study, occurs mainly due to methanol addition, and (iv) in the case of the presence of storage polymers, their formation is lower at higher temperatures (34°C in this study) (Krishna and van Loosdrecht, 1999). When a new value of $Y_{STO,NO}$ was implemented in the batch model, a very good fit of predicted data to measured data was obtained regarding methanol consumption to nitrate conversion, as demonstrated by Figure 6.3. When the new $Y_{STO,NO}$ was implemented in the full-scale plant model, good results were also achieved concerning the nitrate effluent level (decreased from 7.0 to 2.5 mgN/L). However, regarding the denitrification batch rate prediction, the change in $Y_{STO,NO}$ greatly influenced nitrate uptake which was predicted to be faster than the measured one at the lab (Figure 6.3). Hence, in order to fit the denitrification rate in both the batch test and full scale plant model, the anoxic reduction factor (h_{NO}) was calibrated as suggested elsewhere (Roeleveld and van Loosdrecht, 2002). A satisfactory denitrification fit was obtained when h_{NO} was decreased from 0.6 to 0.2 (Figure 6.3). However, during the implementation of this change to simulate the full-scale plant, nitrate in the effluent increased dramatically (to 10 mg/L). Due to the fact that h_{NO} obtained from the feedback calibration of the batch test is only related to a methanol degrading bacteria group (Hyphomicrobium. *sp*), the value of 0.2 is underestimated at the full scale plant (since influent COD is also used in the denitrification process). Hence, to avoid the underestimation of h_{NO} a value of 0.3 was finally used in the batch and the plant model. Finally the nitrate concentrations in the effluent as well as process rates were in accordance with those measured at the plant (Table 6.6). Furthermore, neither COD, NH_4 nor methanol to nitrate consumption were further influenced due to implementation of obtained calibrated parameters during the batch tests calibration step.

It was finally decided that there was no need for further calibration given the inaccuracies associated with data collection, handling, measurements, analytical procedures, etc. (Brdjanovic *et al.*, 2000). A summary of model outputs as predicted using the described calibration procedure is shown in Table 6.6. Additionally, an overview of the adjusted parameters during the calibration procedure is presented in Table 6.7.

Table 6.7 Model parameters for which the value was altered during model calibration in this study.

Parameter	Symbol	Unit	This study	ASM3 default value
Nitrogen content of particulate inert	i_{NXI}	gN/gX_I	0.03	0.02
Autotrophic maximum growth rate	μ_A	1/d	0.50	1.00
Anoxic yield of stored products per Ss	$Y_{STO,NO}$	$gCOD_{XSTO}/gCOD_{Ss}$	0.50	0.80
Anoxic reduction factor	h_{NO}		0.30	0.60

6.5.5 Model validation

Model validation was performed by evaluating its capacity to predict the measured concentrations of NH_4, NO_3 in the liquid phase and TKN of the biomass along the three bioreactors B1, B2 and B3 obtained during sampling campaign (sampling points 3, 4, 5, 6 in Table 6.1). The recorded average daily inflow during sampling (4,871 m^3/d) was used as input for the model. The model provided a close prediction regarding the above mentioned parameters: the TKN content of the sludge was predicted as 176.0 mg/L compared to a measured value of 175.5 mg/L. Likewise, the model provided a satisfactory prediction of the practically complete ammonia depletion at the end of basin B2 (0.5 mg/L) whereas the nitrate removal (2.0 mg/L) was comparable to measured values (Table 6.4). The same statement applies to the end of basin B3 and effluent sampling points.

6.5.6 Performance evaluation

Based on routine monitoring as well as detailed observation during this study, it is clear that in terms of the removal of suspended solids, organic materials and nitrogen compounds, the selected wastewater treatment plant of the refinery is performing well and complying with current effluent discharge regulations. Analyses of operational data recorded in 2005 showed that the average percentage of removals of COD, NH_4 and NO_3 were 98, 89 and 91%, respectively. The high NO_3 removal efficiency at the plant is attributed to the addition of methanol as carbon source (Nyberg *et al.*, 1992; Purschert and Gujer, 1999). According to reported data in the literature (Metcalf and Eddy, 2003), the denitrification process with external carbon addition is capable of achieving effluent nitrogen levels as low as 3 mg/L which is in agreement with the performance of this plant (2.7 ± 3.4 mgN/L in the effluent).

6.6 Process optimization

For this refinery plant in particular, one of the major challenges is to maintain the required denitrification activity and at the same time optimizing (reducing) the amount of added methanol. In this regard, the model was used to evaluate three different optimization scenarios.

6.6.1 Scenario 1: Implementation of an idle phase

Based on the denitrification batch test, the theoretical amount of required methanol addition for full nitrogen removal was estimated at approximately 2,490 $kgCOD_{MeOH}/d$. Plant records showed that the actual amount of methanol added to the plant during the studied period was 17% (higher than the experimentally obtained required dosage) that neglects the potential contribution of S_S during denitrification. Figure 6.2 shows that the methanol addition starts when the dissolved oxygen (DO) concentration in the basin is still in the range of 2.0 mgO_2/L. Consequently, this means that some methanol may be utilized under aerobic conditions. In order to avoid the potential oxidation of methanol under aerobic conditions, a modified control strategy was proposed consisting of the introduction of two so-called idle phases of 2 minutes at the end of the aerobic period and 1 minute at the end of the anoxic period during which neither oxygen nor methanol is added. This is done to ensure that the total methanol consumption takes place under anoxic conditions and, therefore, it is only used for denitrification. According to model results, a theoretical reduction of 20% in methanol addition could be achieved, while still allowing reaching a good effluent quality regarding nitrate and other evaluated parameters (2.65 $mgNO_3$-N/L).

6.6.2 Scenario 2: Transforming B3 basin from aerobic to anoxic

Simulations showed that the influence of the last aerobic basin (B3) on overall ammonia removal performance was insignificant. The results indicated that virtually complete ammonia removal was already achieved at the end of basin B2, which was confirmed by the results obtained by the sampling campaign. The results also showed that at the outlet of B2, the ammonia concentration was very low (lower than 1 mg/L) and approximately 33 mg/L of COD was entering B3. Model predictions are well in accordance to the observed situation (of less than 1 mgN/L and 32 mgCOD/L, respectively). From a process optimization perspective, B3 was retrofitted from an aerobic to an anoxic tank. Such a change could bring some savings as less aerobic volume is needed resulting in a reduced requirement for pure oxygen of 3.18 tons on a yearly basis. Furthermore, an approximate reduction in methanol supply of 29% could be achieved without affecting the nitrate effluent standards (predicted value of 4.3 mgN/L).

6.6.3 Scenario 3: Combined pre- and post-denitrification with external methanol addition

In order to utilize the COD present in the influent wastewater for achieving good nitrate removal, the plant layout was modified into a combined pre- and post-denitrification scheme. As a consequence, basin B1 was converted to a purely anoxic tank; B2 was kept as an aerobic tank and B3 was modified into an anoxic tank as well. Additionally, an internal recirculation line was implemented from basin B2 to basin B1 to recycle the nitrate-rich mixed liquor to B1 in order to create anoxic conditions and methanol was dosed only to basin B3. According to the model, this modified process configuration could result in a reduction of 67 % of methanol addition while still maintaining satisfactory effluent NO_3 concentrations (3.0 mgN/L). Furthermore, additional savings in oxygen supply could also be achieved as part of the organic matter in the wastewater is oxidized by nitrate instead by oxygen (Henze *et al.*, 1997). As the optimization proposed in this study was made within the existing tank volumes and presently used equipment, no other expenditures except construction of a nitrate return pump and its respective line are foreseen in this scenario. Overall, this makes this option even more attractive due to the apparently low required investments. A summary of the selected operational parameters for the present situation and proposed optimization scenarios is given in Table 6.8.

Table 6.8 Comparison of selected operational parameters of present and alternative scenarios of the oil refinery plant.

Scenario	Aerobic volume (m^3)	Anoxic volume (m^3)	Methanol dose (mg/L)	COD_{total} effluent (mg/L)	NH_4 effluent (mg/L)	NO_3 effluent (mg/L)	Remarks
Present case	3,054	1,226	2.4	41	0.17	4.0	Intermittent system
Scenario 1	3,054	1,226	1.9	40	0.04	4.2	
Scenario 2	2,454	1,826	1.7	39	0.40	4.3	
Scenario 3	2,000	2,280	0.8	38	1.20	3.0	Nitrate recycle flow 1:2 respective to the influent

6.7 Discussion

Based on the result of this study, it can be stated that the calibration using the Dutch standard procedure for domestic wastewater characterization (STOWA protocol), proved to be appropriate for modeling an industrial plant treating the effluent from an oil refinery. The attributes of the protocol allowed to have a practical approach and reduced the need of additional sampling and measuring. This is noteworthy because of the fact that the STOWA protocol has been originally developed primarily for its application to municipal wastewater treatment plants, and only very limited experience has been reported so far on industrial applications.

Regarding the choice of using static versus dynamic modeling, the common practice readily applied within the industry of constructing equalization/buffering tank(s) at the beginning of an industrial treatment plant to level out hydraulic peaks, equalize shock loads and isolate eventual toxic discharges from the production processes, certainly contributed to the minimization of variations in both flow and load to the plant. However, due to a lack of dynamic influent data, it was not possible to quantify the extent of such contribution. Nevertheless, based on the experience from previous modeling studies (Brdjanovic, 1998; van Veldhuizen *et al.*, 1999) it was assumed that the presence of equalization systems and primary treatment units created satisfactory pseudo-state conditions to support the use of static data (average daily flow proportional sample values) as was the case in this modeling exercise. However, regarding the plant operation mode, the dynamic (cyclic) operation of the first two biological basins (alternate aerobic/anoxic conditions with regular switching between oxygenation and methanol addition) was fully taken into account by the model. Such an inclusion of internal operation dynamics was necessary for the successful description and interpretation of the system.

Concerning the execution of the batch test, they proved to be useful for the determination of the biomass activity as a part of the iterative process of model calibration. In the case of denitrification, the fact that methanol is regularly added to enhance the denitrification capacity resulted, somehow, in a deeper investigation on its effect on the overall plant performance. As it has been reported (Nurse, 1978; Purschert *et al.*, 1996), if methanol is supplied for denitrification purposes, after an adaptation period, an enrichment of the required bacteria will occur. Among such bacteria, Hyphomicrobium *sp.* has been observed to be the dominant microorganism under anoxic-methanol conditions being their occurrence mainly attributed to methanol presence. The population composition can be tracked by measuring the degradation rates of methanol (Layton *et al.*, 2000). Moreover, Hyphomicrobium *sp.* is a slowly growing organism with a biomass yield of around 0.40 $gCOD_{XB}/gCOD_{MeOH.}$ The yield found in this study

($0.45\ gCOD_{XB}/gCOD_{MeOH}$) is in agreement with the values reported for the biomass yield of Hyphomicrobium *sp.*

The calibration procedure applied in this study comprised the evaluation of the ability of the model to satisfactorily describe the effluent plant concentrations and biomass activity. In this regard, a sludge, that had different characteristics obtained from the simulations of the full-scale treatment plant, was used to simulate the nitrification and denitrification batch tests. The final calibration of the nitrification batch test showed the need to reduce the value of μ_A from 1.0 to 0.5 1/d which agrees with previous findings from full-scale municipal and industrial plants modeling studies (Brdjanovic *et al.*, 2000; Moussa *et al.*, 2004). The choice in the change of the growth rate of autotrophs may be supported by occurrence of toxicity. Some studies (Eckenfelder *et al.*, 1986) reported toxicity as a recurrent problem from petrochemical and refinery industries, due to the presence of heavy metals, phenols, etc., that may inhibit kinetics of growth by reducing the maximum specific autotrophs growth rate (Henze *et al.*, 1997). Similar to nitrification tests, denitrification batch model results showed the need to recalibrate $Y_{STO,NO}$ and h_{NO} in order to not only predict nitrate effluent concentrations but also to satisfactorily describe the denitrification rate and the consumption of methanol for nitrate removal. The choice in the selection of the above mentioned parameters to be adjusted in the denitrification batch model were principally governed by the influence of methanol addition to the biomass. As mentioned previously, this involves the presence of a specialized methanol degrading bacteria group (Hyphomicrobium *sp.*) and the fact that, although some influent wastewater COD is present during denitrification, the majority of denitrification organisms use methanol as electron donor in this particular plant. ASM3 does not make any distinction between either the heterotrophic bacteria group species or carbon sources, but estimates the bacterial growth based on the presence of substrate under anoxic or aerobic conditions. Hence, for practical issues (*e.g.* plant optimization or getting a better understanding of the system) the approach followed in this study is considered to be satisfactory. However, if a more intensive study concerning the nitrogen conversions is needed then the implementation of a separate module to describe the activity of methanol degraders is recommended (Purschert *et al.*, 1996; Purschert and Gujer, 1999). This can be necessary for the determination of the denitrification potential especially during the start up of treatment plants when a period of inoculation of bacteria has to be established and the denitrification potential of the plant needs to be estimated both in terms of methanol and the COD of raw wastewater.

With the use of a stepwise calibration and ASM3 as the applied model, it was possible to obtain a satisfactory simulation of a specific oil refinery plant. It is noteworthy that only four model parameters needed to be adjusted ($Y_{STO,NO}$, h_{NO}, m_A and iN_{XI}) in order to reasonably match the observed selected characteristics of both liquid phase and the biomass and those predicted by the model.

The model was demonstrated to be a powerful tool for the evaluation and optimization of plant operations. The study provided a better understanding of the plant operation and treatment process involved. The potential of the model was illustrated in the evaluation of the adjustment and upgrading of the plant configuration. As, in general, the plant is performing satisfactorily and complies with current regulations in place, the optimization focused on exploring the possibilities to reduce the operational costs. During this study, the denitrification potential of the activated sludge system was examined in particular as it is directly related to the external addition of carbon source. From that perspective, the results of the three scenarios indicated that some minor modifications in process configuration, combined with an optimized operational strategy regarding oxygen and methanol addition, could lead to a more stable and economically sustainable plant performance. The simulations results suggested that by modifying the existing plant to a combined pre- and post- denitrification system, large savings on methanol dosage and oxygen supply can be achieved. Additional benefits of modeling can be achieved through its use as a decision support tool by the oil refinery treatment plant operators.

6.8 Conclusions

Although ASM3 was originally developed for its application on domestic wastewater at a temperature range between 10 and 20°C, it proved to be capable of describing the performance of an industrial wastewater treatment plant treating the effluent from an oil refinery operating at 34°C. Likewise, the STOWA influent characterization and calibration procedure proved to be suitable for its application to this plant. As far as the sludge batch tests are concerned, they proved to be useful and the model was able to describe both nitrification and denitrification processes satisfactorily. This modeling exercise was not only carried out to gain more experience about the practical application of ASM3 within the refining industry, but it was also beneficial to obtain a better understanding of the plant operation and the conversion processes involved. It was also demonstrated that the model can be used for optimization of the plant performance regarding operational costs.

References

Barañao, P.A. and E.R. Hall (2004). Modelling carbon oxidation in CTMP pulp mill activated sludge systems: calibration of ASM3. *Water Science and Technology*, 50(3): 1-10.

Boursier H., Beline F. and Paul E. (2004) Activated sludge model No 1. Calibration for piggery wastewater tretament using respirometry. *Water Sci. Technol.*, 49, 389-396.

Brdjanovic D., van Loosdrecht M.C.M., Hooijmans C.M., Alaerst G. J. and Heijnen J. J. (1997) Temperature effects on physiology of biological phosphorous removal systems. *ASCE J. Environ. Eng.*, 123, 144-154.

Brdjanovic D. (1998) *Modelling Biological Phosphorous Removal in Activated Sludge Systems.* Balkema Publishers, Delft.

Brdjanovic D., van Loosdrecht M.C.M., Veersteeg P., Hooijmans C.M., Alaerts G. J. and Heijnen J.J. (2000) Modelling COD, N and P removal in a full-scale WWTP Haarlem Waarderpolder. *Water Res.*, 34, 846-858.

Chudoba J., Albokova. J. and Cech J.S. (1989) Determination of kinetic constants of activated sludge microorganisms responsible for degradation of xenobiotics. *Water Res.*, 23, 1431-1438.

Coen F., Vanderhaegen B., Boonen I., Vanrolleghem P.A. and Vanmeenen P. (1997) Improved design and control of industrial and municipal nutrient removal plants using dynamic models. *Water Sci. Technol.*, 35, 53-61.

Eckenfelder W. (1986) Operation control and management of activated sludge plants treating industrial wastewaters. *Proc. of a seminar sponsored by Vanderbilt University.* Tennessee.

Gujer W., Henze M., Mino T. and van Loosdrecht M.C.M. (1999) Activated Sludge Model No. 3. *Water Sci. Technol.*, 3, 183-193.

Hellinga C. and van Loosdrecht M.C.M. (1999) Model based designed of a novel process for ammonia removal from concentrated flow. *Math. Comput. Modell. Dyn. Systems*, 5, 351-371.

Henze M., Grady C. P. L., Gujer W., Marais G.v.R. and Matsuo T. (1987) *Activated sludge model No 1.* IAWPRC Scientific and Technical Report No. 1. IAWQ, London.

Henze M. (1992) Characterization of wastewater for modeling of activated sludge processes. *Water Sci. Technol.*, 25, 1-15.

Henze M., Harremoes P., Jansen J.L.C and Arvin E. (1997) *Waste water treatment biological and chemical processes.* Springer, Berlin.

Hulsbeek J.J.W., Kruit J., Roeleveld P.J. and van Loosdrecht M.C.M. (2002) A practical protocol for dynamic modelling of activated sludge systems. *Water Sci. Technol.*, 45, 127-136.

Koch G., Kuhni M., Gujer W. and Siegrist H. (2000) Calibration and validation of activated sludge model No. 3 for Swiss municipal wastewater. *Water Res.*, 34, 3580-3590.

Krishna C. and van Loosdrecht M.C.M. (1999) Effect of temperature on storage polymers and settleability of activated sludge. *Water Res.*, 33, 2374-2382.

Layton A.C., Karanth P.N., Lajoie C.A., Meyers A.J., Gregory I.R., Stapleton R.D., Taylor D.E. and Sayler G.S. (2000) Quantification of *Hyphomicrobium* populations in activated sludge from an industrial wastewater treatment system as determined by 16S rRNA analysis. *Appl. Environ. Microbiol.*, 66, 1167-1174.

Meijer S.C.F., van Loosdrecht M.C.M. and Heijnen J.J. (2001) Metabolic modelling of full-scale biological nitrogen and phosphorus removing WWTP's. *Water Res.*, 35, 2711-2723.

Meijer S.C.F., van der Spoel H., Susanti S., Heijnen J.J. and van Loosdrecth M.C.M. (2002) Error diagnostics and data reconciliation for activated sludge modelling using mass balances. *Water Sci. Technol.*, 45, 145-156.

McCarty PL. (1965) Thermodynamics of biological synthesis and growth. *Adv. Water Pollut.*, 2,169-199 (1965).

Metcalf and Eddy (2003) *Wastewater engineering: Treatment, disposal and reuse.* McGraw-Hill, New York.

Moussa M.S., Rojas A.R., Hooijmans C.M., Gijzen H.J. and van Loosdrecht M.C.M. (2004) Model-based evaluation of nitrogen removal in a tannery wastewater treatment plant. *Water Sci. Technol.*, 50, 251-60.

Nyberg U., Aspegren H., Anderson B., Jansen J.L.C. and Villadsen I.S. (1992) Full scale application of nitrogen removal with methanol as carbon source. *Water Sci. Technol.*, 26, 1077-1086.

Nurse G. (1978) Denitrification with methanol: microbiology and biochemistry. *Water Res.*, 14, 531-537.

Nowak O., Franz A., Svardal K., Muller V. and Kunh V. (1999) Parameter estimation for activated sludge models with the help of mass balances. *Water Sci. Technol.*, 39, 133-120.

Orhon D. (1998) Evaluation of industrial biological treatment design on the basis of process modeling. *Water Sci. Technol.*, 38, 1-8.

Printemps C., Baudin A., Dormoy T., Zug M. and Vanrolleghem P.A. (2004) Optimization of a large WWTP thanks to mathematical modelling. *Water Sci. Technol.*, 50, 113-122.

Purschert I., Siegrist H. and Gujer W. (1996) Enhanced denitrification with methanol at WWTP Zurich-Werdholzli. *Water Sci. Technol.*, 33, 117-126.

Purschert I. and Gujer W. (1999) Population dynamics by methanol addition in denitrifying wastewater treatment plants. *Water Sci. Technol.*, 39, 43-50.

Roeleveld P.J. and van Loosdrecht M.C.M. (2002) Experience with guidelines for wastewater characterisation in the Netherlands. *Water Sci. Technol.*, 45, 77-87.

Sin G., Vanhulle S.W.H., De Pauw D.J.K., van Griensven A. and Vanrolleghem P.A. (2005) A critical comparison of systematic calibrating protocols for activated sludge model: A SWOT analysis. *Water Res.*, 39, 2459-2474.

Timmermans P. and Haute A.V. (1983) Denitrification with methanol. Fundamental study of the growth and denitrification capacity of Hyphomicrobium sp. *Water Res.*, 17, 1249-1255.

van Veldhuizen H.M., van Loosdrecht M.C.M. and Heijnen J.J. (1999) Modeling biological phosphorous and nitrogen removal in a full scale activated sludge process. *Water Res.*, 33, 3459-3468.

Warner A.P.C., Ekama G.A. and Marais G.v.R. (1986) The Activated sludge process-IV: Application of the general kinetic model to anoxic-aerobic digestion of waste activated sludge. *Water Res.*, 20, 943-958.

Wichern M., Obenaus F., Wulf P. and Rosenwinkel K. H. (2001) Modelling of full-scale wastewater treatment plants with different treatment processes using the activated sludge model No. 3. *Water Sci. Technol.*, 44, 49-56.

Figure 6.5 WWTP Shell Godorf oil refinery, Godorf, Germany (Photo: Pinzon A.).

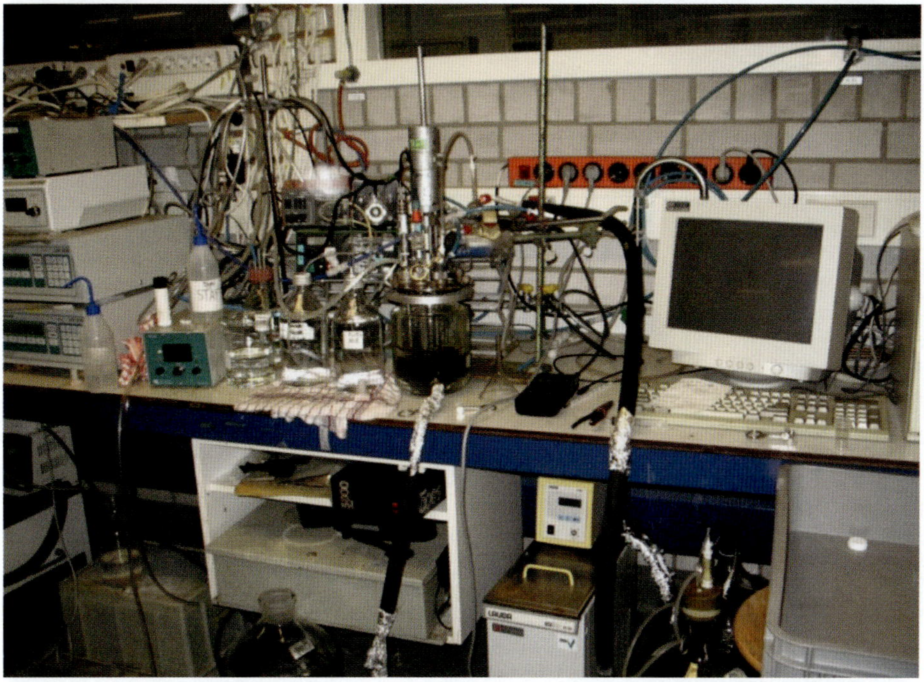

Figure 6.6 Experimental equipment used for sludge activity tests at WWTP Shell Godorf, Germany (Photo: Brdjanovic D.).

Figure 6.7 Sampling campaign at WWTP Shell Godorf, Germany (Photo: Brdjanovic D.).

Chapter 7: WWTP Walcheren, the Netherlands

Model-based evaluation of a novel upgrading concept for N removal

Salem S., Berends D.H.J.G., van Loosdrecht M.C.M. and Heijnen J.J.

This chapter is based on "Model-based evaluation of a new upgrading concept for N-removal" Salem S., Berends D.H.J.G., van Loosdrecht M.C.M., Heijnen J.J. (2002). Wat. Sci. Tech. 45, 6, 169-176.

7.1 Introduction

Activated sludge is the most widespread wastewater treatment process today. For almost a century, it has been successfully utilized as a system for carbon removal from sewage. For the last few decades its potential to remove nutrients has been explored and used. The growing public concern for environmental protection and stricter regulations directed research to improve the removal efficiency of the treatment systems. The main difficulty with activated sludge systems is their limited potential to nitrify at low temperatures (<15°C), making large volumes of aeration tanks needed. The large volume is due to the low growth rate of nitrifiers requiring long aerobic SRT. Besides nitrification, denitrification needs often to be achieved too. In large aerobic tanks also a significant heterotrophic COD-oxidation occurs limiting denitrification potential.

Increasing nitrification in the mainstream aerobic reactor can be obtained by several approaches. One approach is adding nitrification capacity in the aerobic compartment by addition of immobilized nitrifiers (e.g. the Pegasus process; Tanaka *et al.,* 1996) or by adding a carrier material on which a biofilm can grow (e.g. Chen *et al.,* 1995). Adding immobilized nitrifiers is however a relative expensive process. A second alternative is to augment the sludge with extra nitrifiers grown in separate fermentation equipment leading to a lower minimal SRT requirement (e.g. Rittmann *et al.,* 1996).

Inoculation of external grown nitrifiers is expensive and might lead to introduction of a strain that is not active at the prevailing process conditions. Using the sludge treatment flow with a high N-content to boost the nitrifying bacteria in a side stream tank has been proposed by DHV-Water under the name BABE® process (Zilverentant *et al.,* 1999). BABE® is an acronym for Bio-Augmentation Batch reactor Enhanced. By maintaining a short solid retention time in this side stream process predominantly the endogenous nitrifying population will be enhanced. The potential of this process has been extensively evaluated by Salem *et al.* (2003).

Mathematical modelling has become an operational tool for the design and operational procedures of activated sludge systems. Operational difficulties together with nutrient removal greatly increased the demand for process modelling. Considering the design, modelling has led to the identification of procedures to estimate the optimal or near optimal design configuration, reactor sizes and operational parameters (e.g. sludge age) and an estimation of the expected response (Henze *et al.,* 1987). With regard to the application of new treatment concepts, models can also be used to decrease the time and costs of scaling-up these new processes.

Traditionally a new application is tested extensively on lab scale, and then a pilot scale test period is used after which finally a full-scale application will be constructed. Usually it takes long and expensive periods especially for pilot plant phases. Mathematical models can help to bridge this gap between lab and full-scale application. Microbial conversions can be well studied at lab-scale, as they are scale independent. Based on laboratory experiments a proper model for the description of microbial processes as influenced by the microenvironment can be made. By modelling the well-known mass transport and transfer processes in a large reactor the micro-scale biological model can be coupled to a macro-scale model. This can give adequate prediction of the full-scale behavior and sensitive process parameters can be rapidly evaluated and eventually further tested at lab-scale. This approach has been successfully applied in the scale-up of the SHARON® process (a 2 liter lab-scale was scaled up to a 1,500 m^3 full-scale reactor, Hellinga *et al.,* 1998) and is used here for the BABE process. We used the WWTP Walcheren, The Netherlands as an example. Application of the BABE concept and traditional upgrading by expansion to meet a total N-content in the effluent of 10 mg N/L were compared.

7.2 Materials and methods

7.2.1 Walcheren wastewater treatment plant

The Walcheren WWTP in the Netherlands was designed for 140,000 PE and treats a total flow of 43,215 m^3/d. Wastewater is treated in two parallel lanes (Figure 7.1).

Figure 7.1 WWTP Walcheren, The Netherlands.

First, the water flows through 2 pre-sedimentation tanks followed by 2 selectors (anoxic tanks) each of volume 340 m^3. In the pre-sedimentation chemical P-precipitation occurs, this is maintained in the upgrading. The pre-settled water is treated in 2 aeration tanks each one with a capacity of 2,660 m^3 followed by 4 final clarifiers. The treatment plant operates at an average temperature of 14.5°C and a sludge age of 4.8 days. The treatment plant has an influent TKN of 46 mgN/L and influent COD of 236 mgCOD/L. The TSS is maintained at 2.8 gTSS/L in the aeration tanks and is 4.2 gTSS/L in the return sludge. The DO in the aeration tanks is controlled at 0.8

mgO_2/L. The return sludge flow is two times the influent flow and no internal recirculation of mixed liquor was present. The return water coming from digested sludge filtration (the so-called reject-water) has the following characteristics: a total flow of 324 m^3/d, an average temperature of 26°C and an NH_4-N concentration of 636 mgN/L. This reject-water is pumped back upstream the treatment plant.

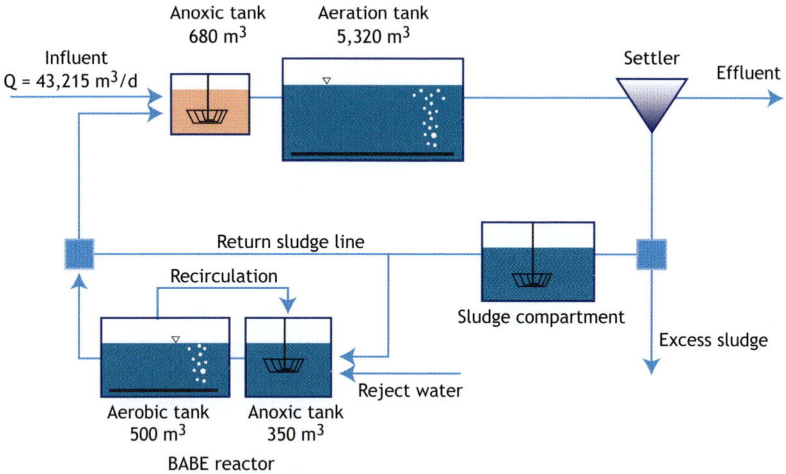

Figure 7.2 Process scheme of the WWTP Walcheren used for modeling.

The AQUASIM 2.0 software with the TUDP model (Hao *et al.,* 2001) was used to simulate the treatment plant. This simulation environment allowed the simulation of the new BABE concept and to evaluate its effect on the performance of the treatment plant. Only the activated sludge tanks and the settlers were considered in the simulation model. The settler was modelled with no processes inside but in the return sludge line a non-aerated tank of 1,650 m^3 was used to simulate the sludge blanket in the settler (Figure 7.2). This volume was based on the observed denitrification in the settler during summer periods as evaluated based on nitrate measurements in the effluent and in the return sludge. The model was calibrated to obtain correct effluent parameters according to the measured data reported from the treatment plant. The daily variations in influent composition were small therefore only the daily flow variations needed to be taken into account.

7.2.2 Wastewater characterization

The influent composition was evaluated (Table 7.1) according to STOWA method (Roeleveld and Kruit, 1998). The model was calibrated according to a procedure described by Meijer *et al.,* 2001. In the calibrated model, all model parameters remained at their default values (Hao *et al.,* 2001) except for the half-saturation/inhibition coefficient for oxygen for growth of autotrophic biomass (K_{NO}). It was changed from 0.5 to 0.3 to obtain a correct effluent prediction for ammonium and nitrate.

Table 7.1 Influent composition of the WWTP Walcheren, The Netherlands.

Measured influent composition	Concentration	Model fractions	Concentration
Total influent COD	236 mgCOD/L		
Soluble influent COD (S_F+S_A+S_I)	160 mgCOD/L	S_F	100 mgCOD/L
		S_A	0 mgCOD/L
		S_I	60 mgCOD/L
Particulate influent COD (X_I+X_S)	76 mgCOD/L	X_I	24 mgCOD/L
		X_S	52 mgCOD/L
Influent TKN	46mgN/L	S_{NH4}	34.7 mgN/L
		S_{NO3}	0 mgN/L
Alkalinity		S_{ALK}	8 mmol/L

7.2.3 The BABE reactor

The sludge line nitrification was introduced as a nitrification/denitrification system. Denitrification was implemented to control the alkalinity of the side stream process. The design was based on a general model-based evaluation of the process (Salem *et al.,* 2003). The anoxic compartment has a volume of 350 m^3 for denitrification and the aerated compartment has a volume of 500 m^3 (Figure 7.2). Internal recirculation between the two tanks was at a ratio of about 15 times the total flow to the BABE reactor. The source of ammonia to be treated in the BABE reactor was the warm reject-water (T=26°C, 636 mgN/L) resulting from filtration of the digested sludge. The load of ammonia sent to the side-stream reactor was subtracted from the influent ammonia composition of the WWTP. Alkalinity in the sludge liquor flow was based on the assumption that the molar ratio between NH_4 and HCO_3 is 1:1, (636 $mgNH_4/L$ or 46 mmol/L alkalinity). Alkalinity inside the BABE reactor was controlled by means of denitrification. The DO in the aerated compartment was set at 2 mgO_2/L. The sludge liquor was mixed with a portion of the return sludge of the main line of the treatment plant. The effluent from the BABE reactor, return sludge enriched with nitrifiers, was mixed with the remainder of the return sludge and then introduced in the selector compartment of the treatment plant. For comparison, two ratios of reject-water flow to return sludge flow (1:1 and 5:1) were used. Thereby two different operating temperatures occurred in the BABE reactor, 22°C and 28°C respectively.

7.3 Results and discussion

The treatment plant operates at a relatively low DO level (0.8 mg/L). Consequently, the effect of increasing the DO in the aeration tanks to improve the effluent quality was first evaluated. Then, the effect of applying the BABE concept on the existing treatment plant effluent was estimated. Finally the system (with or without the BABE concept) was upgraded to meet the effluent requirements.

7.3.1 Increasing the DO in the aeration tanks

The effect of increasing the DO level in the aeration tanks of the treatment plant was evaluated at an average (14.5°C) and winter (9°C) temperatures (Figure 7.3). The DO was increased from 0.8 mg/L to 1.5 mg/L and the corresponding changes in the effluent N parameters were calculated with the calibrated model.

Increasing the DO at winter temperature has no effect on the effluent N parameters of the treatment plant, as the system is still unable to nitrify. At the yearly average temperature (14.5°C), increasing the DO level has a beneficial effect on the performance of the treatment plant but this is to a certain extent and then it levels off. Increasing the DO from 0.8 to 1 mgO_2/L could decrease the effluent NH_4-N with about 60% from its original value. The effluent NO_3-N was increased concomitantly. Under summer conditions full nitrification occurs.

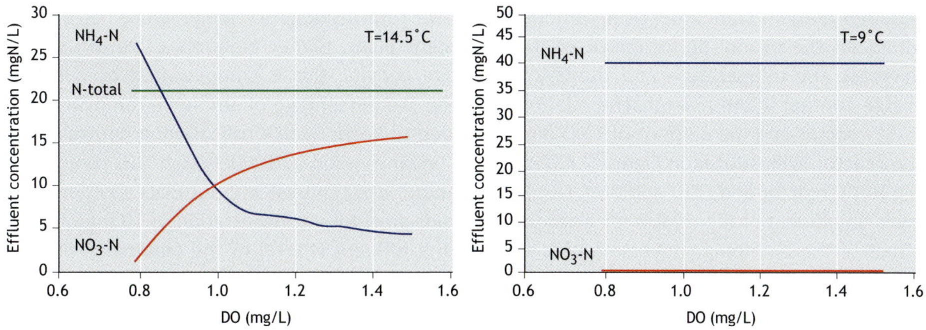

Figure 7.3 The effect of DO in the main aeration tank of the WWTP on the effluent N-parameters (at 14.5°and at 9°C).

7.3.2 Upgrading of the WWTP by the BABE concept

The treatment plant was upgraded through simulation by applying the BABE concept (Figure 7.2). The approach was introduced as a nitrification/denitrification system. Two tanks were implemented in the return sludge line. The first is anoxic where denitrification takes place and the second is aerated for converting all the ammonia coming with the reject-water. The results are presented in terms of the concentrations obtained in the aeration tanks of the main stream (with and without the BABE process) and the side-stream process. This was done for two ratios of reject-water flow and return sludge flow. The results are illustrated in Table 7.2. In order to prevent a limiting alkalinity (too low pH) extra COD was added to the process. To allow full nitrification this accounted for 1,166 kg/d (11.4 % of the input COD load to the treatment plant). In principle it would be possible to enlarge the denitrification zone to allow more endogenous denitrification in the side stream reactor. The choice for larger tanks or adding extra COD will depend on the specific site and costs.

Table 7.2 The effect of applying the BABE concept on the performance of the treatment plant

Parameter	Concentration without BABE process	Concentration with BABE process			
		Reject-water/RAS			
		1/1		5/1	
		T_{BABE} 22°C		T_{BABE} 28°C	
		Main stream	BABE reactor	Main stream	BABE reactor
DO	0.8	0.8	2.0	0.8	2.0
NH_4-N	25.4	9.0	7.7	10.0	37.0
NO_3-N	2.2	7.0	18.8	6.6	27.5
X_{AUT}	80	154	155	148	87
X_{AUT}/X_{TOT}	2.5%	4.7%	4.7%	4.2%	4.0%

The results illustrate the positive effect of bio-augmentation on improving the effluent N-parameters. The total N concentration in the effluent was reduced by around 40%. Part of this improvement is due to the reduction in the load of ammonia to the main stream of the WWTP (about 40%) and the rest (60%) is due to the inoculation of nitrifiers cultivated in the side-stream process and introduced into the main stream. The amount of nitrifiers present in the main stream process has been approximately doubled. By applying bio-augmentation the effluent parameters were improved but still the system was not able to fully nitrify or to meet the required effluent standards. Using different ratios of return sludge to sludge liquor has a small effect on the overall performance of the treatment plant. Higher amounts of return sludge decrease the temperature and thereby the sludge activity, this is compensated by a higher sludge content in the reactor. The relation between the percentage of ammonia removal in the BABE reactor and the amount of COD (kg/d) needed to keep, by denitrification, a neutral pH in the reactor is illustrated in Figure 7.4. An almost linear relation exists between NH_4 removal in the BABE reactor and COD addition, resulting in about 40% removal without COD addition. This gives an effluent of the treatment plant that contains an N-total concentration of 19 mgN/L. The effect of the ammonia conversion efficiency on the effluent N-total of the treatment plant is relatively small. It should be remembered that this is without any change in the main WWTP. It is clear that the side stream process should not be optimized on maximal N-conversion in the side stream, but on the effluent N of the upgraded WWTP and the costs of extra COD addition. Such complex evaluations can only be performed by simulation models.

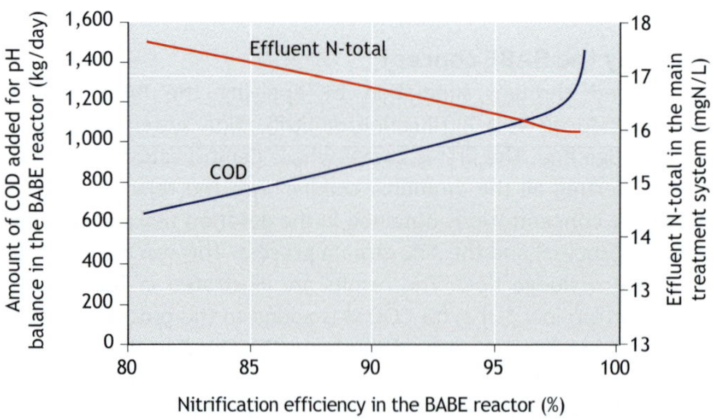

Figure 7.4 The relation between the ammonia removal efficiency in the BABE reactor and the amount of COD added (for pH balance in the reactor) and the subsequent effect on the effluent N-total of the treatment plant (at 14.5°C).

7.3.3 Modification of the WWTP Walcheren to meet the effluent requirements

Only implementing a side stream nitrification and augmentation of the WWTP Walcheren was not sufficient to reach the required effluent standard of 10 mgN/L. We evaluated the needed increase of the main stream treatment plant with and without the BABE process to obtain a N-total in the effluent of 10 mgN/L at the average operating temperature. The volumes of the aeration tank and the anoxic tank in the main stream were increased; also an internal recirculation was introduced. The following modifications were needed if the BABE process is applied (for the option 1:1 reject-water: return sludge flow):

- The volume of the aeration tanks needed to be increased by 22% from their original volumes.
- The anoxic tanks needed to be extended by 370% from their original volumes.
- An internal recirculation flow between the anoxic and aerated tanks in the main stream at a ratio of 4 times the influent flow rate was introduced.
- The SRT in the system was modified to fit the total suspended solids in the system (from 4.8 to 7.8 days).

On the opposite, if the Walcheren WWTP is upgraded following the traditional approach of extending the aeration basins and the anoxic tanks to achieve an N-total concentration in the effluent lower than 10 mgN/L, then the following modifications would be needed:

- The volume of the aeration tanks needed to be increased by 88%.
- The anoxic tanks needed to be increased by 1,300% from their original tank volumes.
- An internal recirculation flow between the anoxic and aerated tanks in the main stream at a ratio of 4 times the influent flow rate was introduced.
- The SRT in the system was extended to describe the total suspended solids in the system (from 4.8 days to 20 days).

7.3.4 Comparison of the upgrading strategies for the Walcheren WWTP

A comparison between upgrading the treatment plant by the conventional approach (of only extending the aeration and anoxic basins) and by applying the BABE concept is illustrated in Table 7.3. It is clear that a large saving in area requirements for the treatment reactors (approximately 50 %) can be obtained by the bio-augmentation concept. The aerobic volume can be reduced because of the reduced minimal aerobic SRT needed. The anoxic volume can be reduced because a lower denitrification is needed in the main stream of the treatment plant, and the aerobic COD removal is also lower because of the shorter aerobic SRT applied.

Table 7.3 Comparison between conventional upgrading and upgrading including bio-augmentation in the return sludge with reject-water of the Walcheren WWTP to meet effluent N-total of 10 mgN/L.

Item	Upgrading the treatment plant by a conventional method	Upgrading the treatment plant by applying the BABE concept
	Increase (additional volume required / original volume) (%)	
Aeration volume	88	22
Anoxic volume	1,300	370
Side stream reactor	0	14
Total increase	225	75

The cost effectiveness can be roughly compared, keeping in mind that site specific aspects might give a large change in these calculations. Here we compare the costs based on the standard cost analysis as used by DHV-Water. Using the BABE concept reduces the construction costs, for upgrading of the WWTP Walcheren to comply with a N-total effluent concentration of 10 mgN/L, by approximately EUR 750,000 because of the much smaller tank volumes required. The costs of land use are neglected. The addition of external COD in the form of methanol in the BABE concept costs approximately 35,000 EUR/yr. This addition of methanol and the lower SRT lead to a higher sludge production before digestion. This higher sludge production generates extra energy in the methane digestion process leading to a saving in energy costs of 70,000 EUR/yr. The slightly higher amount of sludge after digestion costs 30,000 EUR/yr for treatment and disposal. The difference in net yearly costs can be calculated based on an interest rate of 6% and 30 years depreciation for the civil engineering works. The net savings per year amount 115,000 EUR.

7.3.5 Use of modelling

In most modelling studies at Dutch WWTP it was shown that the default parameter set did not need to be calibrated in order to describe the process performance adequately. Since the concept proposed here is not strongly different we expect that the results will match future practical results. Modelling studies compared with pilot scale tests are relatively cheap and give the opportunity to evaluate all kind of process aspects in a systematic manner. This can strongly shorten the scale-up time of new treatment concepts. Based on the model results it has been decided to construct a BABE process at the WWTP Garmerwolde of 300,000 PE capacity. The planning of the upgrading of the Walcheren WWTP was already in a too far stadium to allow changes in the project planning.

7.4 Conclusions

The BABE concept was shown to be a cost-effective method for upgrading activated sludge systems, which do not meet the effluent requirements. It allows treatment plants to work at shorter aerobic SRTs which is compensated for by the nitrifiers grown in the return sludge tank and digester effluent or other ammonium waste sources. The side stream process should not be optimized on maximal N-conversion in the side stream, but on the effluent N of the main treatment system. The mathematical modelling can efficiently bridge the gap between lab-scale tests and full-scale application of new technologies saving time and money.

Acknowledgements

This project was financed by the Dutch organization for water research (STOWA). S. Salem was supported also by the Netherlands organization for international co-operation in higher education (Nuffic).

References

Chen JM, Hao OJ, Al-Ghusain IA, Lin CF (1995). Biological fixed-film systems. *Water-Environ. Res.* 67, 450-459.

Hao X, van Loosdrecht MCM, Meijer SCF, Heijnen JJ, Qian Y (2001). Model-based evaluation of denitrifying P removal in a two-sludge system. *Enviromental Engineering*, 127,2, 3, 112-118.

Hellinga C, Schellen AAJC, Mulder JW, van Loosdrecht MCM, Heijnen JJ (1998). The SHARON process: An innovative method for nitrogen removal from ammonium-rich wastewater. *Wat. Sci. Tech.* 37, 9, 135-142.

Henze M, Grady CPL, Gujer W, Marais GvR, Matsuo T (1987). Activated sludge model No.1, IAWQ Scientific and Technical report. No. 1, IAWPRC, London. ISSN: 1010-707X.

Lei Yang (1997). Investigation of nitrification by co-immobilized nitrifying bacteria and zeolite in a batchwise fluidized bed. *Wat. Sci. Tech*, 35, 8, 169-175.

Meijer SCF, van Loosdrecht MCM, Heijnen JJ (2001). Metabolic modelling of full-scale biological nitrogen and phosphorus removing WWTPs. *Wat. Res.* 35, 11, 2711-2733.

Rittmann BE (1996). How input active biomass affects sludge age and process stability. *J. Env. Eng.*, ASCE, 122, 1, 4-8.

Roeleveld PJ and Kruit J (1998). Guidelines for wastewater characterisation in The Netherlands. Korrespondenz Abwasser. 45, 3, 465-468.

Salem S, Berends D, van Loosdrecht MCM, Heijnen JJ (2003). Bioaugmentation by nitrification with return sludge. *Wat. Res.* 37, 8, 1794-1804.

Tanaka K, Sumino T, Nakamura H, Ogasawara T and Emori H (1996). Application of nitrification by cells immobilised in polyethylene glycol. *Prog. Biotechnol.* 11, 622-632.

Zilverentant A (1999). Process for the treatment of wastewater containing specific components e.g. ammonia. Patent PCT/NL99/00462, WO0005177.

Chapter 8: WWTP Anjana, India

Coupling models for integrated and plant wide modelling

Lopez-Vazquez C.M., Moussa M.S., Mithaiwala M., Van Loosdrecht M.C.M. and Brdjanovic D.

This chapter is based on "Use of Plant-wide Modeling for Optimization and Upgrade of a Tropical Wastewater Treatment Plant in a Developing Country" (2007) Brdjanovic D., Moussa M.S., Mithaiwala M., Amy G. and M.C.M. van Loosdrecht (2007) *Water Science and Technology*, 56(7), pp 21-31., and "Coupling ASM3 and ADM1 for wastewater treatment process optimization and biogas production in a developing country: Case-study Surat, India" Lopez-Vazquez C.M., Mithaiwala M., Moussa M.S., van Loosdrecht M.C.M., Brdjanovic D. (2013) *J. Water, Sanitation and Hygiene for Development*, doi: 10.2166/washdev.2012017.

8.1 Introduction

In developing countries, most of the large centralised wastewater treatment facilities are designed to accomplish the removal of suspended solids (SS) and biodegradable organic matter (often measured in terms of the biochemical oxygen demand - BOD). Moreover, in areas where the treatment plant effluent is reused (e.g. for irrigation), usually no legal requirements are imposed regarding nitrogen and phosphorus removal. However, N-removal starts to attract attention since the excessive concentrations of nitrogenous compounds can lead to adverse effects in agriculture (Khassab, 2009). Also, an increasing number of developing countries and countries in transition are strengthening their environmental regulations. Consequently, their existing treatment facilities (will) need to be upgraded and new sewage treatment plants need to incorporate biological nutrient removal processes. It is noteworthy that under the tropical conditions often found in most of developing countries, the environmental (air) temperature varies from moderate to relatively high (Arceivala, 2000). This is reflected in higher wastewater temperatures (from 20 to 30°C and even higher than 35°C) that enhance the kinetics of biological conversion processes. The latter combined with fact that usually the plants do not operate at their maximum designed capacity (having available sufficient reactor volume and aeration capacity) favours that, although originally designed for suspended solids and organic matter removal only, such plants can exhibit partial or even full nitrogen removal (nitrification and denitrification). In recent years, sludge waste management has attracted increasing attention. Considering that large activated sludge plants in developing countries usually include anaerobic sludge digesters (especially those with a tropical climate), anaerobic sludge digestion becomes an attractive technology for biogas production and energy recovery, contributing to reducing the energy costs while minimising the waste sludge generation. However, their overall performance depends on the efficiency of both the wastewater and sludge treatment lines, where their interaction affects the effluent quality, biogas generation and energy recovery.

Mathematical modelling of wastewater treatment plants has become a reliable and popular practice in developed countries. However, it has not been extensively applied in developing countries and countries in transition which, consequently, do not benefit from its application.

Furthermore, coupling activated sludge (Henze *et al.*, 2000) and anaerobic digestion models (Batstone *et al.*, 2002) can be a promising approach towards optimising the wastewater and sludge treatment processes and evaluating potential upgrade options in developing countries in a cost-effective manner (Salem *et al.* 2002). So far, successful practical applications of such an approach, involving a municipal wastewater treatment plant from developing countries, are fairly limited (Fall *et al.* 2009; Meneses *et al.*, 2011). Therefore, it is interesting from both a practical and a modelling perspective to further investigate the interactions between the wastewater treatment and the sludge treatment lines such as (i) the effects of modifying the return sludge flows and/or sludge wasting rate, (ii) the impact of upgrading the plant to perform N-removal on the anaerobic sludge digestion and (iii) the influence on high strength return flows from sludge treatment on the activated sludge system.

In this study, the Anjana wastewater treatment plant (WWTP), located in the city of Surat, India, was subject to modelling using a combination of the Activated Sludge Model No. 3 (ASM3) (Henze *et al.,* 2000) and the Anaerobic Digestion Model No.1 (ADM1) (Batstone *et al.*, 2002). The main goals of the study were (1) to assess the interactions between the wastewater and sludge treatment lines by the integration of ASM3 and ADM1, (2) to evaluate the applicability of ASM3 and ADM1 as well as the suitability of the Dutch Foundation for Applied Water Research (STOWA) wastewater characterisation protocol to describe the operation of a full-scale wastewater treatment plant, working under tropical conditions, from a developing country (Hulsbeek *et al.,* 2002), (3) to optimise the performance of the sewage plant in terms of effluent quality, sludge production and biogas generation, (4) to investigate different operational, technological and infrastructural requirements to meet new effluent requirements concerning nitrogen removal by mathematical modelling and (5) to assess the effects of the recirculation of the high strength reject flow from the sludge treatment to the wastewater treatment line on the performance of the plant.

8.2 Materials and methods

8.2.1 WWTP Anjana

Anjana WWTP is located in the city of Surat in India, Province of Gujarat (Figure 8.1).

Figure 8.1 Wastewater treatment plant in Anjana.

The ambient temperature during summer oscillates between 37 and 44°C and from 10 to 16°C in winter. The raw sewage temperature ranges from 23 to 35°C with an annual average of 30°C. Sewage is collected and conveyed in a separate sewerage network, serving approximately 200,000 people and associated industry (mostly dyeing, printing and weaving industries). The plant was constructed in 1958 with a design capacity of 25,000 m^3/d with only primary treatment, with the settled sewage used for irrigation. In 1981, the plant capacity was extended to 76,000 m^3/d. Meanwhile, due to rapid expansion and urbanization of agricultural land, the local authorities decided to dispose of the plant effluent (settled sewage) to nearby Koyli Creek, which ultimately discharges into the nearby Arabian Sea. To meet the effluent disposal requirements regarding organic matter, the plant was upgraded in 1996 by biological sewage treatment facilities (activated sludge system) and by facilities for anaerobic sludge digestion. Thus, since 1996 the plant has a maximum installed capacity of 82,500 m^3/d. It is based on a conventional configuration comprised of pre-treatment facilities, 2 primary settlers, 12 biological tanks with surface aeration (activated sludge system), and 3 final clarifiers (one secondary clarifier serves as a stand-by unit). The sludge retention time (SRT) is controlled by wasting excess sludge from the bottom of the final clarifiers. Sludge treatment consists of 3 sludge thickeners, 3 sludge digesters, a biogas collection tank, and sludge drying beds (Figure 8.2).

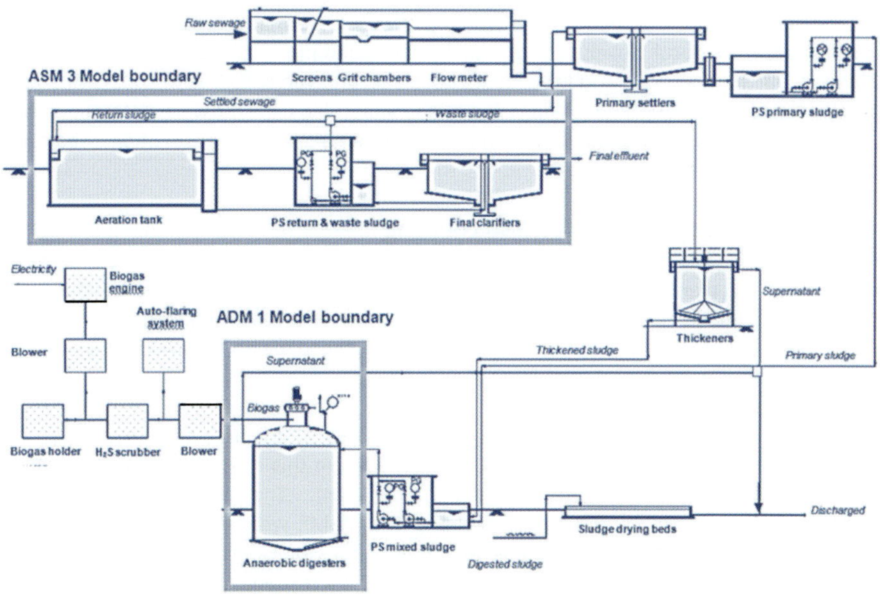

Figure 8.2 Simplified plant process lay-out.

In 2004, a biogas-based power plant (capacity 0.5 MW) was installed. The supernatant from the thickeners is returned and mixed with settled sewage; meanwhile the anaerobic digesters supernatant and filtrate from sludge drying beds are discharged into the sewerage network. Sludge digesters operate as low-rate units without mixing and the biogas plant generates an electric power of 0.2 MW since the hydraulic load (40,000 m^3/d) is approximately at 50% of its design capacity (82,500 m^3/d). Future operational scenarios contemplate the return of these reject effluents to the beginning of the treatment plant.

The treated effluent discharges into a nearby river. Local effluent disposal standards are: 5-day biochemical oxygen demand (BOD_5) ≤ 20 mg/L, suspended solids (SS) ≤ 30 mg/L and chemical oxygen demand (COD) ≤ 100 mg/L (GPCB, 1999). In the near future, the discharge limit for total nitrogen concentrations is expected to be 10 mg/L.

8.2.2 Sampling program and analytical methods

To monitor the plant operation, samples of raw sewage, settled sewage and effluent were collected on a daily basis. For modelling purposes, a tailor-made sampling programme was executed from October 24th to November 8th, 2004, under stable weather conditions (dry weather with both sewage and air temperature around 30ºC). The sampling points, sampling frequency and selection of the parameters of interest were determined based on the content of existing records and requirements of the mathematical models (ASM3 and ADM1). All analyses were performed in accordance with Standard Methods (APHA, 1996). The influent flow records and internal flow rates were obtained from the plant logbooks and on-site calibration of the pumps. The methane content of the biogas was measured by the on-line analyser of the plant. The average values of routinely collected data during the period October-November 2004 and from data collected during the sampling campaign are displayed in Table 8.1.

8.2.3 Wastewater and sludge characterization

The characteristics of the settled sewage (Table 8.2) were determined following the wastewater characterisation protocol developed by the Dutch Foundation for Applied Water Research (STOWA in Dutch) (Roeleveld and Kruit, 1998; Roeleveld and van Loosdrecht, 2002). Meanwhile, the characteristics of the primary sludge fed to the anaerobic digester were determined based on data measured in the plant and assuming the same sludge fractionation reported by Blumensaat (2002) (Table 8.2).

8.2.4 Model building and ASM3-ADM1 coupling

The sewage plant was described in AQUASIM (Reichert, 1998) where ASM3 and ADM1 were incorporated. The plant was modelled according to the hydraulic and operational parameters displayed in Table 8.3. The ASM3-to-ADM1 converting interface was used to couple the models and particularly to express the characteristics of the thickened waste activated sludge (WAS) as ADM1 parameters, prior to its treatment in the digester (Figure 8.3).

In recent years, different approaches have been proposed to couple activated and anaerobic models (Grau *et al.,* 2007; Nopens *et al.,* 2009). However, in this study, ASM3 and ADM1 were coupled following a simplified approach (Table 8.4). The soluble and particulate unbiodegradable organics (S_I and X_I, respectively) from ASM3 were directly associated with S_I and X_I for their input in ADM1 following the observations of Ekama *et al.* (2007). Also, the complex particulate COD fraction (X_C) was assumed to correspond to the volatile suspended solids fraction (X_{VSS}) from the sludge waste (Ekama *et al.,* 2007). The inorganic carbon and inorganic nitrogen concentrations (represented as S_{IC} and S_{IN} in ADM1, respectively) were assumed to correspond to alkalinity (S_{ALK}) and free and saline ammonia concentrations (S_{NH}), respectively. These relationships relied on the assumptions that carbonates and carbon dioxide (CO_2) were the main contributors to alkalinity and that free and saline ammonia (FSA) were practically the only inorganic nitrogen compounds in the sludge waste. Direct measurements and the model outcomes were used to validate these assumptions.

Table 8.1 Average wastewater characteristics obtained from the sampling program (October 24th to November 8th, 2004)

	Sampling point	Parameter[3]															
		COD_{tot}[1] mg/L	COD_{filt} mg/L	BOD_5 mg/L	TKN mgN/L	NH_4 mgN/L	NO_3 mgN/L	NO_2 mgN/L	TSS[1] mg/L	VSS[2] mg/L	VSS/TSS	T[1] °C	DO[1] mg/L	Gas[1] m^3/d	CH_4[1] %	pH[1]	Alk[2]
1.	● Before primary settling tank	853		3,812	43.2[2]	21.9[2]	1.7[2]	0.1[2]	781	550	0.70	29.5	0.87			7.2	433
2.	● After primary settling tank	357	1,561	1,361	37.2[1]	21.8[1]	1.9[1]	0.1[1]	142	104	0.73	29.5	1.81			7.2	387
3.	◆ Aeration tank (middle)	1,449				17.3[2]	1.8[2]	0.8[2]	1,434	926	0.65	27.7	0.5[4]			73	
4.	◆ Aeration tank (end)	1.39				16.2[2]	1.8[2]	0.8[2]	1,401	801	0.57	27.6	1.02			7.3	
5.	● After secondary settling tank	78	551	171	18.8[1]	14.7[1]	1.8[1]	0.7[1]	29	16	0.55	27.7	6.03			7.6	
6.	◆ Waste activated sludge	3,509	2,112				0.5[2]	0.2[2]	4,066	2.49	0.61	27.8				7	396
7.	◆ Primary sludge	53,456	7,552			40.2[2]			60,879	34.69	0.57	29.8				6.2	595
8.	◆ Thickened sludge	8,863	2,102			25.3[2]			8,326	5,728	0.69	28.6				6.4	449
9.	◆ Thickener supernatant	103	772						54	42	0.78	27.9				7.3	434
10.	◆ Before digester	32.16	8,772						26,024			29.1				6.1	588
11.	◆ Digester supernatant	9,748	1,862			237.6[2]	0.5[2]	0.1[2]				29.8				6.7	1,47
12.	◆ After digester	62,058	2,372						99,977	43,989	0.44	30.0		3,026	67	6.6	1,267

[1] Daily collected (16 days)

[2] Taken every 3 days

[3] All samples are 24 hrs composite of 12 two hrs grab samples

[4] Measured and corrected/calculated based on the net oxygen consumption and existing aeration capacity

● Routine monitoring

◆ Additional sampling point

Table 8.2 Influent wastewater characteristics determined in accordance to STOWA characterization guidelines (Roeleveld and van Loosdrecht, 2002) and sludge characterization based on sludge degradation data for ADM1 modelling applications (Blumensat, 2002).

Influent characterization required by ASM3		Settled sewage characteristics[4] at 30°C		Equations for determination of the influent characteristics	Parameters needed to be measured for the influent characterization		
Symbol	**Name**	**Unit**	**Value**		**Symbol**	**Name**	**Unit**
Soluble compounds					$COD_{tot,inf}$	Total influent COD	$gCOD/m^3$
S_{O2}	Oxygen (negative COD)[1]	$gCOD/m^3$	1.80	*Soluble compounds*	$COD_{filt,inf}$	Influent COD filtered	$gCOD/m^3$
S_S	Soluble substrates[2]	$gCOD/m^3$	107.36	$S_I = 0,9 \cdot COD_{filt,eff}$	COD_{VFAs}	COD from VFAs	$gCOD/m^3$
S_{NH4}	Ammonium and ammonia N[2]	gN/m^3	31.17	$S_S = COD_{filt,inf} - S_I$	$BOD_{5,inf}$	Influent BOD_5	$gBOD/m^3$
S_{NO3}	Nitrate and nitrite nitrogen[1]	gN/m^3	1.81	*Solid compounds*	NKj	TKN	gN/m^3
S_I	Soluble inert organics[2]	$gCOD/m^3$	49.06	$X_S = \{[BOD_t/(1-e^{-kt}]/(1 - Y_{BOD})] - S_S\}$	NH_4	Ammonium	gN/m^3
S_{ALK}	Alkalinity[1]	eq/m^3	6.44	$X_I = COD_{part,inf} - X_S$	NO_3	Nitrate	gN/m^3
Solid compounds				$X_H = 0$	TSS	Total Suspended Solid	g/m^3
X_I	Particulate inert organic	$gCOD/m^3$	37.56	$X_{AUT} = 0.1$ to 1.0	$COD_{filt,eff}$	Effluent COD filtered	$gCOD/m^3$
X_S	matter[2]	$gCOD/m^3$	162.74	$X_{STO} = 0$	$BOD_{5,eff}$	Effluent BOD_5	$gBOD/m^3$
X_H	Slowly biodegradable substrate[2]	$gCOD/m^3$	0.0	$X_{TSS} = i_{TSSXI} \cdot X_I + i_{TSSXS} \cdot X_S + i_{TSSBM} \cdot (X_H + X_{AUT}) + i_{TSSSTO} \cdot X_{STO}$	ALK	Alkalinity	eq/m^3
X_{STO}	Active heterotrophic biomass[3]	$gCOD/m^3$	0.0	*Other components*			
X_{AUT}	Glycogen[3]	$gCOD/m^3$	0.0	$COD_{tot,inf} = COD_{part,I\,nf} + COD_{filt,inf}$			
X_{TSS}	Active autotrophic biomass[3]	$gTSS/m^3$	142.11	$COD_{filt,inf} = S_S + S_I$			
	Influent suspended solids[2]			$COD_{part,inf} = X_I + X_S + (X_H + X_{STO} + X_{AUT})$			

Sludge characterization required by ADM1		Sludge characteristics[4] at 30°C		Equations for determination of the sludge characteristics	Parameters needed to be measured for the sludge characterization		
Symbol	**Name**	**Unit**	**Value**		**Symbol**	**Name**	**Unit**
Soluble compounds							
S_I	Soluble inert[2]	$kgCOD/m^3$	0.20	The fractionation of soluble degradable COD (i.e. S_{su}, S_{aa}, S_{fa}) were based on Blumensaat (2002). The inorganic carbon and nitrogen were measured as mg/L and then converted into molar units.	$COD_{tot,sludge}$	Total sludge COD	$gCOD/m^3$
S_{su}	Sugars[2]	$kgCOD/m^3$	0.06		$COD_{filt,sludge}$	Total filtered COD	$gCOD/m^3$
S_{aa}	Amino acids[2]	$kgCOD/m^3$	0.10		NH_4-N	Ammonium	gN/m^3
S_{fa}	Long chain fatty acids[2]	$kgCOD/m^3$	0.14		ALK	Alkalinity	$gCaCO_3/m^3$
S_{IN}	Inorganic nitrogen[1]	$mole/m^3$	0.0023				
S_{IC}	Inorganic carbon[1]	$mole/m^3$	0.0053				
Solid compounds							
X_I	Particulate inert[2]	$kgCOD/m^3$	15.74	The X_I and X_c were calculated based on COD particulate degradation data			
X_c	Particulate composites[2]	$kgCOD/m^3$	16.24				

[1] Measured [2] Calculated using given equations [3] Defaults in ASM3 [4] Based on average values obtained from the sampling program 24 October to 8 November 2004

Table 8.3 Average operational data of Anjana wastewater treatment plant during the period of study

Components	Units	Value	Modelled as
ASM3			
Aeration tank	m^3	26,250	Mixed Reactor Compartment
Secondary clarifier	m^3	3,300	Mixed Reactor Compartment (Clarifier)
Sludge compartment [1]	m^3	1,600	Mixed Reactor Compartment (Sludge compartment)
Influent flowrate (Q_{INF})	m^3/d	38,485	
Primary sedimentation tank underflow rate (Q_{UPS})	m^3/d	429	
Sludge recirculation flowrate (Q_{RAS})	m^3/d	21,186	
Internal thickener supernatant flowrate (Q_{TH})	m^3/d	709	
Waste of activated sludge (Q_{WAS})	m^3/d	1,090	
Solids retention time in the activated sludge system (SRT_{AER})	day	7	
ADM1			
Headspace [2]	m^3	3,050	Mixed Reactor Compartment (Headspace)
Digester	m^3	29,790	Mixed Reactor Compartment (Digester)
Reject outlet [3]	m^3/d	758	Mixed Reactor Compartment (Reject Outlet)
Sludge outlet [3]	m^3/d	192	Mixed Reactor Compartment (sludge Outlet)
Solids retention time in the anaerobic digester (SRT_{ANA})	day	35	

[1] Introduced in the model to stimulate the sludge retention in secondary clarifiers. Represents the volume of sludge blanket
[2] Introduced in the model to define the gas collection area in the digester. Represents the volume of gas collection area
[3] Introduced in the model to define the supernatant (reject outlet) and sludge outlet chambers in the digester.

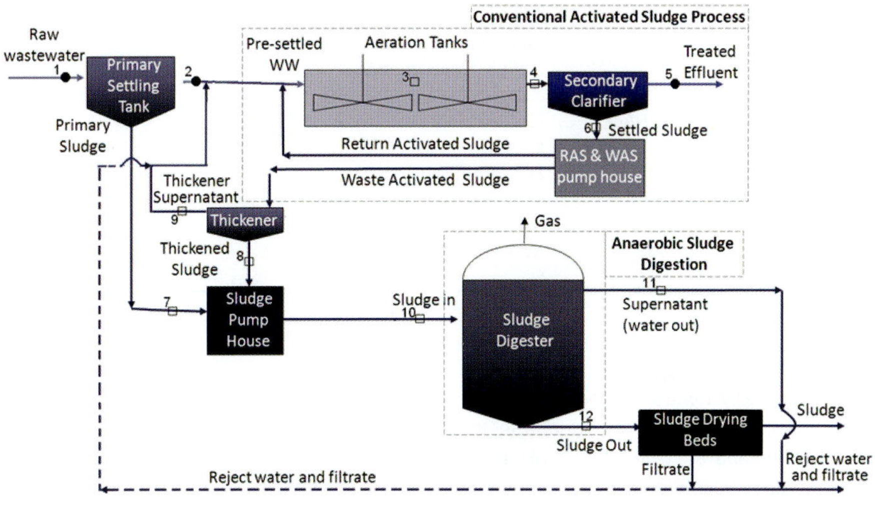

Figure 8.3 Simplified Anjana WWTP process scheme. The numbers indicate the routine monitoring points (closed circles) and additional sampling points assigned during the sampling program (open squares) as described in Table 8.1. The dotted line illustrates the future recirculation of supernatant from the digesters and filtrate from the sludge drying beds to the wastewater treatment line.

8.2.5 ADM1-ASM3 coupling

Although the high-N-strength filtrate from the sludge drying beds is not returned to the plant inlet, an ADM1-ASM3 converting interface was incorporated to describe its eventual return and evaluate its impact on the plant performance (Table 8.4). Like in the ASM3-ADM1 interface, the ADM1-ASM3 converting interface assumes that both soluble and particulate unbiodegradable organics (S_I and X_I, respectively) correspond to S_I and X_I for their input in ASM3. The soluble COD fractions such as aminoacids, sugars and volatile fatty acids (VFA) were considered to conform the readily biodegradable COD fraction (S_S) and the particulate COD assumed to represent the slowly biodegradable COD fraction (X_S). Regarding the nitrogen compounds, the biodegradable organic nitrogen was only associated to the particulate COD fractions (X_S) in the return flow. FSA and ammonium were introduced as such in ASM3 and the inorganic carbon (S_{IC}) assumed to correspond to alkalinity (S_{ALK}). The ADM1-ASM3 interface was incorporated in AQUASIM in the link that connects the filtrate from the sludge drying beds and the aeration tank.

Table 8.4 ASM3-ADM1 and ADM1-ASM3 interfaces.

	ASM3-ADM1 interface	
ADM 1 component/parameter	**Value from ASM3 output**	**ADM1 input**
Complex particulates	X_{VSS}	X_C
Particulate inert	X_I	X_I
Soluble inert	S_I	S_I
Inorganic carbon	S_{ALK}	S_{IC}
Inorganic N	S_{NH}	S_{IN}
Cations or anions	None, estimated from measured pH	S_{cat}, S_{an}
	ADM1-ASM3 interface	
ASM3 component/parameter	**Value from ADM1 output**	**ASM3 input**
Soluble inert COD	S_I	S_I
Readily biodegradable COD	1. If S_A is not used, then ΣS_i, excluding S_I, S_{IN} and S_{IC} 2. If S_A is used, then ΣS_i, excluding S_I, S_{IN}, S_{IC} and S_{VFA}	S_I and S_F
Volatile fatty acids (VFA, mostly HAc)	ΣS_{VFA}	S_S
Particulate inert COD	X_I	X_I
Slowly biodegradable COD	ΣX, excluding X_I	X_S
Free and saline ammonia (FSA)	S_{IN}	S_{NH}
Slowly biodegradable organic N	None, estimated by fitting	S_{ND}
Particulate biodegradable organic N	None, estimated by fitting	X_{ND}
Inorganic carbon	S_{IC}	S_{ALK}

8.2.6 Modelling strategy

First, ASM3 and ADM1 were applied by separate to describe the wastewater and sludge treatment lines, respectively. The boundaries of ASM3 and ADM1 were defined as shown in Figure 8.3. After separate calibration and validation, the models were coupled. For model calibration, data collected during the sampling programme were used.

8.2.7 Model calibration and validation

The ASM3 parameters were calibrated under steady-state conditions by fitting the COD and N model predictions to the data obtained from the sampling programme. The model calibration followed the procedure described by Hulsbeek *et al.* (2002) that included the calibration of: (1) the solids balance with X_I/X_S and mixed liquor suspended solids (MLSS) in the aeration tank, (2) effluent COD, (3) nitrification, and (4) denitrification. After calibration, the model was validated under dynamic conditions using the daily annual data. Annual data did not include nitrogen measurements; therefore, nitrogen data were determined by extrapolation of N/COD ratios, based on the N/COD ratios obtained from the sampling programme.

Two different periods were used for validation: (1) January 2004 at a wastewater temperature of 24°C, and (2) August 2004 at 29°C.

The ADM1 model was calibrated with the data collected during the sampling programme following a similar calibration protocol for ASM3. It included the calibration of: (1) solids balance, (2) methane production, and (3) inorganic nitrogen concentrations in the supernatant from the digester. Since the anaerobic digester has always showed a similar steady-state operation since its construction, the ADM1 could not be validated due to the lack of available data under different operating conditions. Despite this disadvantage, ADM1 and ASM3 were coupled to estimate the impact of the return of filtrate and explore potential control and operation measures.

8.2.8 Scenarios evaluation for process upgrade and optimization

Different upgrade alternatives were evaluated concerning the optimization of the biological nitrogen removal and methane generation under current conditions (with an estimated influent wastewater flowrate of 40,000 m^3/d) but also taking into consideration future operating conditions. In this regard, two scenarios were identified by the local sewage corporation based on medium and high flow increase prognosis for the year 2010 (60,000 m^3/d) and 2020 (80,000 m^3/d), respectively. For the current situation (base case) as well as for each of the two scenarios, a number of actions and adjustments were identified and subjected to simulation using the combined ASM3-ADM1 model (Table 8.5). First, the approach was to check how the existing plant operation could be improved with relatively simple operational adjustments (alternations of flow rates of internal streams, adjustment of oxygenation, creation of anoxic zones, etc.), and then to examine more substantial measures to upgrade the plant by the construction of additional treatment units.

8.3 Results

8.3.1 Model calibration

ASM3 was calibrated in 4 steps:

1. The influent X_I/X_S ratio was adjusted to describe the MLSS concentration in the aeration tank as described by Meijer *et al.* (2001). Initially, the model predicted a MLSS concentration of 1.33g/m^3, compared to the measured concentration of 1.43 g/m^3. After increasing the influent X_I fraction from 0.108 to 0.14, a satisfactory estimation was obtained (1.42 g/m^3).
2. The simulations showed an effluent COD concentration of 77 gCOD/m^3 against the measured value of 78 gCOD/m^3. The difference was assumed to be acceptable (at around 2%) so no additional calibration was necessary.
3. The nitrification process needed to be calibrated since the model predicted full nitrification (e.g. no ammonia observed in the effluent), whereas 14.7 gN/m^3 was actually measured. Thus, the growth of autotrophic organisms was reduced from 1.0 1/d (the default value) to 0.46 1/d. Thereby, ASM3 predicted an effluent ammonia concentration of 14.1 gN/m^3 which was about 4% lower than that measured.
4. Finally, the model predicted a nitrate concentration of 2.3 gNO_3-N/m^3 in the effluent against the measured value of 1.8 gNO_3-N/m^3. These concentrations were considered to be within the same relatively low concentration range and hence there was no need to calibrate the denitrification process further.

Table 8.5 of applied actions and measures regarding the current situation and two development scenarios.

A. Base case: Current situation (Q=40,000 m3/d)	B. Scenario 1: Medium flow increase prognosis (Q=60,000 m3/d)	C. Scenario 2: High flow increase prognosis (Q=80,000 m3/d)
	Applied actions and measures	
A.1. Alteration in return activated sludge flow (RAS) and waste activated sludge flow (WAS) A.1.1. Existing situation A.1.2. Increased RAS from current 50 to 75% expressed as percentage of the influent flow A.1.3. Increased RAS from 50 to 100% A.1.4. RAS at 50% and reduced WAS by ⅓ A.1.5. Increased RAS to 50 to 75% and reduced WAS by ⅓ A.1.6. Increased RAS to 50 to 75% and reduced WAS by ⅔ **A.2. Reduction in sludge withdrawal from the digester to optimize the methane production** A.2.1. Increased RAS to 50 to 75%, reduced WAS by ⅓ and reduced sludge withdrawal from the digester by 25%. A.3. Adjustment of DO level in the entire biological tank from current 0.3 to 1.0 mgO2/L A.3.1. Reduced WAS by ⅓ A.3.2.Reduced WAS by ⅓ and one digester shut down A.3.3.Reduced WAS by ⅓ and two digesters shut down **A.4. Adjusting DO level in the second half of biological tank from current 0.3 to 1.0 mgO2/L** A.4.1. Partial DO increase A.4.2. Partial DO increase and reduced excess sludge by ⅓ **A.5. Creation of anoxic phase for N-removal upgrade within the existing biological tank** A.5.1. Partial DO increase, reduced WAS by 50% and creation of anoxic zone in the last quarter of the existing biological tank A.5.2. Partial DO increase, reduced WAS by 50%, creation of anoxic zone in the first quarter of the existing biological tank, and internal recirculation of 300%	**B.1. Increased COD concentration to biological tank** B.2. Increased primary sludge withdrawal rate to optimize methane production B.3.1. Increased primary sludge withdrawal rate by 50%, and reduced WAS by ⅓ B.4. Creation of anoxic phase for N-removal upgrade in the last quarter of exiting biological tank B.4.1. Increased primary sludge withdrawal rate by 50%, and reduced WAS by ⅓ B.4.2. Increased recirculation from 50 to 100%, increased primary sludge withdrawal rate by 50%, and reduced WAS by ⅓ **B.5. Creation of anoxic phase for N-removal upgrade in the last quarter of exiting aeration tank with increased COD concentration** B.5.1. Increased RAS from 50 to 100%, increased primary sludge withdrawal rate by 50%, and reduced WAS by ⅓, at current COD concentration. B.5.2. Increased RAS from 50 to 100%, increased primary sludge withdrawal rate by 25%, reduced WAS by ⅓ and increased COD concentration by ⅓ B.5.3. Increased RAS from 50 to 100%, increased primary sludge withdrawal rate by 10%, reduced WAS by ⅓ and increased COD concentration by 50%. **B.6. Creation of anoxic phase for N-removal upgrade in the first quarter of exiting biological tank** B.6.1. Increased RAS from 50 to 100%, increased primary sludge withdrawal rate by 50%, reduced WAS by ⅓ and increased internal recirculation 250%	**C.1. Increasing primary sludge withdrawal rate to optimize methane production** C.1.2. Increased recirculation from 50 to 100%, increased primary sludge withdrawal rate by 100% and reduced WAS by ⅓ **C.2. Adjustment of DO level in the biological tank from current 0.20 to 0.80 mgO2/L** C.2.1. Increased DO level in the biological tank from current 0.20 to 0.80 mgO2/L **C.3. Creation of anoxic phase for N-removal upgrade in the last quarter of exiting biological tank** C.3.1. Creation of anoxic zone in the last quarter of the existing biological tank C.3.2. Creation of anoxic zone in the last quarter of the existing biological tank, increased COD load by ⅓, and increased primary sludge withdrawal rate by 65% **C.4. Introduction of additional anoxic volume** C.4.1. Introduction of additional anoxic volume after existing biological tank C.4.2. Introduction of additional anoxic volume after existing biological tank, reduced RAS from 100% to 50% C.4.3. Introduction of additional anoxic volume after existing biological tank, increased COD concentration by 25%, and increased primary sludge withdrawal rate by 60% C.4.4. Introduction of additional anoxic volume before existing biological tank and increased internal recirculation 200%

ADM1 was calibrated in 3 steps, as follows:

1. The solids balance was calibrated based on the X_I/X_C fractions and COD-sludge leaving the digester. Initially, the ADM1 default parameters and the sludge characterisation were used (Table 8.2). It resulted in a COD-non-stabilised concentration in the digester of 22.8 kgCOD/m^3 against a measured concentration of 22.7 kgCOD/m^3. Hence, there was no need for further calibration. However, a COD concentration of 10.6 kgCOD/m^3 was predicted in the supernatant against 9.7 kgCOD/m^3 measured, and the predicted stabilised COD sludge was 61.9 kgCOD/m^3 against 64.6 kgCOD/m^3. Consequently, the X_I and S_I fractions were changed from 0.20 to 0.29 and from 0.10 to 0.01, respectively. This led to COD concentrations in the sludge and supernatant of 64.2 and 9.8 kgCOD/m^3, respectively.
2. After COD calibration, a satisfactory description of the generated biogas (1,526 m^3/d against the measured value of 1,513 m^3/d) and of the percentage of methane in the biogas was observed (67.9% predicted *vs.* 67.4% measured).

3. An inorganic nitrogen concentration in the supernatant of 241.5 gN/m^3 was observed against the measured value of 237.6 gN/m^3. This was considered satisfactory so no further calibration of the model was needed.

After coupling ASM3 and ADM1, the predictions of the integrated model provided a satisfactory description of the plant operation (Table 8.6) and it was assumed that there was no need to calibrate the coupled model.

Table 8.6 Comparison of the results of simulations using the coupled model and separate models.

Parameter	Unit	Measured value	ASM3	ADM1	ASM3 + ADM1
Aeration tank (middle)					
COD	$gCOD/m^3$	1,449	1,451	-	1,458
Ammonia-N	gN/m^3	17.3	14.5	-	14.1
Nitrate-N	gN/m^3	1.8	1.7	-	2.2
MLSS	mgSS/L	1,430	1,419	-	1,415
Wasted activated sludge					
COD	$gCOD/m^3$	3,509	3,558	-	3,670
Ammonia-N	gN/m^3	-	14.5	-	14.1
Nitrate-N	gN/m^3	0.5	1.7	-	2.0
MLSS	mgSS/L	4,066	3,806	-	3,802
Thickened sludge					
COD	$gCOD/m^3$	8,863	-	-	8,871
Ammonia-N	gN/m^3	25.3	-	-	14.1
Nitrate-N	gN/m^3	-	-	-	2.0
MLSS	mgSS/L	8,326	-	-	8,638
Water line effluent					
COD-effluent	$gCOD/m^3$	78.4	76.7	–	74
Ammonia-N	gN/m^3	14.7	14.1	–	14.1
Nitrate-N	gN/m^3	1.8	2.3	–	2.2
Sludge line effluent					
COD-sludge	$kgCOD/m^3$	64.6	–	61.9	63.9
CH_4	%	67.4	–	67.9	67.8
Gas flow	m^3/d	1,513	–	1,526	1,524
Ammonia-N	gN/m^3	237	–	242	249

8.3.2 Model validation

The ASM3 was validated with data from winter conditions (January 2004) and the wet (monsoon) season (August and September 2004). In these periods, different sewage temperatures and flow rates were observed (for instance of 24°C in January 2004 *vs.* 28°C during the sampling campaign, and flow rates around 45,000 m^3/day in September-August 2004 and 36,000 m^3/day in January 2004, compared to 38,000 m^3/day during the sampling programme). Despite these differences, the ASM3 satisfactorily predicted the effluent COD concentrations from the validation periods (Figure 8.4).

However, the concentrations of nitrogenous compounds could not be validated because of the unavailability of historical data concerning these parameters. The ADM1 could not be validated because of the lack of historical data under operating conditions different from those used to calibrate the model, since the anaerobic digesters have always been operated under similar conditions (exhibiting a steady-state operation reflected in a stable sludge production and methane formation).

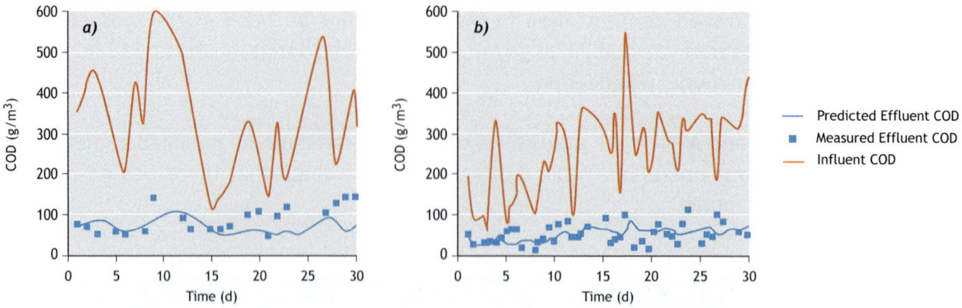

Figure 8.4 (a) COD validation for January 2004 and (b) August - September 2004

8.3.3 Model-based evaluation for process optimization and upgrade

A) Base case: Current situation (Q=40,000 m^3/d)

The coupled model was very capable of predicting the existing situation (A.1.1) (Table 8.7). For example, the predicted effluent ammonia and nitrate concentrations were 14.1 and 2.3 mgN/L, respectively, against average measured concentrations of 14.7 and 1.8 mgN/L. The same conclusion can be drawn based on a comparison of predicted and measured methane production. Simulation results showed that by simple manipulation of return activated sludge (RAS) and waste activated sludge (WAS) flow rates, better N removal could be achieved (A.1.6.: 8.7 gNH_4-N/m^3 and 2.9 gNO_3-N/m^3) with only a marginal effect on methane production (reduction from 3,097 to 3,013 m^3/d). Increased aeration resulted in improved nitrification (A.3.1: 2.7 gNH_4-N/m^3) while overall N-removal deteriorated (22.9 gNO_3-N/m^3). A partial increase in aeration in the first half of the biological volume resulted in improved N removal (A.4.2: 3.8 gNH_4-N/m^3 and 12.8 gNO_3-N/m^3). Conversion of one quarter (6,562 m^3) of the biological volume from aerobic to anoxic resulted in improved N removal efficiency, especially in the case of pre-denitrification (A.5.2.: 3.6 gNH_4-N/m^3 and 6.1 gNO_3-N/m^3, methane production unchanged). According to coupled model predictions, these adjustments did not affect methane production significantly (variations within 3%). However, taking one or two digesters out of operation resulted in a decrease in methane production by 4 and 15%, respectively.

B) Scenario 1: Medium flow increase prognosis (Q=60,000 m^3/d)

The simulation with increased COD load to the activated sludge system by one third, suggested that the existing plant configuration will result in poor N removal efficiency (B.1.1: 31 gNH_4-N/m^3) and increased methane production (22%), (Table 8.7). Assuming that the currently oversized primary settling tank could retain its present efficiency regarding COD removal, further simulations with alternations in RAS and WAS flow rates only, indicated that overall N removal efficiency could be approximately equal to that of the present situation, while methane production could be increased by 47% (B.2). An increase of aeration over the entire biological tank improved nitrification, as expected; however, denitrification was absent (B.3.1). Introduction of post-denitrification within the existing biological volume (last quarter), improved overall N removal (B.5.1), however, the new N removal target of total N $\leq$ 10 mg/L was not achieved assuming the present efficiency of primary clarifiers. Further simulations indicated that increased COD load to the activated sludge system would improve denitrification and that satisfactory N removal could be achieved (B.5.2 and B.5.3). The comparative option with pre-denitrification appeared to be the better solution: the results showed similar N removal efficiency (effluent ammonia and nitrate concentration of 4.6 and 5.3 gN/m^3, respectively). However, the methane production increased by 15 to 24%.

Table 8.7 Summary of simulations of selected actions and measures regarding the current situation and two development scenarios for Anjana plant

Parameter[a)] Scenario	COD_{sol}[b)] g/m^3	NH_4^+ gN/m^3	NO_3^- gN/m^3	DO[c)] gO_2/m^3	O_2-Net gO_2/m^3	SRT[d)] d	MLSS g/m^3	COD_{sludge} kg/m^3	CH_4 %	Gas_{gen} m^3/d	$CH_{4gen.}$ m^3/d
A. Base case: Current situation (Q=40,000 m^3/d)											
A.1.1.	49	14.5	1.8	0.30	141	7.3	1,419	63.9	67.7	4,572	3,097
A.1.2.	49	13.4	1.9	0.30	127	8.6	1,539	63.1	67.9	4,505	3,058
A.1.3.	49	12.9	1.9	0.29	114	9.6	1,598	62.5	68.0	4,457	3,031
A.1.4.	49	11.6	2.4	0.29	152	10.4	1,761	62.2	66.2	4,670	3,090
A.1.5.	49	11.0	2.4	0.28	134	12.2	1,885	61.4	66.3	4,602	3,051
A.1.6.	49	8.7	2.9	0.27	142	21.1	2,536	59.2	64.8	4,654	3,013
A.2.1.	49	11.0	2.4	0.28	134	12.2	1,885	78.6	66.3	4,602	3,051
A.3.1.	49	2.7	22.9	1.00	201	10.3	1,726	62.1	66.5	4,639	3,084
A.3.2.	49	2.7	22.9	1.00	201	10.3	1,726	62.4	66.7	4,458	2,974
A.3.3.	49	2.7	22.9	1.00	201	10.3	1,726	64.6	67.4	4,015	2,705
A.4.1.	49	5.0	12.3	0.26/1.00	173	7.3	1,425	63.7	68.1	4,540	3,090
A.4.2.	49	3.8	12.8	0.25/1.00	181	10.3	1,726	62.1	66.5	4,639	3,084
A.5.1.	49	7.0	7.7	0.95/Anox	170	12.9	1,983	61.2	65.5	4,696	3,074
A.5.2.	49	3.6	6.1	Anox/0.95	171	13	2,003	61.2	65.5	4,699	3,077
B. Scenario 1: Medium flow increase prognosis (Q=60,000 m^3/d)											
B.1.1.	71	31.0	0.02	0.15	168	7.8	3,612	77.8	65.6	5,778	3,792
B.2.1.	49	20.5	0.3	0.23	128	7.3	2,050	98.9	66.0	7,062	4,663
B.2.2.	49	17.5	0.2	0.20	138	10.5	2,519	96.7	66.4	6,862	4,557
B.2.3.	49	17.5	0.2	0.20	138	10.5	2,519	123.8	66.4	6,862	4,557
B.3.1.	49	3.1	21.0	0.90	200	10.4	2,342	96.0	66.7	6,778	4,522
B.4.1.	49	9.7	4.2	0.81/Anox	160	10.2	2,329	96.1	66.6	6,801	4,526
B.4.2.	49	8.5	5.0	0.80	123	14.1	2,509	93.5	66.8	6,585	4,400
B.5.1.	49	8.5	5.0	0.80	123	12.5	2,509	93.49	66.8	6,585	4,401
B.5.2.	65	8.3	1.2	0.69	150	13.5	3,320	78.56	67.1	5,606	3,763
B.5.3.	71	8.1	0.6	0.64	162	13.9	3,690	72.49	67.3	5,198	3,498
B.6.1.	49	4.6	5.3	Anox/0.78	128	15.0	2,790	92.3	67.0	6,483	4,344
C. Scenario 2: High flow increase prognosis (Q=80,000 m^3/d)											
C.1.1.	49	26.4	0.1	0.20	112	6.6	2,700	132.9	65.3	9,336	6,096
C.1.2.	49	23.3	0.1	0.14	92	12.5	3,428	80.5	65.9	8,741	5,762
C.2.1.	49	4.0	18.6	0.80	145	11.9	2,969	79.6	66.2	8,610	5,699
C.3.1.	49	11.1	2.2	0.68/Anox	114	11.8	2,977	79.7	66.1	8,634	5,704
C.3.2.	65	10.9	0.3	0.53/Anox	141	13.0	4,010	66.7	66.2	7,377	4,885
C.4.1.	49	7.3	7.8	0.80	129	14.5	2,800	79.4	66.2	8,587	5,681
C.4.2.	49	8.0	7.3	0.80	167	11.0	2,688	81.5	65.9	8,858	5,839
C.4.3.	65	8.2	2.3	0.80	204	11.8	3,545	69.0	66.0	7,660	5,052
C.4.4.	49	4.7	6.5	Anox + 0.80	172	11.0	2,710	81.6	65.9	8,863	5,840

[a)] Water quality parameters are related to plant effluent.

[b)] In all simulations COD content of the settled sewage was used. Settled sewage COD concentration obtained from the sampling programme (357 mg/L) was used in most of cases assuming that the efficiency of primary settling tanks remained unchanged as result of existing spare capacity within these units. In a few cases (B.1.1., B.5.2., B.5.3., C.3.2. and C.4.3.) the process performance was checked with comparatively higher COD concentrations in settled sewage (33 and 50% increase).

[c)] Since aeration capacity was designed for the organic load associated with influent flow rate of 82,500 m^3/d, the need for aeration supplement in both base cases and future scenarios still need to be verified.

[d)] SRT was calculated taking into account the biomass present in the biological tanks, final settlers, plant effluent and excess sludge stream.

C) Scenario 2: High flow increase prognosis (Q=80,000 m^3/d)

In general, analogous results were obtained in comparison with scenario 1; however, different combinations of adjustments of RAS, WAS, increase in DO, etc., within the existing plant configuration, still could not bring the total N effluent concentration below 10 mg/L (Table 8.7). Simulations indicated that with the existing efficiency of primary settlers, introduction of post-denitrification within the existing volume would not bring the total N concentration in the effluent to a satisfactory level (C.3.1). Reduced settling efficiency would help N removal (C.3.2: 10.9 gNH_4-N/m^3 and 0.3 gNO_3-N/m^3) on the expense of methane production (reduction by 17%).

The best option appeared to be the construction of an additional anoxic volume (6,562 m^3) in front of the existing aeration tanks which showed the best balance between the effluent quality requirements and methane production (C.4.4: 4.7 gNH_4-N/m^3 and 6.5 gNO_3-N/m^3, marginal influence on methane production).

8.3.4 Modelling the return of the filtrate stream

Once the integrated ASM3-ADM1 model described the plant operation, the ADM1-ASM3 interface was added to assess the effects of the return of filtrate from the side-stream sludge handling to the mainstream wastewater treatment line. This doubled the MLSS concentration in the aerobic reactor (from 1.42 to 2.92 g/L) (due to a higher accumulation of inert organics and the potential recycling of biodegradable organics coming from the digester), but kept within the average values for activated sludge systems (Ekama 2008). Moreover, the plant has one secondary clarifier on stand-by which can be used if necessary. The model also predicted higher effluent COD and ammonia concentrations (of 155 mg/L and 22.6 mg/L, respectively). Conversely, the effluent nitrate concentration decreased from 2.3 to 1.0 mg/L, possibly due to a higher availability of biodegradable COD coming from the digester. Due to the recycling of biodegradable COD and the accumulation of inert material in the system, an increase in the biogas production (from 1,526 to 1,740 m^3/d) and a higher production of stabilised sludge (from 64.1 kgCOD/m^3 to 73 kgCOD/m^3) were observed. However, the plant could not comply with the effluent COD and total nitrogen discharge limits of 100 mg/L and 10 mg/L, respectively. The optimal operating conditions found during the process optimization were used as a basis to cope with the return of the filtrate: (a) an anoxic phase in the first quarter of the aerobic tank was included, (b) the internal recirculation from the aerobic to the anoxic stage was increased to 4 times the influent flow rate, and (c) the dissolved oxygen (DO) concentration in the aerobic tank adjusted to 2 mg/L. This contributed to the improvement of the effluent quality concerning ammonia (which decreased from 22.6 to 7.1 mg/L) and nitrate (kept at around 6.7 mg/L), but had no effect on the COD effluent quality (which increased up to 163 mg/L). With regard to the anaerobic digester, the biogas production dropped to 1,505 m^3/d whereas the production of stabilised sludge increased up to 78 kgCOD/m^3. The increase in effluent COD and sludge production can be explained on the basis of the higher accumulation of unbiodegradable (soluble and particulate) COD fractions, meanwhile the drop in biogas generation is a reflection of the higher utilisation of biodegradable COD for denitrification purposes.

The simulation results suggest that the plant will hardly meet the effluent COD and total nitrogen standards. The higher concentration of unbiodegradable soluble COD from the return flow is the limiting factor. It is likely that an advanced oxidation process will be required. On the other hand, the effluent total nitrogen concentration is limited by the lack of biodegradable COD, reflected in an effluent nitrate concentration of 6.7 mg/L. Possibly, the addition of an external carbon source will be needed to enhance denitrification or the implementation of an innovative side-stream nitrogen removal process (van Loosdrecht, 2008). For a more reliable evaluation and further full-scale application, this will require the additional characterisation of the return flow(s).

8.4 Discussion

8.4.1 Influent and sludge characterization

In general, the different COD fractions from the Indian wastewater were similar to those from previous reports (Table 8.8), except for the easily biodegradable substrate (S_S). The difference can be attributed to the nature of the methods: on the one hand, the respirometry-based Hochschulgruppe (HSG) characterisation protocol (Koch *et al.,* 2000; Wichern *et al.,* 2003) and, on the other hand, the more practically-oriented STOWA protocol (Roeleveld and Kruit 1988; Brdjanovic *et al.,* 2000; Meijer *et al.,* 2001; Hulsbeek *et al.,* 2002).

Table 8.8 Fractionation of total COD in the present and previous modeling studies.

Study	Model	Country	S_S	S_I	X_S	X_I	X_A	X_H
Brdjanovic *et al.* (2000)	TUDP	Netherlands	0.32	0.07	0.42	0.19	0	0.00
Meijer *et al.* (2001)	TUDP	Netherlands	0.33	0.07	0.31	0.29	0	0.00
Koch *et al.* (2000)	ASM3	Switzerland	0.06	0.10	0.55	0.20	0	0.09
Wichern *et al.* (2003)	ASM3	Germany	0.19	0.04	0.55	0.12	0	0.10
	ASM3	Germany	0.19	0.08	0.48	0.10	0	0.15
	ASM3	Germany	0.17	0.07	0.45	0.12	0	0.19
This study	ASM3	India	0.30	0.14	0.42	0.14*	0	0.00

*During calibration, the X_I fraction was adjusted from 0.11 to 0.14 to predict the correct MLSS concentration in the aeration tank

As far as ADM1 is concerned, different studies have shown substantial variations in sludge characterisation (Table 8.9), possibly due to the unavailability of a standard protocol. The main difference lies in the particulate composites component (X_C). This can be attributed to the fact that Siegrist *et al.* (2002) and Blumensaat *et al.* (2002) consider the hydrolysis of the particulate composites in the storage tank, leading to higher concentrations of soluble biodegradable organics when compared to the values obtained in this research. The direct determination of particulate and soluble COD measurements and COD degradation, and the acceptable description of the plant, support the approach followed in this study.

Table 8.9 Fraction of total COD in the sludge.

Study	Model	X_c	X_I	S_{su}	S_{aa}	S_{fa}	S_I
Siegrist *et al.*, (2002)	Siegrist	0.300	0.400	0.060	0.090	0.135	0.015
Blumensaat, (2002)	ADM1	0.240	0.430	0.065	0.100	0.150	0.015
This study	ADM1	0.500	0.485	0.002	0.003	0.004	0.006

8.4.2 Model calibration

Through the adjustment of two model parameters in ASM3 (influent inert particulate COD, X_I, and the autotrophic growth, μ_A), a satisfactory description of the operation of the mainstream treatment line was obtained. The value of μ_A of 0.46 1/d is considerably lower than the ASM3 default value of 1.0 1/d (at 20°C), likely due to the presence of inhibitory compounds from the industrial discharges since nitrifying organisms are rather sensitive to these compounds (Ekama 2008). Nevertheless, the adjusted value is in the upper range of those suggested by Ekama (2008).

The ADM1 predicted the correct COD stabilisation, methane generation and percentage in the gas flow. However, the total alkalinity input in the inlet sludge was set at 0.035 eq/L due to the relatively low pH observed in this stream (around 6.3). The 0.035 eq/L value only took into account the contribution of the neutralised fatty acids to the total alkalinity, and neglected the presence of bicarbonate (HCO_3^-) whose concentration in the inlet sludge was indeed considerably low.

The ADM1 was calibrated by adjusting two parameters: the fractions of soluble inert and particulate inert produced during the disintegration step ($f_{XI,Xc}$ from 0.20 to 0.29 and 0.10 to 0.01 for $f_{SI,Xc}$). After calibration, the model provided a satisfactory prediction of the anaerobic digester operation. This suggests that the sludge contained higher X_I and lower S_I fractions, arguably attributable to the activated sludge properties from this plant.

8.4.3 Model coupling

Despite that a simplified approach to couple ASM3 and ADM1 was adopted, the satisfactory description of the performance of the wastewater and sludge treatment lines (either by the separate models or the coupled model) proved that the approach was adequate.

The implementation of an ADM1-ASM3 interface provided a (rough) estimation of the potential effects of returning the filtrate on the performance of the plant. Despite the fact that the simulations suggest that the plant will struggle to meet the discharge standards, the interface proved to be an important tool to assess the maximum capacity of the plant, and could be used as a basis to explore future (upgrade and extension) scenarios and to support the decision-making process whenever it is planned to adopt stricter standards. This is a process that, sooner or later, several emerging and developing countries will likely need to follow.

8.4.4 Plant performance assessment for current and future scenarios

Under the present load and current regulations in place, the performance of wastewater treatment plant Anjana is, in general, assessed to be satisfactorily. Although the plant was originally designed for organic matter removal only, the availability of spare capacity (plant is currently approximately 50% oversized), combined with high sewage temperature, allow for partial nitrification and denitrification. However, partial N removal led to higher oxygen demand, resulting in lower DO levels in the aeration tanks. The operating sludge retention time (SRT) was calculated to be at the low end of that required for nitrification (approximately 7 days). The decreased WAS flow rate (increase in SRT) generally showed improvement in both nitrification and denitrification, while the increased RAS flow rate did not result in major improvement of the plant performance. In addition, the likely presence of "dead" zones within the aeration tank and availability of COD allowed for limited denitrification. Concerning the existing situation, it was shown that limited improvement in N removal efficiency could be achieved by manipulation of WAS and by step-wise increasing the DO level in the aeration tanks. However, future effluent discharge standards seem to be best achievable by introduction of pre-denitrification by creation of an anoxic zone within the existing biological tanks. It is important to note that the surface aeration (turbines) capacity to support N removal still needs to be assessed. Given that the system is designed for a COD load twice as high as the current one, it is likely that a certain increase in DO concentration can be accommodated as well. Further investigation into the current way of operating aerators might bring more possibilities for increasing air input into the biological tanks.

As expected, the requirements for additional N removal are conflicting with the goal to maximize energy recovery by methane production. Since upgrade of the system for N removal requires longer SRT and results in less WAS flow, this will consequently lead to lower methane production. Within the investigated range of sludge age from 7 to 21 days, methane production dropped by a maximum of 14%, which was considered acceptable. It was also shown that, due to spare capacity, taking one digester out of operation will have no impact on anaerobic sludge digestion and taking two digesters out of operation will reduce methane production for only 15%, due to present overcapacity.

In both future scenarios, the model predicted satisfactory COD removal. This was to be expected due to the fact that plant extension from 1996 provided for hydraulic capacity of the activated sludge system of 82,500 m^3/d. The investigation of the performance of the activated sludge plant under most unfavourable conditions (increased flow, reduced efficiency of primary settlers) still resulted in satisfactory COD removal. However, nitrification did not occur. This strongly suggested that the efficiency of the primary clarifier should be carefully considered. Similar to the present case, upgrade of the plant for N removal required introduction of pre-denitrification. Also, the other major findings are analogous to those of the present case. For the maximum plant capacity desired, the targeted N removal was only possible to achieve by construction of an additional anoxic tank in front of the the existing aerobic tanks. The additional aeration capacity required for required N removal in the future still has to be investigated in more detail. Overall, the coupled model indicated that the upgrade of plant for N removal would have marginal negative influence on methane production.

8.5 Conclusions

The main conclusions from this study can be summarised as: (i) The Dutch wastewater characterisation (STOWA) protocol proved capable to be successfully applied under tropical conditions, which suggests its potential to be a reliable tool for defining the wastewater characteristics from tropical developing countries and countries in transition; (ii) The procedure applied to determine the fractionation of the primary and secondary sludge proved to be satisfactory to describe the performance of the anaerobic digester in terms of biogas generation and sludge production by ADM1; (iii) It is noteworthy that the introduction of a simplified ASM3-ADM1 converting interface and a simple calibration procedure (which resulted in the adjustment of only four default values, two for ASM3 and two for ADM1) provided a satisfactory description for the qualitative assessment of the plant regarding the effluent quality, biogas generation and sludge production; (iv) Mathematical modelling can also be a reliable and flexible tool to assess the interactions that take place in sewage treatment from developing countries and countries in transition which can therefore benefit from the availability of advanced modern tools for process optimisation, upgrade and resource recovery.

Acknowledgements

The authors would like to thank Nuffic – Netherlands Organisation for International Cooperation in Higher Education, who sponsored the research of Mayank Mithaiwala, Surat Municipal Corporation and TIFAC-Surat for providing the data and the access to Anjana treatment plant and laboratory facilities, and laboratory staff of UNESCO-IHE.

References

APHA, AWWA, WEF. 1996 *Standard Methods for the Examination of Water and Wastewater*, 19th edition. American Public Health Association, Washington DC.

Arceivala, S.J. 2000 *Wastewater Treatment for Pollution Control*, 2nd edition. Tata McGraw Hill Publishing Company Limited, New Delhi, India.

Batstone, D.J., Keller, J., Angelidaki, R.I., Kalyuzhnyi, S.V., Pavlostathis, S., Rozzi, A., Sanders, W.T.M., Siegrist, H. & Vavilin, V.A. 2002 *Anaerobic Digestion Model No. 1*, ISBN: 1 900222 78 7. IWA Publishing, London.

Blumensaat, F. 2002 *Modelling of two-stage anaerobic digestion using the anaerobic digestion model No. 1 (ADM1)*, PhD Thesis, The University of Queensland, Brisbane.

Brdjanovic, D., Moussa, M.S., Amy, G. & Van Loosdrecht, M.C.M. 2007. Use of modelling for optimization and upgrade of a tropical wastewater treatment plant in a developing country. *Wat. Sci. Technol.* 56(7), 21-31.

Brdjanovic, D., Van Loosdrecht, M.C.M., Veersteeg, P., Hooijmans, C.M., Alaerts, G.J. & Heijnen, J.J. 2000 Modelling COD, N and P removal in a full-scale WWTP Haarlem Waarderpolder. *Wat. Res.* 34(3), 846-858.

Ekama, G.A. 2008 Nitrogen removal. In M. Henze, M.C.M. van Loosdrecht, G.A. Ekama & D. Brdjanovic, (eds.), *Biological Wastewater Treatment: Principles, Modelling and Design*, ISBN: 9781843391883. IWA Publishing, London, UK.

Ekama, G.A., Sötemann, S.W. & Wentzel, M.C. 2007 Biodegradability of activated sludge organics under anaerobic conditions. *Wat. Res.* 41, 244-252.

Fall, C., Loaiza-Navia, J. & Esparza-Soto, M. 2009 Full activated sludge model no. 1 calibration experience at a medium-size WWTP in Mexico. *Wat. Sci. Technol.* 60(12), 3069-3082.

GPCB 1999 *GPCB Norms*. Gujarat Pollution Control Board, Gandhinagar, India.

Grau, P., de Gracia, M., Vanrolleghem, P.A. & Ayesa, E. 2007 A new plant-wide modelling methodology for WWTPs. *Wat. Res.* 41, 4357-4372.

Henze, M., Gujer, W., Mino, T. & van Loosdrecht, M.C.M. 2000 *Activated sludge models ASM1, ASM2, ASM2d and ASM3. IWA Scientific and Technical Report No. 9.* IWA Task Group on Mathematical Modelling for Design and Operation of Biological Wastewater Treatment. IWA Publishing, London.

Hulsbeek, J.J.W., Kruit, J., Roeleveld, P.J. & Van Loosdrecht, M.C.M. 2002 A practical protocol for dynamic modeling of activated sludge systems. *Wat. Sci. Technol.* 45(6), 127-136.

Khassab, G.N. 2009 Combined carbon and nitrogen removal in integrated anaerobic/anoxic sludge bed reactors for the treatment of domestic sewage, PhD Thesis, Wageningen University, Wageningen, The Netherlands.

Koch, G., Kuhni, M., Gujer, W. & Siegrist, H. 2000 Calibration and validation of activated sludge model No. 3 for Swiss municipal wastewater. *Wat. Res.* 34(14), 3580-3590

Meijer, S.C.F., van Loosdrecht, M.C.M. & Heijnen, J.J. 2001 Metabolic modeling of full-scale biological nitrogen and phosphorus removing WWTP's. *Wat. Res.* 35(11), 2711-2723.

Meneses, A., Vargas, D.L., Grosso, J.M., Deeb, A. & Wergara, W. 2011 Biogas availability and its energy use in combined wastewater treatment plants (CWWTP). *Wat. Practice Technol.* 6(2), doi:10.2166/wpt.2011.021.

Nopens, I., Batstone, D.J., Copp, J.B., Jeppssson, U., Volcke, E., Alex, J. & Vanrrolleghem, P.A. 2009 An ASM/ADM model interface for dynamic plant-wide simulation. *Wat. Res.* 43, 1913-1923.

Reichert, P. 1998 *Computer Program for the Identification and Simulation of Aquatic Systems.* Swiss Federal Institute for Environmental Science and Technology (EAWAG), Dubendorf, Switzerland.

Rieger, L., Koch, G., Kuhni, M., Gujer, W. & Siegrist, H. 2001 The EAWAG bio-P module for activated sludge model No. 3. *Wat. Res.* 35(16), 3887-3903

Roeleveld, P.J. & Kruit, J. 1988 Guidelines for wastewater characterization in the Netherlands. *Wat. Sci. Technol.* 45(3), 465-468

Roeleveld, P.J. & Van Loosdrecht, M.C.M. 2002 Experience with guidelines for wastewater characterisation in the Netherlands. *Wat. Sci. Technol.* 45(6), 77-87

Salem, S., Berendsm, D., Heijnen, J.J., van Loosdrecht, M.C.M. 2002 Model-based evaluation of a new upgrading concept for N-removal. *Wat. Sci. Technol.* 45(6), 169-176.

Siegrist, H., Vogt, D., Garcia-Heras, J. & Gujer, W. 2002 Mathematical model for meso- and thermophilic anaerobic sewage sludge digestion. *Environ. Sci. Technol.* 36, 1113-1123.

Van Loosdrecht, M.C.M. 2008 Innovative nitrogen removal. In M. Henze, M.C.M. van Loosdrecht, G.A. Ekama & D. Brdjanovic, (eds.), *Biological Wastewater Treatment: Principles, Modelling and Design,* ISBN: 9781843391883. IWA Publishing, London, UK.

Wichern, M., Lubken, M., Blomer, R. & Rosenwinkel, K. 2003 Efficiency of activated sludge model No. 3 for German wastewater on six different WWTPs. *Wat. Sci. Technol.* 47(11), 211-218.

Figure 8.5 After completion of his Masters at UNESCO-IHE, Mayank Mithaiwala returned to his job at Surat Municipal Corporation and implemented upgrade of WWTP Anjana using results of this modelling study (Photo: Mithaiwala M.).

Chapter 9: WWTP Ecco, the Netherlands

Modelling nitrogen removal from tannery wastewater

Moussa M.S., Rojas A.R., Hooijmans C.M., Gijzen H.J. and van Loosdrecht M.C.M.

This chapter is based on "Model-based evaluation of nitrogen removal in a tannery wastewater treatment plant " (2003) by Moussa M.S., Rojas A.R., Hooijmans C.M., Gijzen H.J. and van Loosdrecht M.C.M. Wat. Sci. Tech. 50, 6, 251-260.

9.1 Introduction

The activated sludge system is currently the most widely used biological wastewater treatment process, treating both domestic and industrial wastewater. The process requires a high degree of operational control and management. In order to obtain the maximum removal efficiency from the activated sludge plant, the operator must have a full understanding of this complex process. Several research studies have been carried out over the last 25 years to understand the behaviour of activated sludge systems using computer modelling. A common language for all modellers in this field regarding concepts and nomenclature is provided by the ASM models developed by the IWA task group (Henze *et al.*, 2000). The ASM models have proved to be a useful tool for the dynamic simulation of activated sludge systems treating domestic wastewater. However, application of these models to industrial wastewater treatment plants remains limited. To apply these models to industrial wastewater treatment plants it could be necessary to extend the ASM models by additional kinetic reactions (Nowak *et al.,* 1995). The physical, chemical and biological properties of industrial wastewater and their variation in flow and composition make the operation more complicated. Moreover, the model could be used as a quantitative way to predict the effect of different production scenarios on their wastewater treatment plant (van Zuylen, 1993), supporting the operator in decision making.

The tannery industry is generating high amounts of polluted water while the profit margin is low. Therefore, purification of the generated wastewater has a high impact on the overall production costs. The work presented here emphasises on modelling of a tannery activated sludge wastewater treatment plant and its practical application. The main objectives of this study are: (1) to modify the activated sludge model ASM1 to satisfactorily describe the COD and N removal in the tannery wastewater treatment plant; (2) to evaluate the plant performance using the modified model; (3) to investigate the required modifications for the plant optimisation and future extensions.

9.2 Materials and methods

9.2.1 Plant and process description

The study covers the wastewater treatment plant of Ecco Tannery Holland B.V., a tannery factory located in Dongen, The Netherlands. The WWTP is in operation since 1987, treating wastewater generated from different steps of the tannery plant and designed for COD and N removal. The configuration of the plant consists of primary and secondary wastewater treatment and sludge treatment.

In the primary treatment chromium (Cr_{total}) from the segregated stream is removed by chemical precipitation. The supernatant liquid is pumped to the next step, where it is mixed with the rest of the generated wastewater in a covered equalisation tank (1,750 m^3). The equalisation tank buffers the dynamic flow generated during the week (5 working days per week) and provides the plant with a minimum flow during the weekend. In the equalisation tank a dose of $Fe(OH)_3$ (iron sludge from a drinking water treatment plant) is mixed with wastewater to remove the S^{2-} compounds. The top gas layer of the equalisation tank is pumped off, washed and used for aeration in the second stage. Eight primary clarifiers (8x50m^3) are used for particulate solid separation at the end of the primary stage.

The secondary treatment consists of a conventional plug flow activated sludge system, which is the main focus of this study. The system involves a plug flow reactor of a total volume of 8,000 m^3. The first part of the reactor (1,000 m^3) is non-aerated to ensure denitrification, while the rest of the reactor is aerated. An internal recycle flow (Q_{int}) supplies the denitrification zone with nitrate. To avoid phosphorus limitation for microbial growth 25L commercially available H_3PO_4 (75%) was dosed in the aerated zone. The outlet of the reactor is connected to a secondary settler of 800 m^3, where the settled sludge (Q_{return}) is pumped to the denitrification zone and the excess sludge (Q_{ex}) is pumped to the equalisation tank of the primary treatment. Finally, the treated effluent is pumped via a 37 km pressure main to the water authority gravity line, which conveys the wastewater to the domestic WWTP Rilland Bath.

The sludge treatment deals with primary and chromium sludge, both produced in the plant. The primary sludge, collected from the primary settlers, is conditioned in a buffering tank (200 m^3), de-watered by a filter press, resulting in a sludge cake with a dry content of 30-35%. This sludge cake and the sludge generated from the chromium removal (chromium sludge) are transported by trucks for final disposal. The rejected water resulting from the primary sludge treatment is pumped to the bioreactor (Q_{in3}). The volumes, the hydraulic and operational collected data are summarised in Table 9.1.

Table 9.1 Operational flow data and reactor volumes of WWTP Ecco Tannery B.V. The values printed italic are obtained from mass balances and are used in the model.

Flow	Average flow rates m^3/d	Reactor	Volume m^3
Influent,Q_{in1}	710	Equalisation tank	1,750
Influent, Q_{in2}	125 (*133*)	Primary settler (8x50)	400
Influent, Q_{in3}	0.025	Unaerated zone	1,000
Efluent, Q_{eff}	755	Aerated zone	7,000
Return Sludge, R_{eturn}	1,920 (*650*)	Total reactor volume	8,000
Internal recycle, Q_{int}	5,760	Secondary settler	800
Excess sludge, Q_{ex}	80 (*90*)	Sludge buffering tank	200

9.2 Measurements

The staff of WWTP Ecco Tannery Holland B.V. provided the routinely collected operational data of the bioreactor and its performance over the year 2000. A detailed sampling and experimental program was conducted in October and November 2000. The pseudo steady state measurements of the WWTP were performed during two sampling runs (11 October and 24

November 2000). During each run samples were collected from the influent, return sludge and effluent. The samples were analysed for T (temperature), pH, Alkalinity, DO (dissolved oxygen), COD_{tot} (Total COD), COD_S (COD of the micro-filtrated fraction, 0.45μm pore diameter), NH_4-N, NO_3-N, TKN (Total Kjeldal Nitrogen), P_{tot} (Total Phosphorus), VSS (Volatile Suspended Solids) and TSS (Total Suspended Solids). In addition, six different sampling points over the length of the bioreactor were defined and sampled during the second run (24 November 2000). These six sampling points were used to describe the hydraulic regime and biological conversion as function of the reactor length. The average values of the routinely collected data for the year 2000 and the average measurements of the sampling program are presented in Table 9.2.

Table 9.2 Measured influent and effluent and the influent composition required for the model of WWTP Ecco Tannery B.V. The yearly average values are presented between brackets. These values were used to calculate steady-state influent composition.

Description	Measurements			Description	Model influent composition			
	Influent	Effluent	Unit		Symbol	Dynamic	Static	Unit
Total COD, COD_{total}	2,525 (290)	167 (190)	$gCOD/m^3$	Soluble compounds				
Soluble COD, CODs	1,785	143	$gCOD/m^3$	Dissolved oxygen	S_{O2}	0.3	0,3	$gCOD/m^3$
Total N-Kj	488 (515)	7 (11)	gN/m^3	Readily biodegrable COD	S_s	840	972	$gCOD/m^3$
Soluble N-Kj	454	6	gN/m^3	Soluble inert COD	S_1	177	205	$gCOD/m^3$
Ammonium, NH_4	438	10	gN/m^3	Ammonium	S_{NH4}	438	448	gN/m^3
Nitrite, NO_2	0	0 (0.1)	gN/m^3	Nitrite	S_{NO2}	0	0	gN/m^3
Nitrite, NO_3	0	99 (50)	gN/m^3	Nitrate	S_{NO3}	0	0	gN/m^3
Total phosporus, P_{total}	0.4	3.3	gP/m^3	Alkalinity	S_{AK}	50	50	$mHCO_3/L$
Ortho phospate, PO_4	0.1	2.8	gP/m^3					
Total chromium, Cr_3	0.23	0.2 (0.1)	gCr/m^3	Particulate compounds				
Calcium, Ca_2	200	320 (400)	gCa/m^3	Inert particulate COD	X_I	454	525	$gCOD/m^3$
Chloride, Cl	8,066	7,960 (7,380)	gCl/m^3	Slowly biodegrable COD	X_s	1,060	1,226	$gCOD/m^3$
Sulfate, SO_4	3,700	3,900 (3,100)	gSO_4/m^3	Heterotropic biomass	X_H	0	0	$gCOD/m^3$
Total suspended solids, TSS	1,593	500	g/m^3	Ammonia oxidisers	X_A	0	0	$gCOD/m^3$
Volatile suspended solids, VSS	255	127	g/m^3	Nitrite oxidisers	X_N	0	0	$gCOD/m^3$
Alkalinity	2,030	570	$gCaCO_3/m^3$					
Temperature	22	22	°C					
pH	9(9.4)	7.5 (7.1)						
Dissolved oxygen, DO	0.3	6.7	gO_2/m^3					

Several biological batch tests were performed at the UNESCO-IHE laboratory to determine the influent and sludge characteristics (Ekama *et al.,* 1986; Orhon *et al.,* 1999a,b). The tests were performed at controlled temperature 20°C, pH of 7.5±0.05 and under aerobic and anoxic conditions. Nitrification batch tests were performed under aerobic conditions, in which $NaNO_2$ and NH_4Cl were consecutively injected (Moussa *et al.* 2003a). This test allows measuring the kinetic parameters of nitrite and ammonia oxidisers separately and was used for model calibration.

9.3 Process model (selection and adjustment)

The simulations of the secondary treatment of Ecco Tannery Holland B.V. WWTP were performed with AQUASIM® (Reichert, 1998), a computer software package used for simulation. The process was modelled according to the flow scheme shown in Figure 9.1, with the hydraulic and operational parameters as presented in Table 9.1. The plug flow condition of the bioreactor was modelled as six completely mixed stirred compartments in series with an internal recycle flow. In the model the secondary clarifier separates solids and water ideally. The amount of suspended solids discharged in the effluent was considered in the sludge age (SRT) calculation. Oxygen concentrations in each compartment were controlled with a PI-controller in accordance to the measured values.

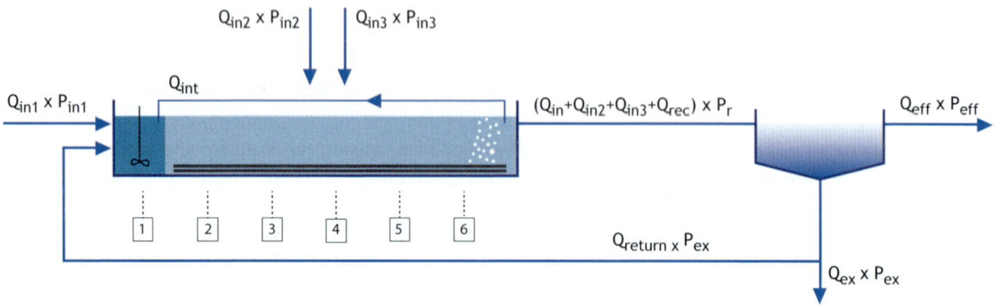

Figure 9.1 Phosphorous and overflow balance over the bioreactor and settler. The mass balances are presented in equation (1)-(3). Concentration and flow are used from Table 9.1 and 9.2. Numbers in boxes indicate sampling points.

A modified version of the activated sludge model No. 1 (ASM 1) proposed by the IWA task group (Henze *et al.*, 2000) was used to calculate the biological conversions in each compartment. The main modification incorporated in ASM1 to simulate the WWTP was that the nitrification was considered as a two-step process (Nowak *et al.*, 1995), because ammonia oxidisers grow faster than nitrite oxidisers at temperatures above 15-20°C (Hellinga *et al.*, 1999). Moreover, inhibiting compounds present in industrial wastewater might lead to an adverse effect on one or both steps of the nitrification process. Thus describing the nitrification in two steps enables the identification of any inhibition and the detection of partial nitrification (NO_2 accumulation).

9.4 Influent measurement and characterization

To simulate the WWTP a detailed wastewater characterisation to determine the model components is required. Laboratory tests involving biodegradation were conducted to determine the influent characteristics. The total influent COD can be described as:
$COD_{total} = S_S + S_I + X_S + X_I$.

The readily biodegradable COD (S_S) was determined by two different approaches using an aerobic and anoxic batch test as described by Ekama *et al.* (1986). The soluble inert COD (S_I) was determined according to the approach suggested by Orhon *et al.* (1999a,b). The influent X_I fraction (X_I/COD_{total}) ratio was estimated as a result of model calibration fitting the solid COD balance as proposed by Meijer *et al.* (2001) and, consequently, the rest will represent the X_S fraction. Average influent measurements and the calculated model influent compositions are presented in Table 9.2.

9.5 Balancing operational data and measurements

9.5.1 Estimation of sludge age, Q recycling and Q_{in2}

For a reliable simulation study the sludge age (SRT) should be known within 95% accuracy (Brdjanovic *et al.*, 2000; Meijer *et al.*, 2001). Therefore a check on the SRT (or sludge production) is strongly recommended. For the evaluation of sludge production the overall phosphorus balance was used as proposed by Nowak *et al.* (1999). Three balances were formulated (see

Figure 9.1): the overall P balance (Eq 9.1), the overall flow balance (Eq 9.2) and the P balance over the settler (Eq 9.3).

$$Q_{in1} P_{in1} + Q_{in2} P_{in2} + Q_{in3} P_{in3} = Q_{eff} P_{eff} + Q_{ex} P_{ex} \tag{9.1}$$

$$Q_{in1} + Q_{in2} + Q_{in3} = Q_{eff} + Q_{ex} \tag{9.2}$$

$$(Q_{in1} + Q_{in2} + Q_{in3}) P_r = Q_{eff} P_{eff} + (Q_{return} + Q_{ex}) P_{ex} \tag{9.3}$$

The yearly average measurements of Q_{in1}, Q_{in3} and Q_{ef} in addition to the two runs average measurements of P_{in1}, P_{in2}, P_{in3} (chemical P addition) and P_{ex} were used to evaluate the Q_{in2}, Q_{return} and Q_{ex} and consequently the SRT. The calculated Q_{in2}, Q_{ex} and SRT (70 days) were in agreement with the recorded values of the plant. However, the balanced Q_{return} (650 m^3/d) was found to be inconsistent with the reported value (1,920 m^3/d). Since the value of the mass balance (650 m^3/d) was in line with the measured sludge concentrations in the reactor and return sludge, we used this value in simulating the treatment plant.

9.6 Model calibration and simulation

After the determination of the main operational parameters and the influent characterisation, the model of the WWTP was calibrated. A step-wise approach was applied as proposed by Meijer *et al.*, (2001). First the solids were fitted (P, COD and TKN) on the basis of yearly average measurements. Next the nitrification and denitrification were calibrated on the basis of yearly average measurements.

9.6.1 Calibration of the solids

The solids balance is a non-conserved balance. An incorrectly assumed COD load or sludge production will generally be compensated by the simulated oxygen consumption of the process. Because the SRT is fixed according to the P_{TOT} balance, the sludge-COD concentration in the process is mainly determined by the influent X_I/COD_{total} ratio (f_{XIin}). Inert COD (X_I) accumulates in the process. Increasing f_{XIin} therefore leads to increasing the COD in the process and vice versa. By adjusting the influent ratio f_{XIin} to 0.18 the model described the measured MLVSS in the reactor (using a conversion factor of 1.4 gCOD/gMLVSS). When f_{XSin} is used to fit the COD balance, all model uncertainties related to the production of X_I and the influent characterisation of X_S and X_I are lumped in the influent f_{XIin} fractionation. As a result of adjusting the f_{XIin} fraction, the model predicted accurately the soluble COD in the effluent and total P and TKN in the bioreactor.

9.6.2 Calibrating nitrification and denitrification

Nitrification and denitrification were calibrated on the basis of yearly average measurements. Adjusting the oxygen half saturation coefficient of the ammonia oxidising organisms ($K_{O2,AOO}$) and heterotrophic biomass ($K_{O2,OHO}$) was used to fit the nitrification and denitrification in the bioreactor according to Meijer *et al.* (2001). To simulate the measured effluent ammonium and nitrate concentration, a value of 1 mgO_2/L and 0.75 mgO_2/L was used for $K_{O2,AOO}$ and $K_{O2,OHO}$, respectively. The calibrated values of $K_{O2,AOO}$ and $K_{O2,OHO}$ are linked to oxygen diffusion limitation within the sludge flocs and oxygen concentration gradients in the tanks caused by non-ideal mixing, processes that are not accounted for in the simulations. Because of differences in mixing intensities, $K_{O2,AOO}$ and $K_{O2,OHO}$ are expected to be different for each compartment, but the same values were used in all compartments for simplicity. Since these values are most sensitive in the anoxic compartment this will not affect the outcome of the simulations significantly.

The nitrification batch tests were used in this study as a last step in the calibration procedure. Ammonia and nitrite oxidising activities in the batch test were not predicted very well by the model; observed ammonia conversion and nitrite conversion rates were approximately 20% lower than predicted. Due to the fact that the WWTP is under-loaded, these differences could not be observed when simulating the full-scale yearly average values. There are three possible reasons for this lower experimental nitrification conversion rates in the batch tests than predicted by the model: either the amount of nitrifiers predicted by the model in the plant is too high or the growth rate in the model is too high or both. This means that re-calibration was required. Since the decay rate is the most uncertain parameter, we choose to calibrate this coefficient. Increasing the decay rates by 30 % (from 0.15 and 0.10 1/d to 0.20 and 0.13 1/d for the ammonia and nitrite oxidisers, respectively) resulted in a good fit of the predicted data to the measurements. When this newly calibrated model was used for the full-scale simulation of the treatment plant, the nitrogen content in the effluent did not change. In a separate test, similar results were obtained when the growth rate was reduced because of the high correlation between the decay and growth rate parameters.

9.7 Model validation

The model validation was performed by validating the capacity to predict the measured concentrations of NH_4^+, NO_2^- and NO_3^- along the bioreactor length using the dynamic influent data. The recorded average daily influent flow during the period 13-26 November 2000 (Figure 9.2) was used as input flow for the model.

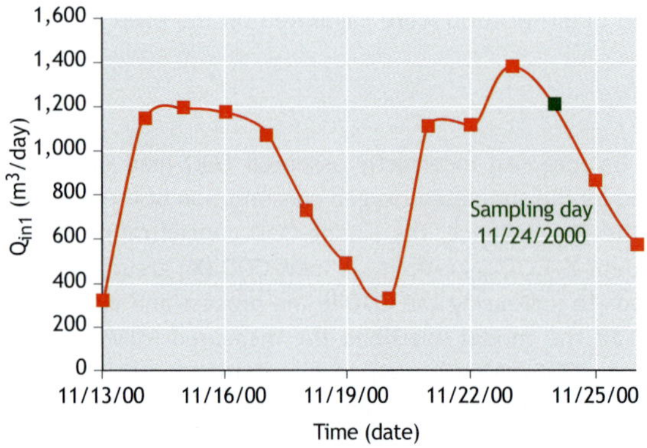

Figure 9.2 Measured influent flow (Q_{in1}) during 13-26 November 2000, on 24 November 2000 samples were taken and used for model validation (see Figure 9.4).

The simulated values of NH_4^+, NO_2^- and NO_3^- on November 24th (the 12th day of the 14 days dynamic simulation period) were compared with the values obtained from sampling over the length of the bioreactor conducted on the same day. The model provides an accurate prediction of NH_4^+ and NO_3^- along the bioreactor (Figure 9.3).

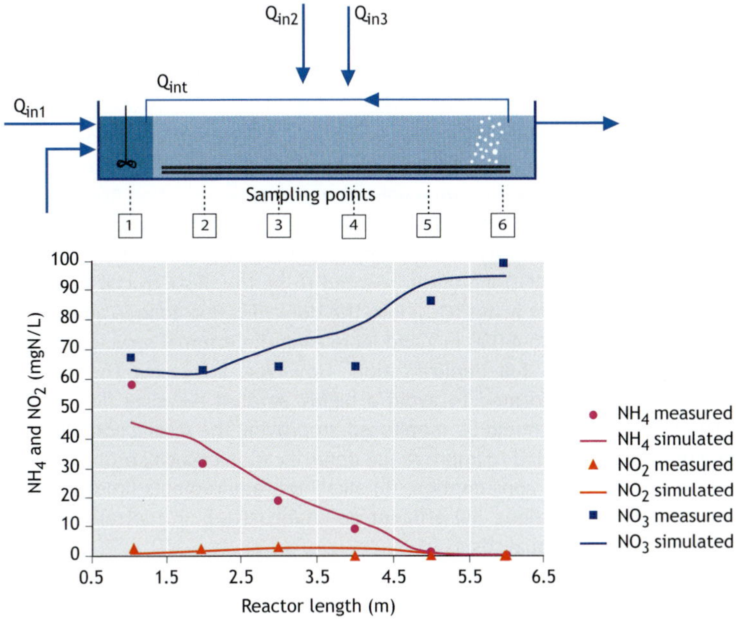

Figure 9.3 Measurement results in terms of nitrogen over the length of the bioreactor of WWTP Ecco on 24 November 2000. The curves are calculated by the calibrated model.

9.8 Model application

This study was conducted in the first year of the new owner of the company. At the same time various viability studies were performed to assess the potential increase of the leather production and application of different tannery processes and operational strategies. The main question was, whether the existing WWTP could cope with these future requirements. Based on the plant performance and effluent quality, an additional volume requirement (25% of the bioreactor volume) was proposed by the operator. The old equalisation tank was selected to be the additional volume, and to be operated as an anoxic reactor. This upgrading aimed to increase the plant capacity and to achieve full denitrification. Evaluating the upgrading concept for the WWTP was the main focus of our modelling study. The validation step demonstrated the capability of the proposed model to describe the WWTP correctly under both steady state and dynamic conditions. Hereafter, the model was used as a tool to obtain insight in the existing plant performance, possible extension and ways of process optimisation.

9.8.1 Evaluation of the existing plant performance and possible extension

The model visualised the existing conditions and the potential capacity of each process involved in the purification system. Nitrification as the most sensitive process was accomplished within 60% of the total bioreactor volume under both steady state and dynamic conditions (Figure 9.4) and denitrification seemed incomplete (only 80%). The same effluent quality was also predicted even up to a twofold load increase. This illustrates that the WWTP Ecco has been under-loaded and that the bioreactor volume is not the limiting factor if the load is increased. It is possible that the aeration capacity and the sludge treatment units will be limiting factors in the treatment process in a possible expansion. Despite the high concentration of NO_3^- in the effluent (50 mgN/L) the present set-up of the plant has a high denitrification potential. Within the existing process configurations, a better effluent quality with lower operational costs could be achieved via process optimisation.

9.8.2 Process optimization

One of the major achievements of plant optimisation is enhancing the denitrification activity to achieve full denitrification in the system. The model was used to quantify the main limiting factors, which hinder the denitrification. Anoxic zone volume, internal recirculation flow and the availability of easily degradable COD were investigated as limiting factors. Different modifications were simulated under steady-state conditions to study their effect on the accomplishment of full denitrification. COD limitation seemed to be the most crucial factor for the denitrification. Introducing the reject water (Q_{in2}) into the denitrification zone in addition to increasing the denitrification zone (up to 4,000 m^3) and increasing the internal recycle flow (up to 10,000 m^3/h) will directly lead to full denitrification. However, increasing the internal recycling flow should be carefully performed to avoid a highly aerated recycling flow to the denitrification zone. Therefore oxygen should be monitored. Improving the mixing conditions in the denitrification zone was recommended to improve the denitrification capacity of the system. This results in a decrease in: (i) oxygen consumption, (ii) alkalinity requirements (implying that lesser external alkalinity must be provided), (iii) effluent pollutant (NO_3^-) concentrations, and, consequently, (iv) in the overall operational costs.

9.9 Discussion

The influent characterisation results of the wastewater do agree with the reported values of the tannery wastewater (Orhon *et al.,* 1999a). The modified ASM1 model proposed in this paper for COD and N removal described the performance of the WWTP well with the adjustment of only two parameters ($K_{O2,AOO}$ and $K_{O2,OHO}$). An accurate and correct description of the system configurations, balancing the operational data with the measurements to accurately calculate the SRT and the use of the stepwise calibration proposed by Meijer *et al.* (2001) simplified the complexity of the model calibration. The use of batch tests for model re-calibration was useful because the plant was under- loaded and therefore the effluent concentrations profiles were not sensitive for the rate parameters. Model validation under dynamic conditions is of great significance in industrial WWTPs, because the plants are usually working under highly dynamic conditions in comparison with domestic WWTPs.

The observed reduction in ammonia and nitrite oxidising activity is attributed to the presence of salt (7.5 gCl/L). This decline in the activity for ammonia and nitrite oxidisers is in agreement with earlier results of salt inhibition effects on nitrifying sludge (Moussa *et al.,* 2003 b). The use of a higher decay rate of nitrifiers to mathematically describe the salt impact simplifies the model description of salt inhibition (only one parameter to calibrate). Other parameters for calibration could have been chosen but the decay rate was considered as the most uncertain coefficient in the model.

The potential use of the model was illustrated in the evaluation of the upgrading of the WWTP. During this evaluation the bioreactor capacity and process configuration were investigated. The model provided a better and quantitative understanding of the plant operation and treatment process. The investigation of several modifications to reach further plant optimisation was very time-effective and economic (Salem *et al.,* 2002). Because it gives quantifiable results, modelling supported the decisions to be taken with respect to the plant extension. It was originally proposed to increase the plant volume (25% with the use of old equalisation tank) to cope with anticipated load expansion. The simulations clearly indicated that with the present system and future increase in load good effluent quality can be reached even combined with an increased denitrification.

Plant managers and staff use the model in a different way, namely as a tool to quantitatively predict the effect of certain decisions on the treatment process (Salem *et al.,* 2002). This application of the model is very useful in case of industrial WWTPs, like the tannery studied here. The cost of treating the wastewater generated during the production process has a crucial impact on the overall production costs. This also increases the awareness of the impact of each pollutant term in each part of the process and stimulates the practice of waste minimisation to have an environmentally friendly production.

9.10 Conclusions

Activated sludge models commonly applied for domestic wastewater treatment plants can also be used for industrial WWTP if the following steps are carefully considered:

- An accurate description of the system configurations;
- Balancing the operational data with the measurements to accurately calculate the main important reactor input parameters (flow rates and SRT);
- Selection of model process and components which are significant and dynamic in this system configuration;
- Complementary analyses to assess the wastewater and the sludge characterisation;
- Stepwise calibration under steady-state conditions and finally model validation under dynamic conditions.

The modified ASM1 model proposed in this paper for COD and N removal proved to be able to describe the performance of Ecco Tannery Holland B.V. WWTP wastewater treatment plant.
The model was successfully used to evaluate and optimise the plant performance. In addition it was demonstrated that the model could be used by the plant manager to support his decisions quantitatively resulting in saving time and money.

References

Brdjanovic D., van Loosdrecht M.C.M., Versteeg P. and Hooijmans C.M., Alaerts G.J., Heijnen J.J. (2000). Modelling COD, N and P removal in a full-scale WWTP Haarlem Waarderpolder. *Wat. Res.*, 34, 846-858.

Ekama G.A., Dold P.L., Marais G.v.R. (1986). Procedures for determining influent COD fractions and the maximum specific growth rate of heterotrophs in Activated sludge systems. *Wat. Sci. Tech.*, 18(6), 91-114.

Hellinga C., van Loosdrecht M.C.M. and Heijnen J.J. (1999). Model based designed of a novel process for ammonia removal from concentrated flow. *Mathematical and Computer Modelling of Dynamic Systems*, 5(4), 351-371.

Henze M., Gujer W., Mino T. and van Loosdrecht M. C. M. (2000). Activated sludge models ASM1, ASM2 and ASM3. Scientific and Technical Report, IWA Publishing, London.

Meijer S.C.F., van Loosdrecht M.C.M. and Heijnen J.J. (2001). Metabolic modelling of full-scale biological nitrogen and phosphorus removing WWTP's. *Wat. Res.*, 35, 2711-2723.

Moussa M. S., Lubberding H. J., Hooijmans C. M., van Loosdrecht M. C. M. and Gijzen H. J. (2003a). Improved method for determination of ammonia and nitrite oxidation activities in mixed bacterial cultures. *Appl.Microbiol. Biotechnol.*, 63, 217-221.

Moussa M. S., Lubberding H. J., Hooijmans C. M., van Loosdrecht M. C. M. And Gijzen H. J. (2003b). Short term effects of various salts on ammonia and nitrite oxidisers in mixed bacterial cultures. *Appl. Microbiol. Biotechnol.*

Nowak O., Svardal K. and Schweighofer P. (1995). The dynamic behaviour of nitrifying activated sludge systems influenced by inhibiting wastewater compounds. *Wat. Sci. Tech.*, 31(2), 115-124.

Nowak O., Franz A., Svardal K., Muller V. and Kuhn V. (1999). Parameter estimation for activated sludge models with the help of mass balances. *Water Sci. Tech.*, 39(4), 113-120.

Orhon D., Ates E. and Ubay Cokgor E. (1999a) Modelling of activated sludge for tannery wastewaters. *Water Environment Research*, 71, 50-63.
Orhon D., Ubay Cokgor E. and Sozen S. (1999b). Experimental basis for the hydrolysis of slowly biodegradable substrate in different wastewaters. *Wat. Sci. Tech.*, 39(1), 87-95.
Reichert P. (1998). AQUASIM 2.0- Computer Program for the Identification and Simulations of Aquatic Systems. Swiss Federal institute For Environmental Science and Technology (EAWAG), Dübendorf, Switzerland.
Salem S., Berends D., van Loosdrecht M.C.M. and Heijnen J.J. (2002). Model-based evaluation of a new upgrading concept for N-removal. *Wat. Sci. Tech.* 45(6), 169-176.
van Zuylen H. J. (1993). From scientific computation to decision support. *Knowledge-based system*, 6, 3-10.

Figure 9.4 Part of the leather treatment process at Ecco Tannery Holland BV (Photo: Moussa M.S.).

Figure 9.5 Salted hides in stock at tannery in Dongen (Photo: Moussa M.S.).

Figure 9.6 Wet blue after tanning (Photo: Moussa M.S.).

Figure 9.7 Primary setttling tanks of WWTP Ecco Dongen (Photo: Moussa M.S.).

Figure 9.8 Activated sudge tank of WWTP Ecco Dongen (Photo: Moussa M.S.).

Chapter 10: WWTP Sarajevo, Bosnia and Herzegovina

Use of modeling for cost-effective reconstruction of urban wastewater infrastructure

Hodzic A., Vojinovic Z., Seyoum S.D., Pathirana A., Drnda N., Meijer S.C.F., and Brdjanovic D.

This chapter is based on "Model-based evaluation of the urban wastewater infrastructure reconstruction options in a developing country: Case study Sarajevo in Bosnia and Herzegovina" by Hodzic A., Vojinovic Z., Seyoum S.D., Pathirana A., Meijer S.C.F., and Brdjanovic D. (2011), 2nd IWA Development Congress, Kuala Lumpur, Malaysia., and "Upgrading Large Wastewater Treatment Plants: Use of Modeling as a Decision-making Tool". Meijer S.C.F. and Brdjanovic D. (2010) IWA Specialist Conference on Water and Wastewater Treatment Plants in Towns and Communities of the XXI Century: Technologies, Design and Operation, Moscow, Russia, June 2010.

10.1 Introduction

Sarajevo is the capital of Bosnia and Herzegovina with 425.000 inhabitants (2010) located in the valley of River Miljacka which flows into the River Bosna at the proximity of the sewage works. Most of the sewage system of Sarajevo is designed as a separate; however, it heavily suffers from cross-connections and numerous operational problems. Sewerage in the historical city center was built about a century ago, while the rest of the network, comprising of main collectors and wastewater treatment plant (WWTP) in Butile, was built in 1984. The WWTP is a conventional activated sludge plant designed for the removal of solids and organic matter from wastewater with anaerobic digestion of primary and secondary sludge. The plant was in operation until 1992 when it was vandalized in the Civil War (1992-1995) and is not operational since. As a result of that, untreated sewage has been discharging into the river Miljacka, just before the convergence with River Bosna, for almost two decades. In the period 1984-1992 the plant effluent has been discharging into the River Bosna downstream of the convergence with River Miljacka. The local government is currently planning to invest in the rehabilitation of the sewerage system and treatment plant. Several main questions arose during the decision-making process regarding the reconstruction of the network and the plant, namely: (i) what is the best framework to understand the interactions between sanitation infrastructure and receiving environment in the city of Sarajevo?, (ii) how can the operations of the sewerage network, particularly from the perspective of the sediment transport, be further improved?, (iii) should the plant be reconstructed in the original state as it was in 1984?, (iv) should it be upgraded by nutrient-removal (nitrogen and phosphorus) facilities to satisfy possibly more stringent effluent requirements (EU standards) in the future?, (v) what is the most appropriate technological process to be applied?, (vi) what are the possible extent and scope of reconstruction, given the available financial resources?, and (vii) what is the impact of sewage and effluent discharges on receiving waters of two rivers?. These questions were addressed in this study which included the combination and application of modern computational packages (simulators) such as Mike Urban (sewerage modeling), BioWin (WWTP modeling), HEC-RAS (recipient modeling) supported by data management tools (GIS) and a digital topographic model (DTM) all integrated into in a tool to support the decision-making process concerning the selection of the most appropriate reconstruction strategy for the sewage network and WWTP. This strategy aims to maximize the environmental and public health protection within the given financial constraints. This chapter describes the model-based holistic evaluation of the urban wastewater infrastructure reconstruction options for Sarajevo City (Figure 10.1).

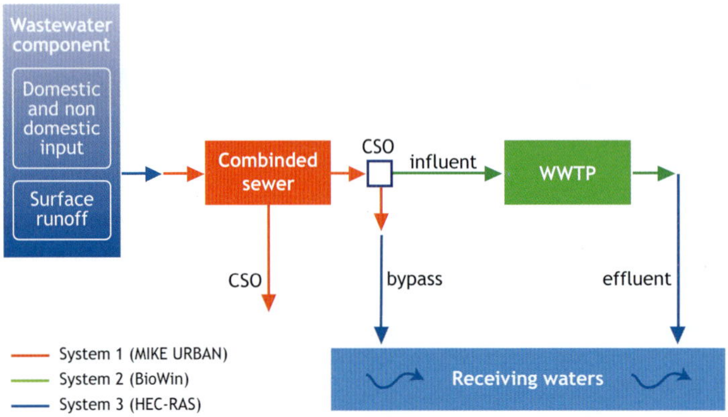

Figure 10.1 Schematic presentations of system subcomponents of concern to this study

10.2 Model of Sarajevo sewerage system

The sewage system of Sarajevo includes a sanitary sewage system and rain water drainage system and consists of an old part covering the oriental city center constructed by the Austrians a century ago and a new part including the main collectors constructed as part of the preparation of the city for the XIV Winter Olympic Games. The main characteristics of the Sarajevo's drainage and sewage collection system are presented in Table 10.1 and Figure 10.2.

Table 10.1 Main characteristics of the Sarajevo sewage and drainage system.

Total area size	7,900 ha	Number of manholes	26,623
Network total length	1,120 km	Total model area size	6,451 ha
Main sewer collector length	52.6 km	Impermeable surface area	1,482 ha
Number of CSO's	24	Average impermeability - city center	35%
Pipe diameter main collectors	0.3-1.8 m	Average impermeability - suburbs areas	15%

An investigation performed in 1999 showed a good structural condition of the main collectors which is likely not the case for the older secondary and tertiary network of pipes. However, besides numerous cross connections in the system, which together with infiltration are the two main reasons for a rather significant dilution of the sanitary sewage, the system regularly experiences problems with sediment depositions within the network as well as at the inlet of WWTP. Because of that, the City's sewerage system is prone to blockages, uncontrolled overflows into the River Miljacka and excessive load to the inlet pumping station of the WWTP.

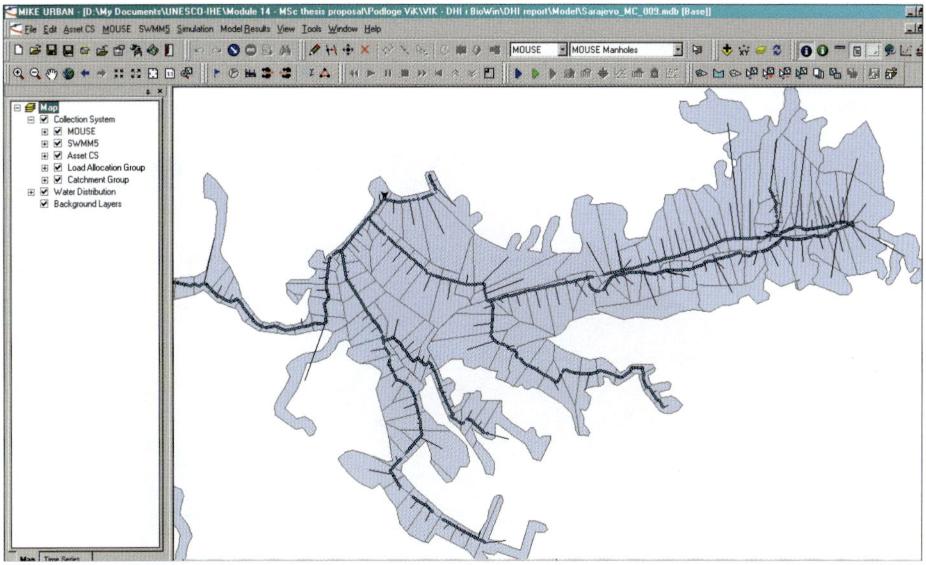

Figure 10.2 Catchment boundaries, CSOs and manholes on the main collectors (Software HEC-RAS)

During the 8-year operations, the WWTP experienced problems, in particular caused by the excessive accumulation of sand, grit and gravel at the inlet to the plant, which resulted in frequent disruptions and interruptions of the plant operation. Possible suspects for such events were intrusion of (i) eroded material via cross connection during the heavy rain events, (ii) sand which enters the system from streets during rain events and periods of intensive snow melting, (iii) illegal dumping of construction materials and other waste into sewers, and (iv) fine stones and gravel by the occasional backflow from the river Miljacka during floods at the location of the plant. Study carried out in early 90's (Peric *et al.,* 1991) recommended solving the problem by the construction of a sand trap in front of the existing WWTP inlet. However, this sort of 'end-of-the-pipe' solution could not address the causes of the problem, and furthermore, the study was rather limited as the computational applications at that time were not so far advanced as they are nowadays. Given the fact that nowadays several excellent numerical models and simulators are available and in use, it was decided to use the sediment transport module of Mike Urban which has enabled to undertake a range of sediment transport simulations from which the 'hot-spots' (those parts of the network which have the highest potential for sediment build up) were identified. With such simulations, the option of constructing several smaller sand traps within the sewer network was evaluated. Furthermore, those options which were considered as preferred options during 90's, were incorporated into the model and compared with other options. From the overall analysis, the preference was given to the option which has identified four critical spots (Figure 10.3).

From an array of different cases, two situations were considered critical from the perspective of sediment transport in the sewer network as well as of the treatment plant operation, namely: (i) the case of dry weather conditions (characterized by dry weather flow: $DWF_{avg,24h}$ = 2,16 m^3/s) and, (ii) rainy conditions (characterized by wet weather flow: $WWF_{avg,5h}$ = 4,20 m^3/s and maximum sewage capacity – Q_{max} = 5,20 m^3/s). The WWF was calculated using the single rainfall event of a 5 year return period with 2 hours duration as approximation as there were no available recorded rainfall time series. In addition, no data regarding the sediment characterization and transport was available resulting in the fact that the simulations were based on default and estimated parameters and consequently that the model results are limited to qualitative analyses of the system functionality and for identification of potential problem locations.

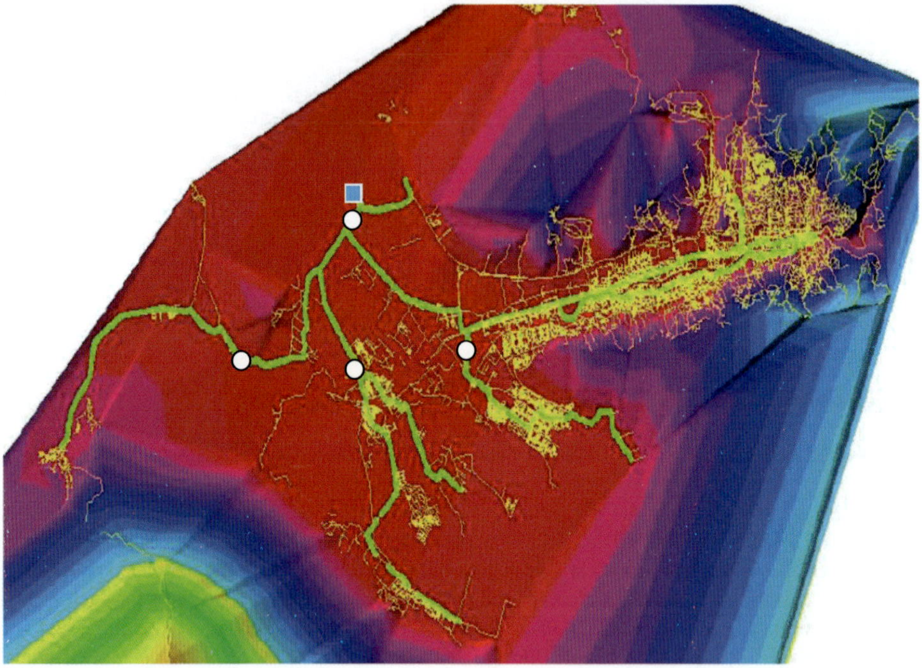

Figure 10.3 An overlay of the Sarajevo's sewage system on a raster DTM displaying locations (white circles) with highest affinity for increased sediment accumulation.

Also, the study was limited to modeling of the main collectors and primary collectors only. Therefore, the present study should rather be seen as a demonstration that the use of sediment traps within the Sarajevo's sewer system can be beneficial in mitigating the sediment transport problems, and to a lesser extent as an identification of an optimal solution to this problem. This study can be considered as a very useful starting point in a decision making process, but one should be cautious in using its results straightforwardly for the final decision on the both number and location of the sediment traps in the system. In conclusion, further more detailed modeling work is necessary to provide solutions that can enable efficient removal of sediments with the minimum number of sediment traps under different flow conditions while maintaining an acceptable depth of sediment bed. The reason why the analysis based on a single rainfall event cannot yield an optimal solution can be explained due to the fact that different sewer sediments (depending on their size, density and settling characteristics) will form deposits under certain flow conditions and they can be re-entrained again under some other flow conditions. Therefore, in order to come up with an optimal solution that can enable efficient removal of sediments given their rate of deposition, characteristics and magnitude, it is necessary to undertake a more comprehensive analysis using series of rainfall events with different characteristics and to assess the system's performance with respect to the frequency and magnitude of sediment depositions. Furthermore, given that the Sarajevo's sewer system has a number of combined sewer overflows (CSOs) structures, its performance should not be measured only on the basis of sediment transport characteristics, but also on the basis of frequency and volume of overflows. Such analysis also requires a more comprehensive study using the series of rainfall events (i.e., long-term model simulations and not only event-based simulations). This work could be then further supplemented with the inclusion of analysis of CSOs impacts on receiving waters along the sewer system (and not only at the location of the WWTP Butile). The local authorities were therefore advised to extend the future work by following additions: (i) develop a proposal for the measuring program within the sewer network concerning the approach for system survey, flow measurements and sediment sampling; the present model can be used to identify the strategic locations for these measurements, and (ii) as the proposed sediment traps were not verified within the model (although the generated model could be used for such verification), it is recommended to perform such verification.

Conclusions of this part of the study can be summarized as follows:

- The study has achieved good outputs considering the amount of data provided/sourced and a very short time frame given for its execution.
- The sewer system model was built with one of the best commercial modeling package.
- The model developed represents a good starting point for the process of assessing sewer system problems and particularly the problems associated with sediments.
- The study demonstrates that the use of sediment traps can be beneficial in managing the sediment transport problem in Sarajevo's sewer system.

10.3 Model of WWTP Sarajevo

Four main questions arose during a decision-making process regarding the reconstruction of the WWTP, namely: (*i*) should the plant be reconstructed in the original state as it was in 1984?, (*ii*) should it be upgraded by nutrient-removal (nitrogen and phosphorus) facilities to satisfy more stringent effluent requirements (EU standards) in the future?, (*iii*) what is the most appropriate technological process to be applied?, and (*iv*) what are the possible extent and scope of reconstruction, given the available financial resources? These questions were addressed by the application of an activated sludge model (ASM) embedded in a decision-making process concerning the selection of the most appropriate reconstruction strategy for this plant. Since the introduction of activated sludge models (Henze *et al.,* 1987, 2000), mathematical modelling has become an irreplaceable tool for the upgrade, retrofit, troubleshooting and optimization of wastewater treatment plants (Van Veldhuizen *et al.,* 1999; Brdjanovic *et al.* 2000; Meijer *et al.,* 2001; Pinzon-Pardo *et al.* 2007, Brdjanovic *et al.* 2007, Henze *et al.,* 2008); however, published information on their use as a decision-making tool in a particular context of a less developed country such as Bosnia and Herzegovina, are scarce. This study attempts to address this deficiency.

10.3.1 WWTP Sarajevo

The Sarajevo sewage system includes a sanitary sewage system and rainwater drainage system (the total length of sewer pipes in the entire system is about 1,000 km with a diameter of up to 1.80 m, Table 10.1). Although in theory separate, the system suffers from cross connections, excessive infiltration and backflow from the river. The WWTP was built as a conventional activated sludge system for the removal of suspended solids (SS) and organic matter (COD) with anaerobic sludge digestion of primary and secondary sludge and biogas reuse for energy production (Figure 10.4). The plant was operational from 1984 until 1992, and severe problems were occasionally experienced because of excessive accumulation of sand, grit and gravel at the inlet to the plant. Therefore, the plant operation was often interrupted, usually during and/or after periods of heavy rain or melting snow (Peric *et al.,* 1991). Several mitigation studies have been carried out so far by consulting firms and all recommend the construction of a large sand trap in front of the existing inlet. However, the inlet is only the place where the problem is manifested but not necessarily the location of its origin. A model-based sediment transport study, using the software package Mike Urban (DHI, Denmark), showed that indeed it may well be feasible to build several smaller sand traps within the sewer network as an alternative to the centralized solution from the '90s.

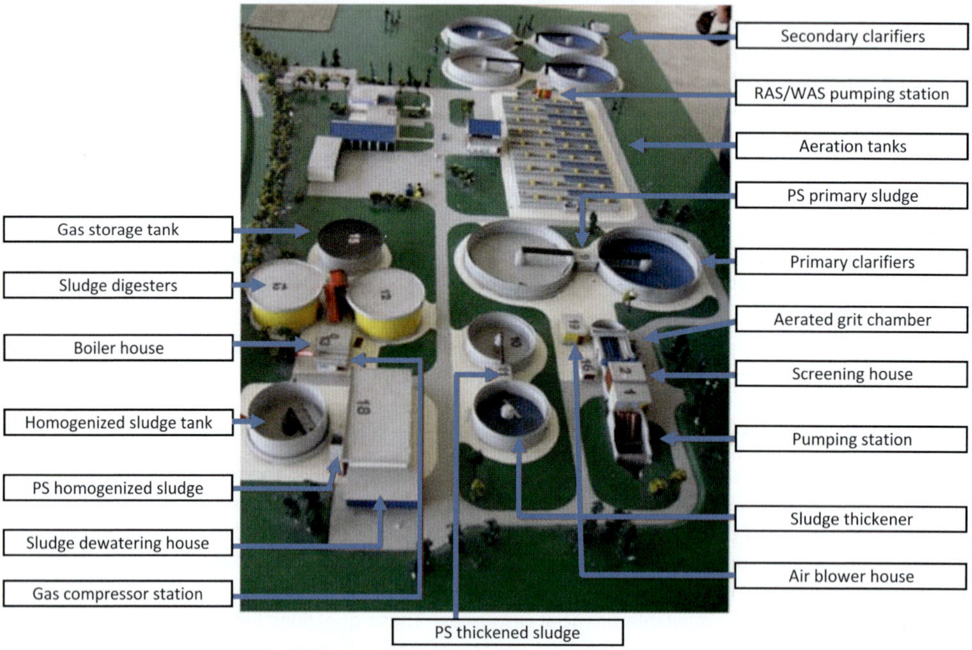

Figure 10.4 Model of WWTP in Sarajevo as built in 1984.

10.3.2 Capacity prognosis

The existing plant has a capacity of 600,000 PE based on specific load of 60 $gBOD_5$/PE.d. At the plant location, sufficient space was left to expand the plant, as planned in the second phase of construction (900,000 PE). The maximum sewer capacity in Sarajevo's main collector at the plant inlet is 5.2 m^3/s. A combined sewer overflow (CSO) located close to plant inlet defines the maximum flow (4.2 m^3/s) which is introduced to pumps, screens and aerated sand traps. Another bypass located before primary settlers limits the flow through to 3.8 m^3/s. Finally, the second bypass allows a maximum flow of settled sewage (2.6 m^3/s) to enter two lines of activated sludge tanks (18 units per lane each equipped with surface aerator – turbine) and later four secondary settlers. The overflow, bypasses and secondary clarifiers discharge into the nearby River Bosna. From the available data, the design assumptions made in 1984 were reconstructed. It was concluded that the average specific wastewater production (310 L/PE.d) and the specific organic load (60 $gBOD_5$/PE.d) values used in the 1984 design are larger than typical design parameters used nowadays. For a more realistic interpretation of present conditions, in this study respectively 250 L/PE.d (source: local sewerage company) and 54 $gBOD_5$/PE.d (or 136 gTOD/PE.d) were used for the calculations. The future population forecast was estimated in one of the studies carried out by a consultancy firm in 1999 using the growth rates (3% p.a. for 1998-2001, 2% p.a. for 2005-2015) based on the 1998 consensus data and population of 374,300. From the prognosis it has been estimated that in 2025 the required plant capacity is 900,000 PE. Accordingly, the capacity of 600,000 PE was reached by the year 2004. However, the prognosis has not been updated yet, and, consequently, not validated and confirmed. Therefore, this study was directed towards the prognosis of the year 2004 (600,000 PE) which is the actual design capacity of the current plant. On top of this the Q_{365} safety factor (125% of the average dry weather flow - ADWF) was applied leading to a design capacity of 750,000 PE. Exceeding this load has large technical implications as the current plant design is limited to 600,000 PE. The results of the model-based study indicate that to meet the EU effluent standards already up to 70,000 m^3 reactor volume is required, threefold the then currently available volume (23,900 m^3); should the design capacity be 900,000 PE in 2025 to meet the same (EU) standards, 6 to 8 times the current reactor volume would be required. Focusing on the current design capacity is therefore justified, as the goal of this project was to study the possibilities to revitalize the current design using as much possible the currently available units/capacity. In addition, based on a few measured flow data from 2007, the activated sludge process is already hydraulically overloaded and therefore, the second bypass is active even during the ADFW. However, because a proper explanation is lacking on the origin of the high measured flows and moreover, because the available wastewater quality values are by far out of the usual range, it was questionable whether the flow measurements from 2007 are realistic and should be used for plant design. Additionally, the 2007 data neither correspond to the original design nor to the PE prognosis. As the 2007 flow data cannot be validated, the flow prognosis was thus based on the original design and key design values from demographic database. In this scenario study, the prognosis in 2004 was used for the design calculations. According to the prognosis the design capacity of 600,000 PE was reached by 2004 (2.1 m^3/s), and for 2025 the prognosis is 900,000 PE and flow of 2.92 m^3/s (Q_{365}). The latter is calculated under the assumption that towards 2025 the sewer is well maintained and the current discrepancy is reduced to acceptable proportions.

10.3.3 Reconstruction of the influent characteristics

From 1992, wastewater produced in the city of Sarajevo flows untreated into the River Bosna. The discharged water quality was monitored in 2007 by the local sewage company. To obtain a reliable wastewater composition that can be used in this study and for future design studies, an influent composition assessment was conducted using different data sources. Based on the measured data in 2007, it was concluded that the Sarajevo's wastewater is characterized as unusually weak. Its concentrations are estimated threefold lower than one would expect in a practically combined sewer system. For example, the average measured concentration for COD in 2007 was 138 mg/L, where a typical combined sewer concentration would be in the range of 300-500 mg/L. In the period prior to 1992, when the plant was in operation, an average COD concentration was 517 mg/L. For the difference to be caused by dilution as a result of infiltration, the average influent flow in 2007 would need to be 3 to 4 times the flow prior to 1992, to result in the measured concentration, which is not a realistic assumption. More likely, the low influent concentrations observed in 2007 are caused by the non-representative operating and sampling conditions (stagnant flow conditions and settling in the measurement channel). That settling has likely occurred, can be concluded from the analysis of the in 2007 measured concentrations compared to the data measured prior to 1992. The settled water quality in 1992 strongly resembles the measured quality in 2007. In 2007, for example, COD and BOD_5 are respectively 138 and 78 mg/L. These values strongly resemble settled influent measured prior to 1992, respectively 173 and 74 mg/L. Prior to 1992, the influent COD and BOD_5 concentrations were 517 and 207 mg/L, which are in the range of those usual for similar conditions/system as in Sarajevo. Therefore in this study, the influent composition was reconstructed based on measured data from 1992, when the plant was in full operation satisfying effluent standards (Table 10.2).

Table 10.2 Model Influent composition as reconstructed from historical sources.

Parameter	Unit	Value	Parameter	Unit	Value
Flow	m^3/day	186,964	Particulate COD	mgCOD/L	349
Total suspended solids	mgTSS/L	358	Filtered COD	mgCOD/L	86
Volatile suspended solids	mgVSS/L	251	Total Kjeldahl nitrogen	mgN/L	24
Inorganic suspended solids	mgTSS/L	107	Ammonia nitrogen	mgN/L	10
Total carbonaceous BOD	mgBOD/L	217	Nitrite and nitrate	mgN/L	0
Filtered carbonaceous BOD	mgBOD/L	46	Total phosphorus	mgP/L	4.5
Total COD	mgCOD/L	435	Soluble phosphate	mgPO_4-P/L	1.5

For determining the influent composition for modelling purposes, the key design parameter was the population prognosis. From the PE equivalents, the BOD loading is determined based on 54 instead of 60 gBOD/PE.day. From the measured COD/BOD ratio the total COD (TCOD) was calculated. From the typical oxygen requirement of 136 gTOD/PE.day, being the sum of oxygen equivalents of COD and TKN (4.57 gTOD/gN-NH_4), the TKN loading was calculated. Although required for modelling, suspended and soluble fractions were not measured by filtration tests. Therefore, the particulate fractions are determined from typical elemental composition numbers; the particulate COD_X/VSS (1.39), and N and P content of organics (respectively 3.3% and 0.9%). These values are in the same range for domestic wastewater, as they arise from organic composition (biomass). Based on these assumptions it was possible to determine all the necessary influent fractions for modelling this plant. Later in the modelling protocol, the above influent characterization was assessed based on typical fractions that are measured in practice. The analysis of the total solids (TSS), inorganic fraction (ISS) and particular organics (VSS) confirmed that the wastewater in Sarajevo has relatively high solids and particulate inorganic material content. All available information confirm a high solid and inert fraction in the raw sewage. This is in line with the observed excessive accumulation of solids in the main collector and plant inlet. Also the observed solids COD and BOD fractions are high. This could be the result of short sewer residence time, probably due to the sloping sewer system of the city of Sarajevo. With short sewer residence time, larger particles are more easily transported and particulate COD (COD_X) does not have time to hydrolyze in the sewer to soluble COD (COD_S). This is expressed by the low soluble COD fraction of 20%, whereas in typical flat anaerobic sewer systems, 30-50% can be expected (typical Dutch conditions). The high solids and inorganic content of the wastewater is expressed in the high TSS removal after settling measured prior 1992 to be 82%. A typical design number is 65-60% TSS removal by primary settling. In the model, it is assumed that all grid and fast settling solids are removed and for remaining TSS, removal is set on 60%. From the high settle efficiency, it is assumed that of 16% of the measured solids is rapidly settling and already will be removed in the grid chamber.

10.3.4 Scenario results

In total a dozen technological and engineering options (Table 10.3) were considered which took into account the original design, the present and future composition and quantity of sewage (up to 2030), the choice of wastewater treatment technology (an existing anoxic-oxic AO system, or a future upgrade to an anaerobic-anoxic-oxic A2O/UCT system or oxidation ditch/Carousel© system), the choice of an aeration system (existing surface aeration by turbines, or an alternative with fine bubbles aeration by air diffusers), and the amount and composition of sludge (and later biogas) generated in the process. These scenarios were directly corresponding to the study objectives listed in the introduction. The total revitalization of WWTP Sarajevo concerns all process units; from preliminary treatment up to sludge processing. This chapter focuses on the design of the biological tanks (the activated sludge process).

Table 10.3 Overview of scenarios used in this study

Nr.	Scenario description	Influent	Biological tanks	Hydraulic regime	Aeration type	Removal
0	Existing reactors for COD removal - installed surface aerators capacity, settled influent	Settled	Existing	Plug flow	Surface	COD
1	Existing reactors for COD removal - new surface aerators capacity, settled influent	Settled	Existing	Plug flow	Surface	COD
2	Existing reactors for COD removal - new bubble aerators, settled influent	Settled	Existing	Plug flow	Diffused	COD
3	Existing reactors for COD removal - new bubble aerators, raw influent	Raw	Existing	Plug flow	Diffused	COD
4	Nitrifying extended volume plug flow - new bubble aerators, settled influent	Settled	Extended	Plug flow	Diffused	COD/NIT
5	Nitrifying extended volume plug flow - turbines and new bubble aerators, settled influent	Settled	Extended	Plug flow	Surface/diffused	COD/NIT
6	Denitrification, extended volume plug flow - bubble aerators, settled influent	Settled	Extended	Plug flow	Diffused	COD/N
7	Extended volume plug flow for COD removal - bubble aeration, raw influent	Raw	Extended	Plug flow	Diffused	COD
7a	Extended volume plug flow for COD removal - bubble aeration, raw influent	Raw	Extended	Plug flow	Diffused	COD
7b	Extended volume plug flow for COD removal - bubble aeration, raw influent	Raw	Extended	MLSS recirculation	Diffused	COD
7c	Extended volume plug flow for COD removal - bubble aeration, raw influent	Raw	Extended	Carrousel	Diffused	COD
8	Nitrifying Carrousel for COD removal - bubble aeration, raw influent	Raw	Extended	Carrousel	Diffused	COD/NIT
9	Denitrifying plug flow AO - bubble aeration, raw influent	Raw	Extended	Carrousel/AO	Diffused	COD/N
10	Denitrifying AO Carrousel - bubble aeration, raw influent (EU standard)	Raw	Extended	Plug flow /AO	Diffused	COD/N
11	A2O plug flow processes - bubble aeration, raw influent (EU standard)	Raw	Extended	Plug flow/A2O	Diffused	COD/N/P
12	A2O Carrousel - bubble aeration, raw influent (EU standard)	Raw	Extended	Carrousel/A2O	Diffused	COD/N/P

However, because the design of the activated sludge plant is directly related to primary settling as well as the maximum loading capacity of the settlers, both these aspects were also taken in account. The current activated sludge system in WWTP Sarajevo is designed for the removal of organic matter (COD or BOD_5). This removal takes place in aerated biological tanks where activated sludge (biomass) and wastewater (substrate) are brought together resulting in COD removal by the biomass and creation of new sludge (biological growth/yield). Due to production of new biomass (waste activated sludge or WAS), the activated sludge system must be reconstructed in parallel with (some kind) of WAS treatment facilities. The long time locally supported reconstruction strategy was simply to turn the current plant into its original condition as of 1984. However, such an approach was recently challenged by expressed doubts whether in 2010 one should go back to the design and technology of ‘80s, having in mind that the new design horizon is the year 2030. Furthermore, as Bosnia and Herzegovina is approaching the EU membership, it is not unlikely that in the (near) future existing effluent standards will become extended to nutrients (N and P) removal. Therefore, it would be pity not to take such a possibility into consideration now. Finally, the scope of a to-be-chosen reconstruction option was not clear, and consequently, no one could say with confidence how far the reconstruction can stretch given the limited financial resources.

Scenarios 0, 1 and 2 are considering turning plant into its original condition (Figure 10.5). In case of implementation of stricter controls for environmental protection purposes is required, the existing BOD-removing activated sludge system shall be retrofitted and extended for biological nitrification (TKN removal) and denitrification (TN removal) (scenarios 4, 5, 6 and 8 to 12), and phosphorus (P) removal (scenarios 11 and 12). Therefore in several scenarios, as explained in Table 10.4, designs are evaluated for different operational conditions.

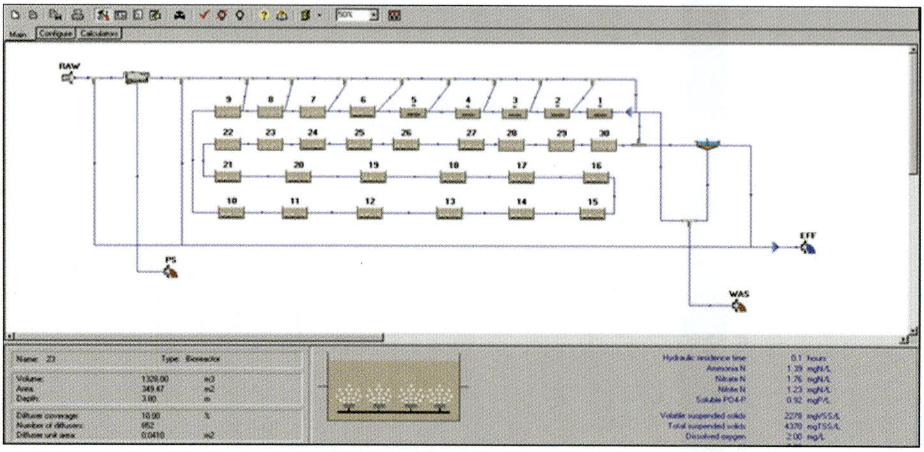

Figure 10.5 Hydraulic plant scheme of WWTP Sarajevo in the BioWin simulator which was used in this scenario analysis to calculate the different WWTP concepts.

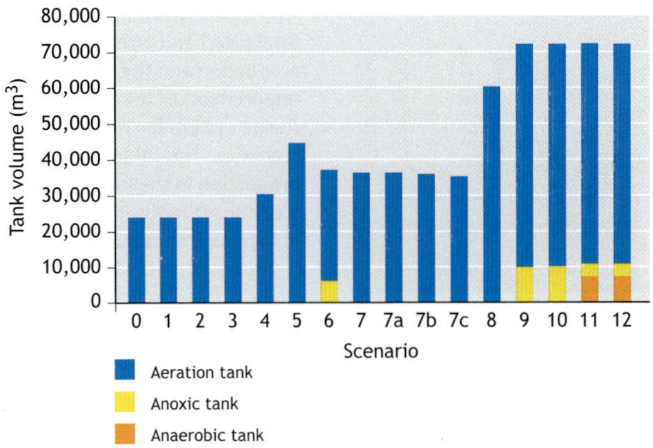

Figure 10.6 Volumes for the aerobic tanks, anoxic tanks and anaerobic tanks belonging to the different studied scenarios. The P-removing scenarios (11 and 12) include an anaerobic tank. For enhanced denitrification an anoxic tank is included. Scenario 5 has semi-anoxic zones due to inefficient aeration by turbines.

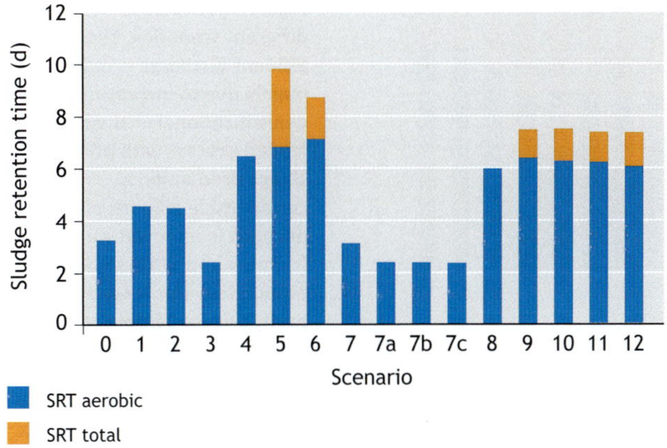

Figure 10.7 SRT of different scenarios. The Figure shows aerobic SRT and the total SRT calculated as aerobic or total sludge mass per WAS produced per day. For full nitrification the minimum aerobic residence time is in the range of 6 to 6.5 days. Scenarios with shorter (aerobic) SRT are mainly BOD-removing systems. Scenario 5 is the combined turbine / bubble aeration scenario. The turbines are incapable of full aeration and therefore the model calculates oxygen shortage and anoxic zones in the inlet of the plug flow configuration. As a result of the turbines, in scenario 5 more volume is required.

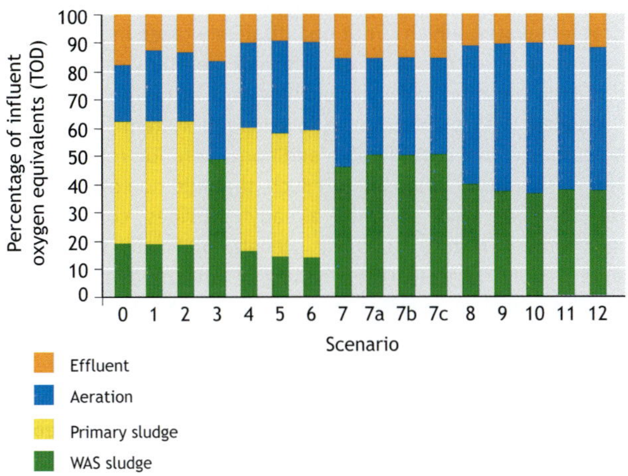

Figure 10.8 Conversion of the influent TOD load being the combined oxygen consumption of influent COD and TKN for the different scenarios. Primary sludge removal shows to be efficient in TOD removal (scenarios 0, 1, 2, 4, 5, 6). In the high loaded activated sludge systems, TOD is mainly removed via the WAS. In low loaded systems (scenarios 8 to 12) more TOD is removed via aeration/oxidation. TOD removal via the effluent is for all scenarios in the range of 10 to 20% and improving when the system is extended.

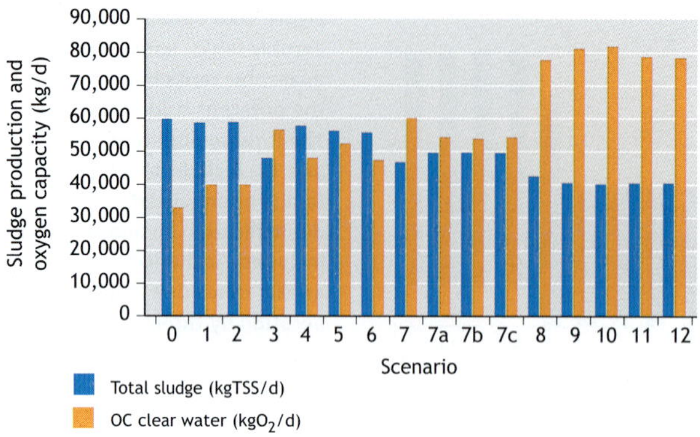

Figure 10.9 Relation between total (WAS and primary sludge) production and the oxygen requirement of the activated sludge system for the different scenarios. Total sludge production in the low loaded scenarios is reduced at the cost of high oxygen requirements.

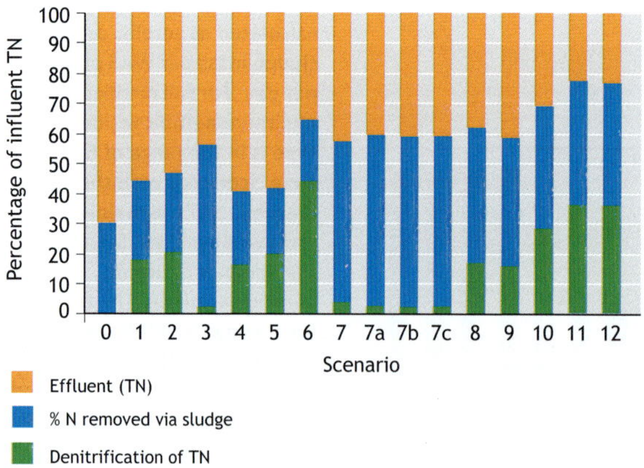

Figure 10.10 Plant performance regarding N-removal for the different scenarios. The effluent discharge is reduced mainly due to increasing denitrification. For the high loaded process with primary sludge production a considerable amount of nitrogen is removed via the (primary and WAS) sludge production. This N-load will be partly released from the sludge when digested which will require additional measures which are not taken in account in this study.

The results show that primary settling is the most efficient method to reduce the pollution load on the plant and thereby that released by the effluent discharge. Moreover, primary sludge can be digested with the biogas partly covering the plant energy requirement. However, sludge digestion holds the consequence of reject water, which is not quantitatively taken in account in this study. Secondly, the results show that introduction of sufficient aeration capacity i.e. replacing the turbines by diffused bubble aeration saves on building cost in (aerobic) volume (compare scenario 5 and 6). For reduced WAS production, a low loaded system is required (Figure 10.9, scenarios 8 to 12). This comes with a considerable additional aeration requirement and additional operational costs. The wastewater composition is such, that addition of a relative small anoxic reactor volume (or semi aerated volume) already provides considerable denitrification and enhanced nitrogen removal (Figure 10.10, e.g. scenario 6). Finally, the results show that the required investments rapidly increase when aiming at an improved effluent quality (e.g. EU effluent standards). This is mainly reflected by increase in reactor volume and required aeration capacity (Figure 10.6 and 10.9).

10.3.5 Scenarios evaluation

The following 10 questions posed by the decision-maker and corresponding answers provided by the authors describe quite nicely the nature and outputs of this study.

1. In the scenario where the WWTP is returned to the original state of 1984 using turbines, can the plant achieve present TSS and BOD/COD effluent standard, a) for the present load, and b) for the future load? Is there sufficient aeration capacity for 2010 and in the future situation?

This is scenario 0 in the study. The pre-war situation results in a highly (over) loaded BOD removing plant for 600,000 PE. Aeration capacity is not (always) sufficient with the current turbines. WAS is not mineralized and possibly requires additional treatment (typically being digestion). Extension of the secondary settling capacity is likely to be achieved by construction of two new settling tanks (recommendation has been made to start with one and to construct the second at the later stage). Sludge reduction is minimal. So it can be concluded that future operation at 900,000 PE is not to be desired due to limited loading capacity of the existing plant.

2. The same scenario: Is replacement of the surface aeration with bubble aeration required and what is the impact of that? By other words, should one stick with turbines or go for diffused aeration with bubbles?

Turbines are only possible in case of restoration of the pre-war conditions. All other scenarios employ replacement with diffused aeration with bubbles which has as a result a volume reduction of the extra built aeration capacity (e.g. volume difference scenario 4 and 5). The turbines do not give the required capacity. Therefore extra aerobic volume of the aeration tanks needs to be built to have full nitrification in winter (in case ammonia conversion is required).

3. The same scenario: What is the amount of sludge (mixed primary and secondary sludge withdrawn from the primary settling tanks - PST) expressed as tons of TSS per day and m^3 of sludge per day (give % DS in mixed PST sludge)?

For a 750,000 PE loading there will be 41,000 kg TSS primary sludge (which is 3,600 m^3/d) and 22,000 kg TSS waste activated sludge (2,500 m^3/d) produced. With gravity thickening, this usually can be brought up to estimated 5% DS (50 kg/m^3, 1,300 m^3/d) (data not given in the study).

4. The same scenario: Can the original plant operate without PST in terms of TSS and BOD removal (standards), and what are the consequences on the aeration tanks volume, aeration capacity, and secondary settling tank size?

Scenario 3 is the current plant without primary settling including diffused aeration with bubbles. Result is an over-loaded BOD removing system that provides some reduction in sludge quantity, however the WAS cannot be expected to be stable and requires treatment (typically digestion). Operation of such a system without digesters is not recommended. From that perspective scenario 0 (or rather scenario 2, including diffused aeration with bubbles) is preferred over scenario 3. The volume-extending alternative without primary settling is in fact scenario 7 (BOD removal). Scenario 8 (nitrification) or scenario 10 (AO process, BOD/NIT/DEN) require respectively extra volumes of activated sludge tanks of 12,000, 36,000, and 48,000 m^3. Average (no design) aeration requirements would be 860,000, 1,100,000, and 1,100,000 Nm^3/d, respectively.

5. So, what is the major general recommendation on this scenario 0 concerning the idea to get the plant in shape as in 1984 concerning the existing effluent standards (with eventual change of aeration system if necessary), shall it work or not for the present and the future load?

In all scenarios, the secondary settling tanks need to be extended by one or two tanks. The current activated sludge tanks volume is already under-designed for BOD removal. If taken in operation again, also primary and sludge digesters need to be included and reconstructed. It is also advisable to replace the total aeration system of turbine aerators with diffused aeration with bubbles (scenario 3). This requires installation of aeration capacity of 870,000 Nm^3/d or 52,000 Nm^3/h (at 142% of pluvial dry weather flow – PDWF and design temperature of 10°C). Alternatives without primary settling and without digestion would require a low loaded activated sludge system and minimal extension - scenario 10: AO system with required volume 48,000 m^3 and installed aeration capacity of 1,100,000 Nm^3/d and design value of 65,000 Nm^3/h (at 142% PDWF and 10°C).

6. There are plans to rehabilitate primary treatment only including PST. These plans do not include rehabilitation of the activated sludge part at this stage. How much sludge will be generated if only PST are rehabilitated and no secondary sludge is present?

At the load of 750,000 PE the estimated primary sludge production is 42,000 kgTSS/day and at 900,000 PE this will be around 50,000 kgTSS/day.

7. What about the idea to rehabilitate primary treatment facilities (and PST) and do "cold digestion" as the first reconstruction step? Or should reconstruction exclude PST and do PST together with activated sludge and SST together at the later stage?

The possible pollution reduction taking in operation PST is 320,000 PE (efficiency of primary settling concerning BOD). From an economical perspective, taking in operation PST is the most cost effective way to reduce pollution discharge. As a rule of thumb, there is 17,000 m^3 digester volume. Regular anaerobic digestion requires 20 to 22 days sludge age. With 5% TSS and 42,000 kgTSS/day of primary sludge, this is 840 m^3/d sludge flow, resulting in a 20 day sludge age which in theory is sufficient for full anaerobic digestion in digesters including energy recovery via biogas production.

8. Considering scenario under which stricter EU standards for nutrients are introduced, can the above plant cope with that? If not, what is necessary to retrofit/upgrade the existing design? What is the effluent quality? What is the amount and quality of sludge?

This is scenario 10. Additional volume of activated sludge of 48,000 m^3 is needed, one additional clarifier, 1,100,000 Nm^3/d of aeration capacity, diffused aeration system with bubbles (at average dry weather flow - ADWF and design temperature of 10°C), primary sludge production of 42,000 kgTSS/day, waste activated sludge production of 22,000 kg TSS/day (of mineralized quality). The effluent would satisfy current EU standards for TSS, BOD, COD, N and P.

9. Would one recommend applying already now N and P removal even if it is not required by the present local laws? What are the advantages and disadvantages?

Nitrogen and phosphorus seem not to be a problem even now. In all extended scenarios N and P requirements are met. This is because of the high BOD/N and BOD/P ratio in the influent. With scenario 10 (AO system), also a good sludge quality is assumed due to anoxic contact time and maybe extension of secondary settling can be reduced to one extra unit. Denitrification will as usual save (some) aeration capacity.

10. Are the existing SST able to cope in scenario with plant as in 1984, for existing standards (now and in the future)? And what will happen if nutrients have to be removed?

Secondary settling tanks need to be extended by one tank possibly two when operated at load of 750,000 PE. The practical extension will depend on the activated sludge quality. This will be better with low loading conditions and anoxic contact time (A/O). For 900,000 PE situation the ADWF hydraulic loading will increase however PWWF is the same because of the maximal sewage capacity. Again, depending on the sludge conditions one or two extra tanks are required.

The model simulation showed that for all the scenarios which involve the adoption of EU standards for nutrient removal, the existing plant will not be able to achieve these standards. Therefore, even for minimal improvement (e.g. nitrification), a large additional volume and aeration capacity is required. In order to reach such standards (nitrification, denitrification and possible bio-P) the sludge age needs to be extended from the designed 2-3 days to 10-15 days and the activated sludge section needs to be 3-5 times the current volume. Furthermore, 150% of the current secondary settling volume will be required. Extension by one and later by another secondary settling tank will be necessary as calculated by five different methods for secondary clarifier design (empirical, ATV, STOWA, WRC, and flux theory; Henze *et al.,* 2008) The study showed that by reconstruction of the primary treatment units, a reduction of an equivalent load of 320,000 PE will be achieved. Reconstruction of digestion to a full anaerobic digestion and biogas production as in the original design will be necessary. Taking into operation the existing aeration tanks with a diffused bubble aeration system will result in BOD, COD and SS concentrations in the effluent of below 25, 125 and 35 mg/L, respectively. It is possible to extend the existing AO plant by an additional tank volume of 48,000 m^3 and retrofit it into an A2O (UCT) process to reach a nutrient level in the effluent of 10 mgN/L and 1 mgP/L (current EU standards for plants larger than 100.000 PE). The study concludes that: (*i*) by reconstruction back to its original state and the replacement of surface aeration by a diffused air system the plant will be able to reach the current local standards regarding suspended solids and organic matter, (*ii*) in case of requirements for nutrient removal the plant will need substantial upgrade in terms of volume and aeration capacity, (*iii*) the existing AO system is considered appropriate for current effluent quality standards; however the most appropriate technological process to be applied to include nutrient removal is the modified UCT process (A2O), and (*iv*) based on the results of this study it will be possible to determine the scope of reconstruction to match available financial resources, given that the costs of preferred options are known. The study clearly showed the usefulness of mathematical modeling in a decision-making process regarding investments in urban infrastructure in which the compromise often has to be sought between the prioritization of investments and environmental benefits, as was the case here.

10.4 Modeling discharge-receiving rivers Miljacka and Bosna

River Bosna originates from the array of strong karstic springs south-west from Sarajevo and is the longest river that flows through Bosnia and Herzegovina. At the source it is of highest quality and as such serves for water supply of the part of the city. However, only few kilometers downstream it is joined by polluted waters of four major tributaries, namely Željeznica, Zujevina, Dobrinja and Miljacka (Figure 10.11).

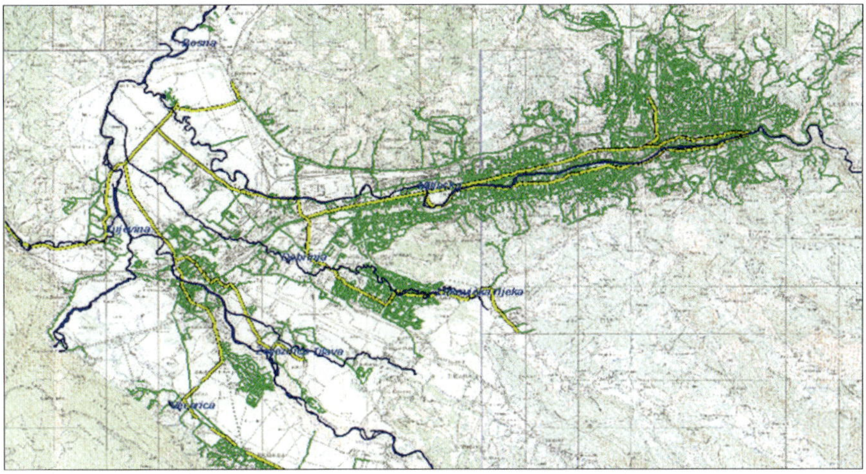

Figure 10.11 Rivers Bosna, Miljacka, Željeznica and Dobrinja

River Miljacka flows through Sarajevo and is recipient of sewage pipes, creeks and drainage channels which are not currently connected to the city's sewerage, as well as of the overflow from 24 CSOs. In addition, just before the convergence with River Bosna, it is receiving for almost two decades untreated sewage of the city which is diverted from the non-functioning WWTP. During the period of plant operation (1984-1992) such discharges were activated only in emergency situations and during its activation as CSO, while the purified wastewater discharging into River Bosna not far downstream of the convergence with River Miljacka.

In order to evaluate the impact of present and future pollution load on the quality of River Miljacka and River Bosna in the surroundings of the WWTP (study area), numerical modeling software HEC-RAS was used. The model was applied for the dynamic flow simulation and water quality analysis in the rivers using GIS information and 149 cross sections and water quality data collected through sampling campaign and regular river water quality monitoring at 7 locations. Catchments of the rivers are delineated from DTM of resolution 20m. For the simulations three scenarios were examined: (i) *Plant out of operation* representing the present situation and discharge of the untreated sewage in river Miljacka, (ii) *Original plant design* in which the WWTP is rehabilitated to resemble situation in 1984 (COD and SS removal only), and (iii) *New plant design* which reflects the possible extension of the plant with N and P removal according to EU standards. From various hydraulic scenarios the DWF situation was selected as this is in fact the worst case scenario (strong sewage with low flow in the rivers).

The following conclusions can be drawn:

- The study shows that application of HEC-RAS on the prediction of water quality in rivers Miljacka and Bosna is able to produce meaningful results. However, in order to verify the model outputs, it is necessary to execute additional sampling campaigns under different scenarios as described by the model.
- In case of the present situation (WWTP non-operational), the negative impact of the untreated sewerage entering river Miljacka is confirmed and described successfully by the model. The impact is clearly observed downstream of the present discharge point and continues on the river Bosna even more than 3.5 km further from the confluence point.
- In case of reconstruction of the WWTP to the original state (BOD and SS removal only), the positive impact of WWTP is obvious on the river Miljacka as the present effluent discharge is not taking place any more, and furthermore, river Bosna is not significantly affected by the effluent discharge in terms of BOD and SS concentrations. However, as the plant has low efficiency concerning N and P removal, the effluent discharge has substantial impact on the river Bosna downstream from discharge point particularly in terms of nitrogen.
- In case of reconstruction and upgrade of the existing WWTP to biological nitrogen removal plant (new plant scenario), the effect of employment of nutrients removal on the recipient is remarkable, particularly in terms of reduction of N concentration in river Bosna as shown in Figure 10.12.
- The study demonstrates the necessity of the reconstruction (and upgrade) of the WWTP, as current self-purification capacity of recipients (especially for river Miljacka) is not sufficient to cope with high load of pollutants discharged into them. The impact of untreated sewage discharge can be seen kilometers downstream even given the fact that the flow of river Miljacka is lower than the flow of river Bosna (so even dilution effect has limited contribution to pollution reduction).
- By using the HEC-RAS model it is now possible to predict and quantify the fate of pollutants in aquatic environment of river Miljacka and river Bosna in the vicinity of the plant at the location Butile as well as the impact on water quality, which is a basis for further EIA study on the aquatic ecosystems (see Figure 10.13).

From the rivers analysis obtained in the research, the following recommendations can be identified to support improvements in the future:

- Data required to calibrate and validate the river model should be collected and available.
- Flow and water level measurements should be made for all rivers.
- The WWTP has significant role in the water quality of rivers; therefore it is recommended that at least the original design scenario should be operational.

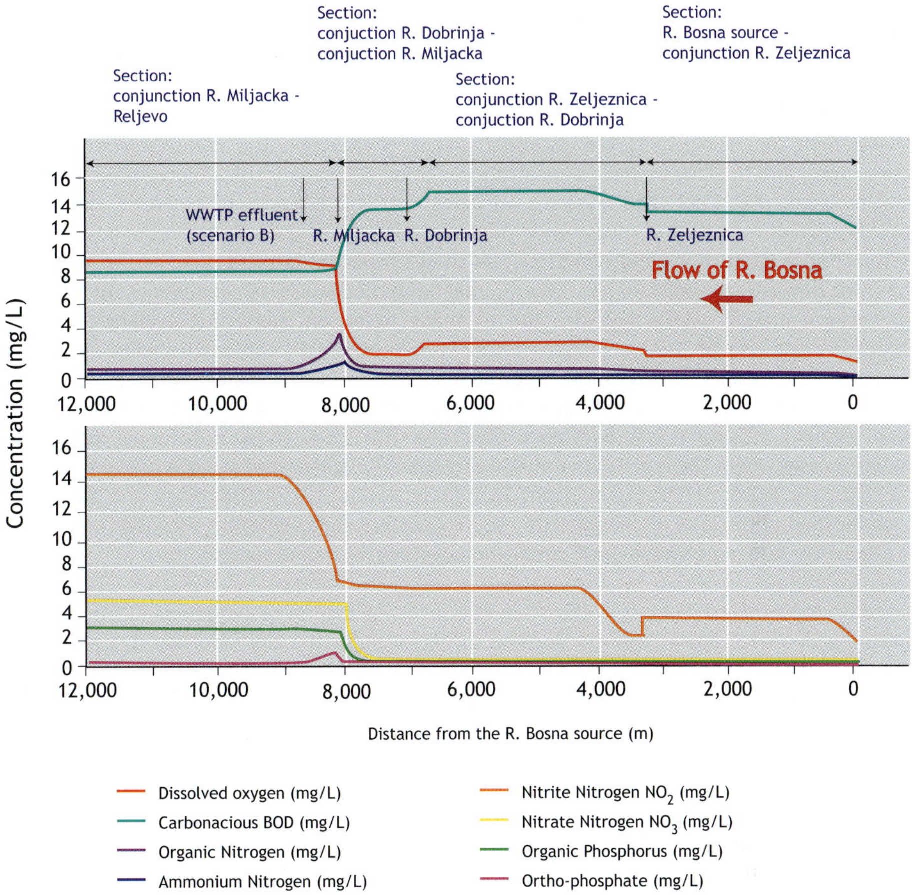

Figure 10.12 Concentrations (mg/L) of DO, BOD, Org. N, NH_4, NO_2, NO_3, Org. P and PO_4 along river Bosna for the reconstruction and upgrade of the existing WWTP to biological nitrogen removal plant (new plant scenario).

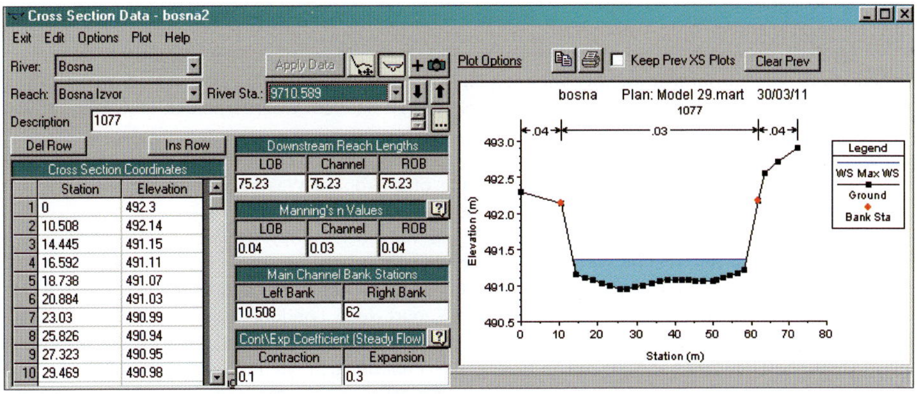

	Station	Elevation
1	0	492.3
2	10.508	492.14
3	14.445	491.15
4	16.592	491.11
5	18.738	491.07
6	20.884	491.03
7	23.03	490.99
8	25.826	490.94
9	27.323	490.95
10	29.469	490.98

Figure 10.13 Cross section of river Bosnia in HEC-RAS.

10.5 Conclusions and recommendations

The study described in this chapter demonstrates the integrated modeling framework used in the research concerning Sarajevo's wastewater system and it shows how such a framework can be effectively applied in a decision-making process. It proves that in a complex urban environment, modeling tools are essential means to describe the complex water-related interactions, and to allow management strategies to be developed. Use of integrated modeling enables connection between systems, development of interaction between them and identification of the effect of measures in upstream segment of the system on the downstream system. These interactions and effects can be shown from the farthest point of the system to the most downstream point of the system. Integrated modeling gives confidence and support for decision making where decisions can be made based on supporting evidence, justified arguments and use of proven modern tools. Even though this approach applies sequential modeling, the results clearly show inter-dependency of urban water systems which are impossible to obtain using conventional, traditional methods and which are therefore often or even regularly missing in less developed countries. This study shows substantial differences between cases where there is treatment or no treatment of generated sewage or even between two different levels of treatment itself. This study develops a methodology and allows for quantification of the load discharges into recipient waters under different scenarios and is able to predict the fate of pollutants (self-purification capacity) of the receiving waters.

Based on the work carried out in this study the following can be recommended: (i) n order to make this approach applicable as a decision making tool for the city Sarajevo, it is needed to upgrade the study with more information regarding each of the three considered systems to obtain the results which are validated and can be taken with more confidence; (ii) having in mind limitations of sequential modeling, this study should be upgraded with its optional part. In other words, the integrated modeling with feedback (i.e. parallel modeling) should be applied, which would allow for a better representation of the real conditions (conditions taking place in reality), and (iii) for the full support decision making process the infrastructure related models should be extended by the cost engineering considerations and the receiving water model should be extended by the description of the impact of wastewater discharges on aquatic ecosystems.

References

ASM Design (2009), Model Design Study WWTP Sarajevo: Model-based process technology selection for rehabilitation of WWTP Sarajevo.

DHI (2009), Preparation of Investment Analysis and modelling study for the sewer collection network in Sarajevo, Bosnia and Herzegovina: Model-based analysis of Sarajevo's Sewer System.

Brdjanovic D., van Loosdrecht M.C.M., Versteeg P., Hooijmans C.M., Alaerts G.J. and Heijnen J.J. (2000) Modelling COD, N and P removal in a full-scale WWTP Haarlem Waarderpolder. *Wat. Res.* 34, 846-858.

Brdjanovic D., M.S. Moussa, M. Mithaiwala, G. Amy, and M.C.M. van Loosdrecht. (2007). Use of Plant-wide Modeling for Optimization and Upgrade of a Tropical Wastewater Treatment Plant in a Developing Country. *Wat. Sci. Techn.*, 56(7), pp 21-31.

Henze M., Grady C.P.L. Jr., Gujer W., Marais G.v.R. and Matsuo T. (1987) Activated Sludge Model No. 1. IAWQ Scientific and Technical Report No. 1, London, UK.

Henze M., Gujer W., Mino T., van Loosdrecht M.C.M. (2000) Activated Sludge Models ASM1, ASM2, ASM2d and ASM3. IWA Scientific and Technical Report No. 9, London, UK.

Henze M., van Loosdrecht M.C.M., Ekama G.A. and Brdjanovic D. (2008) Biological Wastewater Treatment: Principles, Design and Modelling. IWA Publishing, pg 511. ISBN 9781843391883

Meijer S.C.F. van Loosdrecht M.C.M. and Heijnen J.J. (2001) Metabolic modelling of full scale biological nitrogen and phosphorus removing WWTPs. *Wat. Res.* 35, 2711-2723

Meijer S.C.F. and Brdjanovic D. (2010) Upgrading Large Wastewater Treatment Plants: Use of Modelling as a Decision-making Tool, In Proceedings Wastewater Treatment Plants of XXI Century, Moscow, Russia.

Peric M.,Brdjanovic D., and Razdorov P. (1991) Upgrade of the wastewater treatment plant in Sarajevo (in Bosnian), Naša Vodoprivreda, No.5/6, pp.41-51.

Pinzon A., Brdjanovic D, Moussa M., Lopez-Vazquez C.M., Meijer S., Van Straaten H., Janssen A., Van Loosdrecht M.C.M., and Amy G. (2007) Modelling Oil Refinery Wastewater Treatment Plant. Environmental Technology, 29(11).

van Veldhuizen H.M., van Loosdrecht M.C.M. and Heijnen J.J. (1999) Modelling biological phosphorus and nitrogen removal in a full scale activated sludge process. *Wat. Res.* 33, 3459-3468.

Figure 10.14 River Miljacka in Sarajevo (Photo Hodzic A.).

Figure 10.15 River Bosnia at its source (Photo Hodzic A.).

Figure 10.16 WWTP Sarajevo awaiting reconstruction to start in 2015 (Photo: Brdjanovic D.).

Chapter 11: WWTP Illidge Rd., Sint Maarten N.A.

Use of modelling for cost-effective design of wastewater treatment plant

Lake O., Lopez-Vazquez C.M., Hooijmans C.M., Brdjanovic D.

This chapter is based on Steady-state models as cost-effective tools for design and assessment of wastewater treatment systems (2011) by Lake O., Lopez-Vazquez C.M., Hooijmans C.M., and Brdjanovic D. In: 2nd IWA Development Congress. November 21-24th, 2011. Kuala Lumpur, Malaysia.

11.1 Introduction

Mathematical modelling is becoming an integral part of wastewater treatment plant project life cycle, often used for optimization and prediction of process performance, and as a supporting tool for design. However, the costs of simulators and software licences tend to hinder and limit the use and applicability of modelling (e.g. ASM1, ASM2, ASM2d, ASM3) (Henze *et al.,* 2000) particularly in low- and middle-income countries. On the other hand, steady-state models are simple (suitable to implement in commonly used spreadsheets), economical (open source, available in text books and scientific publications), and useful tools for design since they are oriented to determining the important design parameters from performance criteria, becoming also valuable to cross-check simulation model outputs (Ekama, 2009). All these characteristics make steady-state models attractive, reliable and cost-effective tools for design and assessment of wastewater treatment systems in developing countries and countries in transition.

Figure 11.1 Map of the Caribbean Island of Sint Maarten in The Netherlands Antilles.

In the Caribbean island of St. Maarten (Netherlands Antilles), water has a key role with regards to recreation and, by association, to the economy in a tourism-driven market. Tourists partake in numerous aquatic activities and; therefore, pristine and safe surface and coastal waters play an important role. If not properly maintained, the environmental harm can lead to negative economic consequences. As such, surface and coastal waters must be protected from pollution in general but, for the purposes of this research, specifically from wastewater streams exiting the catchment (which introduce high amounts of contamination in the form of organic discharges, nutrients, suspended solids and pathogens). In spite of the importance of adequate wastewater treatment, usually discharge standards are not legislated in tropical areas. Consequently, many simple but robust systems like anaerobic and pond treatment systems are the main solutions applied. While these systems are successful in reducing the organic loads of the wastewater, when stricter discharge standards are imposed with regard to nutrient removal (as in the case of St. Maarten), other treatment technologies are required (like activated sludge). However, in St. Maarten only 20% of the households (approximately) are connected to a domestic wastewater treatment plant (WWTP) where the Illidge Road WWTP treats the wastewater generated by 11% of the households. Most of the connections are in the capital, Philipsburg, and in the Belvedere area. Other a few areas are connected to small plants that do not function properly. Thus, almost 89% of the island is not connected to a treatment system and they make mostly use of septic tanks. When a tank is full, a sewer truck company empties the tank and unloads at the Illidge Road plant. Thus, about 11% of the domestic wastewater is conveyed to a plant through a sewer network originated from the local population (domestic activities), laundry facilities, and tourists. The remaining 89% is delivered with sewer trucks in the form of septic sludge. Daily, 20-40 trucks discharge at Illidge Road. On very busy days however, or when is the content too saline or oily, the trucks unload at the Pond Island instead of the WWTP.

Figure 11.2 Discharge of septage to the buffer tank of WWTP Illidge Rd. (Photo: Brdjanovic D.).

In view of the strong need to achieve appropriate wastewater treatment, the local government in Sint-Maarten has decided to build and operate an activated sludge nutrient removal wastewater treatment plant (a Modified Bardenpho process) to replace the existing Illidge Road sewage treatment plant. The current sewage plant is manly composed of an Imhof tank whose capacity has been exceeded. In the present study, a stoichiometry-based steady-state model was applied to assess the design and operation of the new plant. The main objectives of this modelling study are to illustrate: (i) the importance of wastewater characterization and fractionation as a result of the significant septage component of the wastewater and (ii) the significance and reliability of utilizing steady-state models for technology assessment by comparing the modelling outcomes with those obtained from other simulators and design approaches. Furthermore, based on local space-planning development plans, four different scenarios were evaluated where the effects of the expansion of the sewerage network (up to 85% coverage) and population growth were taken into account to assess their effects on wastewater composition and wastewater treatment plant performance through an estimated 25-year life span of the newly proposed plant.

11.2 Materials and methods

11.2.1 Description of the study area: Cul-de-Sac, St. Maarten

The scope of this study mainly considers the wastewater streams of the Cul-de-Sac and other adjacent sub-catchments in the Dutch site of the Island of St. Maarten (Figures 11.1 and 11.3). Within Cul-de-Sac, there are two main wastewater streams that mix and become the influent of the existing Illidge Road wastewater treatment plant (Figures 11.4 and 11.5).

Figure 11.3 Phillipsburg and Great Salt Pond with the solid waste landfill in the middle (Photo: Teefelen J.).

Figure 11.4 Old WWTP Illidge Rd.: biological treatment unit (Photo: Brdjanovic D.).

Figure 11.5 Old WWTP Illidge Rd.: final clarification (Photo: Brdjanovic D.).

These include: sewage, which amounts to 11% of the wastewater and, septage, which practically comprises the remaining 89%. Sewage is received from the Philipsburg, Dutch Quarter and Belvedere districts. Septage is brought to the plant from the residential and commercial activities in the Cul-de-Sac, Upper and Lower Prince's Quarters (Figure 11.1). Septage is transported by trucks for treatment at the current treatment plant. Excess sludge generated at the plant is dried at the facility on sludge drying beds and transported to the solid waste landfill for disposal.

The current wastewater treatment plant (Figure 11.6a), with a volume of circa 154 m^3 and basically composed of an Imhoff tank and a secondary settling tank (Figures 11.4 and 11.5), is heavily overloaded. Furthermore, the effluent is discharged to the inland water (Great Salt Pond) (Figure 11.3). To alleviate this problem, a Modified Bardenpho (A2O) process has been proposed to upgrade the plant (Figure 11.6b). It is comprised of a carousel reactor for the anoxic/aerobic recycle (Barnard, 1976). In addition, the design allows for chemical phosphate precipitation as a mean of phosphorus removal.

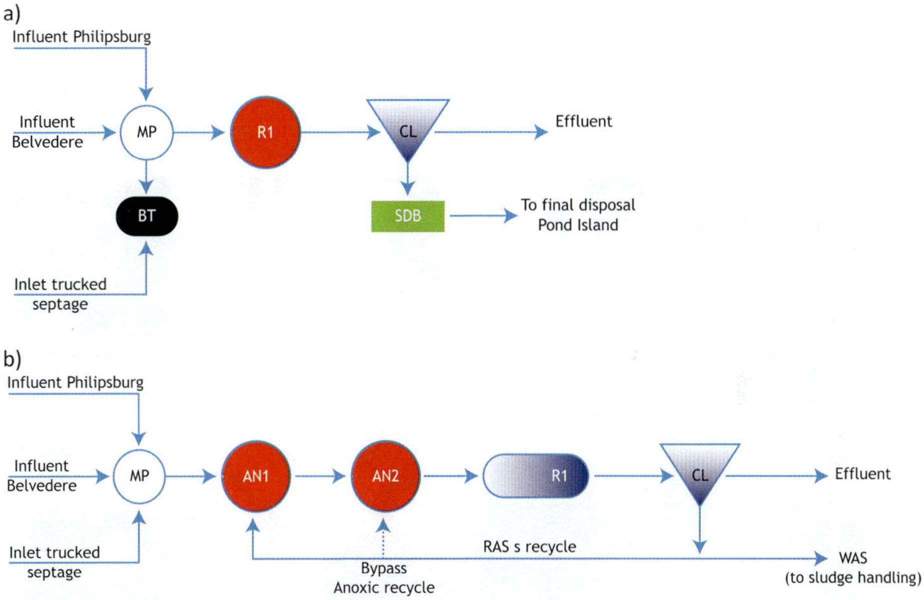

Figure 11.6 Process configurations of the Illidge Road wastewater treatment plant: (a) existing plant configuration and (b) proposed Modified Bardenpho process to upgrade the plant. MP: Mixing pit; BT: Buffer Tank; R1: Imhoff Tank; SDB: Sludge Drying Beds; AN1: Anaerobic Selector; AN2: Anaerobic Tank; AX: Anoxic Zone; AER: Aerobic Zone; CL: Clarifier.

11.2.2 Scenarios of study

Different scenarios were studied to assess not only the life-span of the newly proposed wastewater treatment but also the effects of the expansion of the sewerage network on the performance of the plant. They were defined taking into account the calculated influent flow projected through the design lifetime of the plant based on: (i) projected population dynamics (2.3%), (ii) sewer network expansion (increasing from 11 to 85%), (iii) estimated per capita water consumption (135 L/cap.d) and, (iv) estimated per capita septage production (7 L/cap.d). Thus, four scenarios were studied as a function of the expansion of the sewerage network: 11, 48, 64 and 85%. For each scenario, the composition of the mix influent discharged to the wastewater treatment plant was determined by weighing the concentrations and fractions of the different streams as a function of the expansion of the sewerage network and of the replacement of the septage volumes by wastewater. All scenarios were developed for dry weather flow conditions.

11.2.3 Wastewater characterization and fractionation

The characteristics of the septage and wastewater were determined based on the results of a 3-day sampling campaign carried out in 2008 at the Illidge Road WWTP (Lopez-Vazquez, 2008) (Figure 11.7). The same data were used to assess the efficiency and performance of the existing sewage treatment plant. For the fractionation of the organic matter present in the wastewater influent, the fractions of 0.15 and 0.07, for the unbiodegradable particulate COD fraction (f_{Sup}) and unbiodegradable soluble COD fraction (f_{Sus}) (Ekama *et al.*, 2007), respectively, were used. For the septage component; however, other f_{Sup} and f_{Sus} values were applied. Septage, unlike sewage, has higher fractions of particulate unbiodegradable COD and nitrogen. Using a 'typical' composition for septic sludge (Henze, 2008), the COD and N fractions were recalculated. For the estimation of the unbiodegradable particulate organic nitrogen, a N-to-VSS factor (f_N) of 0.002 mg Org-N/mg VSS was used as proposed by Batstone *et al.* (2000) for the nitrogen content associated to the unbiodegradable particulate organics. This approach was applied assuming that all unbiodegradable organics have the same nitrogen content.

Figure 11.7 Sampling in progress at old WWTP Illidge Rd. (Photos: Lopez-Vazquez C.M.).

11.2.4 Evaluation of the new wastewater treatment plant configurations

The proposed treatment plant (Figure 11.8) was modelled in BioWin (Envirosim Associates Ltd.). This design was compared to the design parameters calculated using a bioprocess stoichiometry based steady-state model for organic matter, nitrogen and phosphorus removal (Ekama *et al.*, 2007). The anaerobic plug flow selector (AN1) was simulated as four unaerated tanks of equal volume connected in series. The anaerobic tank (AN2) was modelled as an unaerated tank. The three anoxic (AX) and aerobic zones (AER) of the carousel reactor were simulated as three unaerated tanks of equal volume in series each separated by on aerated tank circularly connected. Internal flows and recycles were controlled to simulate the design values. An ideal clarifier was used to model the clarifier and solids in the effluent. In the aeration tanks, the dissolved oxygen concentration was set at 2 mgO_2/L. The recycle and waste flowrates were set at the values calculated with the bioprocess stoichiometry based steady-state model and assigned as constant values.

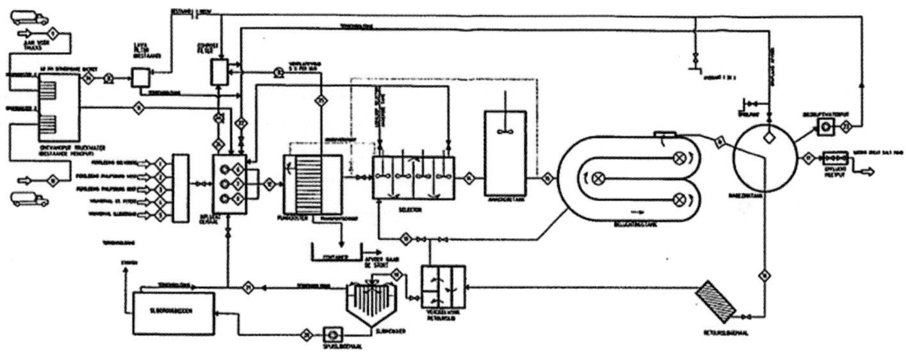

Figure 11.8 Process flow diagram of new WWTP Illidge Rd. (Provided by courtesy of Witteveen & Bos).

11.3 Results and discussion

11.3.1 Evaluation of the current plant performance

The current wastewater treatment plant (Figure 11.6a), with a volume of circa 154 m^3 and composed of an Imhoff tank and secondary settling tank, is heavily overloaded. This was corroborated by the results from a sampling campaign undertook in 2008. As observed (Table 11.1), the plant does not remove a large portion of (COD), nutrients and solids and considerably and continuously fails to meet the discharge standards. Actually, only about half of the total COD and two thirds of the TSS were removed, whereas the total nitrogen and total phosphorus concentrations were practically unaffected by treatment

Table 11.1 Influent and effluent quality of the current Illidge Road wastewater treatment plant including the effluent discharge standards (adapted from Lopez-Vazquez, (2008).

Parameter	Unit	Influent	Effluent	Removal efficiency (%)	Discharge standard (MAC)[1]
COD_{tot}	mg/L	2,011	1,085	46	125
N_{tot}	mg/L	125	115	8	10
P_{tot}	mg/L	22	19	14	2
TSS	mg/L	1,043	393	62	20

[1] Environmental Standards for the Netherlands Antilles for effluent discharges (van Geleuken, *et al.*, 2007).

11.3.2 Scenarios of study

The four scenarios of study were defined mostly based on the gradual increase in population and the potential expansion of the sewer connections. Mainly, they led to an increase in the influent wastewater flowrate from 710 to 2,800, 4,488 and 7,330 m^3/d as the sewer connections increased from 11 to 48, 64 and 88% (Table 11.2) in the period 2010-2035. Among other potential consequences, these flow rates might imply that larger reactor tank volumes could be needed at the wastewater treatment plant as the sewer connections increase.

Table 11.2 Scenarios on projected flows based on sewer connections

Scenario	Sewer connections %	Projected year	Wastewater flow m^3/d	Dry weather flow m^3/d	Total reactor size required m^3
1	11	2010	710	54	3,238
2	48	2016	2,800	16	8,449
3	64	2025	4,488	3	10,953
4	85	2035	7,330	1	13,014

11.3.3 Wastewater fractionation and characterization

Table 11.3 and Figure 11.9a show the fractions and composition of pure septage and sewage. The unbiodegradable COD fraction in septage (f_{Sup}) is the largest fraction (around 0.60), whereas in sewage the biodegradable particulate COD fraction (f_{Sbp}) accounts for the largest portion (0.55). However, as the sewer connections will increase, the fractions of biodegradable organics will tend to increase. The main reason is the larger fractions of biodegradable organics present in raw wastewater compared to those present in septage because septic tank sludge typically has a long storage time with significant stabilization (e.g. years as opposed to days) (Lopez-Vazquez et al., 2014). This leads to the consumption of the biodegradable fractions and the generation and accumulation of unbiodegradable components in septage. On the opposite, wastewater is exposed to relatively shorter retention times in the sewerage network and therefore contains higher fractions of biodegradable organics. As the number of sewer connections increases in time the biodegradable organic fractions increase proportionally (Figure 11.9b). Thus, the fraction of biodegradable particulate organics becomes the highest COD fraction when about 64% sewer connections are built and will reach up to 0.55 at 85% sewerage coverage. Since the biodegradable fractions play a major role for the removal of nitrogen and phosphorus, this is of major importance for the satisfactory performance of conventional biological wastewater treatment systems designed to achieve nutrients removal. Furthermore, the total nitrogen is fractionated a similar manner like septage because the anaerobic degradation of organics leads to the release of inorganic compounds from the hydrolysis of nitrogen bound to the biodegradable organics (Sötemann *et al.*, 2005). This implies that the higher the septage content, the higher the nitrogen concentrations and, the lower the content of biodegradable organics. Consequently, the biological removal of nitrogen cannot be easily achieved in activated sludge systems containing high septage volumes. But, as observed in Table 3 and Figure 11.9b, the expansion of the sewerage network can contribute to increase the biodegradability of the influent treated at the sewage plant. For more detailed evaluation of the impact of septic sludge addition to activated sludge wastewater treatment plants see Strande *et al.* (2014).

Table 11.3 Weighted fractions of unbiodegradable particulate organics (f_{sup}), unbiodegradable soluble organics (f_{sus}) and unbiodegradable soluble nitrogen (f_{nous}) for the mixed influent of the Illidge Road WWTP as a function of the potential expansion of the sewerage network.

Sewer connections %	Weighted fraction particulate unbiodegradable f_{Sup}	Weighted fraction soluble unbiodegradable f_{Sus}	Weighted fraction soluble unbiodegradable organic nitrogen f_{Nous}
11	0.56	0.06	0.13
48	0.39	0.06	0.09
64	0.32	0.07	0.07
85	0.22	0.07	0.05

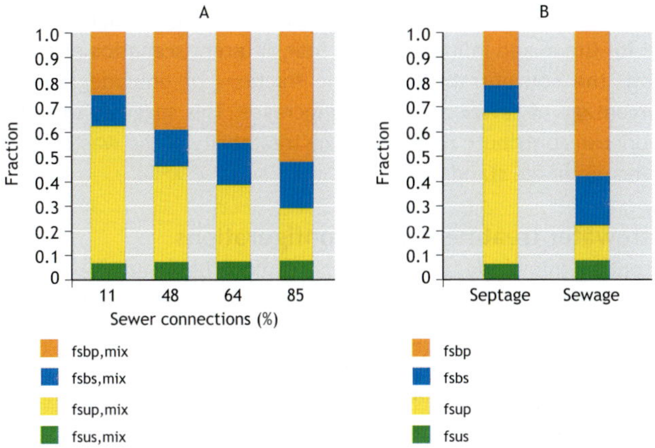

Figure 11.9 Sewage and septage fractionations of organics (measured in terms of COD) as a function of the increase in sewer connections. f_{Sbp}: fraction of particulate biodegradable organics; f_{Sbs}: fraction of soluble biodegradable organics; f_{Sup}: particulate unbiodegradable organic fraction; f_{Sbs}.: fraction of particulate biodegradable organic.

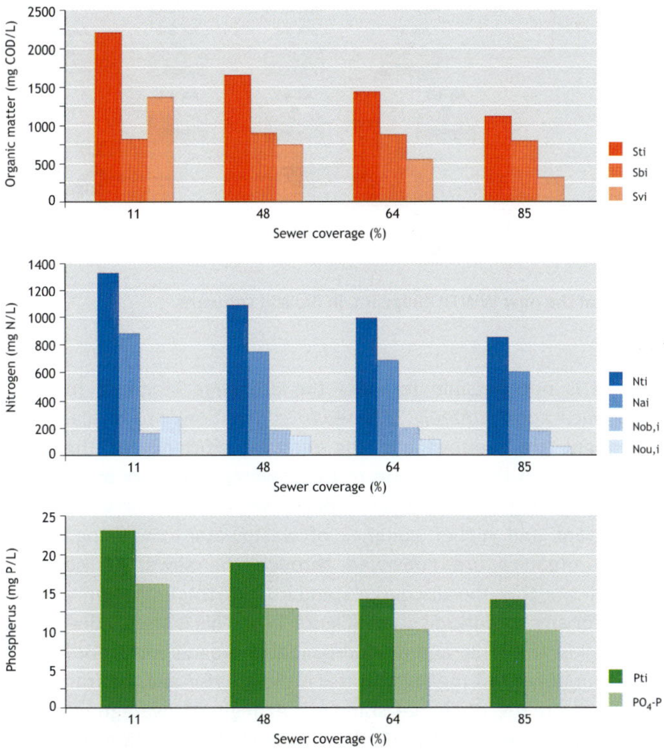

Figure 11.10 COD, nitrogen and phosphorus concentrations in the mixed influent as a function of the increase in sewer connections: (a) COD, (b) nitrogen and (c) phosphorus. Total COD in the influent, S_{ti}; biodegradable COD in the influent, S_{bi}; unbiodegradable COD in the influent, S_{vi}; Total Nitrogen in the influent, N_{ti}; Ammonia in the influent, N_{ai}; biodegradable organic nitrogen in the influent; $N_{ob,i}$; unbiodegradable organic nitrogen in the influent; $N_{ou,i}$.

Using the fractions displayed in Table 11.3 and Figure 11.9, the COD, N and P fractions and concentrations were determined for the mixed influent for the four different scenarios of study (Figure 11.10). As observed, both the concentrations and the fractions of unbiodegradable components decrease as the percentage of sewer connections increase. This suggests that the expansion of the sewerage network can contribute to increase the treatability of the wastewater and the potential compliance of the discharge standards.

11.3.4 Assessment of the wastewater treatment plant configurations

The model comparison carried out in this study showed an agreement between the results obtained with the bioprocess stoichiometry steady-state model and the ASDM model from BioWin (Figures 11.11 and 11.12). The verification of the results from the steady-state analysis was used as a basis to evaluate the proposed treatment plant design using the steady-state simulations in BioWin. In all the cases, as the number of sewer connections increase, the predictions of the models converged.

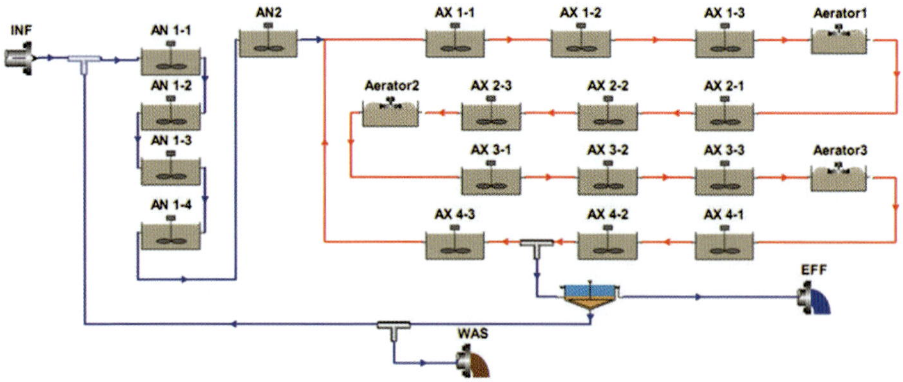

Figure 11.11 Model hydraulic scheme of the new WWTP Illidge Rd. in BioWin software.

With 11% sewerage coverage, it is not possible to meet the discharge standard for COD concentrations (Figure 11.12a). Once approximately 30% sewer connections are in place, the COD concentrations in the influent are reduced and the removal efficiency of the plant increases. This is a direct consequence of the expansion of the sewerage network as both the concentrations and fractions of unbiodegradable organics decrease when lower volumes of septage are discharged at the plant and higher volumes of wastewater containing higher concentrations of biodegradable organics are conveyed through the sewerage network. Unfortunately, the simulations indicate that the discharge standards for total nitrogen will not be met with the proposed design for any scenario of study (Figure 12b). This is largely due to the high concentrations of free and saline ammonia and unbiodegradable organic nitrogen. At 11% sewerage coverage, the septage component of the wastewater is the greater and the fraction of unbiodegradable soluble alone exceeds the discharge standard. At higher sewerage coverage percentages, the simulations indicated that after the nitrification and denitrification processes take place, the effluent ammonia, and to a slightly lesser extent the effluent nitrate concentrations, caused a notable increase in the total nitrogen concentrations in the effluent that do not allow to comply with the standards. One of the factors that do not allow to complying with the standards of 10 mg/L as total nitrogen is the lower availability of biodegradable organics with regard to the nitrate concentrations generated, but also the higher concentrations of unbiodegradable soluble nitrogen (which accounts for about 5 mg/L). Using the model built in BioWin, additional simulations were executed to improve this situation (e.g. by increasing the anoxic retention time). However, not even with the improvement efforts, the discharge standards for total nitrogen could be achieved (data not shown), mostly caused by the high concentrations of unbiodegradable soluble nitrogen. Nevertheless, when reaching about 85% sewer connections, the total nitrogen concentrations in the effluent lie around 12 to 15 mg/L, which suggests that the effluent standards could be possibly met when the island reaches around 100% sewer connections or the discharge of septage to the plant stops.

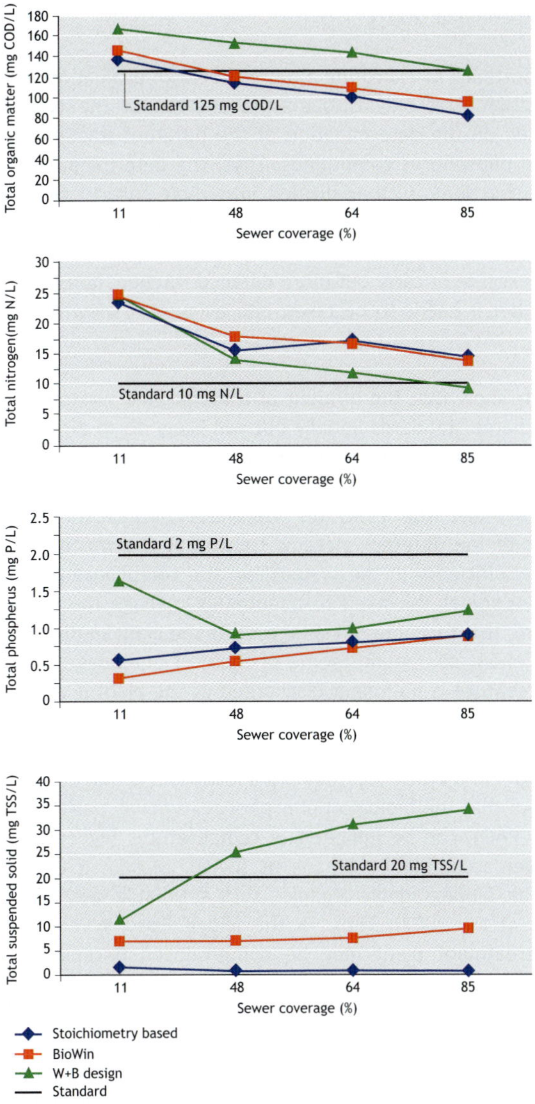

Figure 11.12 Comparison of effluent characteristics (in COD, N_{tot}, P_{tot} and TSS concentrations as a function of sewer connections) for the different models used as a function of the percentage of sewer connections: (a) stoichiometry based steady-state model (♦); (ii) BioWin – ASDM (■); (iii) proposed design tested on BioWin (▲).

Interestingly, all both models applied suggested that the new plant configuration was very suitable to meeting the phosphorus discharge standards throughout its lifetime regardless the coverage of the sewerage network (Figure 11.12c). However, the concentrations provided by the proposed design were higher than those obtained with the stochiometry-based model and BioWin (which were very similar between each other). The compliance of the phosphorus standard can be attributed to the very high concentrations of soluble biodegradable organics which fluctuated around 800 mgCOD/L are far higher than the influent phosphorus concentrations that ranged between 10 and 22 mgP/L. This leads to COD/P ratio of 40 to 80 gCOD/gP that are much higher than the highest value recommended to achieve satisfactory biological phosphorus removal (Henze and Comeau, 2007).

Overall, the present study shows the importance of executing a more thorough evaluation and assessment of the wastewater composition. Not only through the analysis of the typical wastewater components but mostly based on the determination of the different fractions of importance for the organic, nitrogen and phosphorus compounds. Thus, the determination of the biodegradable and unbiodegradable fractions, further divided into their particulate and soluble fractions, can be used as a reliable tool to assess the wastewater treatability and estimate the potential performance of the wastewater treatment technologies applied. However, it is not a common practice to execute or carry out more detailed fractionation of the wastewater components of importance, but it needs to be taken into account in view of the potential benefits.

The results obtained in the present study show that the effluent of the designed wastewater treatment plant will not achieve the discharge standards initially but will likely do so after the sewer connections increase over 30%. Moreover, it appears that some discharge standards will not be met with the current design (e.g. for total nitrogen) until the island reaches 100% sewerage coverage or the discharges of septage stop. First, this underlines the importance of adequately determine the fractionations of the different streams treated at the wastewater treatment plant to estimate the potential efficiency of the system and the compliance of the discharge standards. And, secondly, it underlines the need to adequately manage the waste streams treated at the plant. In this regard, the discharge of septage limits the plant's efficiency due to the high concentrations of nitrogenous compounds some of them of unbiodegradable nature. Thus, it appears that only when septage is no longer discharged at the plant it will be able to fully comply with the corresponding effluent standards.

Despite that the simulations indicate a high EBPR efficiency, they are a few aspects that deserve special attention. For example, it is known that the biological phosphorus removal, performed by phosphorus-accumulating organisms (PAO), can be inhibited at temperatures higher than 20°C due to the proliferation of glycogen-accumulating organisms (GAO). Mostly, it occurs because GAO compete with PAO for soluble biodegradable organics, such as volatile fatty acids, at higher temperatures (Lopez-Vazquez et al., 2007). Thus, in the Illidge Road WWTP the biological P-removal potential of the treatment plant may be compromised because the minimum wastewater temperature is 23°C. Moreover, the growth or proliferation of GAO is not incorporated in any of the models used (Lopez-Vazquez et al., 2009). As such, the actual bio-P removal efficiency could be significantly reduced under actual full-scale conditions which could, in the worst-case scenario, hinder the compliance of the effluent phosphorus discharge standard. Bearing this in mind and following a conservative approach, it can be suggested to apply chemical phosphorus removal, as a back-up strategy (Aguilar et al., 2007), to guarantee the compliance of the phosphorus removal standard.

Overall, the results from the steady-state model were similar to those obtained with commercial software and the design criteria applied by a consulting company. This confirms that the stoichiometry-based steady-state model can be a cost-effective tool for design and assessment of wastewater treatment systems, but also potentially applicable in urban water management to contribute to address potential urbanization changes and evaluate their potential effects on wastewater quality and treatability.

11.4 Conclusions

The bioprocess stoichiometry based model predictions were consistent with those obtained with BioWin indicating that it is a useful low-cost and relatively simple tool for the design and optimization of wastewater treatment plants. The determination of reliable fractionations of the components of concern (COD, nitrogen and phosphorus) are essential. This ensures that the subsequent modelling assessment provides representative and reliable outputs as well. Furthermore, its accuracy is rather important to evaluate the process efficiency of existent wastewater treatment plants and estimate those of new designs. In the present study, due to high load of nitrogen from the septage, the newly proposed plant will initially have difficulties to comply with the discharge limits for total nitrogen (of 10 mgTN/L), as a consequence of the high nitrogen concentrations and low availability of biodegradable organics. This underlines the importance to critically assess and manage the waste streams to comply with the required standards. As such, the discharge of septage must stop to meet the effluent standards or the sewer connections must be carried out earlier than planned.

References

Aguilar MI, Sáez J, Lloréns M, Soler A, Ortuño JF (2002) Nutrient removal and sludge production in the coagulation-flocculation process. Water Research 36: 2910-2919.

Henze, M., Comeau, Y. (2008) Wastewater characterization. In: Biological wastewater treatment: principles, modelling and design. Henze, M, van Loosdrecht, M.C.M., Ekama, G.A., Brdjanovic, D. eds. ISBN: 9781843391883. IWA Publishing. London, UK.

Batstone DJ, Keller J, Angelidaki I, Kalyuzhnyi SV, Pavlostathis SG, Rozzi A, Sanders WTM, Siegrist H, Vavilin VA (2002) The IWA Anaerobic Digestion Model No 1 (ADM1) Water Sci Tech 45(10):65-73.

Barnard J (1976) A review of biological phosphorus removal in the activated sludge process. Water SA 2(3):136-145.

Buiteman J, Kruis H (2009) Quick Scan St. Maarten UNESCO-IHE Institute for Water Education, Delft.

Ekama G.A. (2008) Organic matter removal. In: Biological Wastewater Treatment: Principles, Modelling and Design. Henze, M, van Loosdrecht, M, C, M., Ekama, G, A., Brdjanovic, D. eds. ISBN: 9781843391883. IWA Publishing. London, UK.

Ekama GA, Sötemann SW, Wentzel MC (2007) Biodegradability of activated sludge organics under anaerobic conditions. Water Research 41: 244-252.

Gerardi MH (2006) Watewater Bacteria John Wiley & Sons Inc., Hoboken

Hamersley MR, Howes BL, White DS, Johnke S, Young D, Peterson SB, Teal JM (2001) Nitrogen balance and cycling in an ecologically engineered septage treatment system. Ecological Engineering 18: 61-75.

Henze M (1992) Characterization of Wastewater for Modelling of Activated Sludge Processes. Water Science and Technology 25: 1-15.

Henze, M., Comeau, Y. (2008) Wastewater characterization. In: Biological wastewater treatment: principles, modelling and design. Henze, M, van Loosdrecht, M.C.M., Ekama, G.A., Brdjanovic, D. eds. ISBN: 9781843391883. IWA Publishing. London, UK.

Henze M, van Loosdrecht MCM, Ekama GA, Brdjanovic D (2008) Biological Wastewater Treatment: Principles, Modelling and Design, First edn IWA Publishing, London

Henze M., Gujer W., Mino T, van Loosdrecht (2000) Activated sludge models ASM1, ASM2, ASM2d and ASM3. IWA Task group on mathematical modelling for design and operation of biological wastewater treatment. IWA Publishing. London, UK. ISBN: 1 900 222 24 8.

Hosetti BB, Frost S (1995) A review of the sustainable value of effluents and sludges from wastewater stabilization ponds. Ecological Engineering 5: 421-431.

Lopez-Vazquez, C.M., Song, Y.I., Hooijmans, C.M., Brdjanovic, D., Moussa, M.S., Gijzen, H.J., van Loosdrecht, M.C.M. (2007) Short-term temperature effects on the anaerobic metabolism of glycogen accumulating organisms. Biotech Bioeng 97(3), 483-495.

Lopez-Vazquez, C.M. (2008) St. Maarten Island - Illidge Road Wastewater treatment plant: Wastewater characterization Final Report The Netherlands. UNESCO-IHE, Institute for Water Education. Delft, The Netherlands.

Lopez-Vazquez CM, Hooijmans CM, Brdjanovic D, Gijzen HJ, van Loosdrecht MCM (2009) Temperature effects on glycogen accumulating organisms. Water Research 43: 2852-2864.

Lopez-Vazquez C.M., Dangol B., Hooijmans C.M., Brdjanovic D. (2014) Co-treatment of faecal sludge in municipal wastewater treatment plants. In: Faecal sludge management. Eds: Strande L., Ronteltap M, Brdjanovic D. IWA Publishing. London, UK.

Montangero A, Belevi H (2007) Assessing nutrient flows in septic tanks by eliciting expert judgement: A promising method in the context of developing countries. Water Research 41: 1052-1064

Tchobanoglous G, Burton FL, Stensel DH (2003) Wastewater Engineering Treatment and Reuse Fourth Edition edn McGraw-Hill, New York

Strande L., Ronteltap M, Brdjanovic D. (2014) Faecal Sludge Management: Systems Approach for Implementation and Operation. IWA Publishing, London, UK. ISBN: 9781780404721.

Sötemann, S.W., Ristow, N.E., Wentzel, M.C., Ekama, G.A. (2005) A steady state model for anaerobic digestion of sewage sludges. Water S.A., 31(4):511-527.

UNESCO-IHE (2005) Evaluation of the Proposal for an Integrated Sewage and Drinking Water System on St. Maarten UNESCO-IHE, Delft.

van Geleuken B, Lake O, Montanus P, Holm G, Tearr W, Rooi N, Ras T (2007) Milieunormen Nederlandse Antillen Ministerie van Volksgezondheid en Sociale Ontwikkeling, Willemstad.

Van Loosdrecht MCM, Henze M (1999) Maintenance, endogeneous respiration, lysis, decay and predation. Water Science and Technology 39: 107-117.

WWW.THEDAILYHERALD.COM WEEKENDER May 10, 2014 9

Water, water everywhere

St. Maarten's water gets attention

By Lisa Davis-Burnett

Water, H2O, is considered by many to be the most valuable of all the natural resources on the planet. We humans, as well as every other form of life, need it to survive. It is easy to take water for granted, until you don't have it anymore, then things get deadly serious; for without water, you and I can't last a week. We require water to be clean and reliable, inexpensive and uncontaminated with impurities. In short, we need it managed. That is where water engineers come in; and no one manages water better than the experts in Holland. Historically, the Dutch have shown that sound engineering can reclaim land from the sea, a feat few other countries can match.

That is one of the reasons why, when the United Nations was first interested in establishing an institute where students from all over the globe could train to become experts in water management, they chose Holland's famous city of Delft, specifically the Delft University of Technology. And so that is where we now have UNESCO-IHE or The Institute for Water Education of UNESCO (United Nations Educational, Scientific and Cultural Organization). It is the largest international graduate water education facility in the world offering fully accredited MSc degrees, and PhD degrees. Graduate students from around the world gain specialized technical training in the fields of water purification, water treatment, environmental engineering, sanitation and related topics. Various scholarships are available to those interested in careers in water management or environmental engineering. A visit to the UNESCO-IHE website can provide more information and is recommended for anyone who would consider a career in this important field.

EVERY YEAR, the water situation in St. Maarten is of extreme interest and receives detailed analysis at UNESCO-IHE. As a part of their degree program, they spend four weeks analysing every water-related aspect of the island. This year, some of the lecturers of this program have travelled to see the place they and their students have studied so intently.

Visiting here this week are Giuliana Ferrero, PhD in Water Supply Engineering; Hector Garcia, PhD in Wastewater Treatment Technology; and Loreen Villacorte PhD and MSc in Water Supply Engineering. They are coordinators of the program which examines the case studies of actual hydro engineering situations in St. Maarten. It is an annual event at UNESCO-IHE since 2007.

Ferrero, Villacorte and Garcia serve as leaders to teams of students that tackle real world water problems here on St. Maarten. In an interview with WEEKender, Ferrero mentioned that they hope to be able to extend the work the students do to be more than just academic exercises. To this end, the three have come here to meet governmental leaders; observe water treatment facilities in action and see the place they and their students have studied so intently. The aim is optimization of the wastewater treatment plant, the modelling of the water distribution network or the development of a water quality monitoring plan for the island.

Visiting water experts from UNESCO, left to right: Loreen Villacorte, Giuliana Ferrero and Hector Garcia.

Johan van Teeffelen worked for VROMI for four years from 2004 to 2007 and now he is working as a consultant for UNESCO-IHE and he was the originator of this collaboration. He organized the visiting experts to consult with representatives from GEBE, VROMI, the Ministry of Education, the Health Department and other key players in water management.

The team did a true whirlwind tour, with 18 meetings in four days before flying back to Europe. Only here for four days, they have not had time for even one swim at the beach, but have enjoyed the beauty of the island as they made their way between meetings and sites of interest. "We would like to have had more time, but it is very nice to see the island, it is truly paradise," said Villacorte.

Figure 11.13 The case of St. Maarten gets attention of UNESCO-IHE (Source: Daily Harrard).

THE DAILY HERALD, Tuesday, August 27, 2013 — Islands — 5

Students make strategic plans for water, sanitation and infrastructure

PHILIPSBURG–As part of their International Masters Programme in Municipal Water and Infrastructure, almost forty students at UNESCO-IHE completed module 12, the so-called group work, last Friday, August 23.

UNESCO-IHE – the Institute for Water Education in Delft, the Netherlands – is known in St Maarten for the evaluation study about the sewerage and storm water modelling and improvements in various districts.

With that evaluation study, the UNESCO-IHE professors and lecturers became aware of a lot of water-related and infrastructural problems on the island. As a request from government, UNESCO-IHE was willing and able to integrate a lot of these problems within their curriculum. Because a lot of data was already acquired by UNESCO-IHE, and with help from several Ministries such as Health, VROMI, Economic Affairs, the Fire department, GEBE and the Harbour, these problems could be used in the study of their students.

In cooperation with VROMI, UNESCO-IHE described 10 projects in the areas of storm water, drinking water, sewerage and solid waste as problems to be addressed.

In the group work, students had to make an integrated study about these infrastructural projects. The students were divided into four groups – each of about 10 students – and in a very short period of three weeks, they had to deliver a master plan themed: "Prepare a strategic plan for the water and sanitation sector on St Maarten."

They had to look at the storm water system, the drinking water production and supply system, the sewerage system, the wastewater treatment and the solid waste. Special attention in the subject of storm water was given to disaster management, it was stated in a press release about the initiative. The problems had to be analysed and integrated solutions generated.

This group work module emphasises the process of an integrated approach and teamwork. Each group has to make a full report and, like in real life, they had to make a presentation as consultants trying to get the assignment to solve the problems. They had to present their reports to a panel consisting of Chairman Jan Herman Koster of UNESCO-IHE assisted by Managing Director of Berson UV Paul Buijs, Michael Bentvelsen, a wastewater expert from the Association of Regional Water Authorities, and Johan Department of New Projects and Planning from VROMI.

After each presentation, questions were asked, remarks were made, including very critical ones, by the panel members and the group could try to convince the panel to get the assignment. When all four groups had given their presentations and submitted their reports, the panel had to judge the efforts made by the various groups. It led to discussions and deliberations, but finally the panel came to the conclusion that one group had delivered the best presentation and convinced the panel to give the "assignment" to their company, which they called "Innotech."

Paul Buijs addressed all participants of the group work and gave them some tips and comments about the process, the reports and the presentations. He emphasised making a very good Executive Summary in their future reports and wished them all success in their endeavours to graduate with their master.

Johan van Teeffelen thanked all students for their efforts and reports on behalf of the government of St Maarten and handed the team leader of the winner of this "competition," Group D, and their mentor, Sergio Salinas, a plaque on which the names of the group members will be engraved.

The winning group, "Innotech." From left: Seneshaw Kebede, Grover Mamani, Alida Alves, Luisa Calderon, Chaminda Jayawardana, Clement Ndungutse, Eng Wittike, Mohamed Nazeer, Kazoya Makongoro, Hohee Bang and Abdullah Eidroos.

Winning group "Innotech" receives the winning plaque.

Figure 11.14 St. Maarten is used in a student-centered, problem-based learning at UNESCO-IHE urban Water and Sanitation Masters Programme since 2008 (Source: Daily Harrard).

WWW.THEDAILYHERALD.COM WEEKENDER September 6, 2014 5

Water, sanitation and solid waste

~ UNESCO students make a master plan for St. Maarten ~

As part of their International Master's Program in Municipal Water and Infrastructure, 36 students at UNESCO-IHE completed their group work on Friday, August 22. UNESCO-IHE, the Institute for Water Education in Delft, the Netherlands, is known on St. Maarten for the evaluative study about the sewerage and the storm water modelling and improvements in various districts.

With that study, the UNESCO-IHE professors and lecturers became aware of a lot of water-related and infrastructural issues on the island. At a request from the Executive Council, UNESCO-IHE was willing and able to integrate a lot of these issues within their curriculum. A lot of data was already acquired by UNESCO-IHE and with help from several Ministries, like Health, VROMI, Economic Affairs as well as from the Fire Department, GEBE and the Harbour as well as private companies like Seven Seas N.V., these problems could be used in the study of their students. To update the data and information, a mission, consisting of Dr. Ferrero, Dr. Garcia and Dr. Villacorte, visited the island in May. Therefore, they had meetings with employees of VROMI, GEBE, the Health Inspectorate, SLS laboratory, SevenSeas, SHTA, Indigo Bay, St. Maarten Home and hotels.

So in cooperation with VROMI, UNESCO-IHE described nine projects in the areas of drinking water, sewerage and solid waste as problems to be addressed. The waste water from hotels and an optimization study for the sewerage system for Indigo Bay were also part of the Group work this year.

The students made an integrated study about these infrastructural projects and problems. The students were divided in four groups each with nine students and, in a very short period of four weeks, they had to deliver a Master plan – *Prepare a strategic plan for the water and sanitation sector and solid waste on St. Maarten* – which included the following:

- The vision the recommendations are based on
- An executive summary with the motivations for the chosen solution
- Brief analysis of the current situation and the technical solutions
- Short-, medium- and long-term activities and investments

The causes of the problems had to be analysed, and integrated solutions devised. This group work module emphasized the process of an integrated approach and teamwork. Each group had to make a full report and, like in real life, they had to each make a presentation as consultants trying to get the assignment to solve the problems.

Their reports were presented to a panel consisting of chairman, Dr. Paul Buijs; Managing Director of Berson UV; and Dr. Carlos Lopez-Vazquez; Dr. Mariska Ronteltap and Johan van Teeffelen, representing the Department of New Projects and Planning from VROMI.

After each presentation, questions were asked, remarks were made – some very critical ones – by the panel members as the group tried to convince the panel to get the assignment. When all four groups had given their presentations and submitted their reports, the panel had to judge the efforts made by the various groups. It led to discussions and deliberations, but finally the panel came to the conclusion that one group had delivered the best report and presentation. The panel decided to give the "assignment" to Group C, the company they called "GAIA."

Afterwards, Paul Buijs addressed all participants of the group work and gave them some tips and comments about the process, the reports and the presentations. He emphasized making a very good executive summary in their future reports and wished them all success in their endeavours to graduate for their master and handed the team leader of the winner of this "competition," Group C, and their mentor, Dr. Giuliana Ferrero, a plaque on which the names of the group members will be engraved.

Left to right: Dr. Ronteltap, Dr. Lopez-Vazquez, Ms. Prameshwari, Mr. Buiteman, Ms. Kemboi, Mr. Buijs, Ms. Torres Zambrano, Mr. Barreto Carvajal, Mr. Othello Bendedale, Mr. Herelimana, Mr. van Teeffelen, Ms. Malan Kandage, Dr. Ferrero.

Figure 11.15 Winning group of the St. Maarten groupwork class 2014 at UNESCO-IHE (Source: Daily Harrard).

Figure 11.15 Urbanzing St. Maarten - situation in 60's and in 2014 (Source: St. Maarten Government).

Figure 11.16 Aerial views of the old and new WWTP Illidge Rd. in operation (Photo: Buncamper C.).

Chapter 12: WWTP Dulces-Nombers, Mexico

Model-based evaluation of a full-scale plant hydraulics

Fall C., Flores-Alamo N., Esparza-Soto M. and Hooijmans C.M.

This chapter is based on the paper "Tracer test and hydraulic modeling of a large WWTP" Fall C., Flores-Alamo N., Esparza-Soto M., and Hooijmans C.M. (2012). Water Practice & Technology 7(1).

12.1 Introduction

The setup of the hydraulic model structure of wastewater treatment plants (WWTP) is an important step of the calibration of activated sludge models. In traditional wastewater treatment practice, reactors have generally been designed on the basis of two ideal mixing schemes (Metcalf and Eddy, 2003): the plug flow reactor (PFR) and the continuously stirred tank reactor (CSTR). However, the treatment systems are more often subjected to non-ideal flows, so the actual hydrodynamics of reactors must be known, where there is a need to better predict the performance of the WWTP (De Clerq *et al.,* 1999; Makinia, 2010).

For a series of n equal-size ideal CSTRs (without recirculation, nor dead volume or channeling), the theoretical residence time distribution (RTD) function, $E_n(t)$, may be derived as:

$$E_n(t) = \frac{t^{n-1}}{(n-1)!\,\tau_i^n} e^{-t/\tau_i} \tag{12.1}$$

where *t* is the time variable and τ_i the nominal residence time of each of the individual n tanks.

The ideal n in-series tanks that is equivalent to an actual tubular reactor may be derived (determining n and τ_i) by fitting the field experimental $E_n(t)$ data obtained through a tracer test, to Eq. 12.1 (Olivet *et al.,* 2005). However, the RTD approach is no longer valid (Fall *et al.,* 2007) in the case of high flow variability, unequal CSTRs volumes, and when the quantity of tracer that is returned through the RAS (return activated sludge) is high.

Using a simulator to model tracer tests data is an alternative way for setting up in-series tanks models. The approach can account for the effect of internal flows, returned sludge, dynamic nature of flows and unequal tank sizes (Fall *et al.,* 2007).

In this work, tracer tests and flow measurements were performed at the largest wastewater treatment plant of Mexico (WWTP Dulces Nombres, with capacity of 5 m^3/s), which data were modeled with AQUASIM (Reichert, 1998) to set up the hydraulics model structure of the aerations tanks.

12.2 Materials and methods

12.2.1 The WWTP Dulces-Nombres

The process diagram of the plant is shown in the Figure 12.1. The treatment processes include a grit removal system and 4 primary clarifiers (not shown), as well as 6 secondary settlers (CS_1 to CS_6) and 5 aeration tanks in parallel (R_1 to R_5, with a volume of approx. 20,000 m^3). Each of the longitudinal aeration tanks was built as 4 communicating cells (A to D) separated by 3 baffled walls.

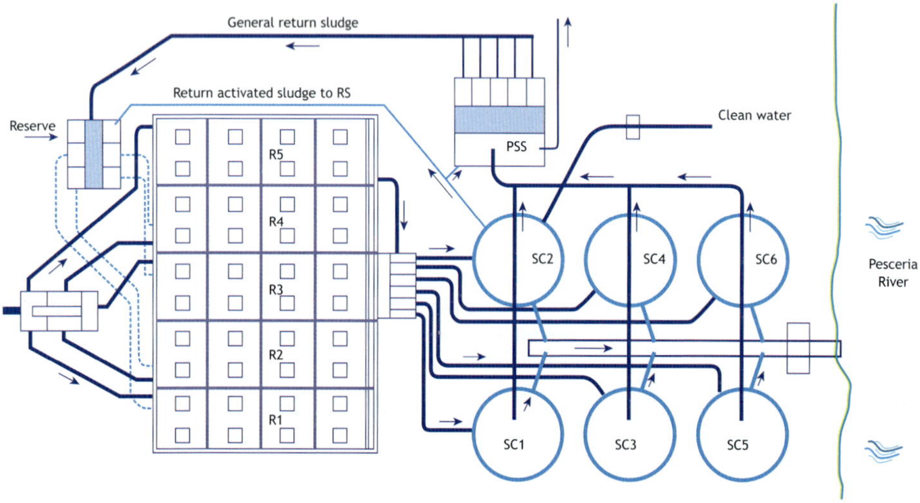

Figure 12.1 Layout of the WWTP Dulces-Nombres (activated sludge part only).

The reactor R5 and one of the clarifiers (CS_2) were isolated from the remaining aeration tanks, to have an almost separated treatment line that produces a better quality of effluent, which is reused in a local industry. The other four reactors (R_1 to R_4) were operating as a same whole lane, receiving the influent from the IDB; their outlet mixed liquors were collected and mixed in a same canal before their redistribution toward the settlers. For the tracer tests, R_1 was selected to represent the R_1-R_4 Lane, while R_5 was evaluated separately.

12.2.2. Preliminary simulation

First, for planning the experiments, a preliminary simulation was carried out with the program AQUASIM (Reichert, 1998). This was to determine the adequate sampling frequency and to predict the volume of dye that was needed. A layout similar to the physical configuration of the plant (4 in-series continuous stirred-tank reactors, CSTRs) was implemented in AQUASIM, with an adequate definition of the different bifurcations and return flows (Figure 12.2). Each reactor (R_1 vs. R_5) was modeled as a series of 4 CSTRs, while the clarifiers and intermediary pits were simply represented as one CSTR each. In the model of R_1 (Figure 12.2b), contrary to R_5 (Figure 12.2a), the others parallel reactors that were not receiving any dye pulse (i.e. R_2, R_3 and R_4) were still kept in the model, to simulate the dilution of the tracer outbreak from R_1. The tracer injection point was at the inlet of the targeted reactors (R_5 vs. R_1). The preliminary simulation was performed with the nominal flows and dimensions of the tanks (Table 12.1), and with the nominal characteristics of the dye (commercial rhodamine WT, 20% w/w active ingredients and a density of 1 g/mL).

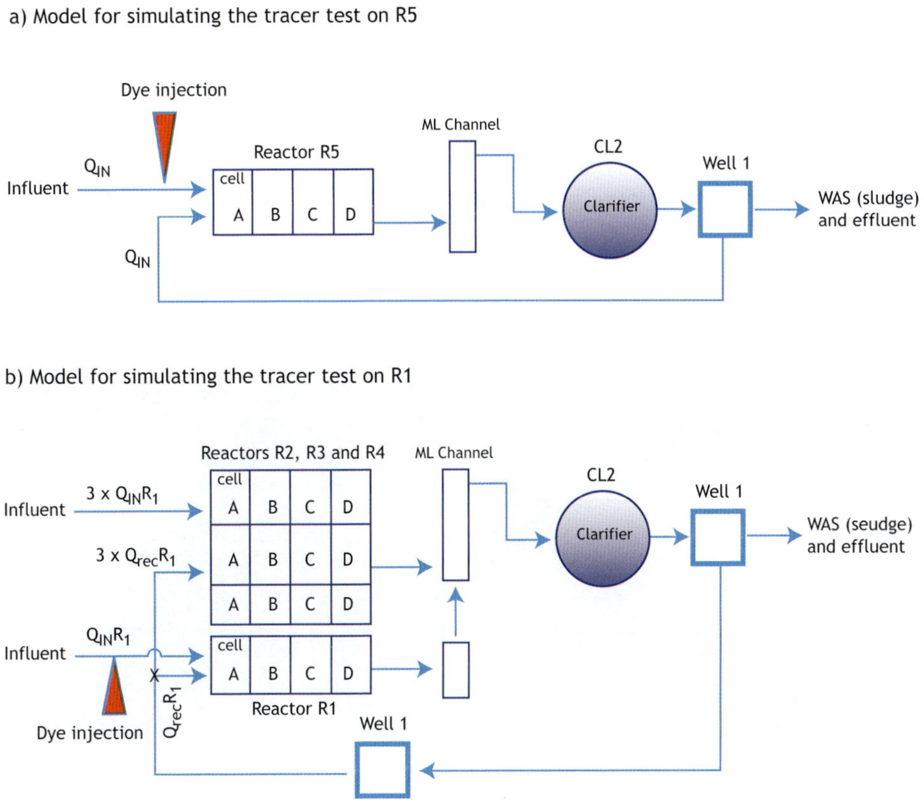

Figure 12.2 Preliminary simulation models implemented in AQUASIM: a) model for simulating the tracer test on R5, and b) on R1.

The injection of the colorant was simulated as a constant mass flow pulse lasting two minutes. The response variable estimated from the pre-simulation was the concentration-time profile of the dye, in the different compartments of the plant, i.e. in each reactor cell, in the effluent of the plant and at the discharge point in the river. The parameter that was varied in the model (trying different values) was the volume of rhodamine injected, seeking for a maximum peak concentration of 100 μg/L at the exit of the reactor (ideal analytical detection range), and no more than 10 μg/L at the discharge point in the river (to limit the visual impact of the red color).

Table 12.1 Nominal operational parameters of the plant

Nomenclature	Description	Value
$Q_{\_in\_tot}$	Total influent flow (1/5 for each reactor)	18,000 m^3/h
$Q_{\_rec\_tot}$	Total RAS flow (1/5 for each reactor)	9,000 m^3/h
$Q_{\_river}$	Flow of the river	17,352 m^3/h
$Q_{\_wasR1\text{-}R4}$	Waste Activated Sludge (WAS) flow of R1-R4.	290 m^3/h
$Q_{\_wasR5}$	WAS of R5	58 m^3/h
$V_{\_R}$	Total working volume of each reactor	19,120 m^3
$V_{\_comp.}$	Volume for each of the 4 compartments per reactor	4,780 m^3
$V_{\_CS.}$	Volume of each clarifier	15,635 m^3
$V_{\_canal\_R1\text{-}4}$	Volume of the mixed liquor canal of R1-R4	200 m^3
$V_{\_canal\_R5}$	Volume of the mixed liquor canal of R5	50 m^3

12.2.3 Flow measurements

The most important flows required for performing the simulations were those of the influent (Q_{inR1} and Q_{inR5}), and of the returned flows (Q_{recR1}, Q_{recR5}). A set of flow gauges was instrumented to know their values. The 4200-flowmeter series from ISCO (ultrasonic probes) were used to estimate the flows of the mixed liquors from the reactors (Figure 12.3b); non-intrusive Doppler-effect based flow-meters were installed above the full-tubes (RAS and influent, Figure 12.3a). A provider was sub-contracted for calibrating the gauges and reporting the measurements. All the flow meters were read at least at each hour (log books), and in the majority of the cases, the data were automatically logged at each 2 min, allowing to calculate the mean daily flows, as well as the dynamic diurnal variations.

Figure 12.3 Flow measurements in closed full tubes (left) and in an open canal (right).

12.2.4 Tracer test and hydraulic modeling

Based on the preliminary simulation results, the tracer tests were performed in-situ, for the reactor R1 and then for the reactor R5. The dye was injected as a pulse in the RAS flow, just before it reaches the reactor. The outlet liquors were sampled at different times and the dye concentrations were measured with a fluoro-meter. The hydraulics data (measured flows and concentrations) were analyzed with a dynamic modeling approach, by using AQUASIM and by representing the different tanks in the plant as a series of CSTRs. The program written in AQUASIM was run to calculate and simulate the outlet dye concentrations from the reactors. Several variants of the hydraulics model were successively implemented in the software and tested (trial-and-error approach). In each variant, the initial layouts (Figure 12.2) were modified by changing the volumes and the number of virtual CSTRs representing the aeration tanks, and by modifying the definition of the inter-connections (Links in AQUASIM) between them. Finally the best hydraulics model was selected as the simplest configuration (lower number of compartments, links and parameters) that was able to adequately fit the tracer data.

12.3 Results and discussion

12.3.1 Preliminary simulation

The preliminary simulation suggested the use of 8L of commercial Rhodamine for performing the tracer test in each reactor. With these quantities, the anticipated maximum dye concentrations at the outlet point of R5 (and R1) were less than the 100-µg/L criterion, which was chosen as the upper limit. The anticipated concentrations at the river discharge point (< 2 µg/L, not shown), were lower than the visibility limit of 10 µg/L. The preliminary simulation results also suggested using the following sampling program: 10 minutes interval for the first six hours, 30 minutes interval for other 2.5 hours and then every hour for the remaining time, up to a total of 24h. The data points (circles) that were measured later on, during the tracer test campaign, are anticipatory represented (Figure 12.4b), showing that the sampling frequency and tracer volumes suggested by the pre-simulation were very adequate. Also, the actual data were very close to the response anticipated with the pre-simulation, suggesting that the 4-in-series CSTRs model used was near the reality. In fact, the aeration tanks at the D-N WWTP were actually very well mixed (too many diffusers units) and the pathway of the water flow between the compartments was very smooth.

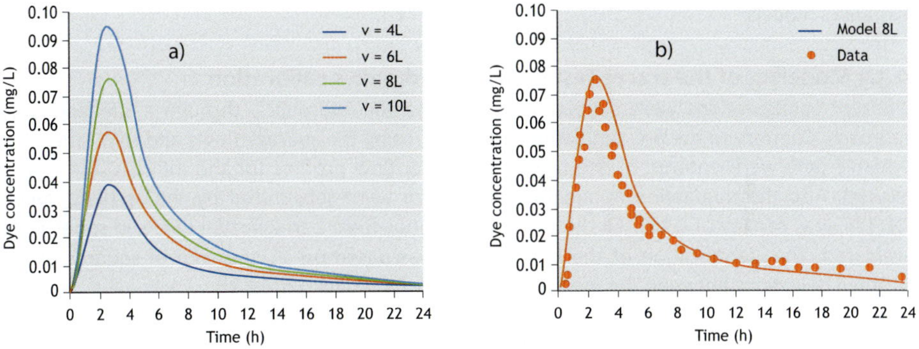

Figure 12.4 Preliminary simulation profiles in R5: (a) for different rhodamine volumes (V = 4 to 10L) and, (b) simulated profile for 8L (solid line) compared with the test-after data (circles).

12.3.2 Flows

Figure 12.5 shows the shape of the diurnal flow profiles at the influent (Q_3) and of the returned flows (Q_{13} for R_5 and Q_{22} for R_1 to R_4). The inflows were low in the mornings and high in the afternoons, with a sudden increase, daily between 11:00 and 13:00 h. At the contrary, the flows of the RAS were stable and continuous in time, as shown. A decision was taken, based on the historical behavior, to start the tracer tests at 13:00 h A.M., so that most of the initial part of the experiment took place during the moments of high flows.

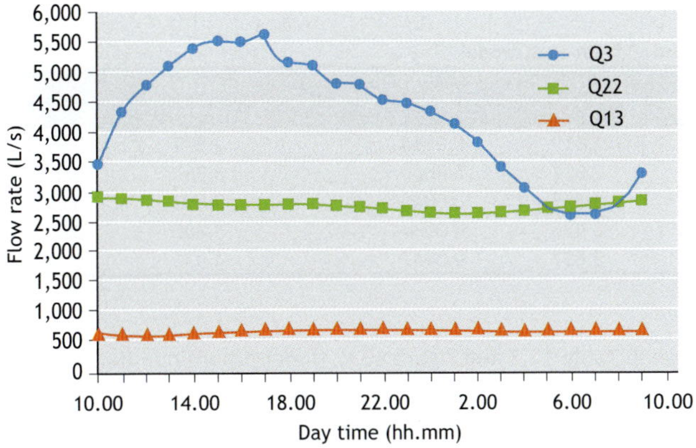

Figure 12.5 Diurnal profiles of the influent (Q_3) and of the RAS flows (Q_{13} and Q_{22}).

The mean flows obtained from the measurements were $Q_{13} = Q_{recR5}$ = 635 L/s (2,286 m^3/h), against Q_{22} = 2,738 L/s (9,857 m^3/h), for the RAS of the R_1 to R_4. The RAS of R_1 (Q_{recR1}) was proportionally estimated to be 2,464 m^3/h. The readings from the influent meters gave an initial estimate of 15,178 m^3/h for the entire plant, which corresponds in average to 3,035 m^3/h toward each reactor. Unfortunately, with the set of gauges that were installed in the mixed liquor canal and in the influent tube, it was not possible to estimate the actual inflow to each reactor. The mass balances showed that the data from theses points were not reliable (calibration issues). Fortunately, the data from the tracer tests (concentration profiles) were another source of information, which could be used to estimate the inflow rate to each reactor, considering that the RAS flows were accurately known (Q_{recR5} = 2,286 and Q_{recR1}= 2,462 m^3/h). The next section deals with the estimation of the inflows, and the identification of the best hydraulics models.

12.3.3 Modeling of the tracer test data and hydraulics calibration

Different configurations were simulated to calibrate the hydraulics, through a trial-and-error approach. Contrary to the pre-simulation, instead of using the average flows and split ratios, the dynamic data were entered in the program (diurnal flow profiles for the influent and sludge recycle). Also the rhodamine nominal characteristics were substituted by the certified values (21.37% ww, density of 1.15). All the possible scenarios from 2 to 7 virtual CSTRs in series were evaluated. The identification of the best models was based on χ^2, the sum of squares of the weighed deviations between the measurements and the model (Eq. 12.2):

$$\chi^2 = \sum_{i=1}^{n} \left(\frac{y_{meas} - y_i(p)}{\sigma_{meas}} \right)^2 \qquad (12.2)$$

where:

$y_{meas,i}$ is the ith measurement of dye concentration;

σ_{meas} is the standard deviation;

y_i is the ith calculated value from the model.

By simulating the models with different scenarios (from 2 to 7 CSTRs), the RAS flows were maintained fixed (2,462 for R_1 and 2,286 m^3/h for R_5, as measured), while the influent flows (Q_{inR1} and Q_{inR5}) were re-estimated by minimizing the errors (χ^2). Table 12.2 represents a summary of the results from this procedure.

Table 12.2: Estimated inflow rates and χ^2 from each model

N° of CSTRs	Reactor R_1		Reactor R_5	
in the models	Q_{in} (m^3/d)	χ^2	Q_{in} (m^3/d)	χ^2
2	2,419	0.0134	2,827	0.0125
3	2,917	0.0066	3,252	0.0059
4	3,233	0.0040	3,561	0.0039
5	3,436	0.0034	3,800	0.0036
7	3,687	0.0047	4,076	0.0062

In both cases (R_1 and R_5), as shown in Table 12.2 and Figure 12.6, the optimum number of CSTRs (lower χ^2) to fit the data was 4 to 5 tanks. The best virtual model was then coinciding with the actual physical configuration of the aeration tanks (reactors with 4 compartments).

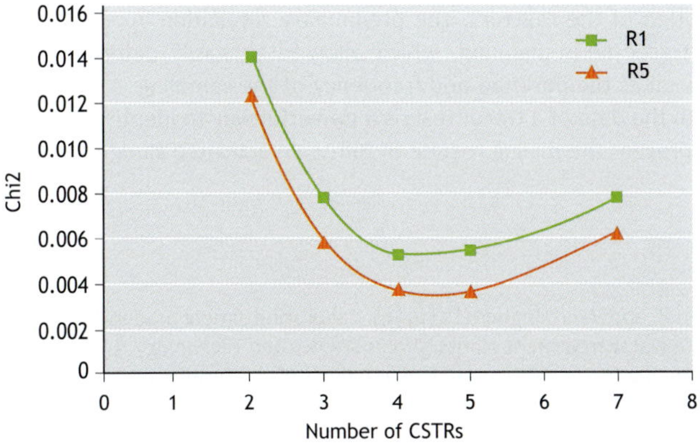

Figure 12.6 Error associated (chi2) with the different scenarios of hydraulics model.

The concentrations profiles fitted with the 4-in series CSTRs model are represented in the Figure 12.7, for R_1 and R_5 respectively. The data were adequately matched by the chosen model. The coincidence between the virtual model and the actual configuration of the tanks (4 CSTRs) suggest that there is not any significant short-circuiting or dead volume in the reactors. In fact, this is consistent with the observations made in the field: very well mixed compartments. This coincidence is far from being the usual case. In a recent compilation, Makinia (2010) reported that between 2 to 12 virtual CSTRs were generally used by the modelers to represent the aeration tanks in different plants. In many cases, poor mixing may occur, which need implementing a virtual hydraulics model that is different from the physical configuration of the reactors. The hydraulics sub-model developed from this work was used later in a WWTP simulator, for calibrating the ASM1 at the plant (Fall *et al.*, 2011).

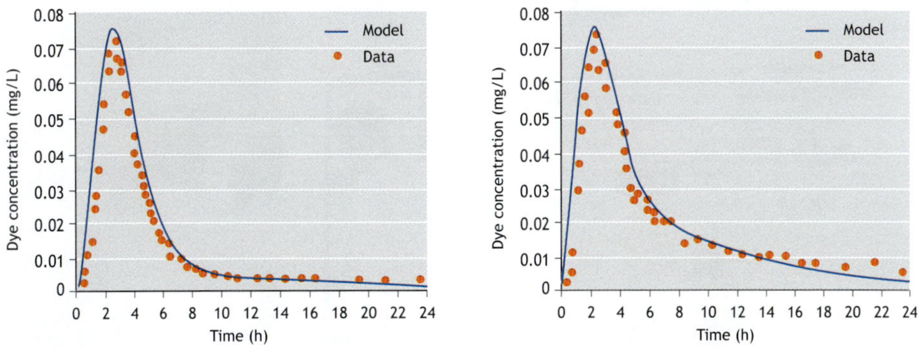

Figure 12.7 Tracer test data of R_1 (left) and R_5 (right) fitted with the 4 CSTRs model

12.4 Conclusions

The hydraulics of the aeration tanks at the examined plant, built each with four well-aerated compartments, was adequately represented by a model of 4 in-series CSTRs. The results showed that the reactors have no dead volumes or short circuits, and, in this case, the hydraulic sub-model that should be implemented in the whole simulator (together with the ASM1) was exactly similar to the physical configuration of the reactors. The preliminary simulation approach was shown to be a good tool for tracer tests planning, which allowed to easily determine the appropriate quantity of dye, as well as the duration and frequency of the sampling. A simulator such as AQUASIM, combined with the data of a tracer test, is a powerful way to identify in-series CSTR models that adequately represent the mixing regime of full-scale activated sludge aeration tanks.

References

De Clercq B., Coen F., Vanderhaegen B. and Vanrolleghem P. (1999). Calibrating simple models for mixing and flow propagation in waste water treatment plants. *Water Science and Technology*, 39(4), 61-69.

Fall C. and Loaiza N. J. (2007). Design of a tracer test experience and dynamic calibration of the hydraulic model for a full-scale wastewater treatment plant by use of AQUASIM. *Water Environmental Research*, 79(8), 893-900.

Fall C., Espinosa-Rodríguez M. A., Flores-Alamo N., van Loosdrecht M. C. M. and Hooijmans C.M. (2011). Stepwise Calibration of the Activated Sludge Model No. 1 at a Partially-Denitrifying Large WWTP. *Water Environment Research*, 83(11), 2036-2048.

Makinia J. (2010). Mathematical modelling and computer simulation of activated sludge systems. IWA Publishing, International Water Association, London, U.K.

Metcalf & Eddy (2003). Wastewater Engineering, Treatment and Reuse. McGraw-Hill, New York, USA.

Reichert P. (1998). User Manual AQUASIM 2.0, Computer Program for the Identication and Simulation of Aquatic Systems. Swiss Federal Institute for Environmental Science and Technology (EAWAG), Dubendorf, Switzerland.

Figure 12.8 Dulces-Nombres, Nuevo Leon, Mexico (Photo: Hooijmans C.M.).

Chapter 13: WWTP Varaždin, Croatia

Use of models for cost-effective planning of plant retrofit and upgrade scenarios

Meijer S.C.F., Brdjanovic D., Šikić T, Matošić M., Deduš B., and Širac S.

This chapter is based on the project *Development and Application of Models and Guidelines to Facilitate Decision making in the Extension of Wastewater Treatment Plants (WWTPs) and to Increase Operational Efficiency of Existing WWTPs* initiated under the Technical Assistance component of the Inland Water Project - Loan 7453/HR in Croatia.

13.1 Introduction

In Croatia, two national strategic projects are being implemented in the field of water pollution control, namely the Coastal Cities Pollution Control Project (The Adriatic Project) in the Adriatic Basin, and the Inland Waters Project in the Black Sea Basin. This study presents part of the project entitled *Development and Application of Models and Guidelines to Facilitate Decision making in the Extension of Wastewater Treatment Plants (WWTPs) and to Increase Operational Efficiency of Existing WWTPs* has been initiated under the Technical Assistance component of the Inland Water Project - Loan 7453/HR. Under this project the performance of all 29 WWTPs that employ secondary treatment in Croatian part of the Black Sea basin was assessed and an upgrade study to tertiary treatment level (EU effluent discharge standards) of five larger plants was carried out using a modelling approach. One of those five plants (WWTP Varaždin) was selected to investigate possible upgrade scenarios by using mathematical modelling at the level of feasibility study and preliminary process design. This chapter summarizes the main findings of this part of the project.

13.2 WWTP Varaždin

The WWTP Varaždin was designed in 1980's and since it has been subject to change, extensions and reconstructions. It is the first WWTP built in Croatia with a capacity larger than 100,000 PE. Situated about one km to the east of the city, at the location of Motičnjak (Figure 13.1), the new extended part of the plant (a secondary treatment) was constructed 1,6 km further from the original location (primary treatment), connected by an open channel. In 2000, the channel was modernized (equipped with aerators, Figure 13.2); biofilm carriers were installed in the plant as well (Figure 13.3). By the most recent reconstruction in 2005, the plant received the present form reaching a capacity of 140,000 PE (from which about 60,000 PE comes from diary industry Vindija), which is occasionally exceeded during peak loads (up to 200,000 PE). Due to various interventions during recent years, this plant became an 'interesting' collection of process-technological solutions. In essence, wastewater treatment plant of the city of Varaždin is a mechanical-biological plant with activated sludge and biofilm technology (a kind of modified AB process) and a separate sludge treatment system.

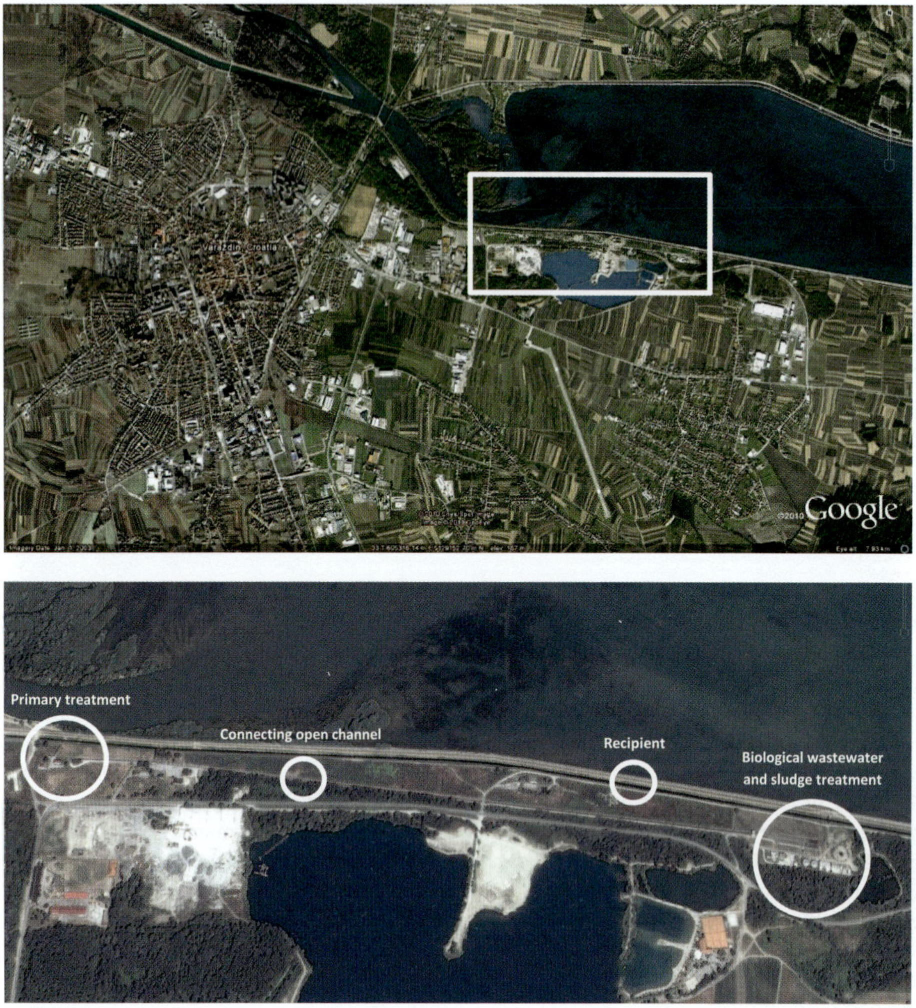

Figure 13.1 Location of WWTP Varaždin (photo: Google Earth).

WWTP Varaždin is treating sewage of the city of Varaždin and a part of sewage from small agglomerations Trnovec, Jalkovec, Biškupec, Hrašćica, Sračinec, and a part of Petrijanec. By planned construction of the collector and the secondary sewerage network, Varaždin WWTP will also receive wastewater from Donji and Gornji Kneginec in the foreseeable future. The sewerage system of the city of Varaždin is a combined system in which household, industrial and rainwater water is collected by a network of total 150 km. There are 10 sewer pump stations in the catchment area, but there are no combined sewer overflows (CSOs) installed. The only release of wastewater from the system takes place via the overflow bypass before the entry to the plant. In the central part of the city, the sewerage network is old and prone to infiltration. Due to flat landscape and low hydraulic gradients in the (mostly gravity) network, accumulations of sediment takes place in the pipes, which is causing problems to the plant operation in case of rain weather. Beside the household sewage, wastewater of industry located in the nearby area also reaches the plant.

Figure 13.2 Channel connecting the primary and secondary treatment at WWTP Varaždin (photos: Meijer S.C.F.).

Figure 13.3 Biofilm careers were introduced to increase efficiency of sewage treatment (photo: Brdjanovic D.).

Industrial effluents make up to 60% of the inflow to the plant. The most important industry connected to the sewerage are Varteks (textile industry), Vindija (food/dairy industry) and Koka (slaughterhouse for poultry). According to the available data, no industrial plant has an efficient pre-treatment. For example, Koka processes about 100,000 of chicken daily, while wastewater from Vindija sometimes contains as much as 50,000 mgCOD/L. Besides that, seepage water from the nearby garbage dumpsite also reaches the plant, as well as the content of septic tanks. Currently, the plant consists of the following facilities: coarse inlet screens, pump station of raw wastewater (6 pumps), flow measuring channel, fine automatic screens (2), sand trap-grease trap with the sand pumping site (2), open trapezoidal connection channel, pump station of the biological part (4), pre-aeration activated sludge tank with the fine diffusers aeration system, pre-aeration tank with the biofilm carriers with the fine diffused aeration system, aeration tanks (2) with fine diffused aeration, pumps for the return of activated sludge (2) pumps for the surplus sludge (2), secondary settling tanks (4), sludge thickener, aerobic sludge stabilizer, compressors and blowers for sludge stabilization, and sludge dewatering with a centrifuge and a system for polymer dosing. Two automatic samplers are installed on the plant, one before the mechanical treatment and another at the effluent outlet. Samples are taken and analyzed daily. The flow measuring device is situated at the plant outlet before discharging to the drainage channel of the power plant Čakovec storage reservoir. Flow data are collected daily (Table 13.1). Influent quality and effluent standards are shown in Table 13.2 and 13.3 respectively.

Table 13.1 Designed and observed flows at WWTP Varaždin.

Flow	Design flows	Measured average flows
Dry Weather Flow (DWF)		
$Q_{d.avg}$	43,200 m^3/d	35,450 m^3/d
$Q_{d.max}$	-	-
$Q_{h.avg}$	1,800 m^3/h	1,440 m^3/h
$Q_{h.max}$	4,680 m^3/h	-
Wet Weather Flow (WWF)		
$Q_{h.avg}$	4,680 m^3/h	-
$Q_{h.max}$	-	25,000 m^3/h

Table 13.2 Designed and observed water quality parameters at WWTP Varaždin.

Parameter	Design values	Measured average values
BOD_5	300 mgO_2/L	280 mgO_2/L
COD	550 mgO_2/L	550 mgO_2/L
TS		140 mg/L
TN		16 mg/L
TP		8 mg/L

Table 13.3 Effluent standards and treatment efficiency.

Parametar	MAC	Design values	Measured average values	Measured efficiency
BOD_5	25 mgO_2/L	25 mgO_2/L	11 mgO_2/L	96%
COD	125 mgO_2/L	125 mgO_2/L	33 mgO_2/L	93%
TS	35 mg/L	35 mg/L	36 mg/L	66%

MAC - maximum allowed concentration

In front of the inlet pump station sewage (about 40,000 m^3/d) passes through the coarse screen and is pumped to fine screening. After screening, sand and grease are separated from sewage in an aerated grit chamber and sewage is transported via an open channel to the inlet of the pumping station of the biological tanks. Because of the plant overload and with the aim to control odors, a part of the open channel is equipped with a pressurized aeration system, however, it is unclear to what extent is this intervention effective.

The biological part of the plant is divided into two parallel but technologically different lines, the so-called North Line and South Line. The North Line employs conventional activated sludge (CAS) systems and receives about 40% of the total hydraulic load. It consists of a pre-aeration tank (stage A), an aeration tank (stage B), and secondary settling tanks. In the pre-aeration and aeration tank, the system with diffused aeration (fine air bubbles) is installed and equipped with automated regulation based on DO measurements. The comparatively more loaded South Line is in fact a CAS system with pre-aeration tank filled up with freely floating biofilm carriers, while the aeration tanks are identical to that of the North Line. To ensure effective settling with respect to the increased load to the South Line, its secondary settling tanks are equipped with inclined (55°) packages of pipes (DN160mm) in the counter-current direction for enhanced settling. The surplus sludge or waste activated sludge (WAS) is pumped to the sludge thickener where the sludge dry content reaches 4-6% DS. Thickened sludge is centrifuged and reaches about 16% DS with the help of polyelectrolyte dosing. Dewatered sludge is mixed with lime, increasing the final dry solids content of the sludge cake to 30% DS. The daily production of dewatered excess sludge is about 20 tons.

After treatment, effluent is dicharged into the drainage channel of power plant Čakovec (Figure 13.4), which flows into the Drava River downstream from the dam. At the inspection shaft, just before dicharge, purified wastewater should meet the discharge values according to the II category rank from the Regulations on Limit Values of Indicators of Hazardous and Other Substances in Wastewater. Throughout the year there is water in the channel (on average 2.5 m^3/s), and hydraulic dilution of the effluent is about 10 times before it reaches the Drava River.

Figure 13.4 Recipient of the plant effluent (photo: Meijer S.C.F.).

A visit to the plant (2011) revealed following observations:

- The plant personnel indicated that the plant is overloaded, and that it is necessary to carry out measures to adjust it to the current and future load. That particularly refers to biological treatment, including the secondary settling tanks as well. Interventions considered by the plant staff include:
  - Reduction of the load by introduction of primary settling tanks, and reconstruction of and leveling the load to aeration tanks;
  - The existing settling tanks are too deep with a small surface and it is necessary to increase the capacity of secondary settling up to the minimum total capacity of 200,000 PE;
  - Introduction of anaerobic digestion of sludge and of the biogas collection and reuse;
  - For the purpose of protection of the recipient (as well as the storage reservoir of the power plant), biological removal of nitrogen and chemical precipitation of phosphorus seems necessary;
- Besides this, several problems related to industrial wastewater are present:
  - Because of the chicken processing industry, a large quantity of fine feathers (down) reaches the plant, forming clusters in the form of balls that pass through mechanical treatment and 'unfold' in the aeration tank, float to the surface, draw clusters of sludge with them and make efficient settling difficult. The problem could be solved by introduction of primary gravitational settling or, maybe even better, by air floatation.
  - Industries do not have (functional) pretreatments of wastewater resulting in large oscillations in the quantity and quality of the inflow.
- Furthermore, the inflow from septic tanks to the plant is considerable; in many cases, households tend to 'stick to' septic tanks and do not wish to get connected to the existing sewerage system.

In order to better understand the plant operation further work was needed, such as:

- To find out what is the total quantity of biofilm carrier filling in m^3 in the pre-aeration tank of the South Line, i.e. the efficiency of the biofilm part of WWTP regarding BOD_5 and TKN;
- To estimate the efficiency of the North Line pre-aeration tank regarding BOD_5 and TKN;
- To assess the aeration capacity (in kgO_2/h or Nm^3/h) per pre-aeration tank (Phase A) and aeration tank (Phase B) ;
- To measure the exact concentration of sludge in all four biological tanks and to calculate the sludge retention time (SRT) in both lines;
- To clarify the reasons for the current practice of the return sludge distribution scheme;
- To determine what is the ratio of the return sludge and the influent flow, and what management control is used for the return sludge flow, particularly at the transition from the dry to the rainy period operating regime;
- To calculate the present hydraulic (surface and overflow) load of the secondary settling tanks in both lines;
- To determine the sludge volume index (SVI) in both lines;
- To explain how and why the sludge surplus pump is operated, and how often does it work;
- To estimate the concentration of settled sludge in both lines;
- To determine the sludge concentration after thickening;
- To explain why the concentration of sludge after the centrifuge is only 16% DS, which is relatively low, and how the polymer dose is determined.

Obviously, this plant, originally designed as a conventional activated sludge (CAS) system for removal of solids and organic matter had undergone serious makover in last decades and gradually become a 'black box' where it is not really clear how and why the technological choices have been made. As the plant is located in the 'sensitive' area of the Black Sea Basin, the plant is subject to new EU and Croatian regulations and consequently, upgrade with nitrogen and phosphorus removal is needed. As this project is the pioneering application of modelling of wastewater treatment plants to Croatia, it was decided to take complex example of Varaždin, where application of modelling is most needed and effective in a view of a planned upgrade. This chapter describes eight model-based upgrade scenarios, all supported by multi-criteria decision approach that include technological as well as financial aspects. Based on a systematic approach, the best upgrade scenario was proposed and design engineering documentation at the level of preliminary design was prepared to facilitate preparation of tender documentation and the start of upgrade works in the near future.

13.3 Upgrade scenarios

The upgrade is governed by the EU standards for COD (125 mg/L), BOD_5 (25 mg/L), TSS (35 mg/L), and removal of nitrogen and phosphorus. For nitrogen, standards require effluent TN concentration <10 mg/L, and full nitrification throughout the year. The other condition is to carry out the upgrade as much as possible within the existing volumes to minimize investments. For phosphorus removal (TP<1mg/L) the chemical option is preferred for three reasons, namely: (*i*) low influent concentration of RBCOD (readily biodegradable COD, mainly VFAs: volatile fatty acids) is not supporting satisfactory biological phopshorus removal (BPR), (*ii*) BPR requires construction of an (additional) anaerobic tank, which is against the phylosophy to maximeize the use of existing volumes, (*iii*) due to the application of 'cold' digestion, application of e.g. iron salts will bind phosphate and thereby avoid internal accumulation of phosphate via recirculation streams. Since the influent phosphorus concentration is relatively low, and due to the fact that portion of phosphorus is utilized by microorganisms as micronutrient, the expected residual concentration of phosphorus is low and the need for chemical dozing is expected to be minimal. Therefore, chemical phosphorus removal is considered as the standard option for all studied scenarios, assuming the effluent concentration of TP will be below 1 mg/L.

This study considers eight upgrade scenarios (S1 to S8); in fact four distinguished scenarios were developed regarding the process layout and two regarding the composition of raw sewage. The main issues arised, namely: (i) whether primary settling should be included and, consequently also mesophilic anaerobic sludge digestion, and (ii) how to increase the capacity of the biological treatment and settling: via conventional way by increasing the tank volumes, or by introduction of more efficient solid-liquid separation system e.g. membrane bio-reactors (MBR)? The two influent composition options include situations based on (i) the measurements conducted in 2011 with observed high concentration of inert COD, and (ii) based on measurements from the same year, which in contrary showed the inert COD concentration in line with more typical values observed at most other plants. The current situation was reffered to as S0 (Table 13.4).

In both scenarios, the existing flow division between the Nort and South Line of 40:60 was corrected to 50:50 as the current unequal distribution is not justified. The main scenarios distinguish between conventional settling and membrane filtration. In both options, two sub-options were considered: with and without primary settling. Primary settling was applied in scenarios S3, S4, S7 and S8, however, it is also possible (seen the existing load to the plant), that the upgrade is executed in several phases, so that primary settling can be introduced in the second upgrade phase, when the load reaches the level when primary settling becomes necessary. The applied activated sludge simulator (BioWin) does not include primary settling design, so sizing of the primary settling tanks was done manually. According to these calculations, two primary settling tanks will be needed (D=24 m, A=452 m^2, V=1,750 m^3 each). The primary thickener is also sized manually (D=16 m, V=785 m^3). The currently applied mechanical treatment is common for all scenarios.

Table 13.4 Overview of scenarios for the upgrade of WWTP Varaždin

Scenario	Scenario description	Influent	Influent particulate COD fraction	Activated sludge tanks	Solids-water separation	Sludge handling	Removal scope
S0	Existing situation	Raw	High inert fraction	2 lane plug-flow	Rectangular SST	Thickening, dewatering	BOD, TSS
S1	Existing biological volume, membrane bio-reactor	Raw	High inert fraction	2 lane plug-flow + MBR	Submerged membranes	Thickening, dewatering	BOD, N, TSS
S2	Existing biological volume, membrane bio-reactor	Raw	Typical inert fraction	3 lane plug-flow + MBR	Submerged membranes	Thickening, dewatering	BOD, N, TSS
S3	Primary settling tank (PST), existing biological volume, MBR, Digester	Settled	High inert fraction	4 lane plug-flow + MBR	Submerged membranes	Thickening, dewatering	BOD, N, TSS
S4	Primary settling tank (PST), existing biological volume, MBR, Digester	Settled	Typical inert fraction	5 lane plug-flow + MBR	Submerged membranes	Thickening, dewatering	BOD, N, TSS
S5	Extended biological volume, new circular secondary settling tanks (SST)	Raw	High inert fraction	Existing + SST volume	New circular SST	Thickening, dewatering	BOD, N, TSS
S6	Extended biological volume, new circular secondary settling tanks (SST)	Raw	Typical inert fraction	Existing + SST volume	New circular SST	Thickening, dewatering	BOD, N, TSS
S7	Primary settling tank (PST), extended biological volume, new circular SST, digester	Settled	High inert fraction	Existing + SST volume	New circular SST	Thickening, dewatering	BOD, N, TSS
S8	Primary settling tank (PST), extended biological volume, new circular SST, digester	Settled	Typical inert fraction	Existing + SST volume	New circular SST	Thickening, dewatering	BOD, N, TSS

13.3.1 Existing situation: S0

Subject to this study is the biological part of the plant, which consists of follwoing units and equipment:

- Influent pumping station to the activated sludge system;
- Pre-aeration tank with fine bubble aeration system (North Line: V = 619 m^3, 32.5 m x 5 m x 3.75 m, aeration capacity 1,800 Nm^3/h);
- Pre-aeration tank with fine bubble aeration system and biofilm carriers (South Line: V = 619 m^3, 32.5 m x 5 m x 3.75 m, aeration capacity 3,500 Nm^3/h);
- Aeration tanks (V = 2 x 2,451 m^3, 77.7 x 8.3 x 3.8 m, aeration capacity 3,000 + 2,500 Nm^3/h);
- RAS pumping station (2 x 25 L/s)
- WAS pumping station(2 x 25 L/s);
- Secondary settling tanks (V = 2 x 2,470 m^3, 38.7 x 16.8 x 3.8 m).
- Sludge gravity thickener (digester 1: V = 1,050 m^3);
- Sludge storage tank (digester 2, V = 1,050 m^3);
- Sludge dewatering by centrifuge (360 m^3/d) enhanced by polyelectrolyte dosing; and
- Lime dosing system.

The main conclusions regarding the current situation are:

- The existing 40:60 split of the activated sludge treatment (North and South Line) is not taken in account in the upgrade as both lines have identical volume;
- Present SRT of the biology is short and about 3 days;
- The South Line performs comperatively less as the thickener supenatant, overflow from the sludge storage tank and reject water from the centrifuge are returnd to this line;
- Only partial nitrification takes place as the SRT is low, influent ammonia concentration is high and plant is not designed for nutrient removal;
- Secundary settling tanks are of insufficent capacity (five most common design approaches were applied and confirmed this statement)

The layout and process flow diagram of the WWTP Varaždin are shown in Figure 13.5 and 13.6, respectively and are used as the basis for development of upgrade scenarios S1 to S8. Some improvements are common for all eight scenarios, e.g. the existing pre-aeration tanks were retrofitted into anoxic (denitrification) tanks, chemical phosphorus removal by dosing of iron chloride in the anoxic tanks was applied, and existing sludge thickener and storage tanks were refurbished into anaerobic digesters.

13.3.2 Scenarios S1 and S2

Scenarios S1 and S2 (Figure 13.5 and 13.6) are in fact identical except that the calculations were performed using two different influent contents of inert COD. The goal of these scenarios was to maximize the use of existing structures and equipment at the plant while maintaining the new (EU) effluent requirements. Consequently, the pre-aeration tanks were turned into anoxic tanks, the main aeration tanks remained as such but the aeration system was adjusted, while the secondary settling tanks were turned into aerobic MBR system to increase the aerobic volume and introduce efficient solids-liquid separation. Surplus sludge from the MBR system is returned to the main aeration tanks. Application of an MBR system allows to operate at higher biomass content in the aeration tanks (10-15 kg/m^3) and it occupies very little space, and consequently is a technologically favourable scenario. However, costs for purchasing membranes and auxilaiary equipment, plus operational costs (energy and chemicals for cleaning and maintaing of membranes) are the offset. On the other hand, MBR systems produce superior and stable effluent quality in comparison with CAS systems.The sludge treatment line was kept within the existing structures; improvements are concerning the recirculation arrangements; surplus sludge is digested and centrifuged. The main additional investment concern the new sludge handling equipment. In both scenarios, the aertaion requirement is substantially higher (365,041 Nm^3/d) in comparison to the existing situation (126,720 Nm^3/d), resulting in the need for the additional compressors, blowers and a system for fine bubble aeration. A part of the existing settling tanks is equiped with sumberged membranes packed in cassettes and the system for chemical cleaning and maintaining of membranes. A chemical dosing system is located in a new building. The sludge treatment line is refurbished with adjusted pipelines for sludge intake, a new system for sludge dewatering, equipment for heating and recirculation of digested sludge, intake and transfer of digested sludge to a dewatring unit, and equipment for intake and incineration of biogas. Energy production from biogas is not included, but this would be possible in the future. All waste liquid streams from the sludge treatment line are introduced to the beginning of the biological treatment and equally distributed over the two lines. Design calculations showed that the difference in the influent quality (inert particulate COD) does not affect the proposed process concepts and layout; the main effect regards a (small) increase in the amount of solids in the internal recirculation Q_{MLSS} (higher in S2), the amount of WAS (smaller in S2), the oxygen requirement (higher in S2), the biogas production and the amount of treated sludge (smaller in S2).

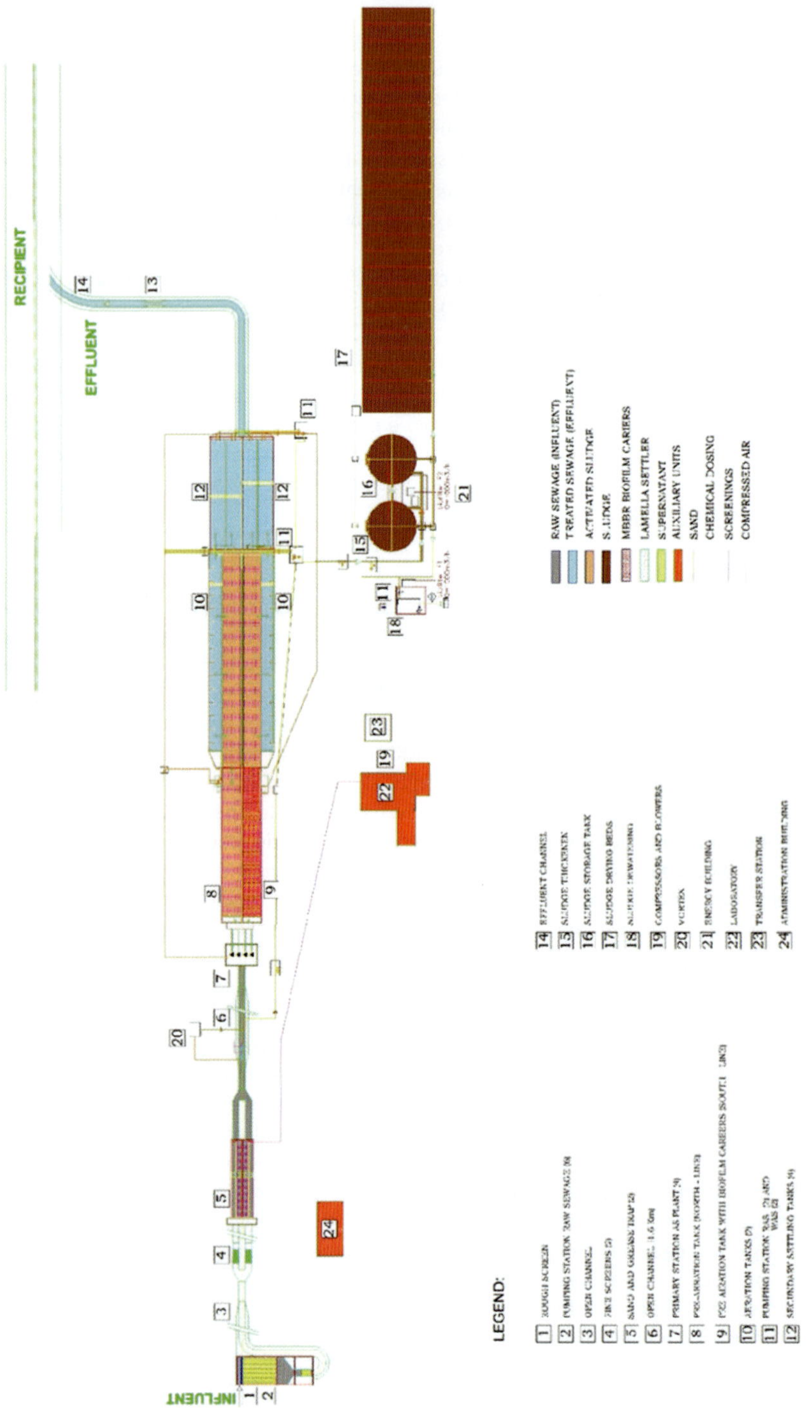

Figure 13.5 Layout of WWTP Varaždin: Current situation S0

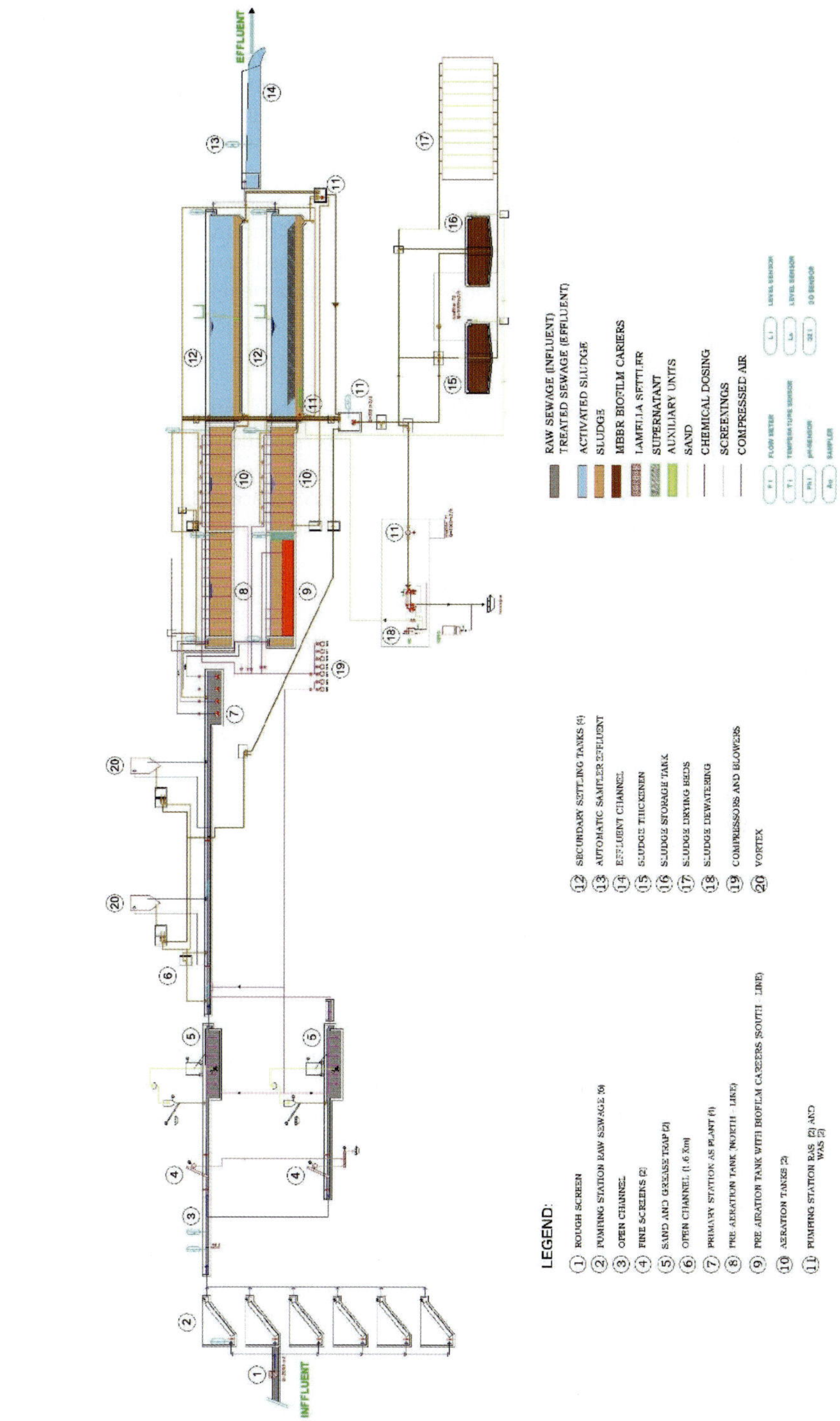

Figure 13.6 Process flow diagram of WWTP Varaždin: Current situation S0

13.3.3 Scenarios S3 and S4

The technological principe of scenarios S3 and S4 are identical to S1 and S2 except of introduction of primary settling (Annex 13.1 and 13.2). Dimensions of tanks are given earlier. In addition, a primary thickener is introduced to receive primary sludge. Introduction of primary settling and mixing primary and secondary sludge induced the need for construction of two additional digesters with a total volume of 2,100 m^3.

Similarly to scenarios S1 and S2, the difference in the influent quality (content of inert particulate COD) does not have any serious impact on the process concept and layout; the main effects are seen regarding the increase in the amount of solids in the internal recirculation Q_{MLSS} (higher in S3), higher amount of WAS (smaller in S4), oxygen requirements (higher in S4), biogas production and amount of treated sludge (smaller in S4).

13.3.4 Scenarios S5 and S6

Scenarios S5 and S6 consider a CAS system with gravity secondary settling without primary settling. Absence of primary settling induced the need for larger aeration tanks in scenarion S5 (high amount of inert COD). Scenario S5 is the only scenario which requires additional aeration volume (3,000 m^3) to satisfy the minimal SRT of 8 days and MLSS concentration up to 5 kg/m^3. Under these conditions the present settling capacity is insuficient, thus four new secondary settlers would be needed (D=37 m).

The sludge treatment remains the same as in the current situation (cold digestion and dewatering). Model calculations show that somewhat higher WAS production as well as more treated sludge is expected in S5.

Scenario S6 does not envisage the need for the additional aerated volume because for the standard influent composition, the minimal required SRT is achieved, aldough it is at the border value. The S6 scenario uses the existing infrastructure to a full extent and requires little additional equipment and rearrangements. However, it requires construction of four new large settling tanks and associated equipment. As it was the case with S1, S2 and S5, the usage of the existing sludge treatment infrastructure is maximized and includes adjustments and upgrade as described earlier. Because the process of S5 and S6 are partially different, these are presented separately (Annex 13.1 and 13.2).

13.3.5 Scenarios S7 and S8

The technological principe of scenarios S7 and S8 are identical as of S5 and S6 except of introduction of primary settling. This results in no need for additional aerated volume, however, new secundary settlers are still needed. The sludge treatment line requires inaddition to new primary thickener, also two new anaerobic digesters due to the production of primary sludge. Comparison of scenarios S6 and S8 shows that for the same influent quality and identical technology applied, except for introduction of additional primary settling in S8, scenario S8 results in somewhat smaller requirement for the area of secondary settling. However, this advantage of S8 is countered in the sludge line where because of the larger sludge production (primary plus secondary sludge), S8 requires additional digesters. Scenarios with CAS systems definitely require larger area in comparison with those with MBR systems, and thus require additional investments for land purchase and landscaping. Because the process of S5 and S6 are partially different, these are presented separately (Annex 13.1 and 13.2).

13.4 Building up the BioWin hydraulic flow scheme

Following the modelling guidelines (Meijer and Brdjanovic, 2012), usual procedure is to first draft the block process flow diagram of a WWTP subject to modelling (Figure 13.7).

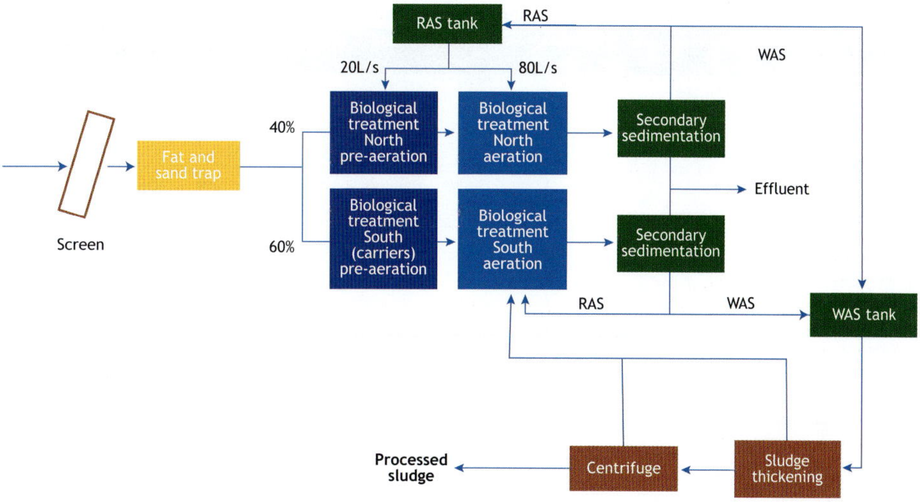

Figure 13.7 Block process scheme of WWTP Varaždin

Based on the block process diagram, design documentation and site visits, the BioWin hydraulic scheme is created (Figure 13.8) with legend given in Table 13.5.

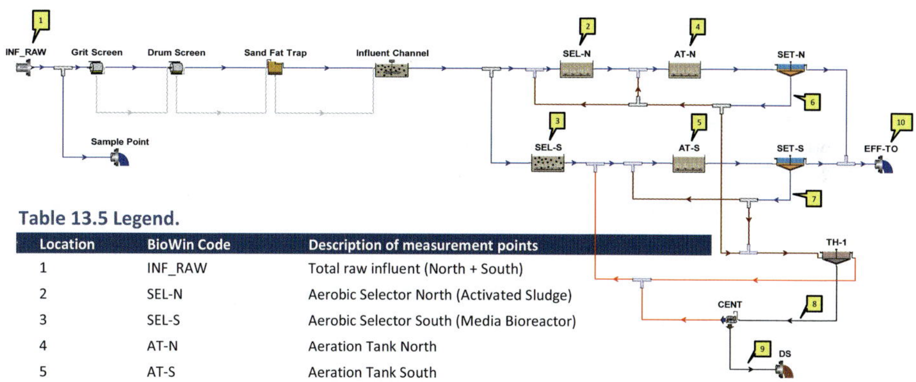

Table 13.5 Legend.

Location	BioWin Code	Description of measurement points
1	INF_RAW	Total raw influent (North + South)
2	SEL-N	Aerobic Selector North (Activated Sludge)
3	SEL-S	Aerobic Selector South (Media Bioreactor)
4	AT-N	Aeration Tank North
5	AT-S	Aeration Tank South
6	SET-N (U)	Return activated sludge (RAS) North (Settler Underflow)
7	SET-S (U)	Return activated sludge (RAS) South (Settler Underflow)
8	TH-1 (U)	Thickened waste sludge (TS)
9	DS	Dewatered Sludge (without lime)
10	EFF-TOT	Effluent (EFF)

Figure 13.8 Schematic hydraulic set-up of the WWTP as introduced in the BioWin simulator. Yellow tabs indicate sampling and measurement points. Table 13.5 Legend belonging to Figure 13.8.

Unit volumes and dimensions used in the hydraulic flow scheme are presented in Table 13.6.

Table 13.6 Reactor volumes and dimensions used in simulations.

Original Design (1996) Process unit	Tanks Units	Depth m	Width[6] m	Radius[2] m	Area[1] m^2	Volume[1] m^3	V total m^3
Sand and grease removal tank[5]	2	3.8	3.75	-	75	285	570
Primary settling tank	-	-	-	-	-	-	-
Aerated influent channel[4)5]	1	1.4	2.8	-	4,448	3,203	3,203
Aerobic selector North	1	3.8	5	-	163	609	609
Aerobic media selector South	1	3.8	5	-	163	609	609
Aeration tank North	1	3.8	8.3	-	645	2,451	2,451
Aeration tank South	1	3.8	8.3	-	645	2,451	2,451
Secondary settling tanks North[3]	2	3.8	4.2	-	163	618	1,235
Secondary settling tanks South[3)7]	2	3.8	4.2	-	163	618	1,235
Gravitational sludge thickeners	1	8.6	-	6.25	123	1,050	1,050
Sludge Storage	1	8.6	-	6.25	123	1,050	1,050

1) Per tank
2) Radius
3) Rectangular
4) V-shaped with average water hight
5) Estimated volume
6) Per tank or aeration lane
7) Equipped with packed tubing to increase sedimentation

Based on the original design documentation and design load of 140,000 PE, the hydraulic loading of the different process units is reconstructed:

- Average mixed flow (indicated as Q_{365}) was estimated from historical measurements (2010) to be 120% of the average dry weather flow (ADWF 33,600 m^3/d).
- Peak dry weather flow (PDWF) was estimated based on inflow over 16 hours per day resulting in a peak flow of 154% of the average dry weather flow being a typical flow for mixed sewage systems with average to short hydraulic residence time in the sewerage.
- Maximum sewage overflow capacity is the (design) capacity where influent bypasses the WWTP (or sewage overflows). A typical design is 300-350% of ADWF resulting from storm water events. In the 1996 design for PWWF 204% of ADWF is available indicating regular sewage overflow at storm water events. This calculation was not verified.
- The installed influent pumping capacity is 417% and much larger than the maximal sewage capacity suggesting considerable reserve pump capacity.

The pre-aerobic tanks and main aerobic tanks are hydraulically designed for the maximum flow. The clarifiers are hydraulically designed for the half of the maximum hydraulic capacity. Therefore, the flow division between the North and South lines should be equal (50:50). In the original design at maximal flow (PWWF), the settling thanks are hydraulically nearly overloaded. This is also expressed by the calculated surface loading in Table 13.11 which is in the range of 2.3 m^3/m^2.h at maximum flow capacity, while the typical design value for (round) settlers (and high SRT) is around 1.6 m^3/m^2.h. From the hydraulic assessment, it can be concluded that the secondary clarifier units are under designed e.g. overloaded.

Following conclusions on the design of WWTP Varaždin can be drawn:

- Most of original design data was readily available or could be reconstructed from the plant information;
- The original design is aerobic CAS system with SRT of 3 days, designed for COD and solids removal;
- The plant was designed for 140,000 PE based on specific BOD load per capita and typical influent load and concentrations;

- More than 70% of the plant loading is of industrial origin;
- In the original design no measures were taken regarding the significant industrial discharge, e.g. additional industrial pre-treatment;
- Hydraulically the plant is overloaded regarding secondary settling capacity;
- The hydraulic capacity of the secondary clarifiers requires equal split of flow between two biological treatment lines;
- Due to a lack of measurements all return sludge flows are estimated;
- The original influent data assume usual influent composition and do not consider dilution by excessive infiltration into the sewer system;
- The installed recilculation activated sludge (RAS) pump capacity is much larger than the observed flow rates;
- The low RAS flow rate resulted in a reduced activated sludge concentration in the pre-aeration tanks which negatively affects the biological removal efficiency;
- The installed influent pumps have 100% spare capacity.

Based on the above presented influent composition, the influent characterization is calculated for all the relevant model fractions (Table 13.7). The calculations are performed according to the BioWin influent characterization procedure (Meijer and Brdjanovic, 2012). The procedure includes the BOD model calculation based on which the biodegradability of the influent is estimated. To fit the model simulation results to the measured data, it is often necessary to adjust the model values for influent biodegradability. The results of the calculation are presented in Table 13.8. Influent fractions are calculated for COD, nitrogen and phosphorus. Influent TSS concentrations are derived from other calculated fractions. In other words, the TSS itself is not a modeled component (does not belong to the model state variables), but is a combined result of the modeled (particulate) COD, N and P compounds (each determined on its weight fraction). Two characterizations are shown: (i) typical raw wastewater (default result of the BioWin influent specifier) and (ii) the characterization that resulted from the calibrated (fitted) activated sludge model. To calibrate the model, it was necessary to adjust several influent parameters mainly to fit the measured activated sludge composition to the model results (nitrogen content of the sludge, the amount of TSS in the sludge and the organic COD/VSS ratio of the sludge).

Following conclusions on the influent composition can be drawn:

- The 2010 data shows a more or less typical wastewater composition;
- It is assumed the influent is diluted by infiltration;
- TKN influent was corrected to a more realistic value based on model simulations and the nitrogen content of the measured activated sludge;
- TSS was corrected to a more realistic value based on model calculations;
- Biodegradability of the influent COD fraction was not estimated with the BOD model but from calibrating the model on the actual measured sludge production being a direct and more reliable method of estimating influent biodegradability;
- To fit the model high inert particulate COD needed to be assumed. This was not validated by measurements. All simulations therefore were also repeated with typical COD wastewater composition.

Table 13.7 Results influent characterization calculation (before and after model calibration).

Code	Component	Unit	Typical	Fitted
	COD concentrations			
	Total COD	mgCOD/L	394.6	394.6
cBOD5	Biodegradable COD of which:	mgCOD/L	322.8	220.2
Sbsc	• Readily biodegradable COD (excluding acetate)	mgCOD/L	199.7	199.7
Sbsa	• Acetate (COD-VFA)	mgCOD/L	0.0	0.0
	Slowly biodegradable COD of which:	mgCOD/L	123.1	20.5
Xsc	• Colloidal slowly biodegradable	mgCOD/L	24.6	17.4
Xsp	• Non-colloidal (particulate) slowly biodegradable	mgCOD/L	98.5	3.1
	Unbiodegradable of which:	mgCOD/L	71.8	174.4
SI	• Unbiodegradable soluble COD	mgCOD/L	20.5	20.5
XI	• Unbiodegradable particulate COD	mgCOD/L	51.3	153.9
	Nitrogen concentrations			
	Total N	mgN/L	34.4	34.4
NO3	Nitrate N	mgN/L	0.6	0.6
	Total Kjeldahl Nitrogen	mgN/L	33.8	33.8
NH3	Ammonia	mgN/L	10.9	10.9
	Organic Nitrogen of which:	mgN/L	22.9	22.9
	• Unbiodegradable Organic of which:	mgN/L	4.2	11.3
Nus	Soluble Unbiodegradable Organic Nitrogen	mgN/L	0.7	0.7
Xin	Particualte Unbiodegradable Organic Nitrogen	mgN/L	3.5	10.6
	• Biodegradable Organic of which:	mgN/L	18.7	11.6
Xon	Particulate Organic Nitrogen (biodegradable)	mgN/L	8.3	5.1
Nos	Soluble Organic Nitrogen (biodegradable)	mgN/L	10.4	6.5
	Phosphorus concentrations			
	Total P	mgP/L	4.3	4.3
PO4	Phosphate	mgP/L	3.0	3.0
	Organic Phosphorus of which:	mgP/L	1.3	1.3
	• Unbiodegradeble Organic of which;	mgP/L	0.3	0.9
Xip	Particulate Unbiodegradable Organic Phosphorus	mgP/L	0.3	0.9
	Soluble Unbiodegradeble (=0)	mgP/L	0.0	0
	• Biodegradeble Organic of which:	mgP/L	1.0	0.4
Xop	Particulate Organic Phosphorus (Biodegradable)	mgP/L	1.0	0.4
	Soluble Biodegradeble (=0)	mgP/L	0.0	0.0
	Solids concentrations			
	Total SS			
TSS	Total Suspended Solids of which:	mgTSS/L	150.8	150.9
ISS	• Inorganic SS	mgISS/L	24.9	24.9
VSS	Volatile Suspended Solids of which:	mgVSS/L	125.9	126.0
	• Slowly Biodegradable Organic of which:	mgVSS/L	82.8	1.9
	Colloidal (not accounted particulate)	mgVSS/L	0.0	0
	Non-colloidal (particulate)	mgVSS/L	82.8	1.9
	• Unbiodegradable Particulate Organics	mgVSS/L	43.1	124.1

All fractions are calculated from BioWin Influent specifier

Fitted - Calibrated/Fitted influent fraction: high inert particulate (used for scenarios S0, S1, S3, S5, S7)

Typical - Scenarios with typical design values (used for scenarios S2, S4, S6, S8)

Table 13.8 Summary table on influent design and characterization.

Code	Description	Unit	Design 2004	Measured 2011	Fitted 2011	Model
Q365	Average mixed flow	m^3/d	28,000	21,808	21,808	21,808
PE	Person Equivalents at 60 gBOD/PE.d	PE	140,000	80,711	56,497	56,319
PE	Person Equivalents at 54 gBOD/PE.d	PE	155,556	89,678	62,775	62,577
PE	Person Equivalents at 136 gTOD/PE.d	PE	141,462	81,163	88,491	88,490
TCOD	Total Chemical Oxygen Demand	mgCOD/L	550	395	395	395
BOD_5	5 day Carbonaceous Biological Oxygen Demand	mgBOD/L	300	222	155	155
TSS	Total Suspended Solids	mg/L	-	121	151	151
TN	Total Nitrogen (TKN + NOx)	mgN/L	30	24.4	34.4	34.4
NH4	Ammonium	mgN/L	-	10.9	10.9	10.9
TP	Total Phosphorus	mgP/L	-	4.3	4.3	4.3
Average influent loads						
TCOD	Total Chemical Oxygen Demand	kgCOD/d	15,400	8,605	8,605	8,605
BOD_5	5 day Carbonaceous Biological Oxygen Demand	kgBOD/d	8,400	4,843	3,390	3,379
TSS	Total Suspended Solids	kg/d	-	2,650	3,297	3,297
TN	Total Nitrogen (TKN + NOx)	kgN/d	840	532	750	750
NH4	Ammonium	kgN/d	-	238	238	237
TP	Total Phosphorus	kgP/d	-	93	93	93
Typical influent ratios						
$TCOD/BOD_5$	Influent COD over BOD ratio	g/g	1.83	1.78	2.54	2.55
Q/PE	Average water use per person equivilent	L/PE	198	269	246	246
TN/TCOD	Nitrogen over COD ratio	gN/gCOD	0.055	0.062	0.087	0.087
TP/TCOD	Phosphorus over COD ratio	gP/gCOD	-	0.011	0.011	0.011
TN/TP	Nitrogen over phosphorus ratio	gN/gP	-	5.7	8.1	8.1

Furthermore, summary of biomass characterization is presented in Table 13.9 and 13.10.

Table 13.9 Measured activated sludge composition.

Symbol	Parameter	Unit	AT-North	Model	AT-South	Model
Particulate						
TSS	Total Suspended Solids	mg/L	2,917	2,897	3,145	3,185
Ash	Ash fraction of TSS	% of TSS	14.8%	14.0%	14.7%	15.1%
VSS	Volatile Suspended Solids (organic fraction)	mg/L	2485	2492	2682	2705
ISS	Inorganic fraction of Total Suspended Solids	mg/L	432	405	464	480
CODx	particulate COD	mgCOD/L	3294	3248	3646	3504
TKNx	Particulate Total Kjeldahl Nitrogen	mgN/L	232	226	255	243
TPx	Particulate Phosphorus	mgP/L	40	40	44	49
Total (MLSS)						
TCOD	Total Chemical Oxygen Demand	mgCOD/L	3,316	3,270	3,668	3,526
TKN	Total Kjeldahl Nitrogen	mgN/L	239	232	262	253
TN	Total Nitrogen (TKN + NOx)	mgN/L	240	233	262	254
TP	Total Phosphorus	mgP/L	41	41	45	51
Soluble						
CODmf	COD 0,45 micron membrane filtered	mgCOD/L	22	22	22	22
NH4	Ammonium	mgN/L	6.9	4.4	6.9	8.4
NOx	Nitrate + Nitrite	mgN/L	1.1	1.0	1.1	0.6
PO4	Ortho-Phosphate	mgP/L	1.1	1.2	1.1	2.0

Table 13.10 Biomass composition.

Parameter	Activated sludge (MLSS) fractions	Unit	AT-North	Model	AT-South	Model
Ash	Inorganic fraction of TSS	% of TSS	14.8%	14.0%	14.7%	15.1%
CODx/VSS	Organic weight to COD ratio	gCOD/gVSS	1.334	1.303	1.368	1.295
VSS/TSS	Organic fraction of total weight	gVSS/gTSS	0.852	0.860	0.853	0.849
N/VSS	Fraction N in VSS	gN/gVSS	0.096	0.091	0.098	0.090
P/VSS	Fraction P in VSS	gP/gVSS	0.017	0.016	0.017	0.018
N/COD	Fraction N in COD	gN/gCOD	0.072	0.069	0.071	0.069
P/COD	Fraction P in COD	gP/gCOD	0.012	0.012	0.012	0.014
Organic N/P	Organic composition	gN/gP	5.80	5.65	5.83	4.95

All design methods show that the current settlers are overloaded, both hydraulically and based on solids loading. Calculations by five settler design models and tests of the input values on its sensitivity show that two additional settlers of 31 meter diameter each need to be built/added to cope with the current (hydraulic) plant loading. Taking also in account a possible future increase of the influent loading, it was recommended to reserve additional space (area) for settling surface in the design footprint (Table 13.11).

Table 13.11 Overview of clarifier design results.

Design Extension	Unit	Empirical	Flux	WRCm	ATV (1976)	STOWA	MEAN
Designed Surface	m^2	1,906	1,860	1,290	3,519	2,387	2,192
Current Surface Available	m^2	650	650	650	650	650	650
Additional Surface Required	m^2	1,256	1,210	640	2,869	1,737	1,542
Extended Surface	% extension	293%	286%	198%	541%	367%	337%
Diameter for Additional Tank	m	31	31	31	31	31	31
Number of Tanks Applied	tank units	1.7	1.6	0.8	3.8	2.3	2.0
At ADWF							
Overflow, QADWF	m^3/h	1,167	1,167	1,167	1,167	1,167	1,167
Overflow rate to surface	m/h	0.6	0.6	0.9	0.3	0.5	0.6
Recycle flow (QRAS)	m^3/h	2,100	700	1,290	1,079	1,079	1,249
Recycle rate to surface	m/h	1.1	0.4	1.0	0.3	0.5	0.6
Recycle ratio (RAS/INF)	-	1.8	0.6	1.1	0.9	0.9	1.1
RAS TSS concentration	kg/m^3	5.4	9.3	6.7	7.3	7.3	7.2
Solids loading rate	$kg/m^2.h$	6.0	3.5	6.7	2.2	3.3	4.3
At PDWF							
Overflow, QPDWF	m^3/h	1,750	1,750	1,750	1,750	1,750	1,750
Overflow rate to surface	m/h	0.9	0.9	1.4	0.5	0.7	0.9
Recycle flow (QRAS)	m^3/h	2,100	700	1,290	1,079	1,079	1,249
Recycle rate to surface	m/h	1.1	0.4	1.0	0.3	0.5	0.6
Recycle ratio (RAS/INF)	-	1.2	0.4	0.7	0.6	0.6	0.7
RAS TSS concentration	kg/m^3	6.4	12.3	8.2	9.2	9.2	9.1
Solids loading rate	$kg/m^2.h$	7.1	4.6	8.2	2.8	4.1	5.4
At PWWF							
Overflow, QPWWF	m^3/h	2,917	2,917	2,917	2,917	2,917	2,917
Overflow rate to surface	m/h	1.5	1.6	2.3	0.8	1.2	1.5
Recycle flow (QRAS)	m^3/h	1,458	1,633	1,290	1,079	1,079	1,308
Recycle rate to surface	m/h	0.8	0.9	1.0	0.3	0.5	0.7
Recycle ratio (RAS/INF)	-	0.5	0.6	0.4	0.4	0.4	0.4
RAS TSS concentration	kg/m^3	10.5	9.8	11.4	13.0	13.0	11.5
Solids loading rate	$kg/m^2.h$	8.0	8.6	11.4	4.0	5.9	7.6

13.5 Scenarios regarding plant layout

Annex 13.2 shows the different plant layouts for the four scenarios as they were introduced in the BioWin model and software. Each of the four plant layouts are tested on the two types of influent (high inert and typical). For simplicity, the design scenarios are all calculated based on a single lane representation. This is allowed when the North and South line are identical loaded and operated and also the internal loading is evenly distributed. In the scenarios where the activated sludge system is extended with a MBR system (S1, S2, S3 and S4), the current secondary settler volume is used to place the submerged membrane units. In the scenario designs with extended conventional settlers (S5, S6, S7 and S8), new round settlers are introduced and the current settling tanks are rebuilt for aerobic extension of the activated sludge system. For all scenarios where primary settling is applied (S3, S4, S7, S8) new tanks would be need to be constructed including new primary thickening tanks. Where possible, the current sludge storage tanks (the former digesters) are used in the digester design. In all cases where digestion is proposed, also a new secondary sludge dewatering unit isrequired.

13.6 Scenarios regarding plant volumes

The building costs of a wastewater treatment plant are for a considerable part expressed in the required reactor volumes and area requirement. For the different scenarios this is presented in Table 13.12 and Figure 13.9 to 13.12. In all scenarios an anoxic volume is included for denitrification (nitrogen removal). The additional anoxic volume is situated in the current pre-aeration tanks. For the scenarios S1 to S4, the membranes are installed in the current secondary settling tanks. As a result of the volume displacement by the submerged membranes not all secondary settler volume (originally 2,400 m^3) is available for the activated sludge process. On the other hand, with the membranes in place, the activated sludge process can be operated at much higher MLSS concentrations (up to 10 to 15 $kgTSS/m^3$). As a result, the scenarios including membranes can handle the design loading of 28,000 PE well within the existing activated sludge (and secondary settler volume). Membranes are therefore rather inexpensive scenario considering the volume extension and use of surface area (Figure 13.11 and 13.12). However, the investment costs for membranes are relative high as well as cost for operation (energy, chemicals, maintenance and replacement). Scenario S5 it is the only scenario that required additional aeration volume (3,000 m^3) to meet the minimal aerobic SRT requirements. Scenario S5 has a high inert influent fraction without primary settling. To maintain the S5 design at a minimal aerobic SRT of 8 days and at the same time a MLSS concentration in the tanks below 5 kg/m^3, it is necessary to introduce additional anaerobic volume. The MLSS concentration of 5 kg/m^3 would be the maximum concentration allowed in the activated sludge tanks to efficiently operate the secondary settling tanks.

Table 13.12 Scenarios and unit volumes.

Volumes from design documentation		Unit	S0	S1	S2	S3	S4	S5	S6	S7	S8
PST	Primary settling tank	m^3	-	-	-	8,000	8,000	-	-	8,000	8000
ANA	Anaerobic tank	m^3	-	-	-	-	-	-	-	-	-
ANOX	Anoxic tank	m^3	-	1,239	1,239	1,239	1,239	1,239	1,239	1,239	1,239
AT	Aeration tank	m^3	6,131	4,912	4,912	4,912	4,912	9,137	6,137	6,137	6,137
MBR	Submerged membrane bio-reactor	m^3	-	1,125	1,125	1,125	1,125	-	-	-	-
SST	Secondary settling tank	m^3	2,470	-	-	-	-	11,775	12,480	11,775	8,925
PPT	Primary thickening tank		-	-	-	1,500	1,500	-	-	1,500	1,500
DIG	Digester	m^3	2,100	2,100	2,100	4,199	4,199	2,699	2,100	4,199	4,199
VAERO	Aerobic volume	m^3	6,131	6,037	6,037	6,037	6,037	9,137	6,137	6,137	6,137
VANOX	Anoxic volume	m^3	-	1,239	1,239	1,239	1,239	1,239	1,239	1,239	1,239
VMLSS	Total activated sludge volume	m^3	6,131	7,276	7,276	7,276	7,276	10,376	7,376	7,376	7,376
VTOT	Total volume waterline	m^3	8,601	7,276	7,276	7,276	7,276	22,151	19,856	19,151	16,301
Volumes extension (reference situation 2011)		**Unit**	**S0**	**S1**	**S2**	**S3**	**S4**	**S5**	**S6**	**S7**	**S8**
PST	Extended volume PST	m^3	-	-	-	8,000	8,000	-	-	8,000	8,000
ANOX	Extended volume ANOX	m^3	-	1,239	1,239	1,239	1,239	1,239	1,239	1,239	1,239
AT	Extended volume AT	m^3	-	-	-	-	-	3,000	-	-	-
SST	Extended volume SST	m^3	-	-	-	-	-	11,775	12,480	11,775	8,925
PTT	Extended volume PPT	m^3	-	-	-	1,500	1,500	-	-	1,500	1,500
DIG	Extended volume DIG	m^3	-	-	-	2,099	2,099	599	-	2,099	2,099

Scenario S6 is identical to S5 however, calculated with typical influent fractions. Because under these conditions less inert material will accumulate in the activated sludge tanks, the current aerobic volume (including secondary settler volume) is sufficient. Thereby, S6 only requires new secondary settling units. Compared to S6, scenario S8 requires a smaller secondary settling tank volume because the activated sludge system can be operated at lower MLSS. This is possible because in S8 primary settling is applied. Besides the smaller secondary settling tanks, in general the advantage of primary settling is a lower total waste sludge production in combination with the possibility of (green) energy production. However, this is at the expense of building primary settling tanks and extending the current digestion units (with energy production facility).

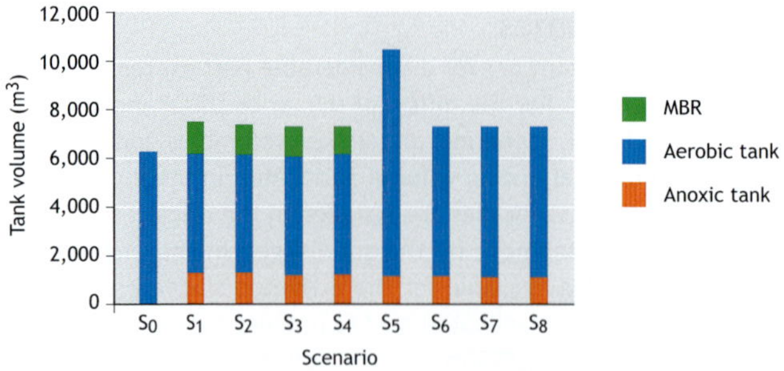

Figure 13.9 Scenario performances regarding the activated sludge units volume (m^3).

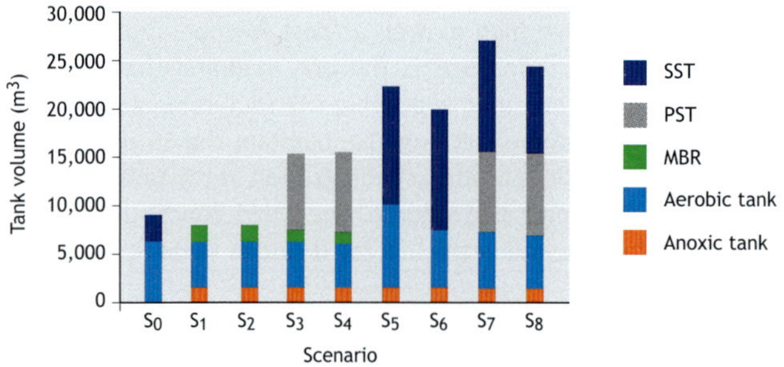

Figure 13.10 Scenario performances regarding the total reactor volume (m^3).

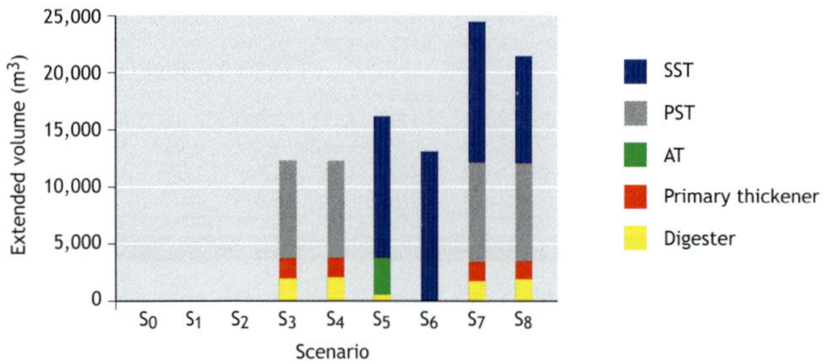

Figure 13.11 Scenario performances regarding the reactor volume extension (m^3).

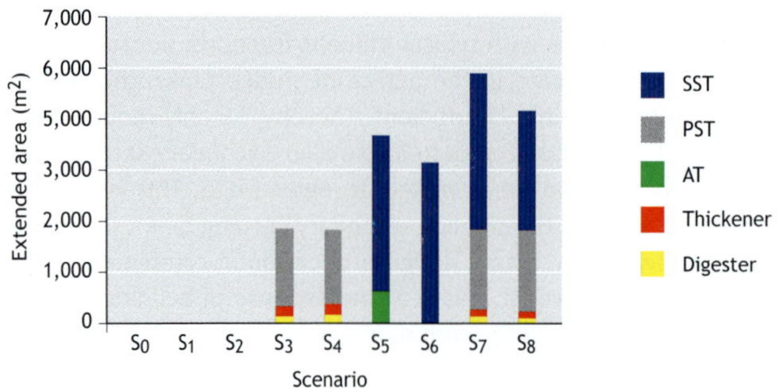

Figure 13.12 Scenario performances regarding the area extension (m^2).

The following conclusions on units volume and area extension can be drawn:

- For all scenarios a relative small anoxic tank, placed in the current pre-aeration tanks, was effective in removing nitrate;
- Scenarios S1 and S2 are most cost efficient from the point of view of volume and area extension. On the other hand, the membranes that are used to save space on itself are less cost efficient;
- Scenarios S1 and S2 do not require extension of the activated sludge system nor primary settling. Secondary settlers are not required and instead submerged membrane units are applied;
- Scenario S5 cannot be realized within the current reactor volumes (including the current secondary volume) and thereby does not meet the design requirements;
- Scenario S7 is the least efficient from the point of view of volume and area extension. The scenario requires both primary settling (including thickening, dewatering, energy production facilities and storage) and large new secondary settling tanks.

13.7 Operational performance of the scenarios

13.7.1 Effluent quality

The main design requirement was to obtain EU effluent standards on nutrient removal for all design scenarios. The simulated results in Figure 13.13 show that all scenarios preform as required according to EU-standards on nitrogen removal (a yearly average TN concentration lower than 10 mg/L). At the same time, effluent ammonium concentrations for all scenarios are at the high side. This is the direct result of the design choice for the upgrade within the existing aerobic activated sludge volume (including the current secondary settler volume). As a result of this design choice, the aerobic volume (aerobic sludge retention time) for all scenarios was limited. This resulted in overall higher effluent ammonium concentrations (calculated at the minimum design temperature of 9.5°C).

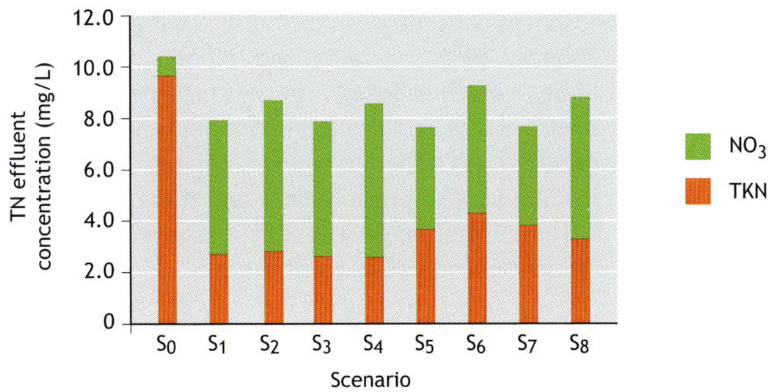

Figure 13.13 Scenario performances regarding the effluent TKN and nitrate concentration.

Figure 13.14 shows the combined effluent ammonium and BOD loads expressed as biological oxygen requirement (where ammonium is expressed in 4.57 $gTOD/gNH_4$). As a result of the upgrade and introducing full nitrification, all scenarios perform better than the current situation (S0). A closer look at the different design scenarios reveals that S6 is the least performing regarding nitrification because of limited aerobic volume. S6 could be improved at the expense of additional aerobic volume to this already in place (3,000 m^3). However, given the fact that for

S6 additional aerobic volume is required, this scenario is out of the scope of the upgrading scenarios. Regarding the scenarios S1 to S4 which include the use of membranes, it can be stated that in general such processes have better performance on removal of particulate BOD and TKN in comparison to conventional secondary settling. The main advantage of the membrane process is that the activated sludge process can be operated at higher and MLSS concentrations and therefore is less sensitive to the aerobic capacity. This results in an overall better performance on ammonium removal for all calculated scenarios involving membranes.

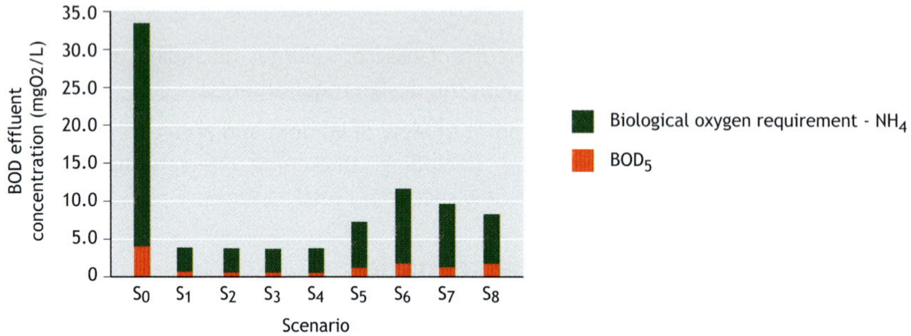

Figure 13.14 Scenario performances regarding the total biological oxygen requirement.

In disregard to the fact that in this study phosphorus removal was not included in the design requirements, effluent phosphate concentrations where also modeled (Figure 13.15). In none of the studied scenarios biological or chemical phosphorus removal was applied. Resulting removal of phosphorus is different as the result of different sludge production for different scenarios. A higher sludge production means that also more phosphorus is removed via the waste sludge, therefore less remains in the effluent and vice versa. Because scenarios S5 to S8 engage slightly lower SRT (to save volume in the secondary clarifier design) comparatively more sludge is produced and, therefore, also the effluent phosphate concentration for these scenarios is lower (S7 and S8). Scenario S6 preforms exceptionally well regarding the effluent phosphate concentration. This is the result of addition of lime which is dosed into the thickener/digester. From the anaerobic digestion model it was calculated that acidification in the thickener/digester unit occurred and therefore lime (calcium carbonate) was added to increase the pH. As a result, phosphorus was removed from the internal loading (due to precipitation) and thereby also the effluent phosphate concentration strongly decreased. Dozing of lime and metals would for all scenarios be equally possible to improve the effluent phosphate concentration.

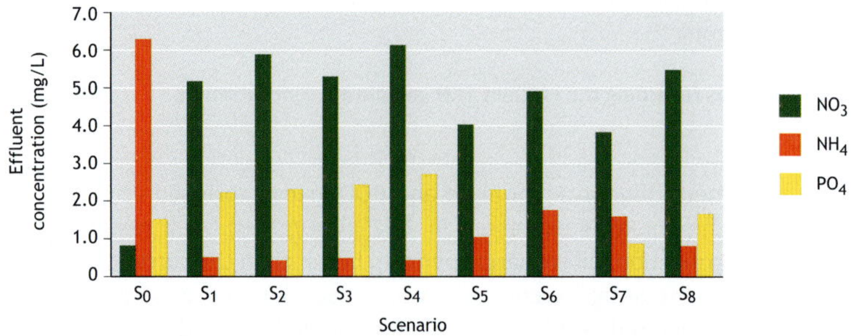

Figure 13.15 Scenario performance regarding the effluent nutrient concentrations.

13.7.2 SRT

To maintain nitrification all year round, including lowest winter temperature (for all scenarios calculated at 9.5°C), a minimal aerobic SRT of 8 days was aimed at for all scenarios. At 8 days SRT nitrification will occur at the most extreme winter conditions (at 8°C). This design requirement does not completely hold for scenarios S5, S6 and S7 as for these scenarios the total SRT including the anoxic zone is 8 days. This results in approximately 6.5 days aerobic SRT (Figure 13.16). As a result, for these scenarios, it would be required to also aerate the anoxic tank during the coldest winter conditions (estimated to occur several days per year). Therefore, in scenarios S5, S6 and S7 the anoxic tank should have standby aeration units installed. Because the planed anoxic (current pre-aeration) tanks already have aeration equipment installed, this would not be an additional investment. For scenarios S5, S6 and S7 during the coldest days of the year denitrification will be scarified for nitrification resulting in temporary higher nitrate (NO_3) effluent concentrations. The yearly average results of this dynamic operation were not calculated in this study. However, based on the average total nitrogen effluent simulations (below 7 mgN/L according to Figure 13.15 for scenarios S5, S6 and S7) and the measured (actual) minimum winter temperature (approximately 11.4°C), it is reasonable to assume that temporary/intermittent aeration of the anoxic tanks would be possible while maintaining the yearly average effluent standards for nitrogen below 10 mg/L.

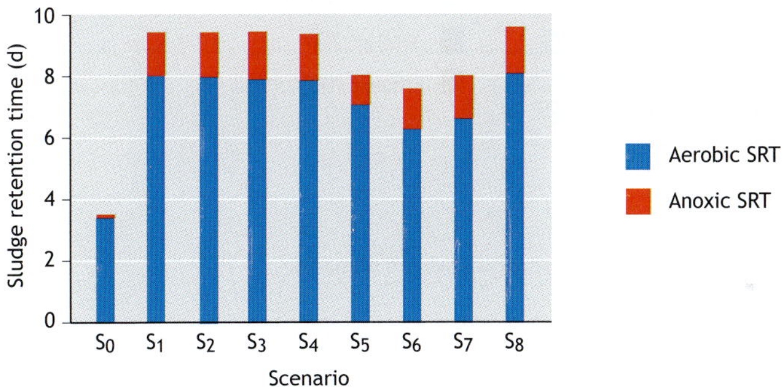

Figure 13.16 Scenario performance regarding the design SRT (aerobic SRT and anoxic SRT).

13.7.3 Sludge production

The main factors in determining the cost of the operation of a waste water treatment plant are energy consumption (mainly for aeration) and sludge production. Depending on the local legislation, processing, transport and disposal of waste sludge can take up as much as 50% of the total operating costs of a wastewater treatment plant. Especially when sludge has to be incinerated and/or transported over longer distances, as often is the case in the Netherlands, the costs of sludge disposal are significant (indicative 80 to 150 euro per ton of wet sludge). In general, there are two ways to operate a wastewater treatment plant to reduce the sludge production; application of primary settling and digestion and/or operating the activated sludge system at longer SRT. Where digestion generally delivers energy from the production of biogas/methane (CH_4), operating at longer SRT will cost energy in the form of additional electricity used to aerate the activated sludge over an extended period (oxidize the sludge to CO_2). On the other hand, investment costs for primary settling, digestion and facilities for energy production out of methane in general are higher than the investment costs needed for an aerobic volume extension. Also, operating a plant at longer SRT is more straightforward than operating a digestion system including all additional process units (sludge dewatering, chemical dosing installations, gas storage and processing equipment, measurements, energy producing facilities, etc.) The choice between the two scenarios depends on the investments costs and cost of energy (electricity) in combination with the availability of knowledge to operate the sludge digestion system. Also the quality of the waste activated sludge in relation to the available opportunities of disposal is decisive in the process selection. For example, for reuse of waste sludge in agriculture in general a different quality is required than when the sludge is transported to be incinerated. Because this study primary focuses on the technical possibilities for upgrading, no further advice is given on this issue. However, these considerations should be taken in account when finally selecting the preferred process.

The overall scenario performance on waste sludge production and the effects of aeration and methane production can best be shown from the plant wide mass balance of Theoretical Oxygen Demand (Figure 13.17).

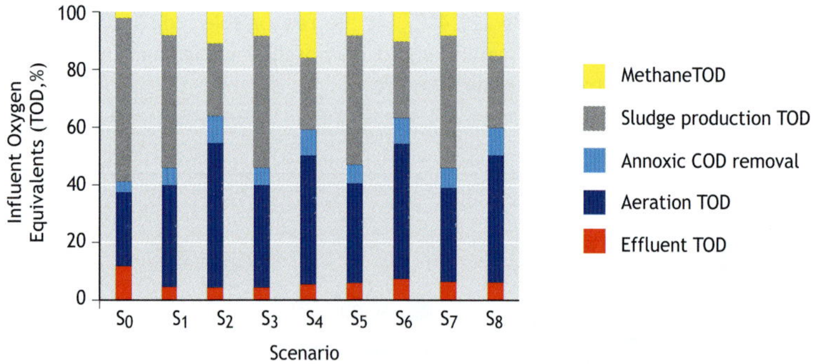

Figure 13.17 Scenario performances regarding the theoretical oxygen demand.

The TOD balance represents all in- and outgoing flows of organic materials, including methane. In the balance aeration (oxidation of organics) and sludge production are reversed proportional; when more aeration is applied (a longer SRT), less (waste) sludge is produced and vice versa. A similar relation applies to the differences between the scenarios with typical wastewater (S2, S4, S6, S8) and wastewater with a high inert COD fraction (S1, S3, S5, S7). Because the inert COD will not be oxidized it will accumulate in the plant. The only outlet for inert particulate COD from the plant is via the waste sludge (and a small fraction via the effluent). Because of this, the scenarios with high inert COD (S1, S3, S5 and S7) have a lower aeration input (34 to 36% TOD) and higher sludge production (45 to 46% TOD). For the scenarios with typical influent composition, more influent particulate COD will be oxidized resulting in higher a higher aeration input (46 to 51% TOD) and subsequently lower sludge production (24 to 26% TOD).

13.7.4 Primary settling and digestion impact on sludge production

Closer look at the TOD of only the sludge production (Figure 13.18) reveals that methane is produced in all scenarios. For the scenarios without digester (S1, S2, S5 and S6) this happens as the result of anaerobic activity in the sludge storage and thickener tanks. Also in the retrofit simulation (S0) it was calculated that the anaerobic activity in the thickener has a large (positive) effect on the operation of the activated sludge system and also actually improves the effluent quality.

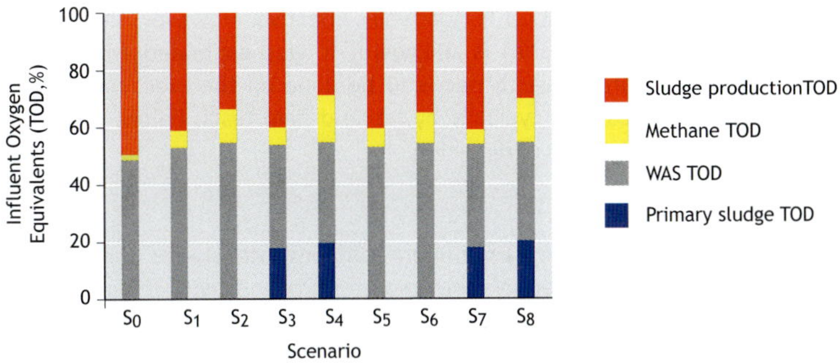

Figure 13.18 Scenario performances regarding the sludge production (expressed as TOD).

The most effective sludge reduction (of primary and secondary (WAS) sludge) and the highest methane production are achieved in scenarios S4 and S8, both operated with primary settling. On the other hand, the scenarios S3 and S7 also including primary settling and digesting hardly produce more methane than scenarios S1 and S3 (without primary settling). This shows that if the influent contains a high particulate inert COD fraction (S3 and S7) applying primary settling and digestion is ineffective from the perspective of reducing sludge (and producing energy). However, from the perspective of volume reduction, applying primary settling has a large positive effect on S7. In the same scenario without primary settling (S5), the aerobic volume needed to be increased by 3,000 m^3. As a result of primary settling in scenario S7, it is possible to obtain effluent EU-standards within the already available volume. This observation underlines the general conclusion that, from the perspective of methane and energy production, it is not effective to apply primary settling and digestion when the influent contains a high fraction of inert particulate COD. On the other hand however it is also shown that primary settling effectively helps to reduce the required aerobic volume of an activated sludge system when the aerobic residence time of the activated sludge system is limiting for nitrification (typically under winter conditions).

13.7.5 Digester design

In the start-up situation, the plant was originally operated with primary settling and two sludge digesters. After the first upgrade to the current situation, the two original primary settling tanks were re-assigned to pre-aeration tanks and the two digesters where reused as sludge storage tank and WAS thickener. In this study, it is assumed that the thickener and sludge storage tanks can be re-assigned to digester. To maintain optimal methane production, the digesters should be operated at approximately 21 days (sludge) retention time. Therefore, the daily mass and volume inflow into the digester of both primary and secondary sludge needs to be controlled. Overloading the digester will reduce the SRT and thereby decrease methane production up to the point where methanogens will be severely affected. Overloading the digester with organic solids (VSS) will result in increasing volatile fatty acids production (VFA) by acidification. When the buffer capacity of the system is reached, this results in a pH drop which causes reduction or even stop of methane production (due to pH toxicity). To control the volume and solids input, for the primary sludge a new gravity thickening unit is included, and for the secondary sludge, mechanical dewatering is applied. The scenarios show that the current digester volume (if 2 tanks are used, 4,200 m^3) is sufficient for most scenarios. This study shows that the production of methane is strongly dependent on the biodegradability of the influent particulate COD. With the high inert influent composition resulting from the retrofit design study (S0), the expected methane production is in the range of 1,000 kgCOD/d. This is a rather small recovery. It could be concluded that under these conditions digestion is not really effective. However, the precise composition of the influent particular material is unknown; it was estimated from other (activated sludge) data. Therefore, more research needs to be done on the exact nature and biodegradability of the influent composition, typically by lab and field tests, before definitive conclusions can be drawn on the effectiveness of digestion.

13.7.6 Nitrogen removal

Figure 13.19 shows the total nitrogen balance (raw influent, effluent total sludge removed). In the figure, the total nitrogen influent loading equals the amount nitrogen leaving the system via the effluent (the total of nitrate, ammonium, soluble and particulate TKN), the amount of nitrogen leaving the system via the sludge (as organic bounds nitrogen) and the amount of nitrogen leaving the system as a result of denitrification (N_2 via the gas phase). As a result of the mass balance, for all scenarios with a higher sludge production also less nitrogen remains to be denitrified. The figure shows how this accounts for all scenarios with a relative high sludge production (high inert influent COD fractions; S1, S3, S5, S7). It is also observed that denitrification is not volume-limited. A relative small anoxic volume is sufficient as the result of the high soluble influent biodegradable COD. The plant is therefore very suitable to introduce denitrification, which does not only improve the effluent quality, but also reduces the aeration energy requirement as the result of anoxic COD removal.

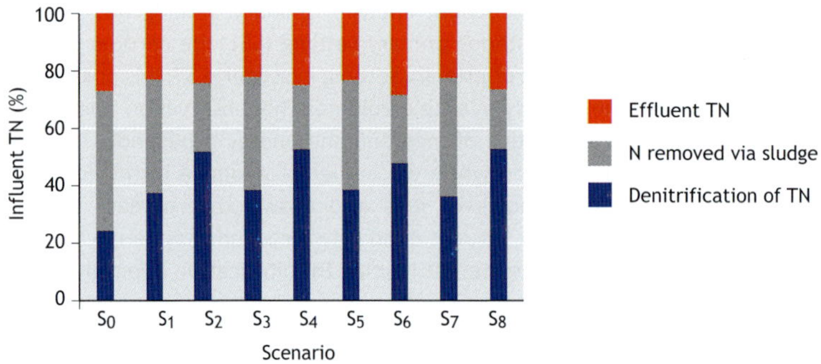

Figure 13.19 TN mass balance evaluation; percentage of TN removed via denitrification (N2), via the effluent and via the sludge.

13.7.7 Internal (nitrogen) loading

Figure 13.20 shows the percentages of nitrification and denitrification relative to the total nitrogen influent load (before primary settling). Denitrification follows on nitrification and therefore is less than (or equal to) nitrification. Nitrification is highest for the scenarios with typical influent (S2, S4, S6 and S8). Figure 13.21 shows that for these scenarios also the sludge production is lower and the aeration requirement higher. All these effects are co-related; with typical influent the biodegradable COD in the influent is higher. Therefore, more COD is oxidized (requiring more air) or digested (resulting in more methane production). When organic material is oxidized or digested, the organic bound nitrogen is released as ammonia (in the activated sludge system or the digester). Eventually, this organic released ammonia is nitrified leading to more air requirement. In other words, high biodegradable influent leads to more organics degraded, more organic nitrogen released, more nitrification, a higher air requirement and less (organic) waste sludge production. Increased nitrification as a result of the degrading organic material is indicated as internal nitrogen (ammonium) loading. The internal N-loading for scenarios S4 and S8 is the highest (also is nitrification, Figure 13.20). This is the result of the most effective sludge reduction by digestion (also the highest methane production). The difference with scenarios S2 and S6 (comparable influent however without digestion but with secondary sludge thickening) is small. This is caused by the fact that also during the thickening anaerobic processes take place, resulting in methane production and release of organic nitrogen. Ammonia release in the digester or thickeners leads to an increased (internal) ammonium loading of the activated sludge system via the internal or dirty water flows. The difference in methane production (a little bit smaller for scenarios S2 and S6 than for scenarios S4 and S8) is also expressed in the aeration requirement being slightly higher for scenarios S2 and S6.

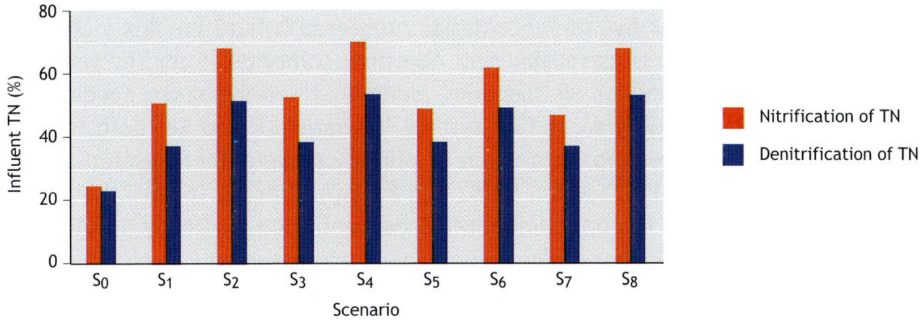

Figure 13.20 Scenario performances regarding nitrification and denitrification as percentage of TN.

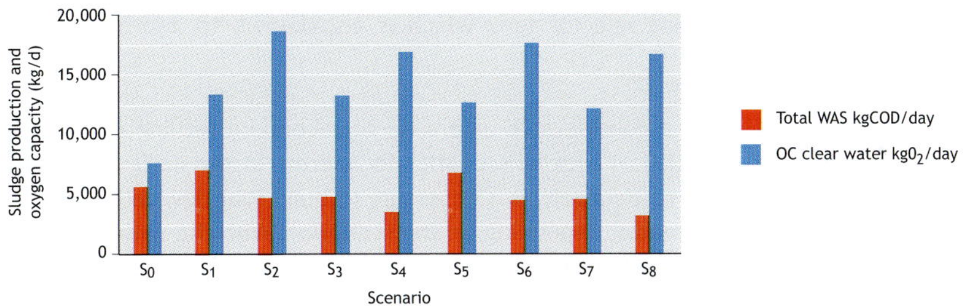

Figure13.21 Scenario performances regarding the sludge production (kg/d) and aeration requirement (kgO_2/d).

For all scenarios the internal loading (dirty water) from the sludge processing (for both thickening and digestion) is considerable regarding ammonia (and also for ortho-phosphate). It confirms that the internal loading needs to be taken in account in the process design. Most effectively, this is calculated using an anaerobic digestion model. In particular, when the internal load from the sludge dewatering is high, the dirty water flow should be correctly taken into account. Firstly, it is necessary to evenly distribute the dirty water flow over the activated sludge lines. This will reduce misbalance in aeration requirement and will improve the effluent quality regarding nutrients. Moreover, the dirty water flow should be redirected to the inflow point of activated sludge system, together with the settled influent and return sludge flow. Often, the internal dirty water flow is redirected into the primary settling tanks with the aim to remove the solids from the sludge dewatering. However, solids from sludge dewatering often will not effectively be removed in the primary settlers. More important, feeding the high nutrient concentrated water from the sludge processing to the primary settling tanks leads to buffering (concentration increase) of nutrients in the primary settlers during (nightly) low influent loading. In the morning, when usually influent flow (DWF) increases, this often results in an ammonium (and phosphate peaks) loaded to the activated sludge system, thereby deteriorating the effluent quality. It is therefore advisable to feed the internal loading (dirty water including TSS) directly to the activated sludge system. Favorably, this should be done during the night when the dry weather flow load is low and activated sludge aeration capacity is not fully used.

13.8 Scenarios regarding secondary settling

Tables 13.13 provides input data for the settler design scenarios. An elaborate explanation on the methodology is found in Meijer and Brdjanovic (2012). Table 13.14 shows the secondary settler design results. The different calculation methods give different results on required settling area. The mean value of the five methods design presented (where the flux method is applied without safety factor) seems as reliable and objective compromise for the required settler design extension. Commonly for all scenarios, additional settling tanks need to be constructed. Out of all scenarios with secondary settling, scenario S7 and S8 (both include primary settling), seem the best scenario. As a result of primary settling, the activated sludge system can be operated at lower MLSS; 5 kg/m^3 for S7 and 4.5 kg/m^3 for S8. To keep the design within the current reactor volume, the operational MLSS concentration is already on the high side. For scenarios S5, S6 and S7 a higher sludge volume index was assumed (SVI of 150 mL/g). This was done based on the lower SRT and the fact that during the winter the anoxic tank would be aerated. Thereby it is assumed this could negatively influence sludge settleability. Scenario S8 was designed at a standard SVI for nutrient removing wastewater treatment plant of 130 mL/g. The lower SVI design (ranging 130 to 150) is justified based on the fact that also chemical phosphorus removal is applied. This will in general improve sludge flocculation and settleability. Moreover, all designs have an anoxic selector tank which is also known to improve sludge settleability.

In general it can be concluded that a considerable extension of secondary settling capacity is required (approximately 3,000 to 4,000 m^3) for all scenarios to operate at the design flow and load of 28.000 m^3/d and 140.000 PE respectively. It is necessary to build four new settling tanks (of diameter ranging from 31 to 37 meter, depending on the scenario). With the extension, the effluent solids concentration is expected to drop to estimated value between 5 and 15 mg/L.

Table 13.13 Clarifier design model input parameters for scenarios S5 and S6, S7, and S8.

General Parameters	Code	Unit	S5 & S7	S7	S8
Measured SVI 30 min	SVI_{30}	mL/g	150	0	0
Sludge concentration	MLSS	kg/m^3	5.2	5.2	5.2
Average Dry Weather Flow	ADWF	m^3/h	1,167	1,167	1,167
Peak Flow - % of ADWF	PDWF	%	150%	150%	150%
Peak Dry Weather Flow	PDWF	m^3/h	1,750	1,750	1,750
Peak Flow - % of ADWF	PWWF	%	250%	250%	250%
Peak Wet Weather Flow	PWWF	m^3/h	2917	2917	2917
Flux theory					
Initial settling velocity	v_0	m/h	6.1	6.1	6.1
Hindered settling parameter	p_{hin}	m^3/kg	0.42	0.42	0.42
Safety factor on area	F_{corr}	-	1.0	1.0	1.0
WRC					
Stirred SVI 3.5 g/L	$SSVI_{3.5}$	mL/g	84	84	84
Safety factor on area	F_{corr}	-	1.0	0.0	0.0
ATV-STOWA					
Diluted SVI	DSVI	mL/g	126	126	126
DSVI*MLSS	DSV30	mL/L	657	657	657

Table 13.14 Overview of the clarifier design scenarios S5 and S6, S7, and S8.

		Scenario S5 and S6						Scenario S7						Scenario S8					
Design Extention	Unit	Empirical	Flux	WRCm	ATV	STOWA	MEAN	Empirical	Flux	WRCm	ATV	STOWA	MEAN	Empirical	Flux	WRCm	ATV	STOWA	MEAN
Disigned Surface	m^2	2,831	4,214	3,156	7,246	3,353	4,160	2,722	3,875	2,920	6,875	3,224	3,923	2,450	2,830	1,907	4,928	2,762	2,975
Current Surface Available	m^2	0	0	0	0	0	0	0	0	0	0	0	0	0	0	0	0	0	0
Additional Surface Required	m^2	2,831	4,214	3,156	7,246	3,353	4,160	2,722	3,875	2,920	6,875	3,224	3,923	2,450	2,830	1,907	4,928	2,762	2,975
Extended Surface	extention	0%	0%	0%	0%	0%	0%	0%	0%	0%	0%	0%	0%	0%	0%	0%	0%	0%	0%
Diameter for Additional Tank	m	37	37	37	37	37	37	36	36	36	36	36	36	31	31	31	31	31	31
Number of Tanks Applied	tank units	2.7	4.0	3.0	6.9	3.2	4.0	2.8	3.9	3.0	6.9	3.3	4.0	3.2	3.8	2.5	6.5	3.7	3.9
At ADWF																			
Overflow, QADWF	m^3/h	1,167	1,167	1,167	1,167	1,167	1,167	1,167	1,167	1,167	1,167	1,167	1,167	1,167	1,167	1,167	1,167	1,167	1,167
Overflow rate to surface	m/h	0.4	0.3	0.4	0.2	0.3	0.3	0.4	0.3	0.4	0.2	0.4	0.3	0.5	0.4	0.6	0.2	0.4	0.4
Recycle flow (QRAS)	m^3/h	2,100	1,103	3,156	2,407	2,407	2,235	2,100	1,050	2,920	2,244	2,244	2,111	2,100	963	1,907	1,551	1,551	1,614
Recycle rate to surface	m/h	0.7	0.3	1.0	0.3	0.7	0.6	0.8	0.3	1.0	0.3	0.7	0.6	0.9	0.3	1.0	0.3	0.6	0.6
Recycle ratio (RAS/INF)	-	1.8	0.9	2.7	2.1	2.1	1.9	1.8	0.9	2.5	1.9	1.9	1.8	1.8	0.8	1.6	1.3	1.3	1.4
RAS TSS concentration	kg/m^3	8.1	10.7	7.1	7.7	7.7	8.3	7.8	10.6	7.0	7.6	7.6	8.1	7.0	10.0	7.3	7.9	7.9	8.0
Solids loading rate	$kg/m^2 \cdot h$	6.0	2.8	7.1	2.6	5.5	4.8	6.0	2.9	7.0	2.5	5.3	4.7	6.0	3.4	7.3	2.5	4.4	4.7
At PDWF																			
Overflow, QPDWF	m^3/h	1,750	1,750	1,750	1,750	1,750	1,750	1,750	1,750	1,750	1,750	1,750	1,750	1,750	1,750	1,750	1,750	1,750	1,750
Overflow rate to surface	m/h	0.6	0.4	0.6	0.2	0.5	0.5	0.6	0.5	0.6	0.3	0.5	0.5	0.7	0.6	0.9	0.4	0.6	0.6
Recycle flow (QRAS)	m^3/h	2,100	1,103	3,156	2,407	2,407	2,235	2,100	1,050	2,920	2,244	2,244	2,111	2,100	963	1,907	1,551	1,551	1,614
Recycle rate to surface	m/h	0.7	0.3	1.0	0.3	0.7	0.6	0.8	0.3	1.0	0.3	0.7	0.6	0.9	0.3	1.0	0.3	0.6	0.6
Recycle ratio (RAS/INF)	-	1.2	0.6	1.8	1.4	1.4	1.3	1.2	0.6	1.7	1.3	1.3	1.2	1.2	0.6	1.1	0.9	0.9	0.9
RAS TSS concentration	kg/m^3	9.5	13.5	8.1	9.0	9.0	9.8	9.2	13.3	8.0	8.9	8.9	9.7	8.3	12.7	8.6	9.6	9.6	9.7
Solids loading rate	$kg/m^2 \cdot h$	7.1	3.5	8.1	3.0	6.4	5.6	7.1	3.6	8.0	2.9	6.2	5.6	7.1	4.3	8.6	3.0	5.4	5.7
At PWWF																			
Overflow, QPWWF	m^3/h	2,917	2,917	2,917	2,917	2,917	2,917	2,917	2,917	2,917	2,917	2,917	2,917	2,917	2,917	2,917	2,917	2,917	2,917
Overflow rate to surface	m/h	1.0	0.7	0.9	0.4	0.9	0.8	1.1	0.8	1.0	0.4	0.9	0.8	1.2	1.0	1.5	0.6	1.1	1.1
Recycle flow (QRAS)	m^3/h	1,458	3,442	3,156	2,407	2,407	2,574	1,458	3,121	2,920	2,244	2,244	2,397	1,458	2,538	1,907	1,551	1,551	1,801
Recycle rate to surface	m/h	0.5	0.8	1.0	0.3	0.7	0.7	0.5	0.8	1.0	0.3	0.7	0.7	0.6	0.9	1.0	0.3	0.6	0.7
Recycle ratio (RAS/INF)	-	0.5	1.2	1.1	0.8	0.8	0.9	0.5	1.1	1.0	0.8	0.8	0.8	0.5	0.9	0.7	0.5	0.5	0.6
RAS TSS concentration	kg/m^3	15.6	9.6	10.0	11.5	11.5	11.6	15.0	9.7	10.0	11.5	11.5	11.5	13.5	9.7	11.4	13.0	13.0	12.1
Solids loading rate	$kg/m^2 \cdot h$	8.0	7.8	10.0	3.8	8.3	7.6	8.0	7.8	10.0	3.8	8.0	7.5	8.0	8.7	11.4	4.1	7.3	7.9

13.8 Conclusions and recommendations on the modelling study

It can be concluded that Varaždin WWTP was modeled successfully. The plant information was checked and the operation could be confirmed by the model simulations with reasonable accuracy. Some information needed assumptions because it was missing or inaccurate. Within this context, the plant retrofit was performed successfully, resulting in meaningful results based on which the project goals could be accomplished and a preferred design scenario was developed to upgrade the plant to EU effluent limits. The proposed scenario was planned in time for the most efficient use of resources and includes; extension of the secondary clarifiers, enlargement of the aeration by using existing infrastructure and finally in the future when the influent loads are to a level this is necessary, introduction of primary settling and sludge digesting. The calibrated model was successfully used to calculate several future scenarios; on influent loading, plant operation and scenario designs. The presented case study serves as demonstration project for model-based plant evaluation, (scenario) design and upgrading scenarios. This case study clearly shows that the application of plant-wide modeling, including sludge treatment and secondary settling, revealed details in plant design decisive for effective plant operation and design. These details would have been missed when only traditional plant design would have been applied and thereby would have resulted (as this case proves it) in operational failure or in plant design with too large safety resulting in increased building costs. Compared to this, the costs of a model-based study are much smaller (see Chapter 1). The results of this study (the calibrated model) can be used for testing other scenarios then here presented and thereby also can help in current and future plant operation of WWTP Varaždin. Moreover, the study indicates several upgrading solutions to obtain EU effluent standards. Possible future operational problems are indicated and solutions are presented to overcome these problems when these should occur. Efficient options for upgrading are suggested optimally using the existing infrastructure thereby reducing development costs and also the proposed improvements were planned in time for the most effective use of resources.

Currently the hydraulic loading of the plant is well under the original design loading of 28,000 m^3 per day (140,000 PE). Upgrading as suggested to load of 200.000 PE does not seem necessary. Regarding the poor current effluent quality, a large positive environmental impact can be obtained by upgrading the plant to 140.000 PE (the current design capacity) however, including two new settlers and full nutrient removal. It is concluded that influent is strongly determined by the industrial discharges, however, for most design parameters the influent is within the range of 'normal' domestic sewage. The exception is a high inert particulate COD fraction that is suspected based on some measurements in 2011 which negatively influences the operation (also after the upgrade). Also there are known problems with grease and fibers in the influent which both can be resolved by better pre-treatment. Therefore primary settling is proposed. However, the most effective upgrade strategy is to start by extending the secondary settlers and introducing aeration in the old setter volume to introduce nutrient removal. Two additional settlers of approximately 31 meter diameter each, need to be built extra for the current (hydraulic) plant loading. The proposed upgrade optimally uses the current tank volumes and is a cost efficient option of upgrading the plant to EU effluent results.

Expectedly, not all information needed for this study was available and sometimes older data were used. Where estimation was needed, this was done on founded reasoning and literature references. Most data were validated using different information sources. Model calculations were used to evaluate the data on consistency however, on several points data inconstancies were found leaving space for interpretation of the here presented results. The presented results need to be judged and used taking this in account. However, within this context, the plant retrofit was performed successfully, resulting in meaningful results leading to the desired project goals and developing a detailed design scenario for upgrading.

Main conclusions regarding the influent flow and measurements can be summarized as:

- Currently the hydraulic loading of the plant is well under the original design loading of 28.000 m^3/day.
- From the influent data it can be concluded that extending the current biological section to capacity of 200.000 PE does not seem necessary in short term. More realistic is the upgrade of the plant to 140.000 PE including settler extension and nutrient removal.
- From the influent analysis it can be concluded that the influent is strongly determined by industrial discharge however, most design parameters fit within the range of 'normal' for domestic sewage.
- An exception is the inert particulate COD level that needed to be increased to fit the model to the measurements. This parameter needs to be further investigated by measurements. Because this parameter in some aspects influences the scenario study, it was incorporated in the design scenarios; all scenarios were calculated for typical and high inert COD.
- Also an exception is the peak wet weather flow which is low relative to the dry weather flow. It needs to be investigated if this is justified for settler design purposes.
- Also it needs to be investigated if currently sewage is bypassed untreated into the recipient during high loading or rain conditions.

Furthermore, some conclusions can be made based on the retrofit study of the current situation:

- The SRT of the current plant is approximately 3.4 days (for both the North and South lane) indicating a typical BOD removing system with partial nitrification in the summer.
- The North and South lanes have similar biomass composition and are completely mixed because of the return sludge flow.
- No evidence was found for the assumed North:South 40:60 influent loading distribution over the two lanes.
- No evidence was found that the South lane is performing better because of the biofilm reactor than the North line using conventional activated sludge.
- All design methods show that the current settler operation is overloaded; both hydraulically and by solids loading.
- Extension of the current secondary settlers is necessary to improve the effluent quality.
- Two additional settlers of approximately 31 meter diameter each, need to be built under the current (hydraulic) plant loading conditions.

Regarding the primary settling and digester design, it can be concluded that:

- The current volume of the sludge thickener and storage (total 4,200 m^3) would be sufficient to apply digestion for most considered scenarios.
- If digestion is applied, also new primary thickening and secondary sludge dewatering need to be applied, together with dewatered sludge storage tanks.
- The expected production of methane is strongly dependent on the biodegradability of the influent particulate COD. This requires further investigation.
- The net additional sludge removal when digestion is applied is limited to approximately 1,500 kgTSS/d.
- For all scenarios, the internal loading from the sludge processing (for both thickening and digestion) is considerable regarding ammonia (and also ortho-phosphate) and therefore should be taken in account.
- Dirty water should be evenly distributed and directed into the activated sludge lines together with settled influent and return sludge flow.

Finally, conclusion on gravity settling versus membrane filtration are:

- In general, it is concluded that a considerable extension of secondary settling capacity is required (approximately 3,000 to 4,000 m^3) for all settling scenarios.
- To operate at the desired design requirement of 140,000 PE (28,000 m^3/d), it would be necessary to build four new settling tanks (ranging 31 to 37 meter).
- All the scenarios using membranes can handle the design loading of 28,000 m^3/d well within the existing activated sludge volume. Applying membranes would be an option when area for settlers is not available.
- For the settler scenarios, the effluent solids concentration is expected to drop to estimated value between 5 and 15 mg/L. Membranes will have even better performance, however on the expense on higher operational costs.

Recommended steps for extension/upgrade can be summarized as follows:

- Extending the secondary settling capacity by addition of two new tanks. A footprint should be secured for two additional settling tanks.
- Re-assignment of the current secondary settling tanks to aerobic activated sludge tanks to increase total aeration capacity for nitrification.
- Re-assignment of the current pre-aeration tanks to anoxic tanks for nutrient removal (denitrification).
- Inclusion of primary settling, primary thickening and sludge digestion.
- Extending the secondary settlers by two additional tanks reaching the full hydraulic capacity of 28.000 m3/d at DWF (140.000 PE). For 140.000 PE the total settler volume should be approximately 4,000 m^3.

13.9 Multi-criteria scenario analysis

Based on the results of modeling study of the current situation (S0) and 8 upgrade scenarios (S1-8) and in order to complete technological comparison with financial analysis, a multicriteria analysis was applied using the follwoing criteria:

1. Removal efficiency;
2. Investment costs;
3. Operational costs;
4. Complexity of the process and operations;
5. Process stability and robustness;
6. Footprint;
7. Upgrade implementation complexcity;
8. Sludge generation.

Here it is to be noted that this is basically a comparison of 4 scenarios for each of the two offered options, i.e. Option 1: High content of inorganic particulate COD fraction in the influent (scenarios S1, S3, S5 and S7), and Option 2: Typical content of inorganic particulate COD fraction in the influent (scenarios S2, S4, S6, and S8).

As the comparison of scenarios of two different options makes no sense, the multi-criteria analysis will be made in parallel for both scenarios.

Each of the proposed criteria is described in short below, including a comment on the specifics of each scenario.

13.9.1 Treatment efficiency

Treatment efficiency is a criterion providing comparison between the achieved level of treatment (expressed as the concentration of selected parameters in the WWTP effluent) calculated under the model and the maximum allowable concentrations in the effluent laid down by the law. The scenario with better overall treatment efficiency has an advantage in selection in the multi-criteria analysis.

The main project requirement is to achieve the EU effluent quality standards in nutrient removal for all the scenarios and options (in addition to the removal of organic matter and suspended solids). The results obtained by the model are presented in Annex 13.3, showing that all the scenarios and options comply with EU standards concerning the removal of organic matter and suspended solids and nitrogen. The EU requirements are met in every scenario. As already mentioned, the application of additional chemical removal of phosphorus in every scenario (where required) has been adopted. It is therefore assumed that the phosphorus concentration in the effluent of less than 1 mg/L is achieved in every scenario. Scenarios S1 through S4 include the application of membrane bioreactors (MBR). This analysis also confirms that their application achieves better wastewater treatment effects compared to conventional processes with secondary sedimentation tanks. The main advantages of the membrane technology are the following: an activated sludge process can be implemented at higher MLSS concentrations, which makes it less restricted in terms of the capacity of an aeration tank; high efficiency and stability of removal of suspended solids. This results in a generally better effectiveness of ammonia removal for every scenario that includes MBRs. However, as the criterion is met by all the scenarios, a slight advantage has been given to MBR scenarios, resulting in the following ranking: Option 1: S3, S1, S5 and S7, and Option 2: S4, S2, S8 and S6.

13.9.2 Investment costs

Investment costs is a criterion providing comparison between the required investment costs with regard to additional construction works and the required investment for additional mechanical and electrical equipment for each analysed scenario. The scenario with the lowest total investment costs has an advantage in selection in the multi-criteria analysis.

Estimate of construction works was made on the basis of the extensions required as presented in Annex 13.3: Additional volume in relation to the current situation S0, as presented in Figure 13.11.

In terms of equipment, the justification is as follows:

- Scenarios S1 and S2: Sophisticated equipment for MBR (modules, washing and maintenance equipment) is foreseen, and additional equipment for existing digesters as well;
- Scenarios S3 and S4: Sophisticated equipment for MBR (modules, washing and maintenance equipment) is foreseen, equipment for primary sedimentation tanks, primary thickener, equipment for existing digesters, and additional equipment for new digesters;
- Scenarios S5 and S6: Additional equipment for aeration (equipping additional tanks for S5) is required as well as additional equipment for new secondary sedimentation tanks, equipment for existing digesters, and additional equipment for new digesters;
- Scenarios S7 and S8: Equipment for primary sedimentation tanks and primary thickener is required, additional aeration equipment, additional equipment for new secondary sedimentation tanks, equipment for existing digesters, and additional equipment for new digesters (the same as for S3 and S4).

Based on the presented information and description, an approximate cost estimate has been prepared for the purpose of comparing the scenarios under the relevant criterion (investments required). Cost estimates for all the analysed scenarios are summarized in Table 13.15.

Table 13.15 Total investment (EUR) for each of the 8 analysed scenarios.

Scenario	S1	S2	S3	S4	S5	S6	S7	S8
Total price in EUR	3,655,430	3,811,430	5,198,880	5,414,880	4,385,130	4,011,130	5,058,130	4,918,130
Ranking	1	1	4	4	2	2	3	3

Ranking under this criterion for Option 1 is as follows: S1, S5, S7 and S3, and for Option 2: S2, S6, S8 and S4.

13.9.3 Operational costs

Operational costs are a criterion providing comparison between the required operation and maintenance (O&M) costs for each analysed scenario. The scenario with lower total operational costs has an advantage in selection in the multi-criteria analysis.

For comparison purposes, only the costs related to biological treatment and sludge treatment have been taken into consideration, because the main reconstruction works were analysed exclusively for that part of the WWTP. It has to be noted that this is an approximate estimate of costs based on the experience from similar WWTPs or similar projects, taking into account the specifics of each scenario. O&M costs are provided in Table 13.16, noting that additional costs were taken into consideration in scenarios S1 through S4 with MBRs, which have to be replaced periodically (every 4-5 years or longer if maintenance is at the optimum level) and are covered by increased percentage for regular equipment maintenance.

Table 13.16 Total O&M costs for each of the 8 analysed scenarios.

Scenario	S1	S2	S3	S4	S5	S6	S7	S8
Total costs (EUR/y)	987,001	1,011,008	1,028,374	1,028,757	961,858	961,466	978,832	989,537
Costs (EUR/PE)	7.05	7.22	7.35	7.35	6.87	6.87	6.99	7.07
Costs (EUR/m^3)	0.119	0.121	0.124	0.124	0.116	0.116	0.118	0.119
Ranking	3	3	4	4	1	1	2	2

Ranking under this criterion for Option 1 is as follows: S5, S7, S1, S3, and for Option 2: S6, S8, S2 and S4.

13.9.4 Technological complexity and maintenance

Technological complexity and maintenance is a criterion providing comparison between the degree of complexity of the applied processes of wastewater treatment and treatment of the generated sludge for each analysed scenario. The scenario with lower technological complexity and need for maintenance has an advantage in selection in the multi-criteria analysis.

The simpler the processes and accompanying equipment, the easier their maintenance, which requires lower qualifications of WWTP staff. On the other hand, the need for maintenance grows as the process becomes more complex. As a rule, due to an insufficient number of qualified staff, conventional plants are in our practice still considered a more favourable solution than newer and more sophisticated plants. Even though these particular solutions can be considered simple because of increasing automation of the treatment processes and possibilities for their full monitoring, this advantage disappears when there is no available staff trained specifically for the management of modern plants. In case of the Varaždin WWTP, the current WWTP staff is thought to be able to manage complex wastewater and sludge treatment processes, with additional training.

The complexity of individual processes for the offered scenarios can be analysed by individual treatment units in the wastewater line and in the sludge line:

- The anoxic section is identical for all the scenarios; comparison does not yield any differentiation;
- In technological terms, there are also no differences between aeration tanks; consequently, no comparison can be made;
- Secondary sedimentation tanks are a relatively simple technological unit, which gives advantage to scenarios S5 through S8; scenarios S1 through S4 with membrane modules are more complex and require more demanding maintenance, regardless of a high degree of automation and stable efficiency. However, if problems occur in secondary sedimentation tanks (such as bulking sludge, denitrification in secondary sedimentation tanks, etc.), then their management or resolution becomes a highly complex task. In such case MBR systems become very interesting because they are not significantly affected by such problems;
- The scenarios which in the technical process also have units for primary sedimentation are by definition more complex since they have an additional technological unit; these are scenarios S3, S4, S7 and S8;
- WWTPs with (in particular heated) anaerobic digesters are generally considered more complex and demanding in terms of maintenance; in this specific case they are used in every scenario, so there are no marked differences between the scenarios in this sense either. A slight advantage can here be given to scenarios S1, S2, S5 and S6, as the remaining scenarios have an additional thickener and additional digesters in the sludge treatment line.

Based on the above, Option 1 scenarios are ranked in order: S5, S7, S1 and S3, and Option 2 scenarios: S6, S8, S2 and S4.

13.9.5 Operating stability and robustness

Operating stability and robustness is a criterion providing comparison between operating characteristics of the applied wastewater treatment processes and equipment and the treatment of generated sludge for each analysed scenario. The scenario with higher operating stability and robustness has an advantage in selection in the multi-criteria analysis.

The process stability criterion takes into account the risk of biological, chemical or physical processes deviating from the desired process conditions or failing completely. For the operation of a WWTP which is foreseen to operate 24 hours a day, more stable processes and equipment contributing to this stability are preferred. While the process stability for conventional activated sludge plants is relatively high, MBR systems have made a step forward, first of all because they operate reliably and with high concentrations of organic matter in reactors, and they are less sensitive to changes in the inflow and quality of incoming wastewater. Systems without primary sedimentation tanks are in principle simpler and more stable in operation, while for plants of higher capacity their establishment is considered an advantage. In this specific case, primary sedimentation tanks can contribute to process stability with regard to the measured influent quality and more than significant share of industrial wastewater in the total inflow to the WWTP.

Anaerobic digestion, which is applied in all the scenarios, shall be carefully managed in order to avoid disturbances in the process. It should be added here that in terms of operating stability, particularly for the sludge line, advantage is given to scenarios S1 and S2, as they leave a possibility of using the existing sludge drying beds in case of failure in the operation of digesters or mechanical dehydration.

Based on the above, Option 1 scenarios are ranked 1: S3, S7, S1 and S5, while Option 2 scenarios are ranked: S4, S8, S2 and S6.

13.9.6 Required space

Required space is a criterion providing comparison between the footprint of the WWTP needed to accommodate additional process units and equipment for wastewater treatment and treatment of the generated sludge for each analysed scenario. The scenario with a lower need for space has an advantage in selection in the multi-criteria analysis.

As the project concerns reconstruction of the existing plant, a very important criterion for the evaluation of scenarios is the need for space. Under this criterion, the most favourable solution is the one with which the required effects can be achieved with maximum use of the existing structures, while the least favourable solutions are those that require additional space for new structures, particularly if that space cannot be provided at the WWTP location and land acquisition is needed. In this particular case, all the land required for WWTP extension is available and owned by the local sewerage company VARKOM.

For each scenario and option, pre-aeration tanks will be used fully as an anoxic tank (required volume for denitrification). Scenarios S1 and S2 are the most favourable in terms of using the available tank volume, and also in terms of the required extension area. So, these scenarios require no additional space for aeration nor for primary and secondary sedimentation, which is the result of application of membrane bioreactors.

Scenario S5 (Option 1) requires the largest additional aeration volume, as well as additional secondary sedimentation tanks, while scenario S7 (Option 1) is the least favourable in terms of requirements for additional space and increased volume. This scenario also includes primary sedimentation tanks (including a primary thickener, additional sludge dehydration, energy generation equipment, and tanks) and additional secondary sedimentation tanks.

Scenarios S6 and S8 (Option 2) are slightly more favourable in relation to S5 and S7 in terms of additional needs for space.

Figure 13.10 and Figure 13.12 clearly present the results in terms of the space required.
Based on the above, Option 1 scenarios are ranked 1: S1, S3, S5 and S7, while Option 2 scenarios are ranked: S2, S4, S6 and S8.

13.9.7 Upgrade implementation complexity

Upgrade implementation complexity is a criterion providing comparison between the feasibility of extension and installation of additional process units and equipment for wastewater treatment and treatment of generated sludge in terms of efforts for continuous WWTP operation during such works for each analysed scenario. The scenario with lower complexity and better implementation method has an advantage in selection in the multi-criteria analysis.

This criterion takes into account whether the proposed treatment process has exceptionally high demands in terms of WWTP reconstruction. This first of all implies all the demands concerning the execution of works of particularly high quality (e.g. high precision, complex procedures, dangerous work, etc.). Here it is important to have in mind that this is a matter of reconstruction of the existing plant which is in continuous operation. Advantage is given to the scenarios which require the lowest scope of construction interventions on the existing structures and which enable phased WWTP upgrade, i.e. make it possible for the WWTP to be partly used even during the execution of works (phased reconstruction by lines).

In that sense, scenario S6 is the most favourable as it implies the lowest scope and the simplest interventions on the existing structures. Every required extension can be done without obstructing the flow of wastewater through the WWTP and its treatment. The situation is also very similar with scenarios S7 and S8, where the most sensitive element is re-connecting the primary sedimentation tanks, as this implies certain works at the entry into the biological part of the plant. The situation is slightly more complex when it comes to S5 due to the extension of aeration tanks. Scenarios S1-S4 are very demanding in this sense when it comes to organization of works, as the major share of the works and interventions is done actually on the existing structures (which is the result of efforts to maximise the use of the existing structures).

Based on the above, Option 1 scenarios are ranked 1: S7, S5, S1 and S3, while Option 2 scenarios are ranked: S6, S8, S2 and S4.

13.9.8 Sludge generation

The quantity of sludge generated in the process is a criterion providing comparison between the generation and quality of sludge generated in the process of wastewater treatment and treated on the sludge line for each analysed scenario. The scenario with lower sludge production has an advantage in selection in the multi-criteria analysis.

Since the problem of final sludge disposal is becoming equally important as the achievement of the desired treatment effect, the sludge quantities generated in the WWTP are important for the analysed scenarios. Even small differences in sludge generation are of big importance here, as the generation of sludge is continuous activity.

Depending on the local legislation, the treatment, transport and final disposal of sludge can account for as much as 50% of total operating costs of the WWTP. Furthermore, when the sludge has to be transported to incineration, to a distance which can be considerable, these costs become even more pronounced (they can range between € 80 and 200 per tonne of wet sludge). In general, there are two ways to reduce sludge generation: application of primary sedimentation and digestion and/or application of an activated sludge system with higher sludge age. While sludge digestion offers a possibility for the generation of energy from biogas, achieving higher sludge age requires additional consumption of energy for extended aeration. In addition, the required sludge quality is related with the available capacities for its final disposal. For example, the desired sludge quality differs in case of sludge use in agriculture as opposed to the quality of sludge intended for incineration. In every analysed Varaždin WWTP scenario, the definite selection includes sludge digestion. Care needs to be taken not only about the generation of sludge, but also about the generation of biogas, for the purpose of future planning of its use for energy. Annex 13.3 was used as the basis for the above considerations. A very indicative Figure 13.17, which was also the basis for the adoption of conclusions and comparison of scenarios, follows below.

Based on the above, Option 1 scenarios are ranked 1: S1, S5, S3 and S7, while Option 2 scenarios are ranked: S2, S4, S8 and S6.

The final overview of analysis of each of the 8 analysed scenarios under 8 different criteria is presented in Table 13.17.

Table 13.17 Overview of analysis of analysed scenarios by adopted criteria.

Criterion	S1	S2	S3	S4	S5	S6	S7	S8
	Scenario							
	Ranking							
Treatment efficiency	2	2	1	1	3	4	4	3
Investment costs	1	1	4	4	2	2	3	3
Operational costs	3	3	4	4	1	1	2	2
Complexity and maintenance	3	3	4	4	1	1	2	2
Operating stability and robustness	3	3	1	1	4	4	2	2
Required space	1	1	2	2	3	3	4	4
Upgrade implementation complexity	3	3	4	4	2	1	1	2
Sludge generation	1	1	3	2	2	4	4	3

13.9.10 Multi-criteria analysis

Each of the 8 criteria mentioned above has been assigned a relative weight in relation to the total available 100% (or 1.00, as presented in Table 13.18). Each scenario was evaluated against each criterion using scores 1 through 5, with 5 representing the most favourable case. This analysis is based on the author's (as much as possible objective) evaluation including the allocation of weights for each criterion. It is possible that representatives of different interest groups have different views concerning the allocation of relative weights among criteria, which may lead to different results (ranking of scenarios) depending on the degree in which their opinions differ. Furthermore, evaluation of each scenario under each criterion is based on the author's experience on similar tasks and their knowledge of the situation and circumstances in the sector in Croatia, even though – by default – a certain degree of subjectivity cannot be fully avoided. When it comes to the first three criteria, they can be assessed pretty objectively as they are based on a technological-economic analysis and realistic indicators and costs, while the remaining criteria are more descriptive, which means that potential deviations in evaluation are probably higher. However, in our analysis the first three criteria already "carry" 60% of the total weight, which reduces the impact of subjectivity to a large extent and eventually ensures a realistic optimum solution. The optimum solution is calculated on the basis of the sum of products of multiplication between the relative weight (0 - 1) and score (1 - 5); the scenario with the largest sum is the best scenario. Care needs to be taken when the total sums for two or more scenarios are close (e.g. differ by up to 5%), because in such case very small changes either in identifying the relative weight or in evaluation may easily change 'the winner'. In such case, additional iteration of such a 'cluster' of optimum solutions is required in order to come to the best solution or a decision should simply be left for the next step in the project's life cycle (preliminary design), where additional information can be collected which will probably contribute to more easily resolve such a dilemma. In our case, the multi-criteria analysis was made in the manner presented in Table 13.18.

Based on the above, it is estimated that for Option 1 the most favourable scenario for Varaždin WWTP upgrade is scenario S1, while for Option 2 the most favourable scenario is S2. As these two scenarios differ only in the concentration of inert particulate COD in the influent, decision has been made to select more critical situation, i.e. Option 1 and scenario S1.

Table 13.18 Multi-criteria analysis of comparison of Varaždin WWTP upgrade scenarios.

Criterion	Weight	S1	S2	S3	S4	S5	S6	S7	S8
		Scenario							
		Ranking[F]							
Treatment efficiency[A]	0,20	5,00	5,00	4,50	4,50	5,00	5,00	4,50	4,50
Investment costs[B]	0,20	5,00	3,00	4,00	3,50	5,00	3,00	4,50	3,50
Operational costs[C]	0,20	4,50	4,25	5,00	4,75	4,25	4,25	5,00	4,50
Complexity and maintenance	0,05	4,50	4,25	5,00	4,75	4,50	4,25	5,00	4,75
Operating stability and robustness	0,10	4,50	5,00	4,25	4,75	4,50	5,00	4,25	4,75
Required space	0,05	5,00	5,00	4,50	4,50	5,00	5,00	4,50	4,50
Upgrade implementation complexity[D]	0,10	4,50	4,50	5,00	5,00	4,50	4,50	5,00	5,00
Sludge generation[E]	0,10	5,00	5,00	4,50	4,50	5,00	5,00	4,50	4,50
Total	1,00	4,78	4,36	4,55	4,44	4,73	4,36	4,65	4,39

(A) Tertiary level of treatment which complies with legal requirements regarding effluent concentrations;
(B) Capital investments for construction of additional units, equipment and retrofit of the existing units of the palnt;
(C) Yearly costs;
(D) Concerning the scope and charatcer of the required extension works as well as the interuption of the regular operation;
(E) Concerning the total slduge production and production of biogas;
(F) Ranking 1 to 5 where 5 is the best.

In line with the above, the upgrade would be done according to the following provisional plan:

1. Works would first be done on one line of pre-aeration tank, and then on the other one;
2. After that, works would start on the reconstruction of one line of secondary sedimentation tank, and then on the other line;
3. In the same way, by lines, works would be done on aeration tanks and lines for excess sludge recirculation and transport;
4. Works on digesters and transport of biogas might be done independently (simultaneously).

13.10 Selected scenario for upgrade of WWTP Varaždin

As presented above, based on the analyses made, scenario S1 has turned out as the most favourable one. Its detailed description follows below, supported with appropriate annexes.

Scenario S1 belongs to Option 1, i.e. influent with a high content of inert COD fraction. It has turned out to be the most favourable scenario for upgrading and improving the operation of the Varaždin WWTP based on the multi-criteria analysis. This technological procedure is based on the basic assumptions of meeting the effluent quality standards, as well as of making maximum use of the existing structures. Scenario 1 has shown itself to be the most favourable in this as well as in other elements of the multi-criteria analysis.

As illustrated by the block diagram, the basic technological units of biological treatment process are the following: anoxic tanks, bio-aeration tanks and membrane bioreactors (MBR). The basic technological units of the sludge treatment line are the following: the line for recirculation of sludge (RAS) from the MBR which returns the sludge to the beginning of the biological treatment process; a line of internal recirculation from the bio-aeration tank to the anoxic tank; a line for surplus sludge (WAS) which is conveyed to mechanical dehydration before its introduction into digesters; digesters; and final mechanical dehydration of digested sludge.

It is beneficial to repeat the main conclusions about the current state of the WWTP, which were highly important for defining the guidelines on the basis of which the design and proposal of measures (scenarios) to improve its performance were made:

- The required data about the current state was available;
- The current technological process is an activated sludge process with full aeration characterised by the sludge retention time (SRT) of 3 days, based on the removal of organic matter (BOD_5 or COD), without nitrification;
- The WWTP is designed for a capacity of 140,000 PE with loads and concentrations typical for urban wastewater;
- However, more than 70% of the WWTP load is attributed to industrial wastewater inflow;
- No special measures are foreseen concerning treatment of industrial wastewater at the WWTP or some other type of its additional pre-treatment;
- In hydraulic terms, the WWTP is under-dimensioned when it comes to the capacity of secondary sedimentation tanks;
- Likewise, the implementation design implies balanced distribution of inflows to North and South lines (50:50), even though the ratio in reality is 40:60;
- Because of the lack of measurement, the sludge recirculation quantities have been estimated;
- The influent load was in the initial design solution calculated using standard pollutant concentrations in the influent, without analysing the possibility of inflow of 'foreign water', i.e. infiltration from the ground and the surface into the sewerage;
- The installed capacity of the pumps for recirculation of activated sludge is much higher than the achieved discharge values;
- Low efficiency of activated sludge recirculation results in a low concentration of activated sludge in the pre-aeration tanks, which negatively affects biological processes of wastewater treatment;
- The installed capacity of the inlet pumping station has 100% spare capacity.

The starting point for designs and preparation of measures to improve the operation of the Varaždin WWTP is to define as precisely as possible the quality and quantity of inflow, with the average dry inflow, the peak dry inflow and the peak wet inflow defined as important inputs. The characteristic inflow to the WWTP was analysed on the basis of calculation of wastewater volumes both from households and from non-households. The calculations are based on actual water consumption data. Likewise, the calculation also takes into account the inflow of 'foreign water' or infiltration into the sewer system. Both the dry inflow and the combined inflow are taken into account in standard designing practice. Based on the measurement made at the Varaždin WWTP, the following characteristic inflows have been identified:

- Average annual dry flow = 21,300 m^3/d;
- Peak dry flow = 25,600 m^3/d (16h inflow);
- Average combined flow = 21,800 m^3/d;
- Average wet flow = 23,900 m^3/d;
- Peak wet flow = 27,000 m^3/d;
- Maximum capacity of the main sewer = 58,320 m^3/d;
- Installed pump capacity (including those on stand-by) = 116,640 m^3/d.

Conclusions made on the basis of inflow analysis:

- The actual hydraulic load of the WWTP is below the designed load of 28,000 m^3/d.
- The difference between the average dry inflow and the average combined inflow is very small (102% in relation to the average dry inflow), indicating a relatively small share of stormwater and a large share of industrial water (up to 72%) in the total inflow.

- The peak wet inflow is 120% of the average dry inflow, which is significantly lower than typical values (150-300%), which may also be the result of a relatively high contribution of industrial water.
- The maximum inflow of 27,000 m^3/d was measured in the year 2010, indicating that the maximum sewer capacity of 58,000 m^3/d has not been achieved, particularly having in mind the fact that no bypasses or overflows through which untreated water would be discharged have been recorded.

13.10.1 Design temperature

Based on the analyses made, the actual values of wastewater temperature have been defined, since this parameter is very important for the nitrification process, as well as for the analysis of aeration requirements. The selected minimum, average annual and maximum temperatures are:

- Average annual influent temperature = 15°C
- Minimum measured influent temperature = 9.3°C
- Maximum measured influent temperature = 22.6°C; when extreme temperature values measured where left out, the adopted maximum temperature is 21.2°C.

The minimum temperature is very important when dimensioning aeration tanks. Namely, in the winter conditions, nitrification takes place with more difficulty because the growth of microorganisms is slowed down. The minimum sludge age in the winter conditions must be sufficient to ensure full nitrification. During design, the minimum sludge age of 8 days which ensures nitrification at 8°C has been adopted. Naturally, many other factors also affect nitrification in reality, but the above two are the most important ones. The maximum temperature is also important for designing aeration capacity, i.e. oxygen input capacity (defining the required capacity of the compressor). It is a known fact that as the temperature of wastewater rises, the solubility of oxygen in the water reduces.

13.10.2 Design influent concentrations

The average concentrations of typical parameters in raw (untreated) wastewater have been identified on the basis of measuring and corrections made based on analyses in this study. On that basis, the following Varaždin WWTP influent quality for the average combined inflow has been adopted:

- TCOD 394.6 mg/L
- BOD_5 155.4 mg/L
- TSS 151.2 mg/L
- TN 34.4 mg/L
- NH_4 10.4 mg/L
- TP 4.3 mg/L

In addition to defining the basic parameters that characterize the quality of raw wastewater, other parameters important for making designs have also been analysed: ratio between biodegradable and non-biodegradable components, and organic and inorganic components. Consequently, in addition to defining the content of total suspended solids, the share of inorganic and organic volatile substances also had to be defined.

Knowing the chemical composition of wastewater is also very important for analysing the applicability of different technological processes. For example, a pH value is an important factor in monitoring the process of wastewater treatment with activated sludge and treatment of sludge in digesters. Besides, this information is important when defining the dose of iron salts for the removal of phosphorus from wastewater.

By applying model-based technological design, the approach to defining the modelling inputs such as the quantity and quality of inflow to the plant is presented in detail. This made it possible to use a BioWin simulator for designs and analyses of different process options.

Conclusions made on the basis of analysis of concentration of typical parameters in raw wastewater:

- Raw wastewater is pretty diluted (probably due to infiltration);
- The measured TKN values have been adjusted based on the simulation model and the measured nitrogen content in activated sludge;
- TSS has been adjusted based on model design;
- Bio-degradability has been defined based on the calibration of actual (measured) generation of sludge;
- In the model, it was also necessary to analyse a scenario with a high load of inert COD fraction, as well as a scenario with its usual (lower) load.

In addition to above, activated sludge was also measured and analysed, resulting in the following conclusions:

- The ash (inorganic matter) content of 15% in the biomass corresponds to the fact that low sludge age of app. 3 days has been achieved;
- The nitrogen and phosphorus contents in activated sludge are within the expected values;
- The ratio between COD and volatile substances in the biomass should be app. 1.42, whereas it usually ranges between 1.29 and 1.37.

13.10.3 Design hydraulic load

The starting assumptions for making analyses in terms of defining the actual hydraulic load and the concentration of substances characterising the quality of raw wastewater were presented in short above. The analysis of incoming loads has already pointed to the usual problems in the current WWTP operation, first of all the low sludge age and insufficient capacity of secondary sedimentation tanks. The designs made show that the sludge age is almost identical for the North and South lines (app. 3.5 days), which indicates even distribution of influent to both lines. The low sludge age is directly related with increased generation of sludge and insufficient nitrification. This confirms the fact that the current WWTP is dimensioned only for the removal of organic matter and suspended solids.

All the five design methods implemented show that the current secondary sedimentation tanks are overloaded both in terms of hydraulic load and SS load. They also show that two more sedimentation tanks 31 m in diameter need to be added in order to achieve good performance in the current conditions, without considering additional WWTP improvement which would require an additional number of new sedimentation tanks. As regards surplus sludge generated in the biological process of wastewater treatment, full (sufficient) sludge stabilization hasn't been achieved in the operation of the WWTP so far. Hence, when the sludge stabilization is not achieved in the process of wastewater treatment, which is the case here, additional process of its aerobic or anaerobic stabilization has to be foreseen.

13.10.4 Design principles

The principles on the basis of which operational improvement of this WWTP is based are explained in the opening part of this chapter. These are: legal requirements (as well as the EU Directive) on effluent quality in terms of nutrient removal (nitrogen and phosphorus), maximum use of the current structures, functionality and provision of the required capacity for all technological units; and achieving the required degree of sludge stabilization in order for the sludge to be suitable for final disposal.

The biological part of the Varaždin WWTP includes a pumping station that sends wastewater into a pre-aeration tank, i.e. a denitrification tank according to the new technological scheme (scenario S1). As the capacity of the current pumps is sufficient, this structure will in the selected option fully retain its original purpose. The activated sludge reaches the pumping station by recirculation, where it is mixed with raw wastewater and conveyed by pumps to the entry into the denitrification tank. The same principle is applied to the return of the supernatant and infiltration water from the sludge treatment process, in order to ensure proper mixing with raw wastewater and activated sludge, as well as balanced distribution of load to both treatment lines (North and South lines).

In this scenario (as well as in all other scenarios), the pre-aeration tank has been improved into the pre-denitrification tank in the full volume of 609 m per tank. Two mixers per tank are installed since the activated sludge requires constant mixing in order for it to be kept in suspension now that there is no aeration to do this. The current aeration system will in the new denitrification tanks be shut down (or, even better, removed), while its continued use in other places in the process (e.g. in the old rectangular sedimentation tank which become additional aeration tanks with MBR) is left as an scenario which has not been analysed any further until its suitability is assessed on site (which, if suitable, can reduce the need for additional investment into the aeration system). With really minor interventions on the removal of the current aeration equipment and additional plastic material (fill material) in the pre-aeration tank on the southern line, and screens, these tanks will fully comply for the required anoxic volume for the denitrification process in the new technological concept of biological treatment (S1). When it comes to mechanical equipment, two mixers will have to be installed into each of these tanks, with the only purpose of the mixers being to enable gentle mixing of water and activated sludge. A simple solution is horizontal submerged mixers with the required mixing capacity of 3 W/m^3. A certain quality of water from the aeration tanks (internal recirculation) is returned to the anoxic tank to establish the contact of raw wastewater and activated sludge after the nitrification process. From the anoxic part the water is released into two bio-aeration tanks where the mixture of water and activated sludge undergoes intensive mixing and aeration, i.e. the required oxygen input for the decomposition of organic matter is done with simultaneous nitrification process. This proposed concept is based on a proven process of pre-denitrification and simultaneous oxidation of organic matter and nitrification (A/O process), where denitrification which takes place in anoxic conditions is designed at the front of the biological treatment in order to be able to use easily degradable organic matter in raw wastewater with mixing with the biologically treated water returned from the aeration tank in a certain share. The volume of the current aeration tanks is 2,451 m^3 per tank, i.e. 4,902 m^3 in total, and these will be used as aeration tanks no. 1. The current aeration tanks will retain their function in full volume, and the same function will also be given to the major share of the current secondary sedimentation tanks, whose role of separating the liquid from the solid phase will be taken over by the membrane modules which will be accommodated in another part of the current secondary sedimentation tanks. The total volume of secondary sedimentation tanks is 4,940 m^3/d, i.e. 1,235 m^3 per tank; they will be used as aeration tanks in the first part and as MBR in the second part. The volume required for MBR is 1,125 m^3, while the remaining part will become aeration tanks no. 2. Diffuse aeration with fine air bubbles is applied here, and the quantity of the required oxygen input is designed through the model. The required oxygen input is almost three times higher than so far (Annex 13.3). No additional construction works on the aeration tanks will be required here, but additional aeration equipment (a diffuse system and compressors) will have to be provided. The diffuse system will also be installed into the current secondary sedimentation tanks from which sludge scraping equipment will have to be removed beforehand, as well as the pipe fill from the southern line of the sedimentation tanks. With this approach, the sludge age is increased to the designed 8 days, which is optimum for systems with separate sludge stabilization (in this case, anaerobic stabilization in digesters). As already mentioned, part of the volume of the current secondary sedimentation tanks (1,125 m^3 in total) will be used for MBR. This means that membrane modules with supports will be placed there. In this case, membranes are packed into modules and are as such submersed into a bioreactor. The module contains membranes, frame, supports, inlet and outlet elements, and other supporting material. The membranes operate under vacuum; the treated water (permeate) passes through the membranes, while particles (activated sludge) remain in the bioreactor. This system can operate properly even with a very high concentration of organic matter in the bioreactor (MLSS as high as 15-25 kg/m^3), but the optimum design range is usually 8-15 kg/m^3. In combination with properly dimensioned aeration, such systems have high efficiency. Membranes undergo backwashing, when filtration is interrupted. Judging from experience, the flux through the membrane fibres ranges 10-15 $L/m^2.h$ with negative pressure of 0.2-0.3 bar, max. 0.5 bar. For example, for real load which, expressed as the average mixed inflow, amounts to 21,800 m^3/d, and for average flux of 800 $L/m^2.d$, the estimated membrane surface is app. 27,250 m^2. This implies installation of 27 modules with membrane surface of 1,000 m^2 and one module of 250 m^2. In any case, in the later stage a special design will have to be made for the specifically selected type of modules and membranes, and systematic optimization will have to be done. The process of membrane washing and maintenance usually consists of three main levels. It starts with air-shaking of membranes (large bubbles); the flow of air through the vertically laid fibres makes the fibres shake and collide. The filtration is stopped every 15-30 minutes, when backwashing with permeate (effluent) is done for 30-45 s. The total time for such backwash is app. 45 minutes per day. Detailed backwash 45-minute process is done two-three times per month using a strong solution of sodium hypochlorite (app. 100 mg/L) or the citric acid. After that, permeate washing lasting for app. 15 minutes is done. Additional permeate washing can continue for 10-15 minutes in order for the system to become completely free from the residual chlorine. A specific solution for washing is sodium hypochlorite (150 g/L of active chlorine), which after optimization amounts to 0.005-0.006 mg per m^3 of wastewater per m^2 of membrane surface. The proposed washing interval is 17-22 days. For the final evaluation of the optimization measures, long-term monitoring of permeability before and after the optimization measures is proposed. The analysis of long-term monitoring, adoption and application of optimization measures show that the permeability and flux of the module achieve a higher level than the comparative module in which the optimization measures were not taken. MBR filtration system are highly efficient since the wastewater biologically treated in the bioreactor is further treated by filtration through membranes with 0.4 µm pores, resulting in very low concentrations of suspended solids, turbidity, BOD_5 in the effluent. Such water can be reused for irrigation. With denitrification ensured in the biological process, an exceptionally high effluent quality can be guaranteed.

The surplus sludge line will also be slightly modified in this selected scenario, S1. In addition to certain interventions on improving the sludge recirculation system, some interventions will be required on the surplus sludge line as well. This concerns primarily equipment for mechanical thickening of surplus sludge in order to release a certain quantity of excess water and reduce the quantity of sludge transported to digesters. Digesters will be built structures used so far as the sludge thickener and tank. As these structures were originally designed as digesters, no demanding construction works will be required for fitting them with equipment. Mechanical works are slightly more demanding in terms of additional equipment and layout of the pipelines. New sludge heating equipment will be installed at the entry line to digesters (as the digesters were originally designed for 'cold' digestion). Digested sludge recirculation, which is standard practice, will be ensured. A mixer for sludge homogenization needs to be installed into digesters. The digested sludge will be collected and conveyed to mechanical dehydration (existing structure) through a system of pipelines. The existing sludge drying fields will not be used in regular operation, but will be left as reserve for potential acceptance of digested sludge, in case of failure of mechanical dehydration. The designed digestion temperature is 35°C; the total volume of the digesters is 2,100 m^3; the sludge retention time in the digester is 23.4 days. During sludge digestion, methane biogas is released. Expected generation of methane biogas is 1,000 kg COD/d. Here it needs to be noted that this study doesn't plan nor design a system for using the energy from biogas, but there are no obstacles to planning its use in the future. For phase I, it is only the installation of equipment for acceptance and transport of methane biogas that is planned. This line is foreseen to be installed with a gas meter and a torch for the combustion of the generated gas.

13.10.10 Estimated investment costs

The total estimated investment costs for scenario S1 (excl. VAT) amount to EUR 3,655,430. This is a relatively rough estimate, the primary purpose of which is to enable comparison of the selected scenarios. As already mentioned, in the total costs, the costs of procurement and installation of electro-mechanical equipment are – as expected – higher than the costs of construction works. The operation and maintenance costs for the selected scenario are estimated at EUR 910,994/year. They include costs related to the facilities for biological wastewater treatment and sludge treatment. The structure of O&M costs consists of: sludge disposal costs, chemicals costs, electric energy costs, costs of regular servicing of equipment, and other minor costs. The sludge disposal costs were calculated on the basis of the quantity of treated WWTP sludge, which for option 1 ranges around 4,366 kg/d, and on the basis of an approximate price of EUR 120 per tonne of sludge (estimated based on similar projects). The costs of chemicals refer to the dosage costs for iron chloride, poly-electrolyte, NaOCl and citric acid. They were estimated on the basis of experience from similar projects. The energy consumption costs were analyzed on the basis of the capacity of equipment to be installed and on the basis of estimated hours of operation during the day. This has given total energy consumption as well as total installed capacity. The price per the unit of measure of these two parameters was assumed, which had no effect on the comparison of scenarios. However, when defining more precisely this type of costs for Scenario 1, the real price for the Varaždin WWTP has to be included. Since it was not possible to analyse other costs occurring in regular operation, such as additional engagement of expert staff for this segment of the plant, the costs have been increased by 5%. Finally, there are also costs of regular servicing and maintenance of electro-mechanical equipment, which in this scenario also include the costs of membrane replacement every 2-3 years. These are estimated at 3.5% of the capital investment.

13.11 Conclusions on the scenario selection

This solution is designed in such a way that it describes the current state of the Varaždin WWTP, makes a detailed analysis in order to define quantitative and qualitative characteristics of wastewater, detailed technological analysis of the current state and all of the 8 proposed scenarios, approximate cost estimate for all the 8 scenarios, approximate O&M costs, technological schemes and bases for all the 8 scenarios, multi-criteria analysis and selection of the optimum solution.

The purpose of the scenario was to improve the current performance of the Varaždin WWTP in accordance with EU standards for COD (125 mg/L), BOD_5 (25 mg/L), TSS (35 mg/L) and nutrient (nitrogen and phosphorus) removal. According to EU standards, the applied limit value for nitrogen is TN < 10 mg/L, including full nitrification throughout the year. In addition, there was a requirement for maximum improvement of the performance of the Varaždin WWTP within the existing volumes in order to minimize the total cost of reconstruction. In short, the specified conditions and requirements are met by adopting scenario S1. The comprehensive study and model analyse in detail all initial data, boundary conditions, current state, different approaches to improvement, and – eventually – the approach to analysing and selecting the solution.

The expected effluent quality after the technological process adopted as scenario S1 – biological treatment system with activated sludge, pre-denitrification and membrane filtration, and separate anaerobic stabilization of sludge – is very high, even with this scenario analysed for an influent with a high share of COD fraction.

The effluent quality is presented in detail, by individual parameters, in Annex 13.3 All the parameters meet the initially set conditions imposed according to the relevant legislation and EU Directive, including those concerning nutrient removal. It has to be stressed here one more time that high phosphorus removal effects can be achieved by dosing iron salts, which was not the subject of the study with model application.

In order to ensure nitrification in the winter conditions when the temperatures are low (in this specific case, 9.3°C), the minimum sludge age of 8 days is required. Under scenario 1, this requirement is fully met, with regard to which this scenario (together with similar scenarios S2-S4) has an advantage compared to other scenarios. When it comes to the requirement concerning the use of existing structures and the requirement for additional space, scenarios S1 has a significant advantage compared to other analysed scenarios. The capacities of the existing structures and equipment are used to the maximum extent. There is no requirement for additional structures and space for them, which reduces the total costs essential for the execution of works, but also for landscaping. The equipment installed under this scenario has the largest share in the total investment.

In addition to the presented advantages of the selected scenario and expected improvement in the WWTP performance, another project benefit is the presentation of a modern approach to wastewater treatment by modelling aimed at finding an optimum solution and making the right decisions.

This solution provides a sound basis and comprehensive information for continuation of activities on the preparation of design documents of higher level of detail.

Reference

Meijer S.C.F. and Brdjanovic D. (2012) A Practical Guide to Activated Sludge Modeling. UNESCO-IHE lecture notes, UNESCO-IHE Institute for Water Education, ISBN 9789073445260, pg 277.

Annex 13.1
WWTP layout and process flow diagram for scenarios S1-S8

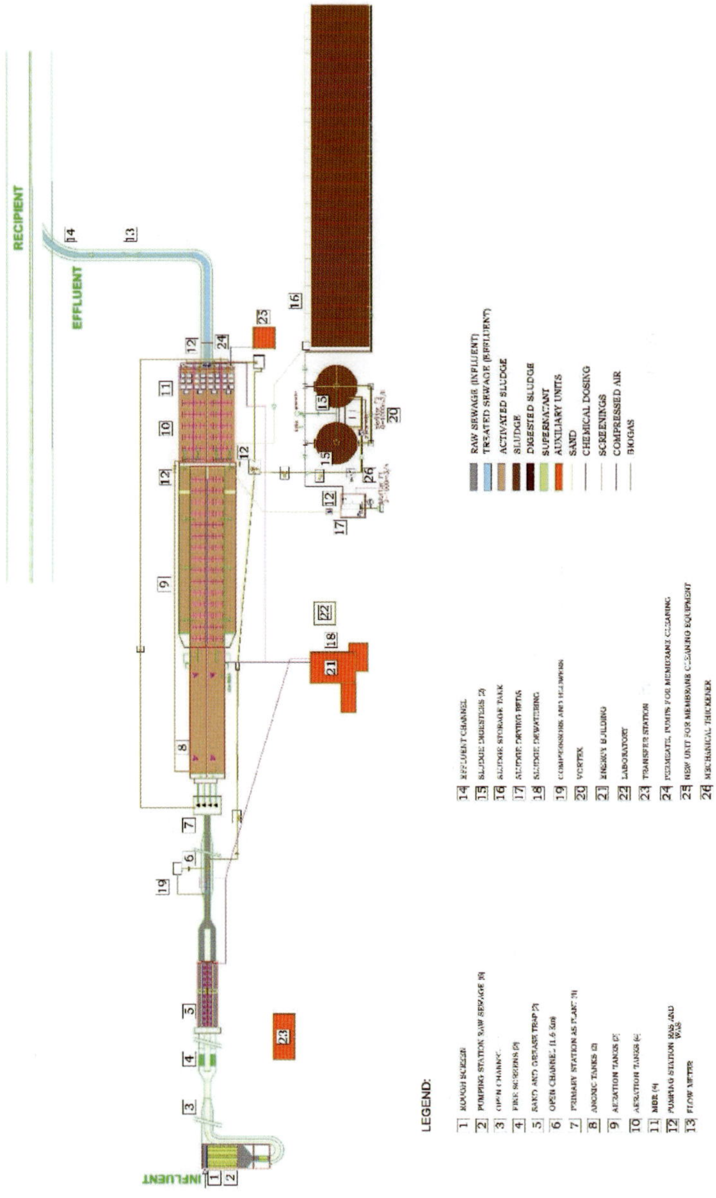

Layout of WWTP Varaždin: Scenarios S1 and S2

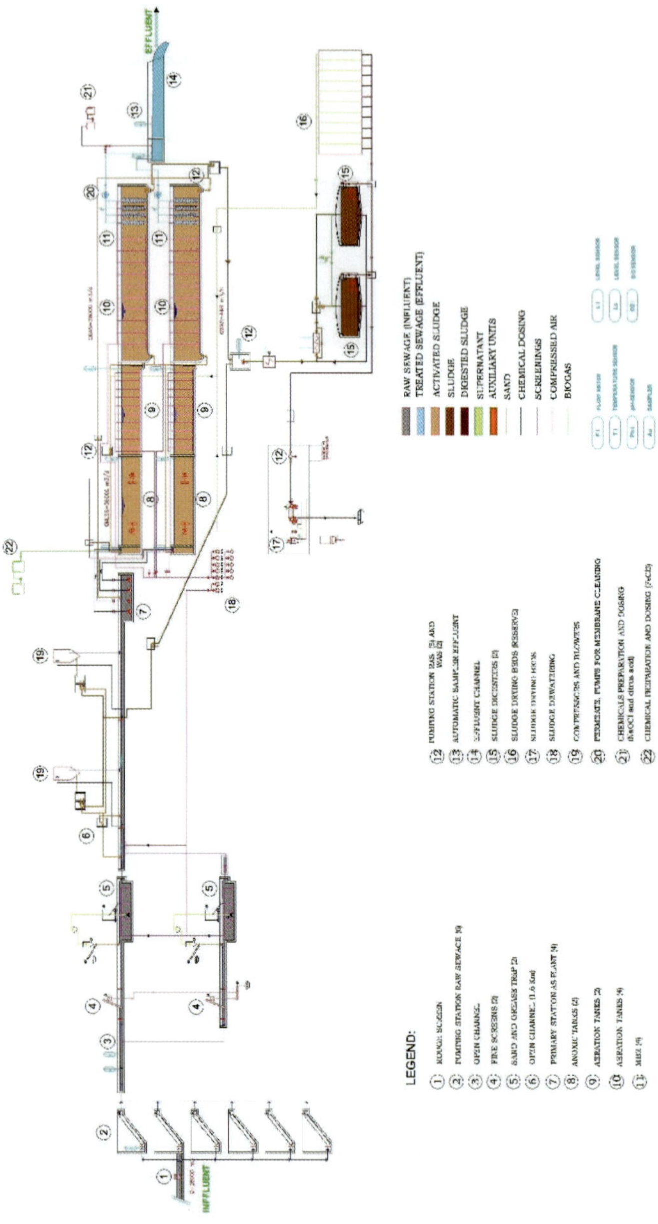

Process flow diagram of WWTP Varaždin: Scenarios S1 and S2

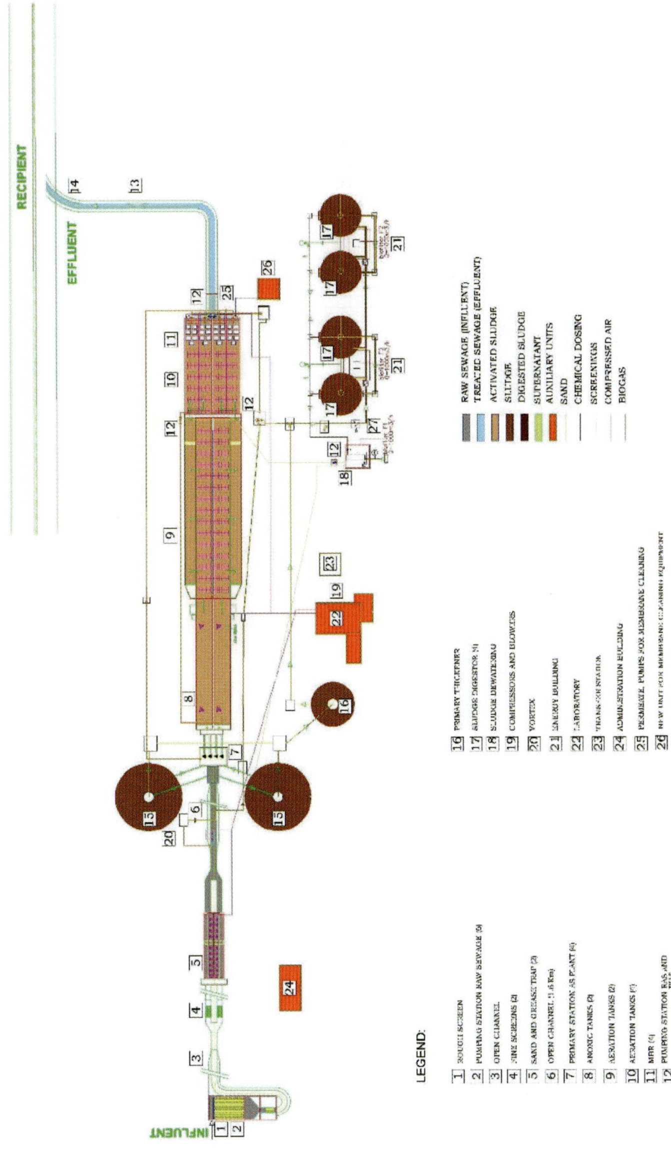

Layout of WWTP Varaždin: Scenarios S3 and S4

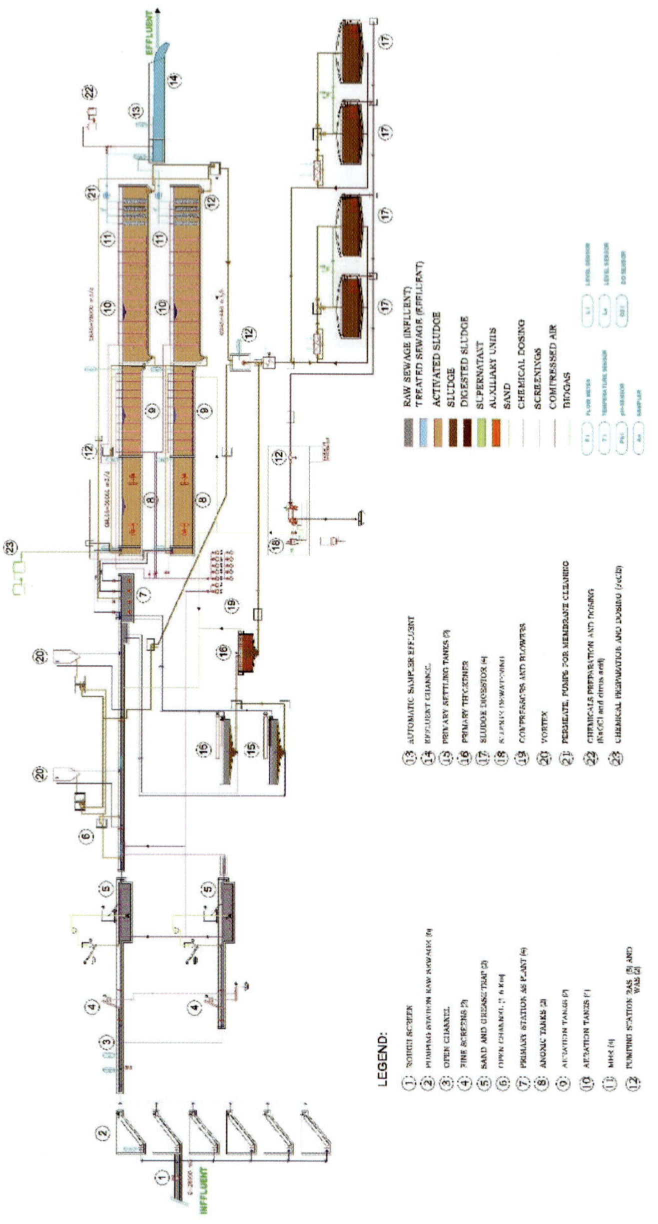

Process flow diagram of WWTP Varaždin: Scenarios S3 and S4

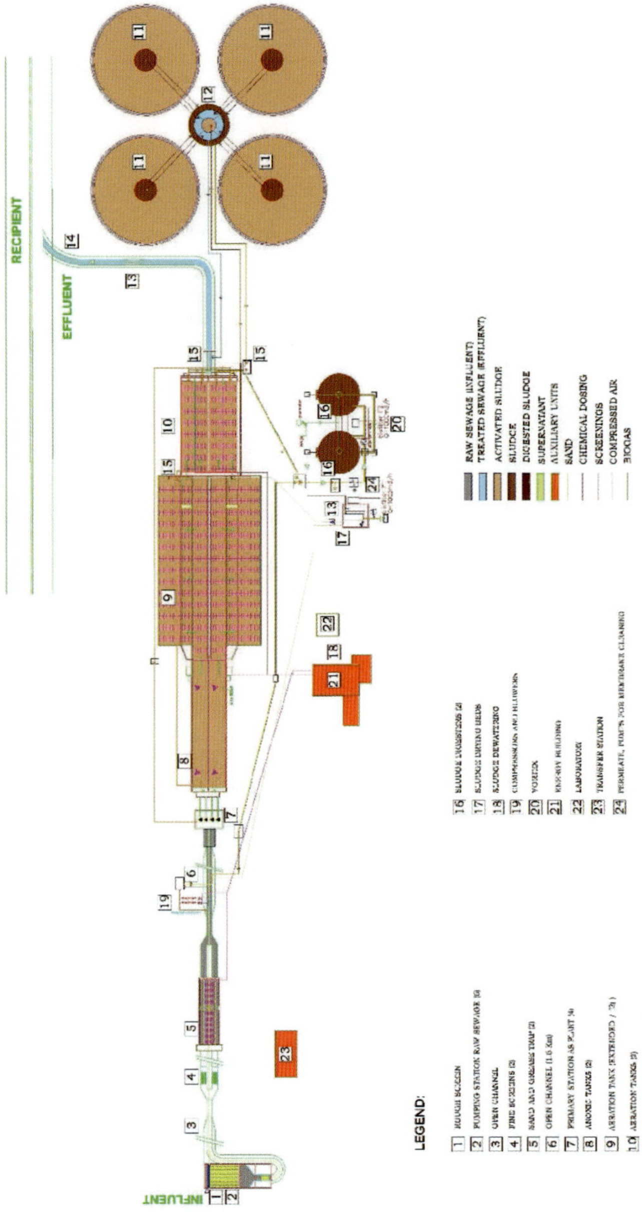

Layout of WWTP Varaždin: Scenario S5

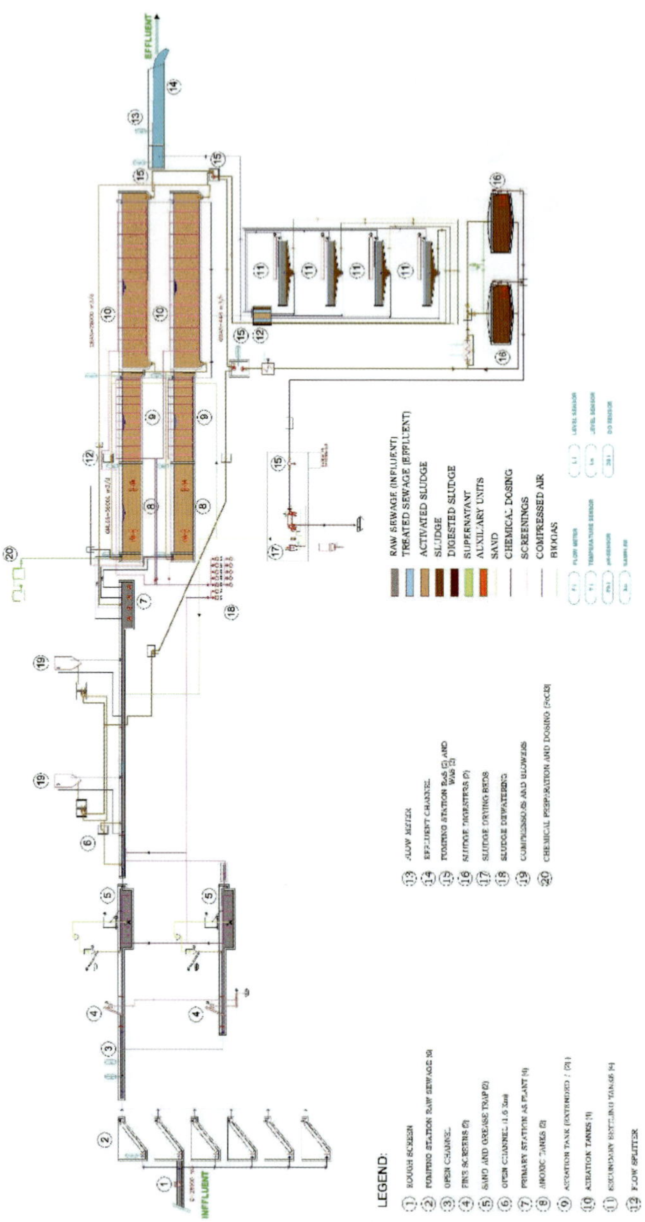

Process flow diagram of WWTP Varaždin: Scenario S5

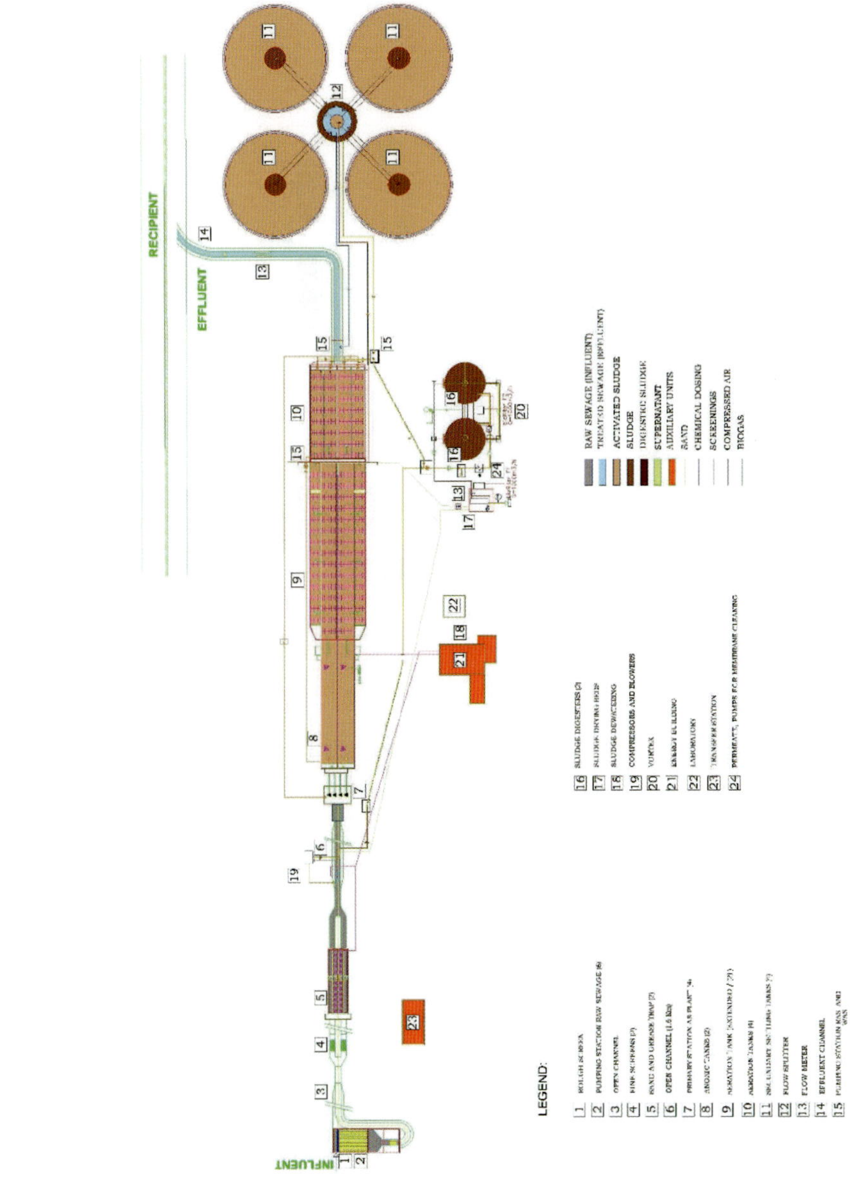

Layout of WWTP Varaždin: Scenario S6

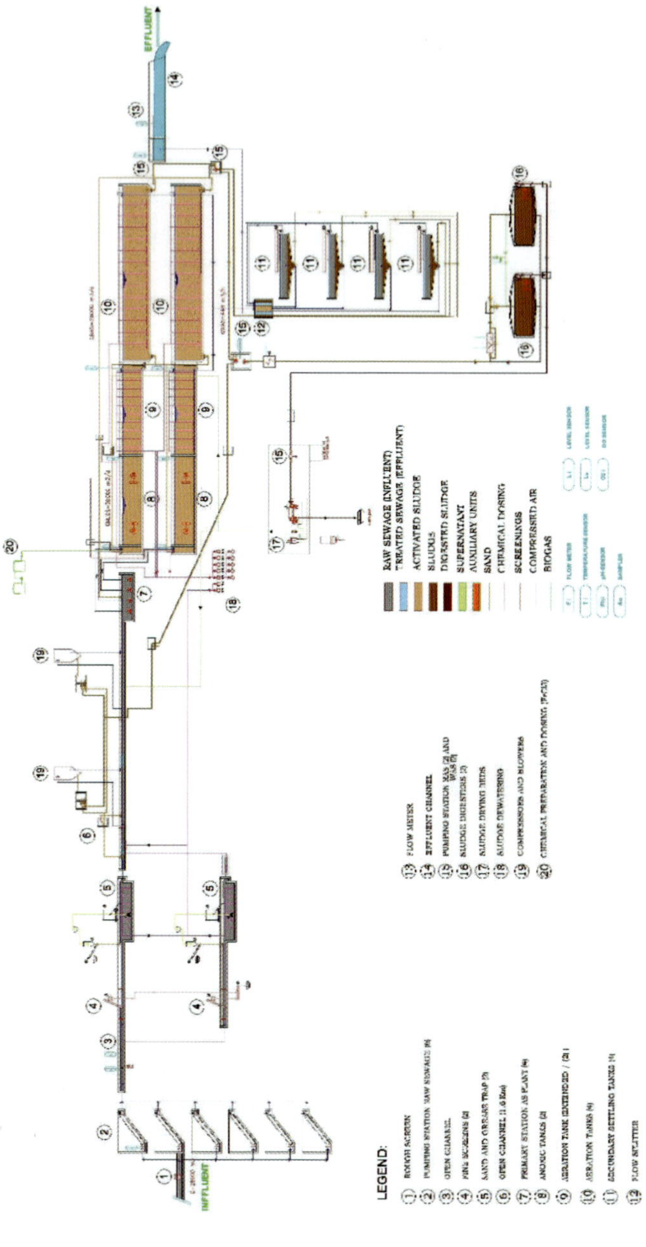

Process flow diagram of WWTP Varaždin: Scenario S6

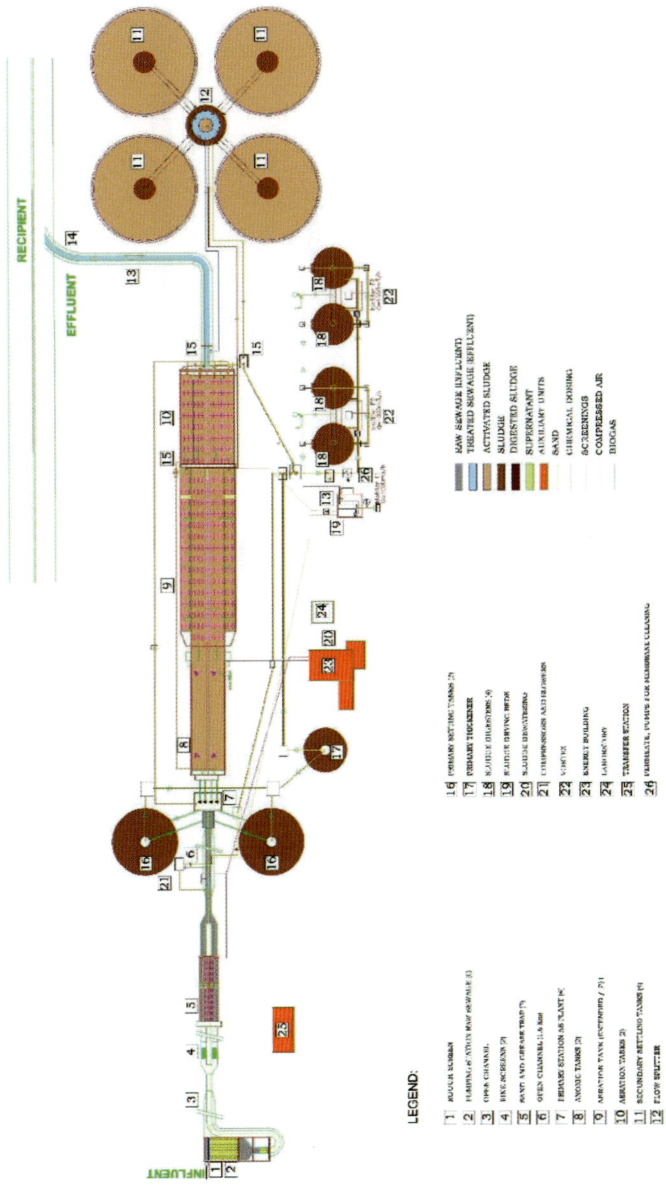

Layout of WWTP Varaždin: Scenario S7

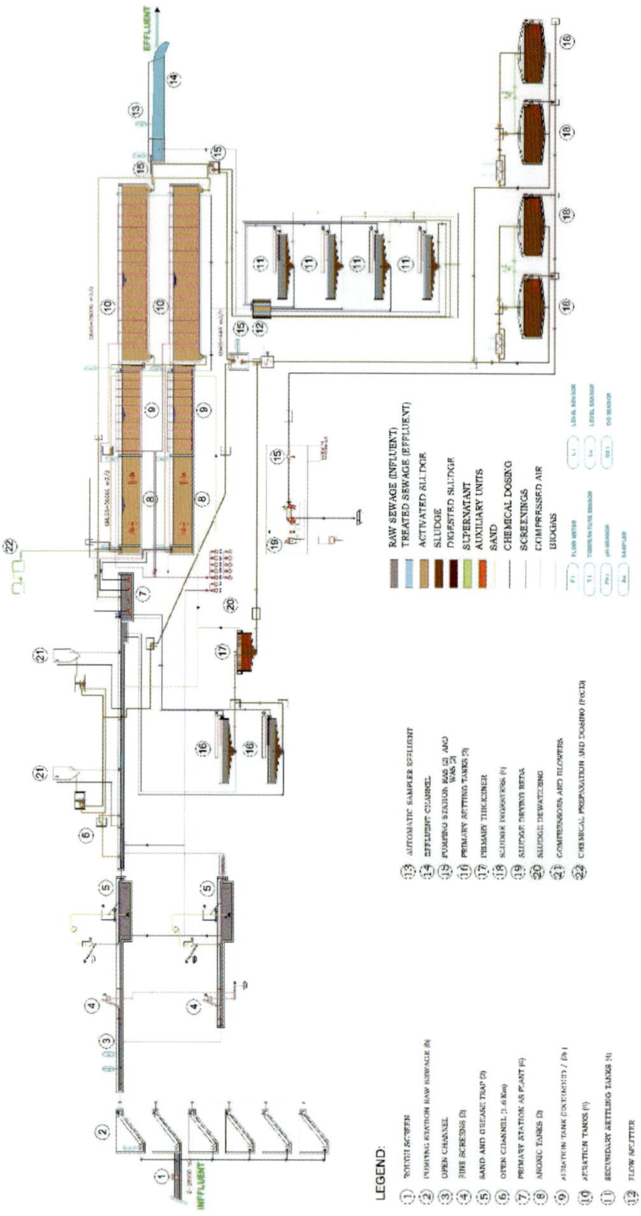

Process flow diagram of WWTP Varaždin: Scenario S7

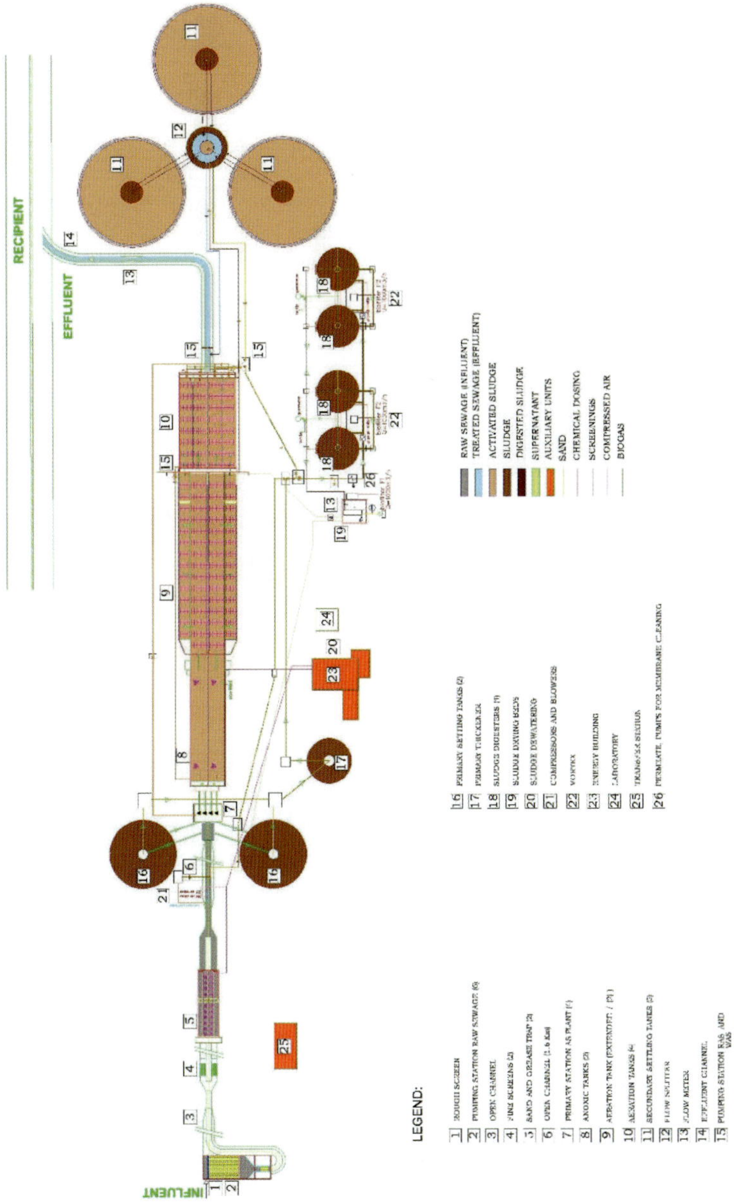

Layout of WWTP Varaždin: Scenario S8

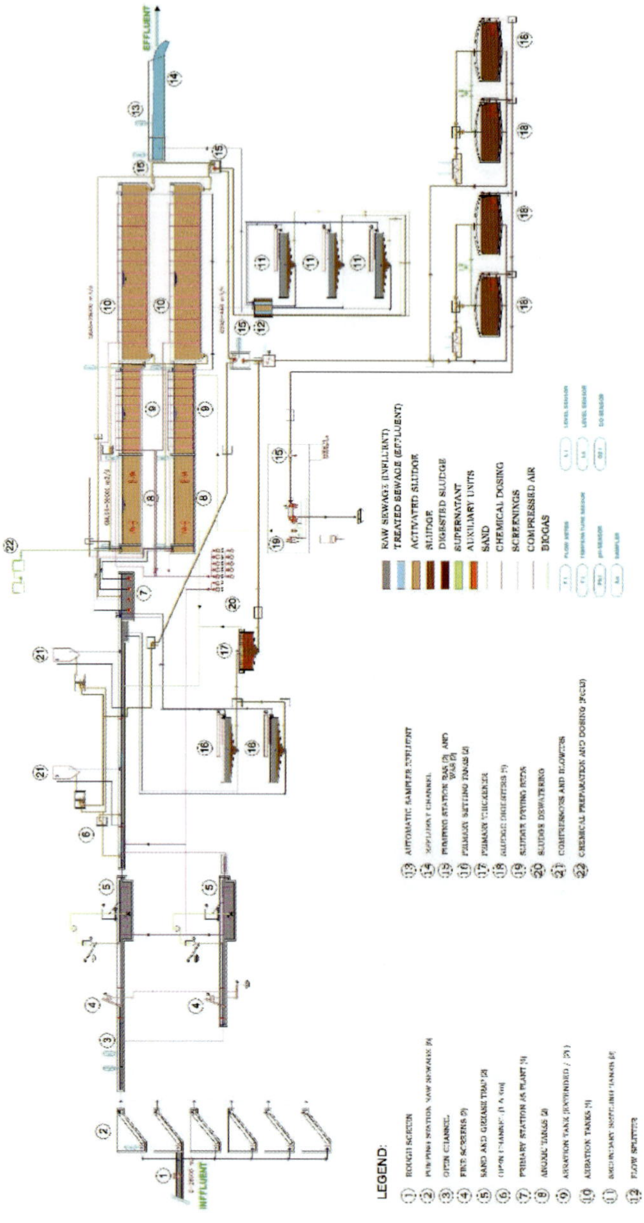

Process flow diagram of WWTP Varaždin: Scenario S8

Annex 13.2
Hydraulic process scheme of the WWTP for scenario S1-S8

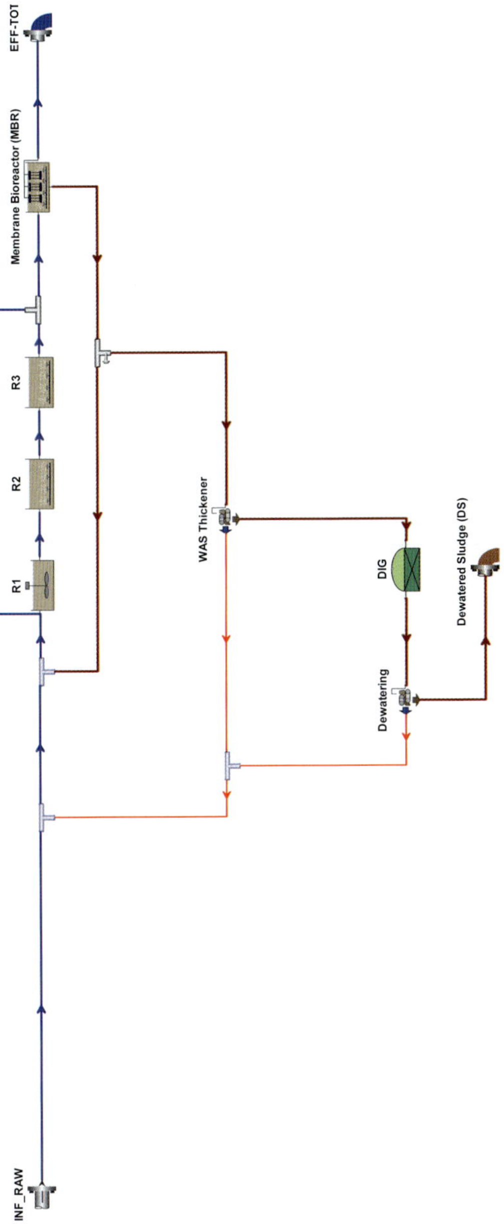

Hydraulic scheme of WWTP Varaždin in BioWIn: Scenario S1

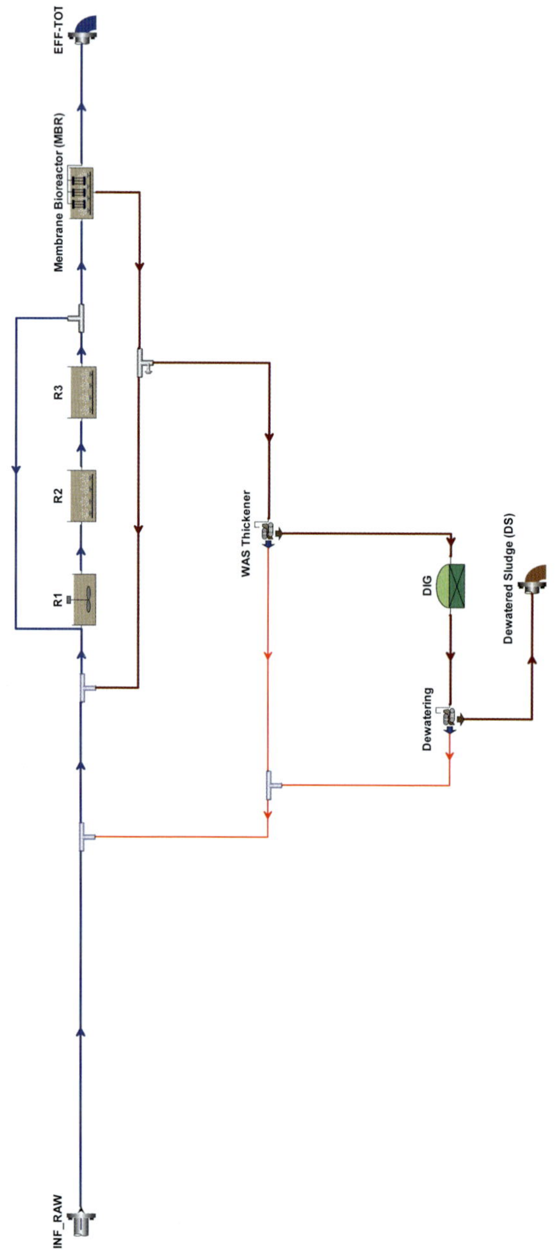

Hydraulic scheme of WWTP Varaždin in BioWIn: Scenario S2

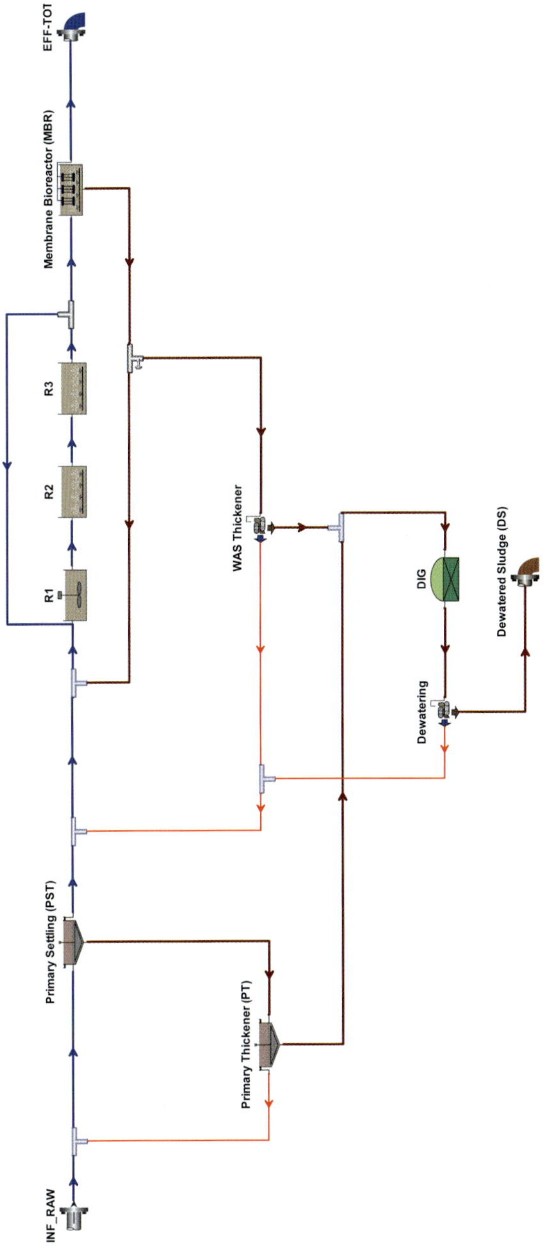

Hydraulic scheme of WWTP Varaždin in BioWIn: Scenario S3

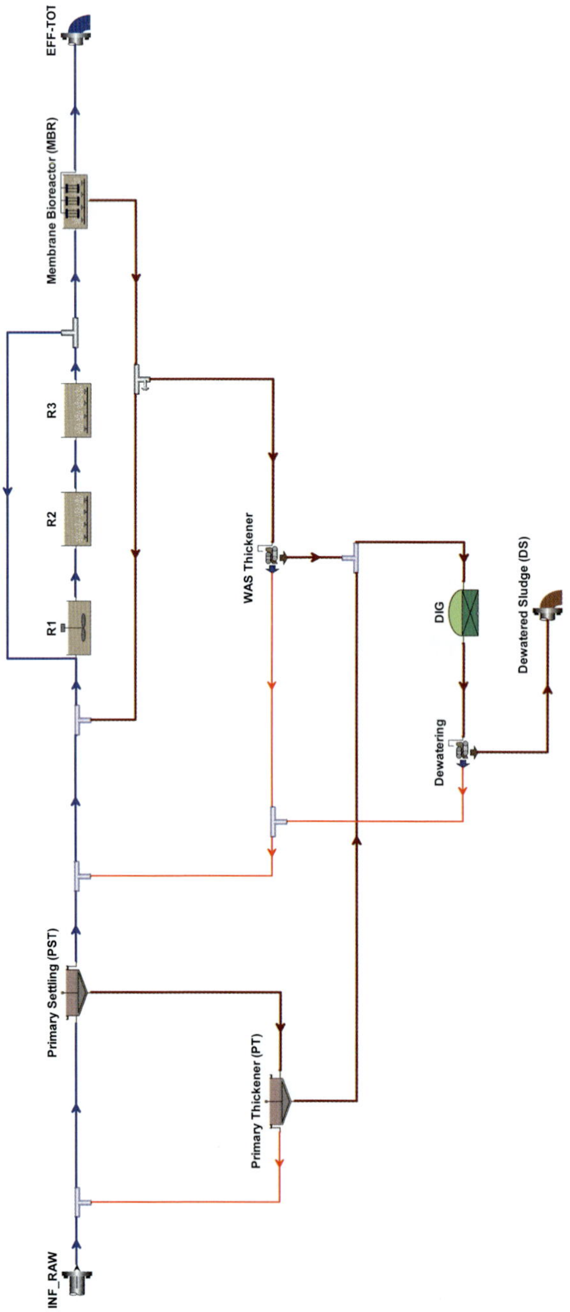

Hydraulic scheme of WWTP Varaždin in BioWIn: Scenario S4

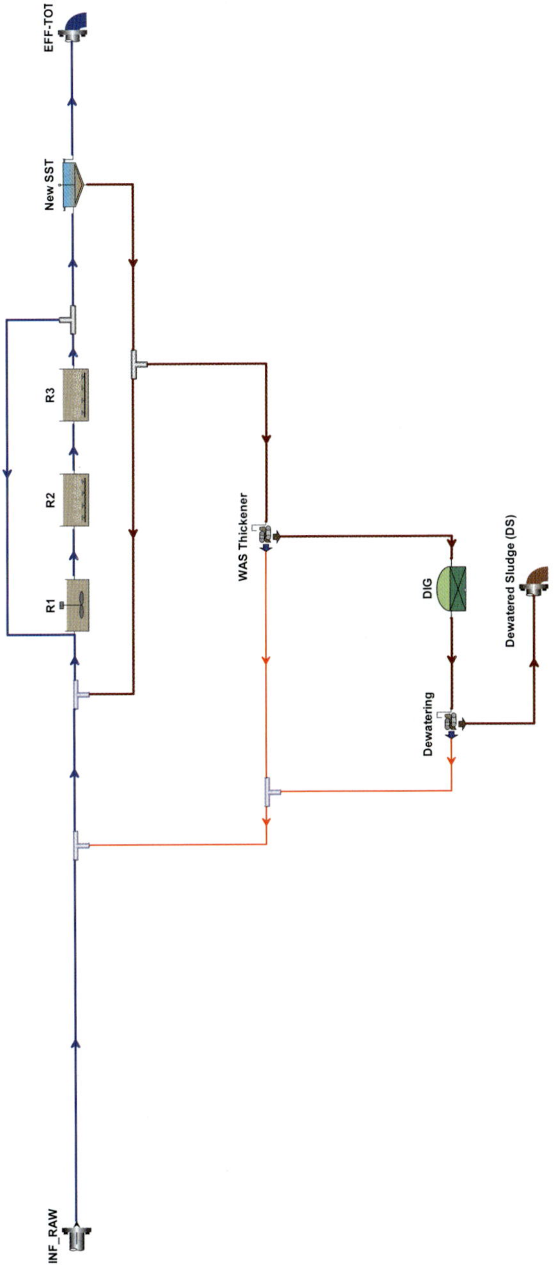

Hydraulic scheme of WWTP Varaždin in BioWIn: Scenario S5

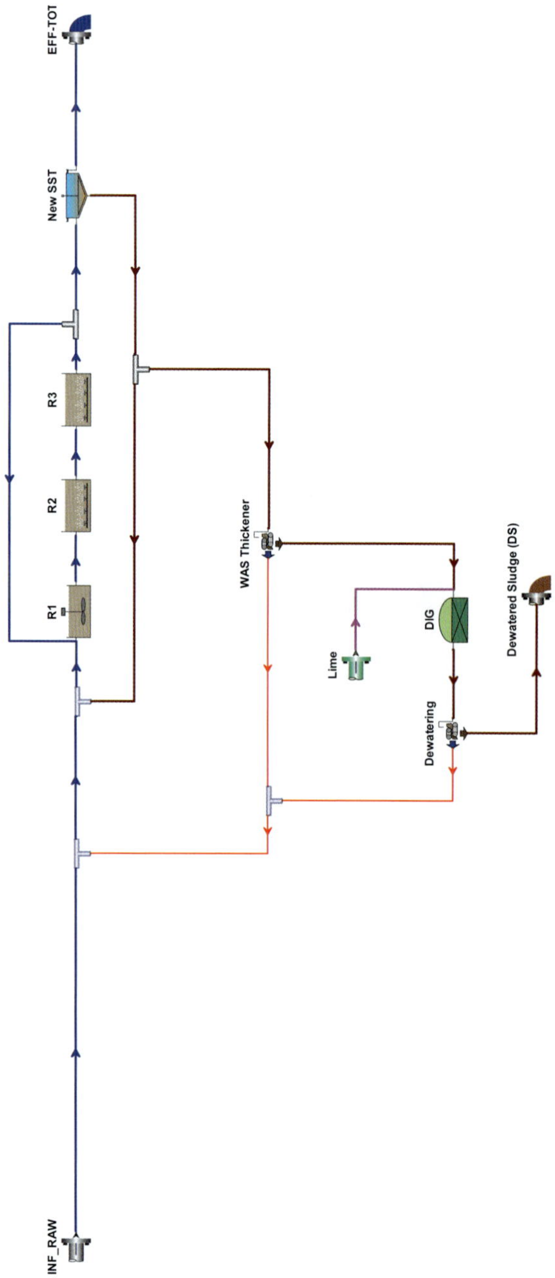

Hydraulic scheme of WWTP Varaždin in BioWIn: Scenario S6

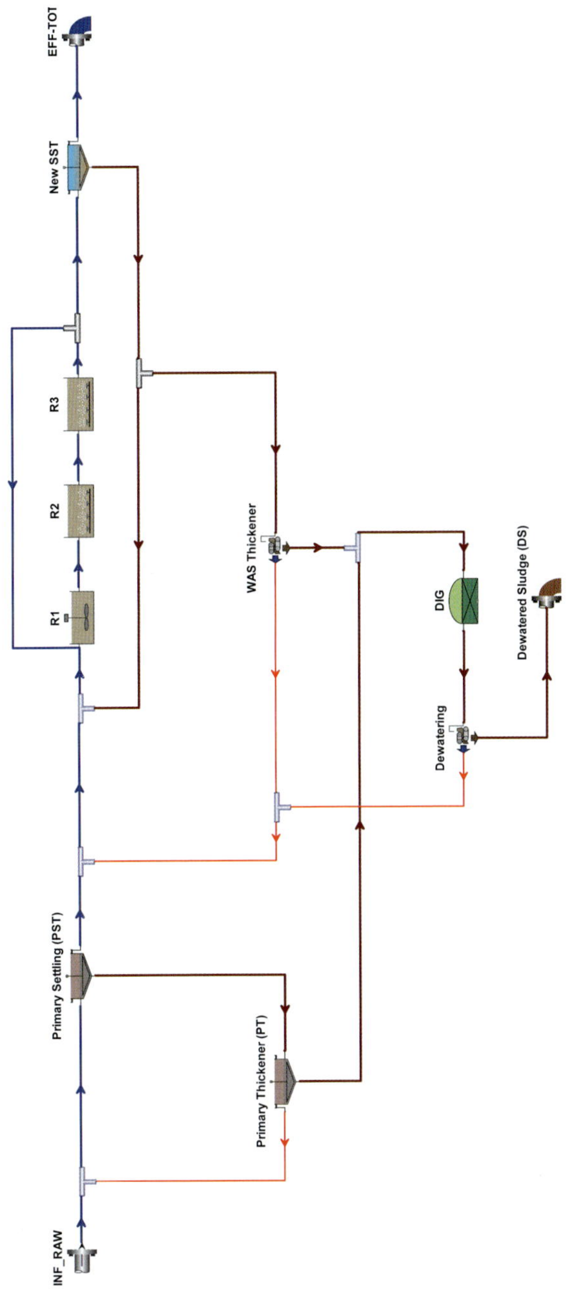

Hydraulic scheme of WWTP Varaždin in BioWIn: Scenario S7

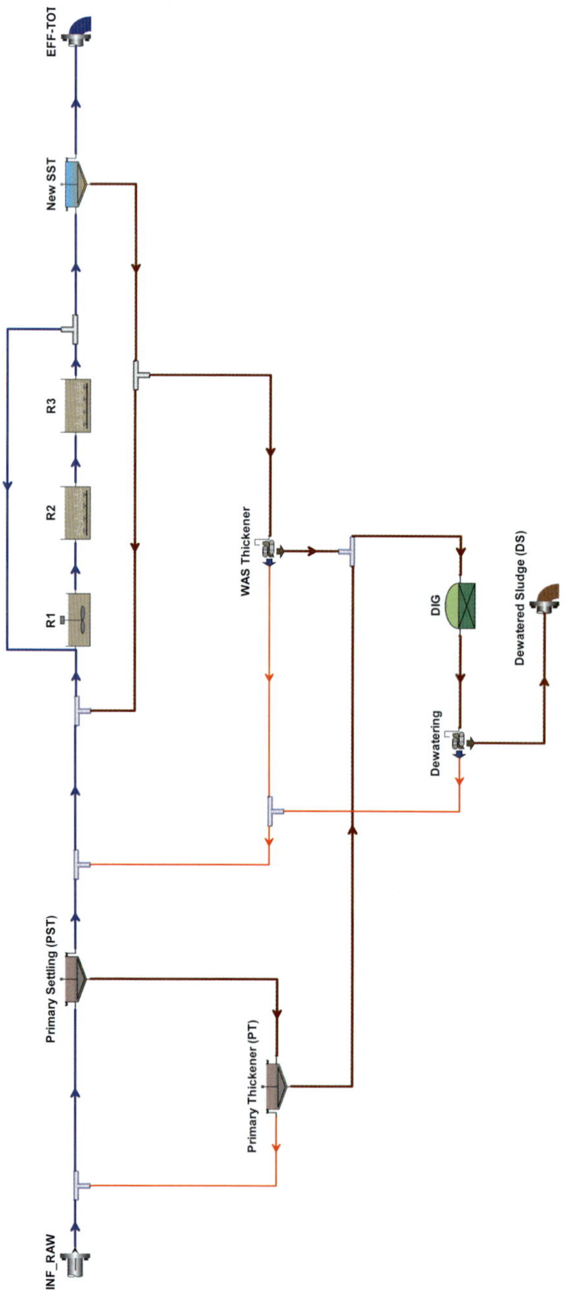

Hydraulic scheme of WWTP Varaždin in BioWIn: Scenario S8

Annex 13.3 Scenario calculations overview table

Item	Description	Unit	S0	S1	S2	S3	S4	S5	S6	S7	S8
Process Type	Aerated, Simultaneous NIT/DEN, Anoxic/Aerobic AO		Aerated	AO	AO	AO	AO	AO	AO	AO	AO
Process Type	MBR or Conventional SST		SST	MBR	MBR	MBR	MBR	New SST	New SST	New SST	New SST
Reactor Type	Carrousel / Plug flow / Combined		Plug flow	Plug flow	Plug flow	Plug flow	Plug flow	Plug flow	Plug flow	Plug flow	Plug flow
Volume	Existing / Extended		Extended	Existing	Existing	Existing	Existing	Extended	Extended	Extended	Extended
Influent Type	Raw / Settled		Settled	Raw	Raw	Settled	Settled	Raw	Raw	Settled	Settled
Aeration Type	Turbine / Bubble		Bubble	Bubble	Bubble	Bubble	Bubble	Bubble	Bubble	Bubble	Bubble
BOD	BOD removing design		BOD	BOD	BOD	BOD	BOD	BOD	BOD	BOD	BOD
NIT	Nitrification design		NIT	NIT	NIT	NIT	NIT	NIT	NIT	NIT	NIT
DEN	Nitrification/Denitrification design		SNIT/DEN	PRE-DEN	PRE-DEN	PRE-DEN	PRE-DEN	PRE-DEN	PRE-DEN	PRE-DEN	PRE-DEN
BIO-P	Bio-P design		-	-	-	-	-	-	-	-	-
Design volume		**Unit**	**S0**	**S1**	**S2**	**S3**	**S4**	**S5**	**S6**	**S7**	**S8**
PST	Primary Settling Tanks	m^3	-	-	-	8,000	8,000	-	-	8,000	8,000
ANA	Anaerobic Tank	m^3	-	-	-	-	-	-	-	-	-
ANOX	Anoxic tank	m^3	-	1,239	1,239	1,239	1,239	1,239	1,239	1,239	1,239
AT	Aeration Tank	m^3	6,131	4,912	4,912	4,912	4,912	9,137	6,137	6,137	6,137
MBR	Submerged Membrane Bio-Reactor	m^3	-	1,125	1,125	1,125	1,125	-	-	-	-
SST	Secondary Settling Tank	m^3	2,470	-	-	-	-	11,775	12,480	11,775	8,925
PTT	Primary Thickening Tank		-	-	-	1,500	1,500	-	-	1,500	1,500
DIG	Digester	m^3	2,100	2,100	2,100	4,199	4,199	2,699	2,100	4,199	4,199
V_{AERO}	Aerobic volume (used for aerobic SRT)	m^3	6,131	6,037	6,037	6,037	6,037	9,137	6,137	6,137	6,137
V_{ANOX}	Anoxic volume	m^3	-	1,239	1,239	1,239	1,239	1,239	1,239	1,239	1,239
V_{MLSS}	Total activated sludge volume	m^3	6,131	7,276	7,276	7,276	7,276	10,376	7,376	7,376	7,376
V_{TOT}	Total volume waterline (including clarification)	m^3	8,601	7,276	7,276	7,276	7,276	22,151	19,856	19,151	16,301
Volume extension (reference situation 2011)		**Unit**	**S0**	**S1**	**S2**	**S3**	**S4**	**S5**	**S6**	**S7**	**S8**
PST	Extended Volume PST	m^3	-	-	-	8,000	8,000	-	-	8,000	8,000
ANOX	Extended Volume Anox	m^3	-	1,239	1,239	1,239	1,239	1,239	1,239	1,239	1,239
AT	Extended Volume AT	m^3	-	-	-	-	-	3,000	-	-	-
SST	Extended Volume SST	m^3	-	-	-	-	-	11,775	12,480	11,775	8,925
PTT	Extended Volume PTT	m^3	-	-	-	1,500	1,500	-	-	1,500	1,500
DIG	Extended Volume Digester	m^3	-	-	-	2,099	2,099	599	-	2,099	2,099

Process flows and internal recirculation flows		Unit	S0	S1	S2	S3	S4	S5	S6	S7	S8
QINF	Total Influent	m^3/d	21,808	28,000	28,000	28,000	28,000	28,000	28,000	28,000	28,000
QPS	Primary sludge	m^3/d	0	0	0	1,777	1,777	0	0	1,777	1,777
QRET	Return Activated Sludge (RAS)	m^3/d	9,023	28,000	28,000	28,000	28,000	28,000	28,000	28,000	28,000
QWAS	Waste Activated Sludge (WAS) to CF	m^3/d	505	448	449	448	448	642	480	450	371
QMLSS	Anoxic MLSS recycle	m^3/d	0	56,000	84,000	56,000	84,000	56,000	56,000	56,000	56,000
QINT	Internal recycle (carousel)	m^3/d	0	0	0	0	0	0	0	0	0
QEFF	Effluent biology	m^3/d	21,778	27,970	27,970	27,995	27,995	27,970	27,970	27,970	27,995
HRT	Total hydraulic residence time WWTP	h	9	6	6	6	6	19	17	16	14
HRT biology	Hydraulic residence time (biology)	h	7	6	6	6	6	9	6	6	6
Raw influent (sewage)		**Unit**	**S0**	**S1**	**S2**	**S3**	**S4**	**S5**	**S6**	**S7**	**S8**
P.E. loading	P.E. loading (136 TOD)	136 gTOD	88,056	113,058	113,058	113,058	113,058	113,058	113,058	113,058	113,058
TSS	Influent suspended solids	kg/d	3,297	4,233	4,226	4,233	4,226	4,233	4,226	4,233	4,226
VSS	Influent volatile solids	kg/d	2,753	3,535	3,525	3,535	3,525	3,535	3,525	3,535	3,525
ISS	Influent inorganic solids	kg/d	544	698	701	698	701	698	701	698	701
BOD_5	Influent BOD Load	kgBOD/d	3,379	4,339	5,894	4,339	5,894	4,339	5,894	4,339	5,894
TCOD	Influent COD loading	kgCOD/d	8,605	11,049	11,049	11,049	11,049	11,049	11,049	11,049	11,049
TKN	Influent TKN loading	kgN/d	738	947	947	947	947	947	947	947	947
TP	Influent TP loading	kgP/d	93	120	120	120	120	120	120	120	120
TOD	Influent TOD loading	kgTOD/d	11,976	15,376	15,376	15,376	15,376	15,376	15,376	15,376	15,376
VSS/TSS	Organic fraction of total solids (0.75-0.85)	g/g	0.84	0.84	0.83	0.84	0.83	0.84	0.83	0.84	0.83
COD/VSS	Particulate organic COD fraction (1.5-1.7)	gCOD/g	1.25	1.25	1.19	1.25	1.19	1.25	1.19	1.25	1.19
TCOD/BOD	COD fraction of BOD (1.9-2.2)	gCOD/gBOD	2.55	2.55	1.87	2.55	1.87	2.55	1.87	2.55	1.87
Fup	Fraction Unbiodegradable Particulate COD of TCOD	gCOD/gCOD	0.39	0.39	0.13	0.39	0.13	0.39	0.13	0.39	0.13
Primary settling (raw influent to settled)		**Unit**	**S0**	**S1**	**S2**	**S3**	**S4**	**S5**	**S6**	**S7**	**S8**
% TSS removed	Removed from raw influent excluding internal loads	%	-	-	-	49%	49%	-	-	49%	49%
% VSS removed	Removed from raw influent excluding internal loads	%	-	-	-	49%	50%	-	-	49%	50%
% ISS removed	Removed from raw influent excluding internal loads	%	-	-	-	49%	49%	-	-	49%	49%
% BOD5 removed	Removed from raw influent excluding internal loads	%	-	-	-	3%	13%	-	-	3%	13%
% COD removed	Removed from raw influent excluding internal loads	%	-	-	-	21%	19%	-	-	21%	19%
% TKN removed	Removed from raw influent excluding internal loads	%	-	-	-	23%	17%	-	-	23%	17%
% TP removed	Removed from raw influent excluding internal loads	%	-	-	-	15%	15%	-	-	15%	15%

Primary sludge (to digester after thickening)		Unit	S0	S1	S2	S3	S4	S5	S6	S7	S8
P.E. loading	Primary sludge after thickening	136 gTOD	-	-	-	23,773	19,705	-	-	23,773	19,706
Primary - TSS	Primary sludge after thickening	kg/d	-	-	-	2,120	1,742	-	-	2,120	1,742
VSS	Primary sludge after thickening	kg/d	-	-	-	1,760	1,393	-	-	1,760	1,393
ISS	Primary sludge after thickening	kg/d	-	-	-	533	535	-	-	533	535
BOD_5	Primary sludge after thickening	kgBOD/d	-	-	-	61	689	-	-	61	689
TCOD	Primary sludge after thickening	kgCOD/d	-	-	-	2,235	1,930	-	-	2,235	1,930
TKN	Primary sludge after thickening	kgN/d	-	-	-	218	164	-	-	218	164
TP	Primary sludge after thickening	kgP/d	-	-	-	18	18	-	-	18	18
TOD	Primary sludge after thickening	kgTOD/d	-	-	-	3,233	2,680	-	-	3,233	2,680
VSS/TSS	Organic fraction of total solids (0.75-0.85)	g/g	-	-	-	0.83	0.80	-	-	0.83	0.80
COD/VSS	Particulate organic COD fraction (1.5-1.7)	gCOD/g	-	-	-	1.25	1.20	-	-	1.25	1.20
COD/BOD	COD fraction of BOD (1.9-2.2)	gCOD/gBOD	-	-	-	36.50	2.80	-	-	36.50	2.80
Loading to activated sludge system		**Unit**	**S0**	**S1**	**S2**	**S3**	**S4**	**S5**	**S6**	**S7**	**S8**
P.E. loading	P.E. loading Biology (136 TOD)	136 gTOD	88,056	113,058	113,058	88,912	91,901	113,058	113,058	88,912	91,901
TSS	Influent suspended solids to biology	kg/d	3,297	4,233	4,226	2,174	2,135	4,233	4,226	2,174	2,135
VSS	Influent volatile solids to biology	kg/d	2,753	3,535	3,525	1,819	1,778	3,535	3,525	1,819	1,778
ISS	Influent inorganic solids to biology	kg/d	544	698	701	355	357	698	701	355	357
BOD_5	Influent BOD Load Biology	kgBOD/d	3,379	4,339	5,894	4,226	5,133	4,339	5,894	4,226	5,133
TCOD	Influent COD loading biology	kgCOD/d	8,605	11,049	11,049	8,764	8,922	11,049	11,049	8,764	8,922
TKN	Influent TKN loading biology	kgN/d	738	947	947	728	783	947	947	728	783
TP	Influent TP loading biology	kgP/d	93	120	120	101	102	120	120	101	102
TOD	Influent TOD loading	kgTOD/d	11,976	15,376	15,376	12,092	12,499	15,376	15,376	12,092	12,499
VSS/TSS	Organic fraction of total solids	g/g	0.84	0.84	0.83	0.84	0.83	0.84	0.83	0.84	0.83
COD/VSS	Particulate organic COD fraction	gCOD/g	1.25	1.25	1.19	1.25	1.19	1.25	1.19	1.25	1.19
TCOD/BOD	COD fraction of BOD	gCOD/gBOD	2.55	2.55	1.87	2.07	1.74	2.55	1.87	2.07	1.74
BOD_5/N	Influent ratio BOD_5/N	kgBOD/kgTKN	4.58	4.58	6.22	5.80	6.56	4.58	6.22	5.80	6.56
TCOD/N	Influent COD/N ratio	kgCOD/kgTKN	11.67	11.67	11.67	12.03	11.40	11.67	11.67	12.03	11.40
TCOD/P	Influent COD/P ratio	kgCOD/kgTP	92.39	92.39	92.39	86.37	87.80	92.39	92.39	86.37	87.80
Sludge Loading	Sludge loading based on BOD_5	kgBOD_5/kgTSS.d	0.18	0.08	0.16	0.12	0.19	0.08	0.16	0.11	0.15
Sludge Loading	sludge loading based on COD	kgCOD/kgTSS.d	0.46	0.21	0.29	0.25	0.33	0.22	0.29	0.24	0.27
Sludge Loading	TKN sludge loading	kgTKN/kgTSS.d	0.04	0.02	0.03	0.02	0.03	0.02	0.02	0.02	0.02

Sludge		Unit	S0	S1	S2	S3	S4	S5	S6	S7	S8
Temperature	Influent w ater temperature	°C	15	10	10	10	10	10	10	10	10
MLSS	Activated sludge in reactors	$gTSS/m^3$	3,081	7,144	5,178	4,847	3,734	4,926	5,153	5,023	4,495
RAS	Return Activated Sludge	$gTSS/m^3$	4,512	14,267	10,322	9,680	7,446	9,829	10,246	10,022	8,971
SRT	Total Sludge Retention Time (anoxic and aerobic)	d	3	9	9	9	9	8	8	8	10
SRT aerobic	Operated Sludge Retention Time	d	3	8	8	8	8	7	6	7	8
SRT aerobic	Maximum Sludge Retention Time	d	3	8	8	8	8	7	8	8	8
Sludge in Effluent	Particulate COD_X in effluent	kgCOD/d	416	0	0	0	0	89	73	85	109
WAS	Waste Activated Sludge	kgCOD/d	5,858	7,056	4,838	4,920	3,670	6,970	4,761	4,791	3,347
Total WAS	Total sludge production (WAS+Effluent)	kgCOD/d	6,274	7,056	4,838	4,920	3,670	7,058	4,834	4,876	3,456
Sludge Loss to Effluent	TSS in effluent	kgTSS/d	375	0	0	0	0	80	76	80	109
WAS	Waste Activated Sludge	kgTSS/d	5,279	6,392	4,630	4,336	3,339	6,307	4,918	4,513	3,324
Total WAS	Total sludge production (WAS+Effluent)	kgTSS/d	5,654	6,392	4,630	4,336	3,339	6,387	4,994	4,593	3,433
Total WAS + Primary	Total sludge production before digestion	kgTSS/d	5,279	6,392	4,630	6,456	5,081	6,307	4,918	6,632	5,066
Digester design		**Unit**	**S0**	**S1**	**S2**	**S3**	**S4**	**S5**	**S6**	**S7**	**S8**
Temperature	Local digester temperature	°C	15	35	35	35	35	35	35	35	35
Volume	Digester volume	m^3	2,100	2,100	2,100	4,199	4,199	2,699	2,100	4,199	4,199
HRT	Total hydraulic residence time	d	4	23	23	21	21	21	22	20	21
CH4	Methane production	kgCOD/d	105	1,000	1,490	1,052	2,227	964	1,293	903	1,984
VSS-destruction	Sludge (organic) reduction	kgTSS/d	220	721	1,033	741	1,475	700	923	646	1,321
VSS-destruction	Sludge (organic) reduction	% VSS	0.15	0.15	0.32	0.14	0.37	0.14	0.29	0.12	0.35
Total Sludge	Sludge production after digestion	kgCOD/d	4,528	5,238	2,801	5,247	2,847	5,192	2,905	5,262	2,778
Total Sludge	Sludge production after digestion	kgTSS/d	4,021	4,866	2,980	4,882	3,028	4,809	3,309	5,074	3,122
Conversions in WWTP (from mass balance calculations)		**Unit**	**S0**	**S1**	**S2**	**S3**	**S4**	**S5**	**S6**	**S7**	**S8**
COD to CO_2	Total oxidized COD	kgCOD/d	3,027	4,214	6,156	4,152	5,375	4,207	6,081	4,142	5,523
COD to Sludge	COD removed w ith sludge	kgCOD/d	5,874	5,238	2,801	5,247	2,847	5,192	2,905	5,262	2,778
COD oxidized	COD oxidized	kgCOD/d	2,450	3,187	4,727	3,084	3,892	3,130	4,730	3,113	4,054
COD denitrified	COD denitrified	kgCOD/d	577	1,027	1,429	1,068	1,482	1,077	1,351	1,028	1,469
Nitrified N Load	Total ammonium nitrified	kgN/d	206	489	648	506	673	474	597	452	652
Denitrified N load	Total nitrate denitrified	kgN/d	201	358	498	372	516	375	471	358	512
Nitrogen removed in Sludge	Organic bound nitrogen	kgN/d	411	385	226	371	204	374	233	389	203
OC_{NIT}	Theoretical oxygen demand for nitrification	kgO_2/d	943	2,233	2,960	2,314	3,077	2,165	2,727	2,063	2,979
OC_{DEN}	COD conversion using nitrate	kgO_2/d	577	1,027	1,429	1,068	1,482	1,077	1,351	1,028	1,469
OC_{COD}	COD conversion using oxygen	kgO_2/d	2,450	3,187	4,727	3,084	3,892	3,130	4,730	3,113	4,054
OC_{TOT}	Total (stoichiometric) oxygen demand	kgO_2/d	3,388	5,560	7,827	5,538	7,109	5,291	7,430	5,163	7,007

Aeration design		Unit	S0	S1	S2	S3	S4	S5	S6	S7	S8
Type	Turbine or Bubble aerators	-	Bubble	Bubble	Bubble	Bubble	Bubble	Bubble	Bubble	Bubble	Bubble
sp_O2	Average oxygen concentration in AT	gO_2/m^3	1.20	2.00	2.00	2.00	2.00	2.00	2.00	2.00	2.00
Oxygen deficit	Cs/(Cs-Cact) oxygen deficiency correction	-	1.13	1.21	1.21	1.21	1.21	1.21	1.21	1.21	1.21
OC dirty w ater	Total dirty w ater oxygen input capacity	kgO_2/d	3,845	6,735	9,482	6,709	8,612	6,410	9,001	6,255	8,488
Alpha factor	Correction for oxygen transfer in dirty w ater	-	1	1	1	1	1	1	1	1	1
OC clear w ater	Total clear w ater oxygen input capacity	kgO_2/d	7,691	13,470	18,963	13,418	17,224	12,819	18,002	12,509	16,976
Air input	Total practical blow er input	Nm_3/d	126,720	365,041	530,283	359,610	468,867	188,953	263,738	172,698	258,937
Air input	Total pow er input	kW	-	-	-	-	-	-	-	-	-
Aeration depth	Air bubble rising height	m	4	4	4	4	4	4	4	4	4
Aeration efficiency	Oxygen transfer efficiency	$gO_2/Nm^3 \cdot m$	16	10	9	10	10	18	18	19	17
Effluent concentration		**Unit**	**S0**	**S1**	**S2**	**S3**	**S4**	**S5**	**S6**	**S7**	**S8**
NO_3	Nitrate	$mgN\text{-}NO_3/L$	0.8	5.3	6.0	5.4	6.2	4.1	5.1	3.9	5.6
NH_4	Ammonium	$mgN\text{-}NH_4/L$	6.4	0.6	0.5	0.6	0.5	1.2	1.9	1.7	1.0
TKN	Total Kjeldahl	mgN/L	9.9	2.6	2.6	2.5	2.5	3.5	4.2	3.8	3.3
TN	Total nitrogen	mgN/L	10.7	7.9	8.6	7.9	8.7	7.6	9.3	7.7	8.9
PO_4	Ortho-phosphate	$mgP\text{-}PO_4/L$	1.6	2.3	2.4	2.5	2.8	2.4	0.1	1.0	1.7
TP	Total phosphorus	mgP/L	1.9	2.3	2.4	2.5	2.8	2.5	0.2	1.1	1.9
COD_{MF}	Soluble COD	mgCOD/L	21.7	21.3	21.5	21.4	21.4	21.4	24.9	23.5	23.4
BOD_5	Biodegradable COD	mgCOD/L	4.0	0.5	0.7	0.5	0.6	1.0	1.5	1.1	1.8
TSS	Particulate COD	mgCOD/L	17.2	0.0	0.0	0.0	0.0	2.9	2.7	2.9	3.9
TOD	Theoretical Oxygen Demand	mgTOD/L	86.0	33.4	33.5	32.6	32.8	40.6	46.7	43.9	42.3
TCOD	Total COD	mgCOD/L	40.8	21.3	21.5	21.4	21.4	24.5	27.5	26.5	27.3
Removal efficiency (effluent vs influent)		**Unit**	**S0**	**S1**	**S2**	**S3**	**S4**	**S5**	**S6**	**S7**	**S8**
TSS removal	Influent TSS in relation to effluent	%	97%	100%	100%	100%	100%	98%	98%	98%	97%
COD removal	% influent COD to effluent	%	93%	95%	95%	95%	95%	94%	93%	93%	93%
TN removal	% influent TN to effluent	%	44%	77%	75%	77%	74%	77%	73%	77%	74%
TKN-removal	Influent in relation to effluent	%	45%	92%	92%	93%	93%	90%	88%	89%	90%
TP removal	Influent TP in relation to effluent	%	94%	45%	43%	41%	34%	42%	96%	75%	56%

Plant performance		Unit	S0	S1	S2	S3	S4	S5	S6	S7	S8
Aerobic sludge fraction	Aerobic sludge mass over total sludge mass	%	100%	83%	83%	83%	83%	88%	83%	83%	83%
Anoxic sludge fraction	Anoxic sludge mass in relation to total sludge	%	0%	17%	17%	17%	17%	12%	17%	17%	17%
Anaerobic sludge fraction	Anaerobic sludge mass in relation to total	%	0%	0%	0%	0%	0%	0%	0%	0%	0%
Anoxic COD utilization	In relation to COD_{BD}	%	19%	24%	23%	26%	28%	26%	22%	25%	27%
Aerobic COD utilization	in relation to COD_{BD}	%	81%	76%	77%	74%	72%	74%	78%	75%	73%
% Nitrification of TN	nitrification over influent TKN load	%	25%	52%	68%	53%	71%	50%	63%	48%	69%
% N removed via sludge	% N removed via WAS	%	49%	41%	24%	39%	22%	40%	25%	41%	21%
Effluent TN	% N removed via effluent	%	28%	23%	25%	22%	24%	23%	27%	21%	25%
Denitrification of TN	Denitrification of TN influent	%	24%	38%	53%	39%	55%	40%	50%	38%	54%
Oxygen balance		**Unit**	**S0**	**S1**	**S2**	**S3**	**S4**	**S5**	**S6**	**S7**	**S8**
Raw Influent TOD	Total Theoretical Oxygen Demand	kgTOD/d	11,976	15,376	15,376	15,376	15,376	15,376	15,376	15,376	15,376
Effluent TOD	Total Theoretical Oxygen Demand	kgTOD/d	1,876	934	937	914	917	1,137	1,308	1,229	1,184
Aeration TOD	Total Theoretical Oxygen Demand	kgTOD/d	3,388	5,560	7,827	5,538	7,109	5,291	7,430	5,163	7,007
Anoxic COD removal	Total Theoretical Oxygen Demand	kgTOD/d	577	1,027	1,429	1,068	1,482	1,077	1,351	1,028	1,469
Methane TOD	Total Theoretical Oxygen Demand	kgTOD/d	105	1,000	1,490	1,052	2,227	964	1,293	903	1,984
Sludge Production TOD	Total Theoretical Oxygen Demand	kgTOD/d	7,738	6,995	3,834	6,945	3,780	6,903	3,968	7,041	3,706
Primary sludge TOD	Total Theoretical Oxygen Demand	kgTOD/d	0	0	0	3,233	2,680	0	0	3,233	2,680
WAS TOD	Total Theoretical Oxygen Demand	kgTOD/d	7,754	9,308	6,393	6,496	4,854	9,204	6,300	6,330	4,426
Raw Influent	% of Total Influent TOD	%	100%	100%	100%	100%	100%	100%	100%	100%	100%
Effluent TOD	% of Total Influent TOD to EFF	%	16%	6%	6%	6%	6%	7%	9%	8%	8%
Aeration TOD	% of Total Influent TOD oxidized	%	28%	36%	51%	36%	46%	34%	48%	34%	46%
Anoxic COD removal	% of Total Influent TOD denitrified	%	5%	7%	9%	7%	10%	7%	9%	7%	10%
Methane TOD	% of Total Influent TOD to CH4	%	1%	7%	10%	7%	14%	6%	8%	6%	13%
Sludge Production TOD	% of Total Influent TOD to WAS	%	65%	45%	25%	45%	25%	45%	26%	46%	24%
Primary sludge TOD	% of Total Influent TOD to PS	%	0%	0%	0%	21%	17%	0%	0%	21%	17%
WAS TOD	% of Total Influent TOD to WAS	%	65%	61%	42%	42%	32%	60%	41%	41%	29%

Chapter 14: WWTP Amsterdam West, the Netherlands

Use of models to explain deterioration of effluent quality under wet weather conditions

Meijer S.C.F. and Piekema P.

This chapter is based on "Model evaluation of biological phosphorus removal under full-scale influent and storm conditions" by Meijer S.C.F. and Piekema P., presented at New Developments in IT & Water Conference, Amsterdam 4-6 November 2012.

14.1 Introduction

In the field of domestic waste water treatment there is an increasing requirement to improve effluent quality for the benefit of receiving surface waters. Additionally, there is a requirement to minimize energy consumption and the use of chemicals in the treatment process. Van Loosdrecht *et al.* showed that biological phosphorus removal in wastewater treatment contributes to a more efficient nitrogen and phosphorus removal resulting in lower energy consumption and reduced use of chemicals (van Loosdrecht *et al.,* 1998).

Enhanced Biological Phosphorus Removal (EBPR) is widely used in full-scale activated sludge wastewater treatment and also applied at Waste Water Treatment (WWTP) Amsterdam West. Operational experience points out that EBPR is sensitive to rain storm events causing effluent ortho-phosphate to increase and sustain for longer periods. Determining the principles behind this limitation has shown to be difficult using traditional plant evaluation methods. Each rain event is unique and depends on varying factors like temperature, the amount of rainfall, paved surface area, weather history, sewer history, type of transport mains etc.. Underpinning Rain Weather Flow (RWF) simulations with measured data is devious and costly when considering the amount of continuous measurements required for recording such an event. A more efficient and cost effective alternative is studying rain storm conditions in the WWTP by model simulation, as is demonstrated in this research.

The activated sludge model (ASM) used in this study is the combined ASM2d and TUDP-model (Van Veldhuizen *et al.,* 1999) developed in the framework of the International Association for Water Quality (IAWQ). This model includes a description of the cell-internal metabolism of the EBPR process and is especially equipped to assess the EBPR processes and detect possible biological limitations in the operation. The model is applied to assess the EBPR and effluent performance of WWTP Amsterdam West under rain storm conditions. For this (qualitative) assessment the model is calibrated based on measured plant data. A simplified sewer model is developed to simulate the rain storm event. A tracer simulation is performed to better understand the hydraulic and mixing characteristics of WWTP Amsterdam West. Finally, several biological process limitations are investigated using the model to better understand the underlying principles of the full-scale EBPR performance and how this is influenced by operation and plant design.

14.2 WWTP Amsterdam West

14.2.1 Plant and process description

The plant lay-out of WWTP Amsterdam West is presented in Figure 14.1. The design is according to the UCT layout (Ekama *et al.*, 1984). Table 14.1 presents the operational data and reactor volumes of WWTP Amsterdam-West.

Table 14.1 Measured operational data (flows) and reactor volumes of WWTP Amsterdam West

Stream	Flow (m^3/d)	Reactor	Volume (m^3)	Area (m^2)	Depth (m)	No.[1]
Total raw influent	171,700	Primary settling (PST)	6,675	1,935	N.A.	4
Primary sludge	4,200	Total primary settling	26,700	7,740	-	4
External primary sludge (AWP)	1,820	Anaerobic tank (ANT)	2,590	-	8	6
Thickened primary sludge (PS)	800	Denitrification tank (DNT)	4,350	-	8	6
Anaerobic recycle	139,350	Facultative aerated tank (FAC)	4,350	-	8	6
Anoxic recycle	429,410	Aerated tank (AT)	5,800	-	8	6
Facultative Recycle	720,000	Degassing tank	460	-	4	6
Return activated sludge (RAS)	202,820	Total bioreactors[1]	105,300	-	-	6
Waste activated sludge (WAS)	4,830	Secondary settler (SST)[1]	7,625	2,210	N.A.	12
External WAS AWP	1,590	Total secondary settlers	118,780	30,950	-	12
Thickened WAS (SS)	820	Primary thickening (PT)	1,570	523	N.A.	2
Centrate from digesters	2,010	Sludge digesters[2]	11,500	-	N.A.	3
Iron chloride	3.5					

[1] During the time of measurement 6 out of 7 activated sludge lines and 12 out of 14 SSTs were operational
[2] Digesters are not shown in flow scheme Figure 14.1.

At the time of measurement, the total treatment capacity of WWTP Amsterdam-West was approx. 778,000 person equivalents (PE) with a total average Dry Weather Flow (DWF) of 172,000 m^3/d. The plant was designed for COD, nitrogen (N) and Enhanced Biological Phosphorus Removal (EBPR). Figure 14.1 shows the plant layout. The combined sewage (mains North, South, East, West) is fed to four Primary Settling Tanks (PST) and after settling, is mixed and evenly distributed over seven parallel activated sludge treatment lines (of which 6 lines were in operation during the time of measurement). Each line consists of an Anaerobic Tank (ANT), Denitrification Tank (DNT), Facultative diffused bubble Aerated Tank (FAC), diffused air Aerated Tank (AT) and a Degassing Tank (DGT) allowing gas bubbles to escape from the water before being divided to the 14 Secondary Settling Tanks (SST). Each activated sludge line has two SSTs and the Return Activated Sludge (RAS) of the separate lines is not mixed. By a controlled valve system Waste Activated Sludge (WAS) is tapped from the SSTs and collected in a Secondary Sludge Buffer (SSB) and mixed with WAS from WWTP Amsterdam Westpoort (Figure 14.1, SS AWP). After Mechanical Belt Thickening (TH) the Secondary Sludge (SS) is digested (not in layout) together with the thickened Primary Sludge (PS) from the primary thickeners (and PS from WWTP Amsterdam Westpoort). Centrate from the sludge digestion and dewatering is collected and mixed with the raw influent (24-hour sampling was not operational).

The 7 activated sludge lines are 8 m deep circular tanks with in each tank the anaerobic, anoxic, facultative and aerobic zones configured in horizontal flowing concentric rings with horizontal propelled mixing. The internal flow in these tanks is estimated based on a horizontal flow velocity of 0.25 m/s (approx. 800,000 m^3/d per tank) which in theory is minimally required to keep activated sludge in suspension. Additional to the EBPR process, phosphate is removed chemically. Therefore iron is dosed in the second flow divider (FD2) leading to the biology. Denitrification is achieved by two internal anoxic recycle flows; one recycling nitrate rich activated sludge from the AT to the Denitrification Tank (DNT), and the other recycling nitrate rich activated sludge to the Facultative Aerated Tank (FAC) during unaerated periods to stimulate additional denitrification. An anaerobic recycle is used to feed low nitrate concentrated activated sludge from the DNT to the Anaerobic Tank (ANT) providing the EBPR process appropriate conditions for anaerobic substrate uptake and storage. Additional chemical phosphorus removal is used to bring Total Phosphorus (TP) in the effluent below 1.0 g P/m^3 by chemical precipitation. For Total Nitrogen (TN) the effluent discharge limit is 10 g N/m^3, for TSS 30 g/m^3, BOD_5 20 g/m^3 and for COD 125 g/m^3.

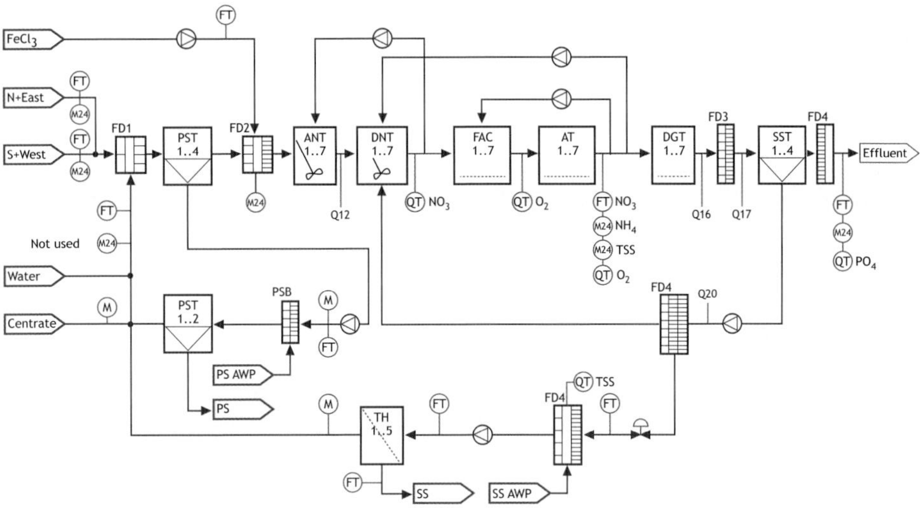

Figure 14.1 Plant lay-out of WWTP Amsterdam-West. PST: primary settling tank, ANT: anaerobic tank, DNT: denitrification tank, FAC: facultative aerated tank, SST: secondary settling tank, TH: sludge thickening belt, SSB: sludge storage buffer, PT: primary sludge thickening, FD: flow dividers. QT: on-line quality transmitter, FT: flow transmitter. M24: 24-hour flow proportional measurement.

Several measurements are performed in the process by on-line Quality Transmitters (QT) and Flow Transmitters (FT). Indicated in Figure 14.1 on-line quality measurements are; nitrate (NO_3), ammonium (NH_4), total suspended solids (TSS), Dissolved Oxygen (O_2). In the figure, M24 indicates a 24-hour refrigerated flow proportional automated sampling machine.

14.2.2 Measurements

The average performance of WWTP Amsterdam West is calculated from operational data and analytical measurements measured during 4 September 2006 until 24 April 2007 (Tables 14.2 and 14.3). The average temperature is 16.4°C. Some additional measurements were performed of the activated sludge composition (measured at the end of the aerobic zone), settled influent and centrate. The influent and effluent are 24-hour composite samples, whereas the centrate and process internal measurements are grab sampled during dry weather peak influent flow. The samples are analysed on TCOD (total COD), COD_{MF} (COD of the micro filtrated fraction, measured with a 0.45 μ pore diameter without applying flocculation), BOD_5 (biological oxygen demand), condensation residue (measuring TSS and soluble salts), ash fraction of the condensation residue (by TSS incineration), VFA (volatile fatty acids), NH_4-N, NO_3-N, TKN (Total Kjeldahl nitrogen), TP (total phosphorus), Ortho-phosphate (PO_4-P) and TSS (total suspended solids). Average pump flow-rates (Table 14.1), mixer/aerator running-time and energy consumption, and process control measurements were recorded daily.

Table 14.2 Average influent measurements (measurement period 4 September 2006 to 24 April 2007).

Measured/calculated influent	Parameter	Unit	Measured	Fraction in the influent[1)]	Fraction in the centrate[1)]
Flow [2)]	Q	m^3/d	28,612	26,163	2,469
Person equivalents, 136 g O_2	P.E.	PE	137,351	116,769	22,711
Total influent COD	TCOD	g COD/m^3	392	392	392
Influent COD micro-filtered	COD_{MF}	g COD $/m^3$	186	186	186
COD from VFA	COD_{VFA}	g COD $/m^3$	47	47	47
Influent BOD_5	BOD_5	g BOD/m^3	145	145	145
Total Kjeldahl nitrogen	TKN	g N/m^3	57	47	188
Total nitrogen	TN	g N/m^3	57	47	188
Ammonium	NH_4	g N/m^3	36	25	166
Nitrate + Nitrite	NO_X	g N/m^3	0.0	0,0	0,0
Total phosphorus	TP	g P/m^3	9.4	6.8	39.8
Ortho-phosphate	PO_4	g P/m^3	4.2	1.6	34.6
Total suspended solids	TSS	g/m^3	155	155	155
Measured effluent	**Parameter**	**Unit**	**Measured**	**Fraction in the influent***	**Fraction in the centrate***
Effluent COD micro-filtered	COD_{MF}	g COD/m^3	25	25	25
Effluent BOD_5	BOD_5	g BOD/m^3	0.3	0.3	0.3

[1)] For the purpose of this study the measured settled influent was divided in a centrate and influent fraction. This was calculated for TKN and TP based on a mass balance over the primary Settling Tanks (PST). [2)] Single lane (in total 6 of 7 lanes were operated).

Table 14.3 Average measurements and simulated results of effluent and the activated sludge (in g/m^3).

Components	Activated Sludge in aerobic tank		Effluent	
	Measured	Simulated	Measured	Simulated
TCOD	5,640	5,664	33.4	33.7
[1)]COD_{TSS}	-	5,639	8.6	8.9
TP	236	234	0.7	0.9
PO_4	-	0.5	0.4	0.5
$FePO_4$	211	155	0	0
TKN	352	352	2.1	2.1
NH_4	-	0.7	0.5	0.5
NO_X	-	9.1	5.0	6.1

[1)] Particulate fraction of COD

14.3 Model construction

14.3.1 Process model description

The SIMBA© simulator (Alex *et al.*, 1997) was used working with Matlab/Simulink©. The plant was modelled as a single line thereby assuming all lines were evenly loaded with settled influent. In SIMBA© plant hydraulics are modelled by connecting Continuous Stirred Tank Reactor-models (CSTRs). Biological conversions in the activated sludge were calculated based on the ASM/IAWQ-type TUDP-model using model parameters as presented by Van Veldhuizen *et al.* (1999) and model adaptations proposed by Meijer *et al.* (2001). The reactor and flow layout of WWTP Amsterdam West was modelled according to Figure 14.1 and the plant data in Table 14.1.

The model includes a simplified sewer and PST model only modelling hydraulic residence time. This is done to better predict the plant hydraulic behaviour under storm (RWF) conditions. Primary settling of particulates is not modelled and the settled influent quality is based on direct measurements in the overflow of the PSTs. The PST is modelled using three CSTR's in series which are fed with settled influent and only representing the top water layer in the PST. Mixing and Hydraulic Residence Time (HTR) in the sewer is modelled by a single plug flow system with an estimated hydraulic volume. The anaerobic tank (ANT), denitrification tank (DNT), facultative tank (FAC) and aeration tank (AT) each are modelled as four CSTRs in series, thereby assuming plug flow characteristics in the concentric rings of the horizontal flowing activated sludge tank. Each of these round flowing tanks (as in a carrousel) are simulated by introducing an internal recycle pump with a flow based on the vertical tank section (height × with) and a horizontal flow velocity of 0.25 m/s resulting in an internal flows of approximately 800,000 m^3/d per circular tank. The degassing tank (DGT) is modelled as one CSTR. The hydraulic mixing property of the secondary settling tank (SST) is modelled as a single CSTR for the water layer and as three CSTRs in series for the return activated sludge flow (RAS) to introduce denitrification in the SSTs.

In the model simulations, particulate COD (COD_X, COD_{TSS}) in the AT is set-point controlled by regulating the waste activated sludge pumps (WAS). Thereby the activated sludge model is maintained at a constant COD solid concentration (Table 14.3) and Sludge Retention Time (SRT) of 14.5 days (measured 15.7 days).

Aeration of both the facultative (FAC) and aeration tank (AT) is modelled using master-slave set-point control. DO in each of the modelled aerated CSTR's is Proportional-Integral (PI) controlled by regulating the diffused air flow rate. The measured DO concentration thereby is maintained on a set-point value by the slave controller. The master controller maintains the ammonium concentration in the AT on a constant set-point of 1.5 mg N-NH_4/l by constantly adjusting the local DO set-points between 0 and 3.0 mg O_2/L according to a linear relation. At low influent (TKN) loading conditions the controller reduces aeration of the FAC to save the use of energy and induce denitrification.

The modelled secondary clarifier (SST) separate solids and water ideally. Suspended solids in the effluent are modelled by introducing a percentile loss of solids from the settlers (0.085 % of the recycle sludge load). Some denitrification in the settlers was observed and therefore also modelled by introducing three CSTRs in series in the return sludge flow.

Chemical phosphorus removal ($FePO_4$) is applied by dosing Fe(III)Cl_3 in the second flow divider (Figure 14.1, FD2). This is done additional to biological phosphorus removal by the EBPR process. During the measurements, a constant Iron dosage was applied (193 mg Fe(III)/L solution dosed at 3.5 m^3/d).

14.3.2 Data reconciliation

Three adjustments on the raw plant data were required to close the mass balances and be able to effectively model the WWTP:

(i) The water balance over the WWTP needed to be corrected because, according to the mass balance, the measured influent and effluent flow were not corresponding. It was concluded that the effluent flow measuring device was the most accurate. To close the water balance, the measured influent flow was corrected by 6% based on the average influent flow measurements. In the reconciled data this resulted in 6% higher influent loads.

(ii) To close the solids (TSS and COD) balance over the secondary settlers (SST), the Return Activated Sludge Flow (RAS) was reduced by 13%. Thereby the WAS concentration and WAS production were increased relative to the raw measurements (Table 14.4, WAS COD).

(iii) Finally, a gap of 4,000 m^3/d in the internal water balance (including the centrate) was closed by assuming this flow consisted largely of clean process water.

Table 14.4 Measured (average values) and simulated loads. The loads are presented in a mass balance matrix with positive signs indicating loads entering the balance or being produced and negative signs indicating loads leaving the system or being converted.

Balances →	Flow	TP	TKN	NO_3	COD	O_2
Loads ↓	(m^3/d)	(kg P/d)	(kg N/d)	(kg N/d)	(kg/d)	(kg O_2/d)
Combined inflow biological tanks	28,611	252	1,541	0	10,581	0
Simulated	28,632	276	1,690	0	11,230	0
Effluent	-28,072	-18	-55	-134	-879	-8
Simulated	-28,052	-24	-58	-171	-946	-1
Waste activated sludge	-540	-234	-347	-3	-5,559	0
Simulated	-580	-252	-377	-2	-6,060	0
Nitrified load			-1,139	1,139		-5,203
Simulated			-1,255	1,255		-5,737
Denitrified load				-1,002	-2,875	
Simulated				-1,083	-3,108	
Aerobic COD conversion					-1,268	-1,268
Simulated					-1,116	-1,116
Stoichiometric oxygen demand						6,479
Simulated						6,854

14.3.3 Model influent characterization

The measured settled influent (Table 14.2) is characterised to the model input requirement according to the standardized Dutch STOWA method (STOWA 1996 and Roeleveld *et al.,* 2001). The (calibrated) model influent results are presented in Table 14.5. The centrate with high phosphorus and nitrogen content is modelled as a separate input and calculated from the total influent. The influent characterization determines de biodegradable COD fraction from BOD_5 measurements and a BOD model. This method can be used to determine an initial estimate of the model influent. In the calibration procedure (Meijer *et al.,* 2001 and 2002) the COD balance is fitted on the (reconciled) plant data by adjusting the influent parameters $X_I/(X_I+X_S)$, $S_I/(S_I+S_F)$ (and all model fractions i_N) to exactly reproduce the COD and N mass balances over the plant (Table 14.4). The calibration procedure is explained in the following section.

Table 14.5 Model influent characterization according to the STOWA method. Model parameters that are not in the table are assumed zero.

Calibrated model influent	Model	Unit	Measured settled influent	Settled influent	Centrate
Soluble compounds					
Readily biodegradable organics	S_F	g COD/m^3	115	115	115
Volatile fatty acids	S_A	g COD/m^3	47	47	47
Ammonium + ammonia nitrogen	S_{NH4}	g N/m^3	37	26	167
Inorganic soluble phosphorus	S_{PO4}	g P/m^3	6.6	3.7	37
Soluble inert organic matter	S_I	g COD/m^3	25	25	25
Alkalinity	S_{ALK}	g COD/m^3	7.0	7.0	7.0
Particulate compounds					
Particulate inert organic matter	X_I	g COD/m^3	171	171	171
Slowly biodegradable substrate	X_S	g COD/m^3	35	35	35
Mixed liquor suspended solids	X_{TSS}	g MLSS/m^3	195	195	195

14.4 Model calibration

Calibration is performed according to the method described by Meijer *et al.* (2001, 2002). Thereby the plant is simulated to steady state at average flow (one lane 28,611 m^3/d) and an average temperature of 16.4°C. The calibrated parameters are presented in Table 14.6. In the calibration procedure, the total phosphorus concentration (TP) in the aeration tank is fitted by controlling the WAS flow. The COD concentration in the aeration tank is fitted by controlling the influent fractions X_I and X_S (Table 14.5) resulting in the calibrated fraction $X_I/(X_I+X_S)$ (Table 14.6). Thereby accurate reproduction of the activated sludge composition and effluent results is obtained according to Table 14.3. Also, this resulted in the simulated mass balances according to Table 14.4 and a SRT of 14.5 days (raw data resulted in 15.7 days SRT).

Table 14.6 Calibrated model parameters. The parameters are fitted to the balanced plant data according to Meijer *et al.* (2001, 2002).

Calibration[1)]	Value	Default	Unit	Description	Purpose
$X_I/(X_I+X_S)$	0.829	-	g COD/ g COD	Fraction inert influent particulate	COD mass balance
$S_I/(S_I+S_F)$	0.176	-	g COD/ g COD	Fraction inert influent soluble	COD mass balance
i_N	0.063	0.03	g N/ g COD	Fraction N in activated sludge	TKN mass balance
K_{NH4}	0.50	1.0	mg N/L	Nitrification sensitivity for NH_4	Nitrification
K_{O2}	0.35	0.2	mg/L	Sensitivity for dissolved oxygen	Denitrification
g_{PP}	1.00	0.22	-	P-uptake	EBPR/Bio-P

[1)] The calibrated parameter $X_I/(X_I+X_S)$, $S_I/(S_I+S_F)$ and i_N relate to settled influent

In the calibration procedure, the model is fitted on the measured activated sludge composition and effluent concentrations for these measurements are considered most reliable. The model and data errors are distributed over the waste sludge production and the influent loading as these two measurements were considered less reliable. Calibration resulted in a model accuracy in the range of 10% within the measured mass balances (Table 14.4, comparing raw versus reconciled data). This model accuracy is sufficient for the purpose of this qualitative scenario study and the simulation results should be judged within this context.

14.5 Modelling dynamic influent conditions

14.5.1 Modelling dry weather flow

DWF is simulated based on a 24-hour dynamic flow and dynamic concentration profiles of COD, TKN and TP (according to Figure 14.2).

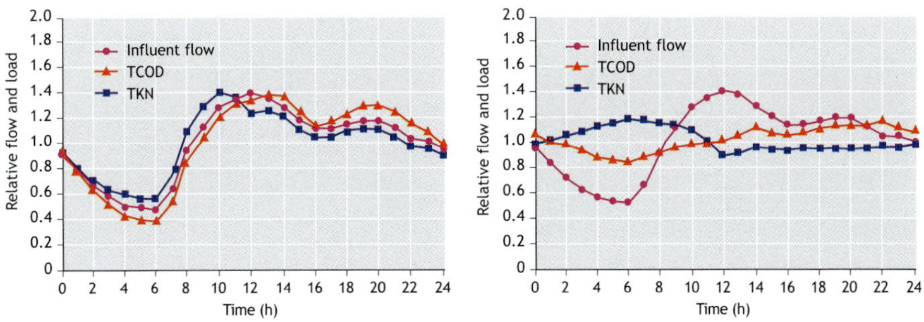

Figure 14.2 Typical 24-hour influent DWF and concentration profile used for the dynamic assessment. The influent data was originally measured at WWTP Hilversum and adapted for simulation of WWTP Amsterdam West.

This dynamic influent profile was originally measured at WWTP Hilversum and also representative for WWTP Amsterdam West. DWF typically peaks in the morning and again in the late afternoon and is low during the night. The time of the morning flow peak (PDWF) depends on the sewer transport delay. In Amsterdam West this delay is relative long due to flat surface and long (pumped) transport mains. The morning typically can be recognised by higher TKN and TP loads and the afternoon by a higher (particulate) COD load.

14.5.2 Modelling rain weather flow (RWF)

RWF is simulated using the SIMBA RWF simulator allowing storm water with a different composition of pollutants. Typically RWF transports more inert and less biodegradable organic particulate (X_I and X_S) material to the WWTP. This is mixed with the 24-hour DWF influent profile. In the model of WWTP Amsterdam West, RWF is fed to the simplified plug flow sewer model without side mains, conversions or back mixing. This model approach resulted in a typical first flush peak loading of the WWTP consisting of influent of DWF quality rapidly transported to the plant including some additional organic particulate material (COD_X) originating from RWF. Figures 14.3 and 14.4 show that during RWF the influent concentrations drop by approx. 50% while the influent loading initially becomes twice as high as a result of the first flush and rapidly emptying sewer. The figures also show that RWF carries little soluble phosphorus (S_{PO4}) and as well as soluble nitrogen (not shown here). As a result ammonium and ortho-phosphate concentrations are diluted following the initial first flush.

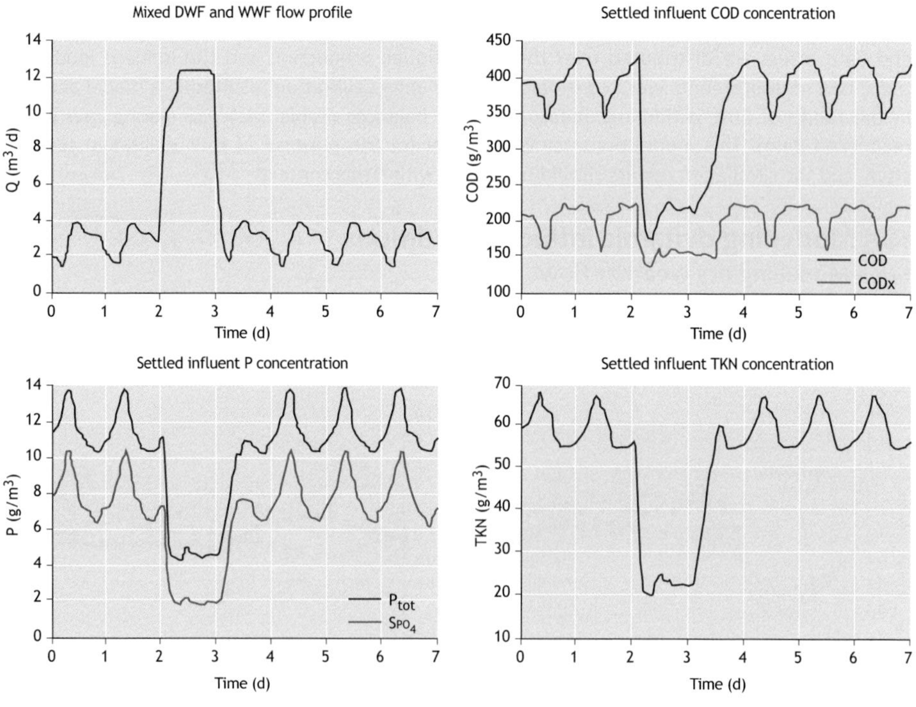

Figure 14.3 Constructed influent flow (Q) and concentration profile; total COD (COD), particulate COD (COD_X), total phosphorus (P_{tot}), ortho-phosphate (S_{PO4}) and TKN. The simulated influent profile is used for the dynamic model assessment of WWTP Amsterdam West.

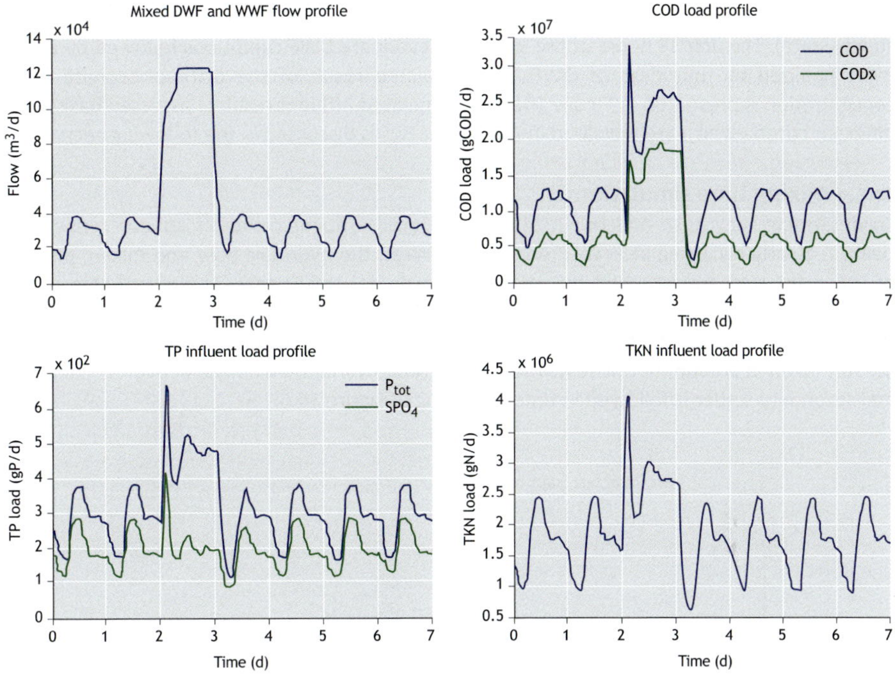

Figure 14.4 Constructed influent flow (Q) and load profile; total COD (COD), particulate COD (COD_X), total phosphorus (P_{tot}), ortho-phosphate (S_{PO4}) and TKN. The simulated influent profile is used for the dynamic model assessment of WWTP Amsterdam West.

14.5.3 Dynamic effluent simulation results

For the purpose of studying the dynamic effect of RWF on EBPR performance and effluent ortho-phosphate, dynamic conditions are applied to the calibrated steady state model. Therefore a hypothetical combined DWF/RWF influent event was simulated with a total duration of 7 days and according to the influent profiles in Figures 14.3 and 14.4. In this simulation no iron is dosed in the model to investigate the effect of RWF on the EBPR process.

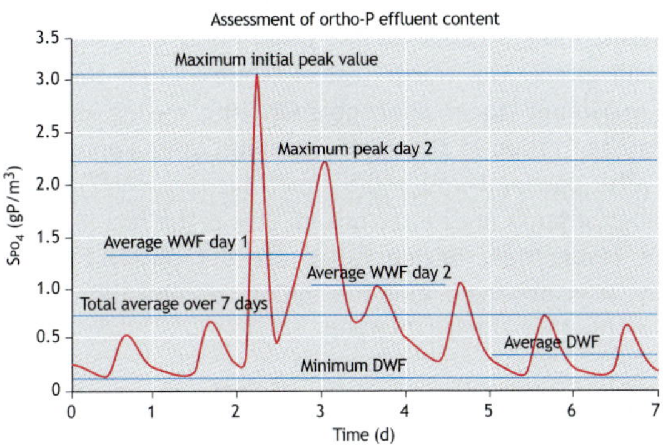

Figure 14.5 Simulated effluent ortho-phosphate performance as the result of the dynamic influent presented in Figures 14.3 and 14.4.

The influent event started from average DWF simulated to steady state for 1,000 days (the initial steady state). The first 48 hours of the simulated event are DWF conditions followed by 24 hours RWF, finalized by four days of DWF. The simulated result of the ortho-phosphate effluent concentration is presented in Figure 14.5. The simulated effluent result is assessed based on the different criteria which are indicated in the figure. This is discussed in the following sections.

14.5.4 Mixing time simulation

To determine the cause and origin of the measured effluent ortho-phosphate concentration peaks regularly occurring at WWTP Amsterdam West, the hydraulic flow and mixing properties of the plant also are assessed by performing a model tracer test. Therefore, at a constant influent flow of 28,610 m^3/d, a peak of inert soluble material (S_I) is introduced in the model to the biological tanks (15.0 mg/L dosed in FD2, Figure 14.1). The sewer and primary settler models are not taken in account in this test. The concentration variation over the activated sludge tanks and secondary settler is followed in time according to Figure 14.6.

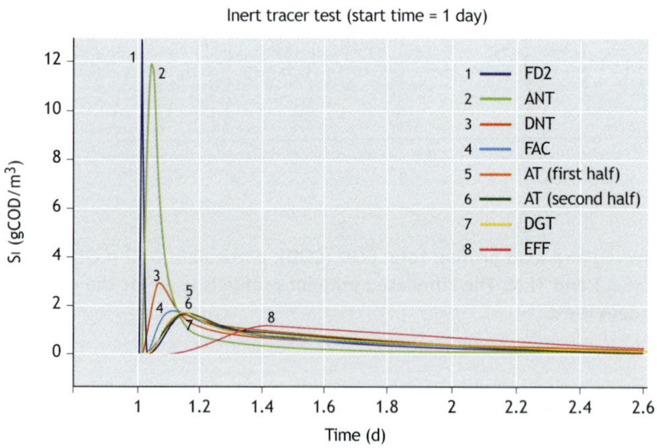

Figure 14.6 Hydraulic mixing characteristics of the calibrated model of WWTP Amsterdam West. At day one an inert tracer is dosed as a spike to FD2 (15.0 mg/l) and the concentration was followed in time in the following tank reactors; FD2: flow divider, ANT: anaerobic tank, DNT: denitrification tank, FAC: facultative aerated tank, AT: aerobic tank (separated in two tanks), EFF: effluent after secondary settling.

The tracer simulation (Figure 14.6) shows how peak loadings are reduced in the WWTP by dilution. Thereby ortho-phosphate is buffered in the process. On the other hand, concentration buffering results in a relative strong tailing effect. The tracer test also shows that the time needed for the effluent to rise to the maximum tracer concentration (the mixing time) is approximately 11 hours. The maximum concentration in the AT is observed after approx. 3.5 hours. The tracer peak rapidly decreases after passing through the anaerobic tank (ANT) which indicates that the anoxic and aerated biological tanks are better mixed. This is the result of the high RAS and anoxic recycle flows (the respectively mixing ratios are 1.1 and 2.1 times proportional to the influent flow). Finally, it is observed that the initial concentration spike results in an elevated effluent concentration (tail) for at least 38 hours.

14.6 Plant assessment and discussion

14.6.1 Considerations on the model and simulations

Each rain event is unique and depends on varying factors like temperature, the amount of rainfall, paved surface area, weather history, sewer history, type of transport mains etc.

Underpinning RWF simulations based on measured RWF data is devious and costly when considering the amount of continuous measurements that are required to record such an event. Studying RWF by model simulation, as is demonstrated in this article, is an efficient and cost effective alternative giving insight in the plant performance as the result of RWF. The quantitative results of this model study may not be fully accurate however, the results are explaining the main principles which occur under dynamic RWF loading events.

The operational experience of WWTP Amsterdam West is that after RWF often an increased ortho-phosphate effluent concentrations is measured with a concentration 2 to 3 times higher than average (results are not shown here). In general the RWF simulations here presented largely reproduce the actual plant behaviour (Figure 14.5). A difference is however, that in practice at WWTP Amsterdam west RWF typically shows only one (distinguishable) ortho-phosphate peak in the effluent. This could be caused by the modelled inflow dynamics (e.g. by an exaggerated simulation of the first flush) or by inaccurate modelled plant or sewer mixing properties. First flush conditions in WWTP Amsterdam West may not be as strong as modelled in this study. In practice this effect will very much depend on the type of rain storm and the weather history. However, for the purpose of this study the modelled extreme RWF allowed a better observation and assessment of the different effects of RWF on EBPR and effluent ortho-phosphate performance. The modelled mixing characteristics could be improved by reducing the amount of CSTRs in series thereby obtaining a more completely mixed system. However, for the purpose of this study the current model was acceptable.

14.6.2 Effluent performance under dry weather conditions

Simulation of DWF shows a reoccurring variation of the effluent ortho-phosphate concentration. Effluent ortho-phosphate peaks approximately 6 hours after DWF (compare Figures 14.4 and 14.5). If this effluent peak would origin from the ANT, a minimal mixing time of approx. 10 hours would be expected (Figure 14.6). If this effluent peak would origin from the influent, approx. 11 hours would be expected. This indicates that the effluent ortho-phosphate variation most likely origins from the anoxic (DNT) or facultative tank (FAC). This points to a possible limitation of EBPR during DWF. Occurrence of this possible biological process limitation is discussed further on.

The average effluent ortho-phosphate concentration simulated at constant influent flow (steady state simulation) is similar to the average result of the dynamic DWF simulation. This indicates that buffering the influent flow and thereby reducing DWF dynamics as a measure to improve the effluent ortho-phosphate performance, most likely is not effective. On the other hand, RWF causes the average effluent ortho-phosphate concentration to increase from approx. 0.35 mg/l (Figure 14.5; Average DWF) to 0.75 mg/L (Figure 14.5; Total average over 7 days). This concentration increase is considerable taking in account the ortho-phosphate effluent discharge limits of WWTP Amsterdam West. Therefore buffering rain water is would most likely improve the effluent ortho-phosphate performance. Also internal peak loadings with ortho-phosphate originating from the digesters could have a considerable effect on the average effluent performance and by reducing these peaks the effluent performance may be improved. However, in this study this possibility to improve the effluent performance was not further evaluated.

14.6.3 Effluent performance under rain weather conditions

The resulting RWF effluent peak (Figure 14.5) can be explained in vive steps: (i) In the first instance, concentrated ortho-phosphate (> 12 mg/L) is hydraulically flushed from the anaerobic tank (ANT); (ii) This is followed by the first flush, rapidly pushing concentrated DWF (approx. 4 mg P-PO_4/L) from the sewer and primary settling tanks (PST) into the biological tanks (ANT). Here, the increased water load leads to a shorter HRT and mixing time; (iii) After an initial maximum effluent peak, relative clean rain water with little ortho-phosphate (and ammonium) is entering the plant resulting in a drop of the ortho-phosphate effluent concentration; (iv) The first effluent peak then is followed by a second and longer lasting peak. This second peak points to a possible biological limitation of the EBPR process; (v) finally, a lagging effluent ortho-phosphate tail extends over 1 to 2 days. The effluent tailing very much resembles the tracer simulation (Figure 14.6) which was simulated by introducing a similar anaerobic peak concentration (12 mg/L). The sustaining high effluent concentration indicates that little EBPR is taking place during the period following RWF. The tracer simulation (Figure 14.6) shows that effluent peak loadings are reduced by dilution. The trade-off of the dilution effect is that ortho-phosphate is buffered in the process resulting in the tailing effect. In theory tailing could be reduced by increasing the internal mixing flows (RAS and the anoxic recycle flow) following RWF for 1 to 2 days. However, this only would be effective when also the EBPR process is fully operational and recycled ortho-phosphate can be biologically removed. The simulated tail in Figure 14.5 indicates that EBPR is limited during the occurrence of tailing. A possible explanation is given by the metabolic TUDP model and discussed in the following section.

14.6.4 Model-based assessment of the EBPR process

The combined ASM2d and TUDP-model used in this study describes the EBPR process from the perspective of cell internal metabolism of Phosphorus Removing Organisms (PAO) responsible for EBPR. The model is used to examine limitations in the EBPR process during storm events in WWTP Amsterdam West. Figure 14.7 schematically explains the normal DWF EBPR cycle in the anaerobic, anoxic and aerobic reactor zones. Limitation of the EBPR process will occur as the result of three possible conditions; (i) under anaerobic conditions, when there is shortage of VFA (mainly acetate or propionate) or when the PAO do not have enough poly-phosphate (PP) or glycogen (GLY) stored cell internal, (ii) under anoxic conditions when there is shortage of nitrate (NO_3) or Poly-Hydroxy-Alkanoate (PHA), (iii) under aerobic conditions when oxygen (O_2) or PHA is limiting.

Limitation of EBPR due to shortage of VFAs

Figure 14.7 shows the simulated result of the TUDP model during RWF simulation. It shows that RWF initiates the decrease of anaerobic stored PHA; between approximately day 2.2 and 3.2 PHA becomes limiting for uptake of ortho-phosphate in the activated sludge process (EBPR). This biological limitation is the main cause of the second occurring effluent ortho-phosphate peak (Figure 14.5 and also Figure 14.7 left).

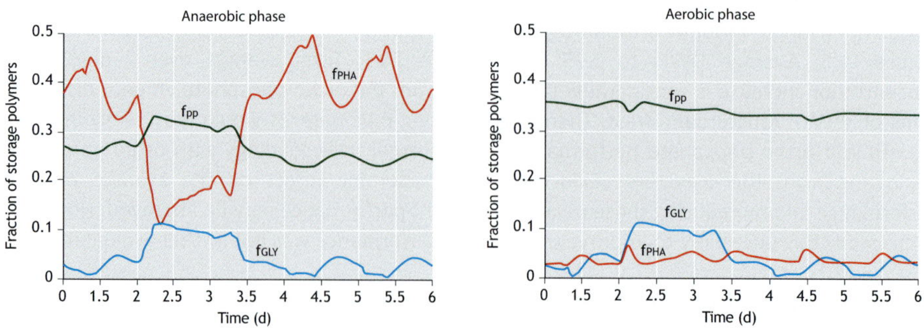

Figure 14.7 Fractions of PAO cell internal storage of glycogen (GLY), poly-hydroxy-alkanoate (PHA) and poly-phosphate (PP) relative to the PAO concentration followed in time, from day 1 to day 6 of the simulated RWF event. The left figure shows the modelled storage intermediates under anaerobic conditions (in the ANT). The right figure depicts the situation under aerobic conditions (in the AT).

The simulations show that PHA is depleted (oxidized) and not restored because (i) there is a shortage of VFA in the RWF influent and (ii) because the residence time in the anaerobic tank (ANT) has become too short for formation of VFA by hydrolysis and fermentation of organic material. Because VFA uptake is not occurring, also glycogen is not converted and therefore accumulates in PAOs (Figure 14.7, GLY). When RWF stops and the influent flow returns to DWF at approximately day 3.2, also hydrolysis and fermentation are restored and VFAs anaerobically produced in the ANT from the fresh influent supply of influent organics. Approximately 12 hours after RWF, the PHA limitation is lifted completely and from that respect EBPR is no longer limited. However, a second limitation remains which is discussed below.

Limitation of EBPR due to shortage of electron acceptor

Following day 3.2, PHA and VFA are no longer limiting for EBPR. However, ortho-phosphate uptake still is lagging resulting in the tailing effluent concentration (Figures 14.5 and 14.9, top). Further model investigation shows that during RWF and nightly (low) DWF there is a general shortage of electron acceptor (oxygen and nitrate) in the process resulting in a reduced anoxic and aerobic EBPR activity. Nitrate in the anoxic tank (DNT) is limiting for anoxic EBPR from day 2 up to approx. day 4.7 (Figure 14.8). During low nightly DWF conditions there is shortage of nitrate as well as oxygen in the facultative tank (FAC) (Figure 14.8). During all times the AT is fully aerated and no limitation of oxygen for EBPR occurred there (not shown in Figure 14.8).

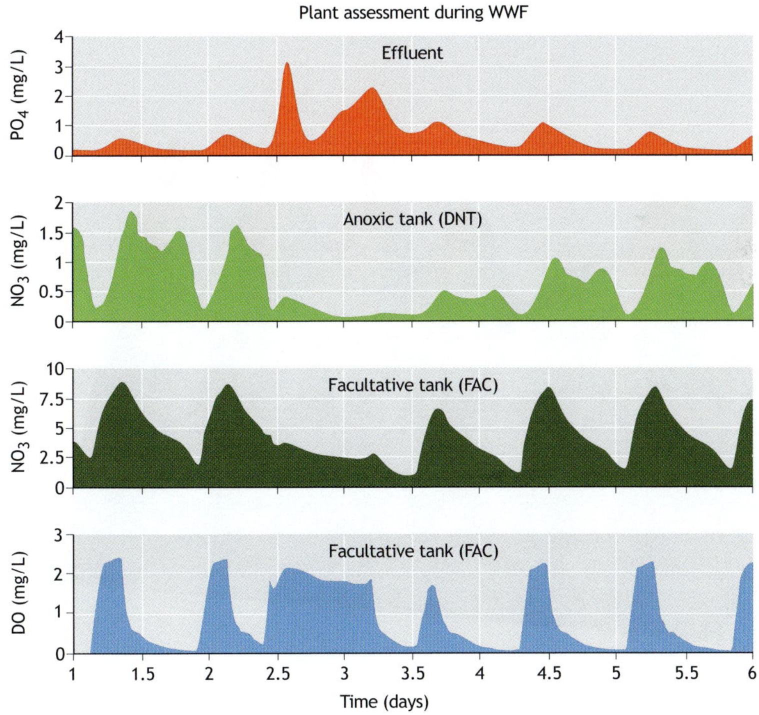

Figure 14.8 Overview of limiting electron acceptor conditions occurring in the anoxic (DNT) and facultative tanks (FAC). In the nights and after RWF, nitrate is limiting for anoxic EBPR in both the DNT and FAC tanks. Also during the night oxygen is limiting in the FAC tank. This is the combined result of low influent loading and process control setting typically applied in most WWTPs as well as Amsterdam West.

EBPR requires an electron acceptor (oxygen or nitrate) to take up ortho-phosphate (Figure 1.13). Figure 14.9 shows the simulated plant-wide nitrate profile of WWTP Amsterdam West; it is observed that limiting anoxic EBPR conditions occur typical following RWF and during the night when DWF is low. Because the TKN loading is low during the night, the master-slave dissolved oxygen (DO) controller measures little ammonium in the effluent. The DO controller automatically reduces the aeration of the facultative tank (FAC) and eventually also the aeration of the AT. This is a logical control action from the perspective of saving energy and to induce denitrification in the FAC and AT and thereby improve the overall removal of nitrate. As a result the nitrate concentration in the process drops because less nitrate is formed by nitrification and more nitrate is denitrified. Nightly shortage of electron acceptor (nitrate and oxygen) however results in an increasing effluent ortho-phosphate concentration. That a limitation of EBPR also occurs during DWF is observed from Figure 14.7. Small PHA peaks are observed during the night (at 1.5, 3.5, 4.5 and 5.5 days) coinciding with low nitrate (and oxygen) concentrations in the plant (Figure 14.8 and 14.9). These nitrate (and oxygen) dips are followed by effluent ortho-phosphate peaks in the morning (Figure 14.8). PHA accumulates aerobically because there is a lack of electron acceptor in the form of oxygen or nitrate. The time that it takes between the process becoming limited and the time that the ortho-phosphate peak appears in the effluent corresponds with the mixing time of the process (Figure 14.6). This is approximately 8 hours, corresponding with the time delay observed from Figure 14.8; this is the time between peak effluent and the minimum NO_3 and O_2 concentrations in DNT and FAC.

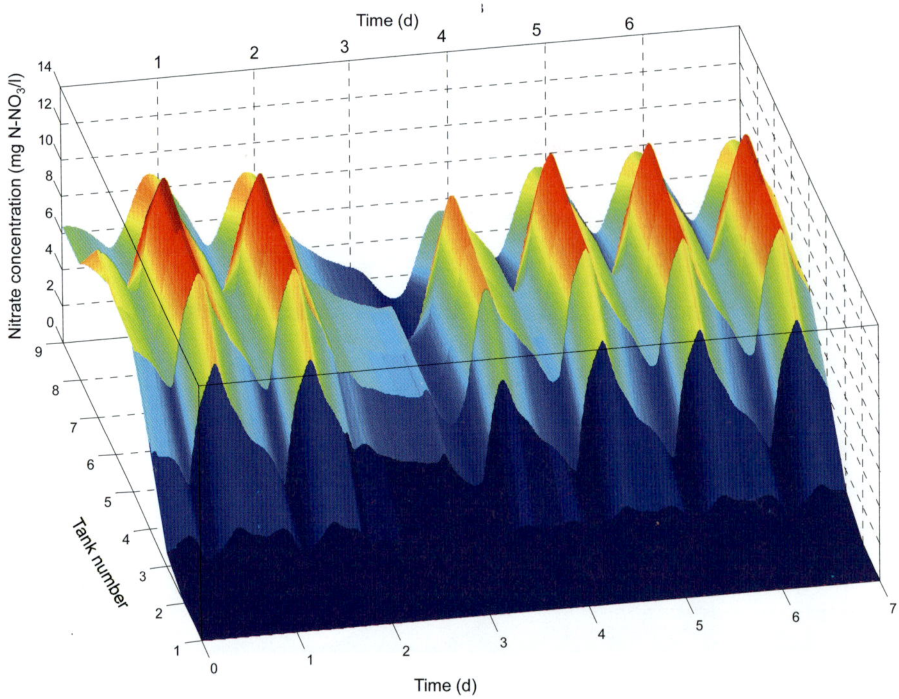

Figure 14.9 Simulated nitrate profile over the entire plant during the DWF and RWF simulation. Tank numbers represent: (1) FD2, (2) ANT, (3) DNT, (4) FAC, (5+6) AT, (7) DGT, (8) SST, (9) effluent. During RWF and low DWF in the night, nitrate is limiting in the process for anoxic EBPR. As a result of the HRT, peaks in the reactors move towards the effluent in approximately 11 hours.

14.7 Conclusions

We were successful in constructing a model of the activated sludge process of WWTP Amsterdam West was including a relative accurate model of the actual dynamic process aeration control. The model was calibrated based on accurate and reconciled plant data within ± 10% accuracy of the established mass balances. We successfully used the combined ASM2d TUDP model for dynamic plant performance. A rain storm event was modelled based on a simplified sewer model. This model predicted typical occurrence of RWF, first flush and influent dilution. The WWTP and influent model were used for qualitative analysis of the EBPR and effluent performance under steady state, dynamic DWF and dynamic RWF conditions. The obtained effluent simulation results are exemplary for typical dynamic DWF and RWF effluent measurements including typical observations of ortho-phosphate peak behaviour at WWTP Amsterdam West during and after rain events. From the model it was concluded that phosphorus peaks in the effluent during DWF are caused by anoxic and aerobic limitation of the EBPR process arising during the night. When the influent TKN loading and effluent ammonium are low, automatic aeration control reduces aeration in the process to save energy and induce denitrification. As the result of this typical DO controller action, too little oxygen and/or nitrate is available for biological phosphate removal. This causes the effluent ortho-phosphate concentration to increase. Also the typical observed tail in the effluent concentration following RWF, which in the model sustained up to 38 hours, is caused by a biological limitation of the EBPR process as the result of the DO control action during the night. Following RWF, as the result of dilution effluent ammonium is low. Automatic aeration control responds to this measurement by reducing the aeration resulting in a limiting availability of oxygen and as well as nitrate for phosphate removal. From the model it is concluded that the initial effluent peak during RWF is a combined effect of; (i) an hydraulic and mixing time effect by the increased RWF influent flow rapidly transporting ortho-phosphate from the influent and anaerobic tank towards the effluent and (ii) a PHA limitation of the EBPR process caused by disruption of anaerobic VFA formation and uptake and RWF influent dilution.

It is concluded that reducing DWF peak flow dynamics in the influent is not useful to improve the phosphate effluent performance. On the other hand, buffering rain water is expected to result in an improvement of the effluent phosphate performance.

It is concluded that to improve the effluent ortho-phosphate performance of WWTP Amsterdam West, sufficient aeration of the facultative and aeration tank is required at all times, also when the plant is low loaded or when (effluent) ammonia is already low. These conditions typically occur in the night and after RWF. On top of the additional aeration, the Return Activated Sludge and internal recycle flows can be increased to reduce the effluent tailing of ortho-phosphate after RWF.

From this study it is shown that limitations of the EBPR process can be explained very well form TUDP model simulations. The study shows that the limitations partly can be resolved by adapting an appropriate control strategy for the aeration system. Typically used master-slave DO control relying on effluent ammonium concentration measurements causes limitations of EBPR in the night and after rain storms. By obtaining a sufficient aeration also during these conditions better effluent performance is expected. This would be at the cost of a higher overall energy consumption. This conclusion seems to be in contradiction with the previous conclusions of van Loosdrecht et al. (1998) indicating that introduction of EBPR results in a net reduction of energy. Further research is required to investigate this effect and come to better aeration control strategies adapted to nitrification, denitrification as well as EBPR.

References

Alex J., Tschepetzki R. and Bitter U. (1997) SIMBA Simulation of biological waste water treatment. User's manual, Institute of Automation and Communication e. V., Magdeburg.

Ekama G. A., Marais G. v. R. and Siebritz I. P. (1984) Biological excess phosphorus removal. Theory design and operation of nutrient removal activated sludge processes 7, 1-32. Water Research Commission, Pretoria, South Africa.

Meijer S. C. F., van Loosdrecht M. C. M. and Heijnen J. J. (2001) Metabolic modelling of full-scale biological nitrogen and phosphorus removing WWTP's. *Water Res.* 35(11), 2711-2723.

Meijer S.C.F., van der Spoel H., Susanti S., Heijnen J.J. and van Loosdrecht M.C.M. (2002) Error diagnostics and data reconciliation for activated sludge modeling using mass balances. Water Sci. Tech. 25(6), 145-156.

Meijer S.C.F. (2004) Theoretical and practical aspects of modeling biological nutrient removal in activated sludge systems (ISDN 90-9018027-3) Ph.D. Thesis, Delft University of technology, The Netherlands (205p.)

Roeleveld P.J and van Loosdrecht M.C.M. (2001) Experiences with guidelines for wastewater characterization in The Netherlands. *Wat. Sci. Tech.* 45(6), 77-87.

Henze M., Gujer W., Mino T., Matsuo T., Wentzel M. C. and Marais GvR. and van Loosdrecht M. C. M. (1999) Activated Sludge Model No. 2d, ASM2d. *Wat. Sci. Tech.* 39(1), 165-182.

Van Veldhuizen H. M., van Loosdrecht M. C.M. and Heijnen J. J. (1999) Modelling biological phosphorus and nitrogen removal in a full scale activated sludge process. *Wat. Res.* 33(16), 3459-3468.

STOWA (1996). Methods for wastewater characterisation, inventory and guidelines (in Dutch), STOWA report 97–23, Hageman Fulfilment, Zwijndrecht, The Netherlands.

Van Loosdrecht M. C. M., Brandse F. A. and de Vries A. C. (1998) Upgrading of wastewater treatment processes for integrated nutrient removal – The BCFS process. Water Sci. Technol. 37(9), 209-217.

Figure 14.10 WWTP Amsterdam West during construction (Photo: Waternet)

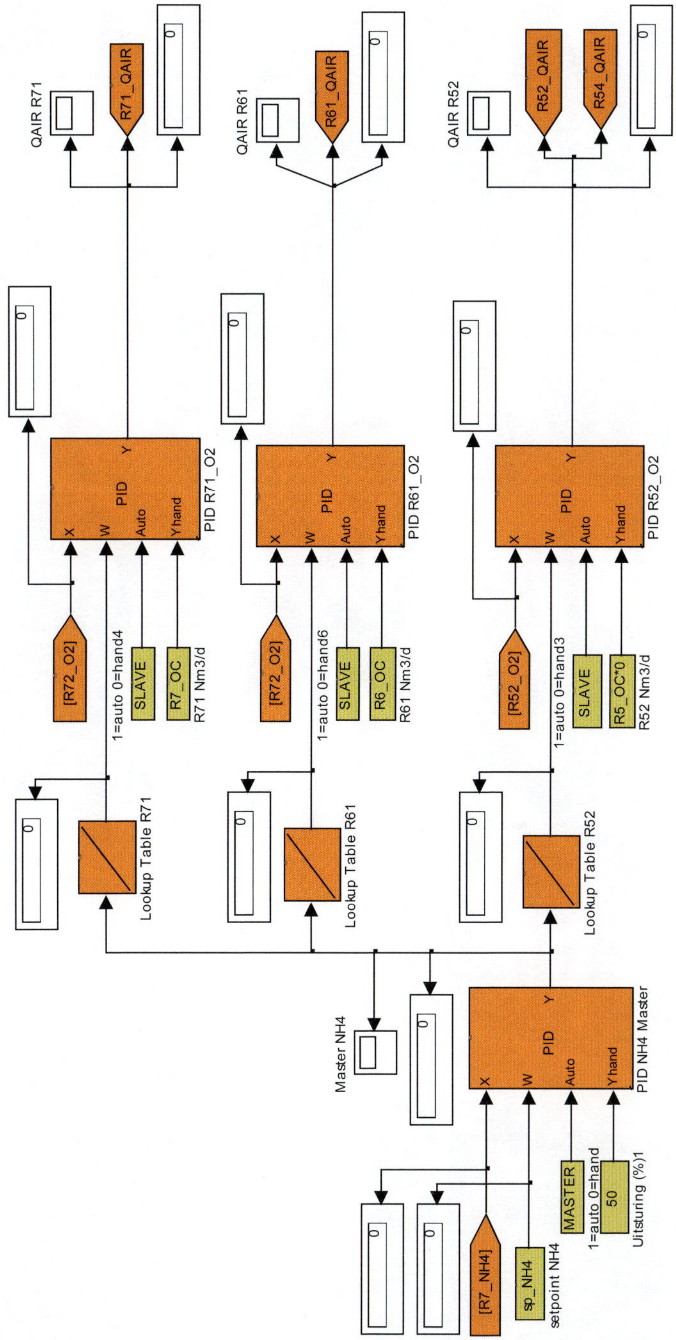

Figure 14.11 Simulated master-slave cascade controller applied in WWTP Amsterdam West. The effluent ammonia concentration is measured using an on-line measurement device. A master controller calculates an actual oxygen set-point. The slave controller regulates the aeration valves to obtain the preset DO set-point in the aerated tanks.

Figure 14.12 WWTP Amsterdam West (Photo: Waternet)

Chapter 15: WWTP Houtrust, the Netherlands

Plant upgrade using big-data and reconciliation techniques

Meijer S.C.F., van Kempen R.N.A. and Appeldoorn K.J.

15.1 Introduction

This chapter describes the performance evaluation of wastewater treatment plant (WWTP) Houtrust which was not performing as expected after refurbishment in 2008 (Figure 15.1).

Figure 15.1 Aerial photograph of WWTP Houtrust in The Hague, The Netherlands. The Houtrust treatment plant is largely built underground. Below the parking lot are the primary setting tanks, selector, pre-denitrification, anaerobic and anoxic tanks. Eight activated sludge tanks are arranged in four lines and are covered (at the bottom of the photo). The thirty two rectangular secondary clarifier tanks are stacked in a double layer arrangement (middle section). The plant includes digestion and sludge processing (top section).

In 2003 the water authorities of Delfland water authorities signed the Design, Build, Finance, Operation and Maintenance (DBFO) contract and the special purpose company Delfluent concerning two plants in the Netherlands, namely: WWTP Houtrust and WWTP Harnaschpolder under a Public Private Partnership (PPP) framework. The contract concerns one of the largest wastewater projects in Western Europe up to date. In this framework, Delfluent Services (DSBV) is responsible for the operational activities for a period of 30 years of the two plants and the accompanying transport systems with 19 sewage pumping stations. Delfluent has sub-contracted the design and construction to BAHR through the EPC contract, and operation and maintenance to DSBV through the operation and maintenance contract (Figure 15.2).

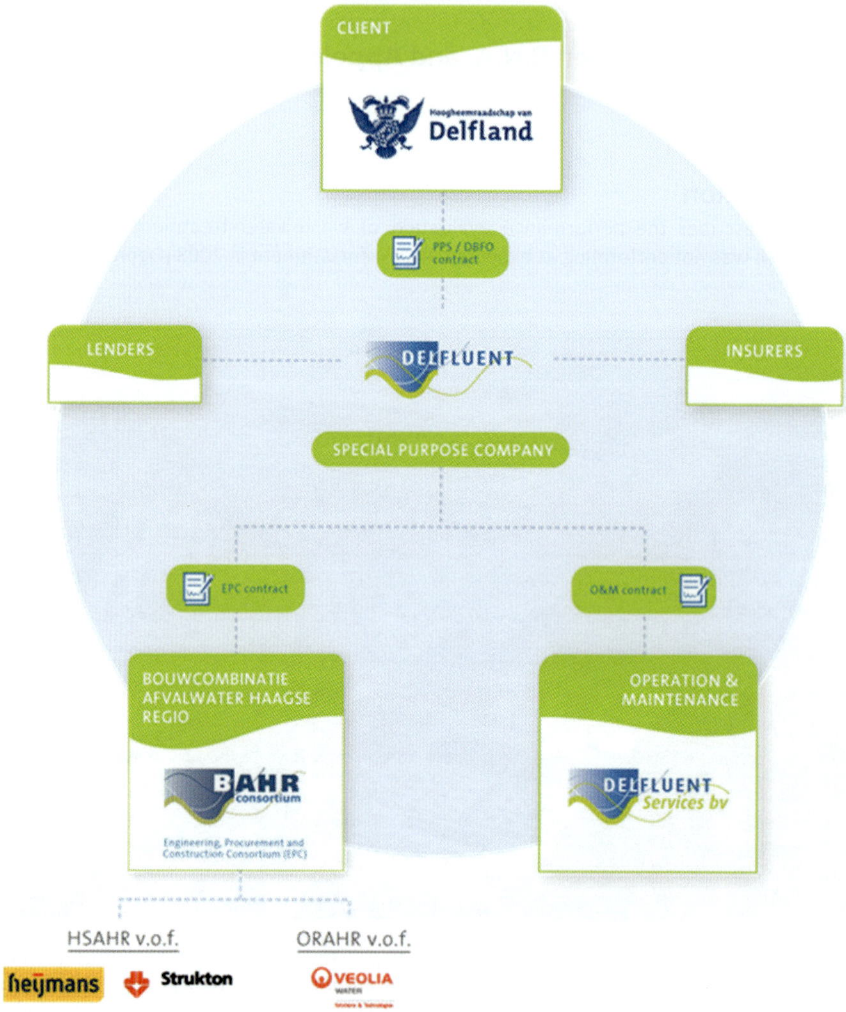

Figure 15.2 The operational framework of WWTP Houtrust and WWTP Harnaschpolder (Courtesy of Delfluent Services BV).

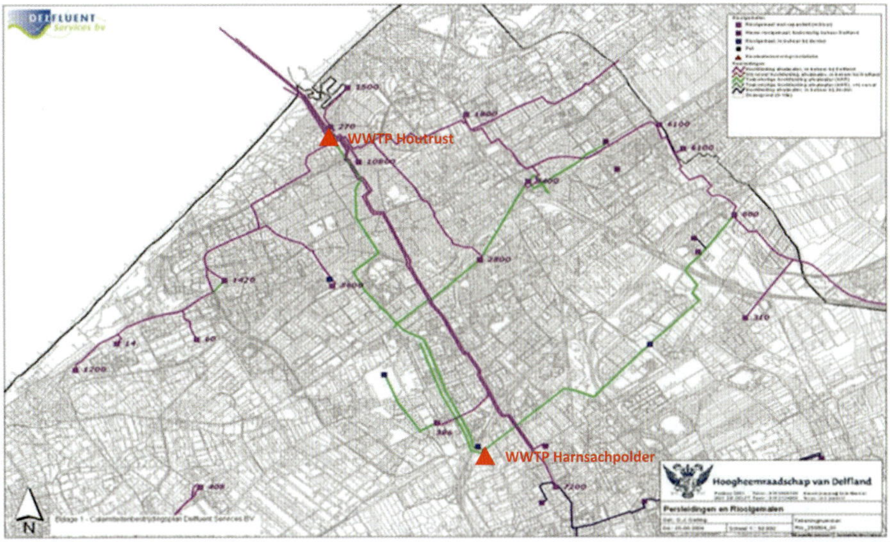

Figure 15.3 Sewage collection area of WWTP Houtrust and WWTP Harnaschpolder with the main pumping stations (squares) and sewer transport mains. The main effluent discharge is in the North Sea. Sewerage is built in a ring arrangement thereby creating redundancy in pumping capacity. Sewage can be transported back and forth from Houtrust to Harnaschpolder.

15.2 Performance assessment of WWTP Houtrust

According the DBFO contract in 2008, the plant in Houtrust was refurbished and retrofitted by a "three stage Phoredox process" or A2/O design. Prior the refurbishment, WWTP Houtrust was a BOD-removing and partly nitrifying plant with primary settling and combined digestion of primary and secondary sludge. Plant effluent is discharged in the North Sea. Under the former discharge regulations there were no requirements for nutrient removal. The construction of the new and comparatively much larger WWTP Harnaschpolder (1.3 million PE) including a new transport system in 2008, made it possible to redirect 75 % of the Houtrust influent towards WWTP Harnaschpolder. This new arrangement made it possible to upgrade and retrofit WWTP Houtrust towards a low loaded COD and nutrient (nitrogen and phosphorus) removing activated sludge plant. The upgraded WWTP Houtrust, usually loaded at 70-80% of the design capacity, was equipped to perform according to the contractual agreements and effluent requirements at full design loading. In the first year of operation of WWTP Houtrust it was concluded that the plant was underperforming based on a higher than expected methanol dosage to meet the total nitrogen (TN) effluent requirement, and also higher than expected iron dosage to meet the total phosphorus (TP) effluent requirement. It was decided to assess the plant performance at full design loading using a model simulation study, and further, to find out whether the plant can meet the contractual effluent requirements at the forecasted design loading of 104% which includes a 4% contractual safety requirement.

The decision to make use of activated sludge modelling choice was only made in 2010, after performing a more traditional consultancy design retrofit. The conclusion of this consultancy study was that more precise and advanced (model-based) design methods are required to draw conclusions with more confidence on the observed and future plant performance. It was also concluded in this study that the treatment data (measured in 2008 and 2009) included inaccuracies and errors and therefore in its raw form data were not very useful for an accurate plant performance assessment. Furthermore, it was decided to introduce another model study with aim to improve the quality of the measured treatment data. These two studies together were named "N-tot study WWTP Houtrust" (N-tot study).

15.3 N-tot study WWTP Houtrust

The goal was to specify the Total Nitrogen (TN) removal performance of WWTP Houtrust at full design loading. This specialist modelling task was performed by ASM design B.V. (currently Yuniko B.V.). Initially, three phases were planned in the study; (phase 1) checking and assessing the measured historical plant performance data, (phase 2) development of a simulation model, and (phase 3) testing the plant performance at full design loading by model simulation. Data evaluation is performed using state-of-the-art reconciliation techniques to detect and correct possible errors in the plant performance data. The main goal of this project was to demonstrate the methodology and effectiveness of these methods for reliable model based assessment and design. After the initial three phases, the N-tot study was extended over a period of 3 years, starting in 2010 and finalizing in 2013 (Figure 15.4).

Figure 15.4 Time planning and overview of the different phases of the N-tot study. The data evaluation and calibration method is discussed in this chapter based on the 2010 phases 1 and 2 of the N-tot study (green boxes).

This study resulted in a partly redesign and fully optimized performance in 2013. The 2010 data evaluation study confirmed that the plant data measured in 2008-2009 was indeed not reliable and included major errors and inaccuracies leading to a misinterpretation of the actual performance of WWTP Houtrust. From the data evaluation it was partly possible to repair and improve the information using statistical analysis and mathematical reconciliation techniques. Analysis of the 2008-2009 measurement plan pointed out that the data collection was not complete and critical treatment information was missing which making it impossible to draw reliable and decisive conclusions. To fil the critical information gaps, additional data was collected in 2010. By this route, a complete data set was obtained, representative for the average plant performance. In a modelling study this data was accurately retrofitted using a calibrated activated sludge model. This model was extrapolated to predict full influent loading conditions (Figure 15.4, phase 3). The results confirmed that at full design loading the plant would was unlikely to meet the contractual performance and that the performance was (very much) limited in its denitrification capacity. A high methanol dosage was predicted to meet the contractual TN effluent limit, which was in disagreement with the contractual requirements. The simulation also predicted that bioP-removal was underperforming despite the fact that the hydraulic residence time (HRT) in the anaerobic tank was more than sufficient. The Dutch (STOWA) design guidelines recommend 0.5h as minimal HRT taken into account local conditions, whereas one hour HRT was available (STOWA, 2001-15, 'Handbook biological phosphorus removal'). The results of the modelling study show that the observed limitations of the EBPR process where most likely caused by excessive amount of nitrate recycled to the anaerobic tank (caused by limitations of the denitrification capacity) and the most likely insufficient mixing and sludge retention in the anaerobic tanks leading to a disruption of the EBPR process.

Based on these preliminary conclusions, a follow up of the N-tot project was initialized with the goal to improve the N and P removal of the plant by redesign and eventually, to establish the improvements in a final reference measurement. This led to several design adjustments, all of which were calculated by the activated sludge model. In 2010, a new volume layout of the biological tanks was implemented, thereby enlarging the anoxic zones at the cost of anaerobic volume. Assumed mixing problems in the anaerobic and anoxic tanks were addressed and several other process optimizations were carried out, under which improvement of the anoxic and sludge return recycle flows and improvements of the measurement and control system of the biological reactors. As result, the plant performance improved considerably. In April 2013, a final plant evaluation was performed. To make a straight comparison possible between the previous and new plant performance, both model assessments were performed using the same methodology (Meijer and Brdjanovic, 2012). This chapter demonstrates the use of this methodology and thereby goes into detail in the data evaluation procedure. Thereby project phases 1 and 2 are used as an exemplary case study.

The final assessment convincingly showed that the redesign was a success. A large improvement of the nutrient removal capacity was established, concerning both nitrogen and phosphorus removal (Figure 15.5 and 15.6).

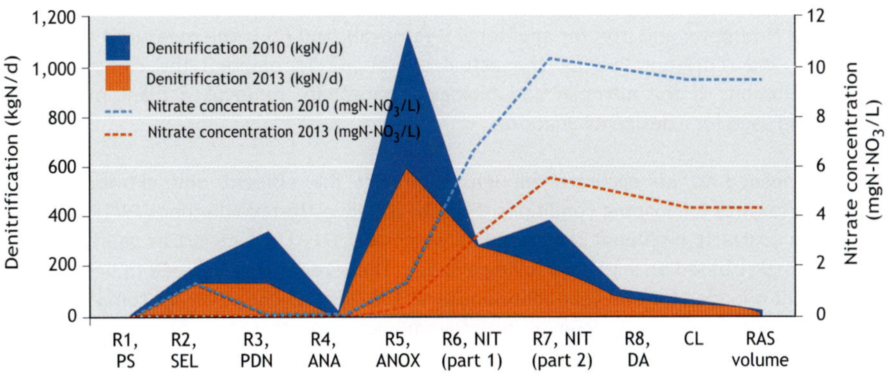

Figure 15.5 Denitrification and nitrate profile of WWTP Houtrust in 2010 (blue) and 2013 (red). Effluent concentrations are measured in the outflow of the secondary clarifier (CL).

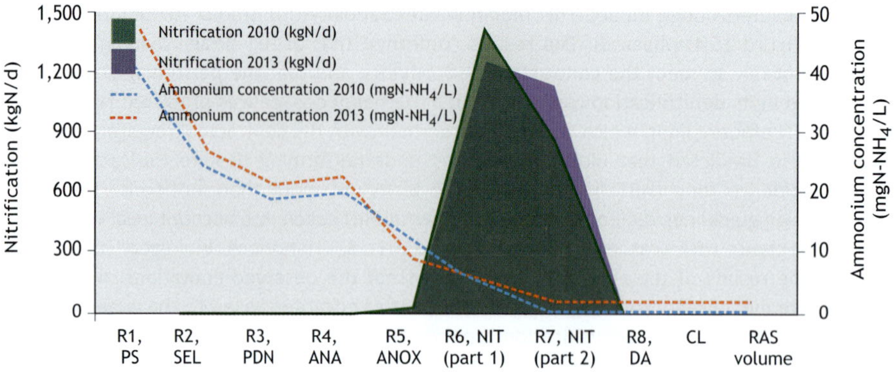

Figure 15.6 Nitrification and ammonium profile at WWTP Houtrust in 2010 (green) and 2013 (purple). Effluent concentrations are measured in the outflow of the secondary clarifier (CL). To improve the TN removal performance in the 2013 effluent ammonium was controlled at 2.0 mgN-NH_4/L. Partial nitrification took place in the anoxic tank as the result of oxygen recycle.

As a result of the plant redesign and optimization, the removal performance of WWTP Houtrust improved by 33% for total nitrogen removal (nitrate in the effluent decreased from 12.4 to 8.3 mgN-NO_3/L). For TP the effluent improved by 50% (ortho-phosphate in the effluent decreased from 2.2 to 1.1 mgP-PO_4/L).

This study convincingly demonstrates that the re-designed plant is capable of meeting the operational and contractual performances. The N-tot project was finalized in 2013 and excellent performance of the plant has maintained up to this date. The study is described in more detail in the sections below.

15.4 Problem definition and goals

Based on the observed plant performance it was assumed that the refurbished plant (in 2008) was limited concerning its nutrient removal capacity. Two questions needed to be answered: (1) can plant meet the contractual effluent requirements at full loading under constraint of robust operation with sufficient design safety and limited use of additional chemical dosage (methanol for additional N-removal and iron for additional P-removal), and (2) is the measured treatment data reliable and representative for the actual operation? To pinpoint the causes for the observed limitations in the nitrogen and biological phosphate removal, a plant model was developed and used for a design evaluation.

The model needed to be accurate enough to predict the effluent and extrapolate this performance towards full loading conditions. Thereby it is investigated whether WWTP Houtrust will meet the contractual effluent requirements which are: TN < 13.4 mgN/L as yearly average and TP < 3.0 mgP/L based on a 10 days rolling average. These requirements need to be met, also when the plant is loaded for one year continuously at 104 %, at a yearly average temperature of 16.2°C. The nitrification capacity (conversion of ammonium) was tested, at full design load and minimal design temperature of 10°C, to establish whether the 2008 design has sufficient (aerobic) safety (typical 10-20%) to compensate for possible design shortcomings and model inaccuracy. According to the contract, WWTP Houtrust should fully nitrify at 10°C for 30 days continuously without losing nitrification capacity. There are no contractual limits regarding dosage of iron to meet the TP < 3.0 mgP/L effluent requirement. However, the contract prescribes that the activated sludge process relies on biological removal of phosphorus (EBPR) and that iron only is to be used additionally. Thereby reference is made to the documents 'Handbook chemical phosphorus removal' (STOWA, 93-06) and 'Handbook biological phosphorus removal' (STOWA, 2001-15). Because the plant was also underperforming on this aspect, this was tested in the modelling study.

With the model it was demonstrated whether a robust operation is possible. Therefore the safety margins were determined for effluent total nitrogen (TN) and total phosphorus (TP). Other operational margins which were determined are the Mixed Liquor Suspended Solids (MLSS) recycle, the Sludge Residence Time (SRT), the aeration capacity and requirement of additional chemicals (methanol and iron). To convincingly calculate these design margins reliable and accurate model results are required. Thereby application of data reconciliation proved decisive.

The study investigated (based on phases 1 and 2 of the N-tot study), how far a reliable model can be developed using mass balance evaluation and statistical data reconciliation. To do this accurately, the treatment data was checked and improved in a data evaluation study. Thereby errors in the treatment data are identified, removed, and corrected, thereby improving the overall quality of the data. This resulted in a more accurate calibration and model simulation results. The case study shows that after performing a proper data evaluation, modelling becomes an efficient and effective tool for operational trouble shooting, plant design and performance improvement. Much of these result already are obtained in the early data evaluation and model calibration phases. The case study demonstrates the practical use of the protocol by Meijer and Brdjanovic (2012) and shows the route towards time and cost efficient upgrade of biological wastewater treatment.

15.5 Plant inventory

According to the first step of the modelling protocol (Meijer and Brdjanovic, 2012), the modelling study started with an inventory of all possible information. Based on the inventory the project was planned. The inventory provides an early project overview and general idea of the availability and quality of the design information and thereby the required investment in time and resources to obtain this information. Acquisition of trustful and reliable plant information is an absolute prerequisite for a successful modelling study. Based on the preliminary data collection, the current state of the WWTP can be assessed, and it can also be determined whether the plant is appropriate subject to modelling. A minimum required quantity and quality of data is necessary for a meaningful execution of a modelling study and achievement of the study goals. Therefore, in the inventory phase, all possible available information for plant modelling and design is collected, screened, (pre)assessed and filtered. Collection was not limited to critical (technological) design information. Also additional information were collected so that later they may be useful to support expert judgment in the case that assumptions needed to be made on the process operation. Typically this includes information on the plant location, operational history, maintenance records, design history, construction and many other operational issues.

For this study up-to-date design documents, technical drawings and process diagrams were used. This included a reported reference measurement of the installed aeration capacity. Information on operation and process control settings were obtained from interviews with plant operators and also verified from the automated monitoring and process control system (Supervisory Control and Data Acquisition: SCADA). Performance data and quality measurements were retrieved from several databases (MIS and SAXO). For the purpose of the modelling study a specific measurement plan was developed largely based on analytical grab samples. All collected quantitative information were organized in a spreadsheet environment (MS Excel). This data set is indicated as raw data because it is not (yet) checked on gross errors and validated by (statistical and mass balance) evaluation. Raw data often contains errors and conflicting information which needs to be corrected before commencing on design calculations and modelling. The final goals of the inventory were to collect: (1) all critical (mostly quantitative) information for the model design study (*critical information*); (2) additional quantitative information that allows to crosscheck the critical data with the goal of improving the accuracy of the study (*redundant information*) and, (3) all other quantitative and qualitative information that later can be possibly used to support expert judgment in the case that assumptions are required on the process operation (*supporting information*).

15.6 Structured approach towards valid model results

When plant data is collected correctly and also checked and validated using sound evaluation methods, the final checkpoint for the data consistency is the model retrofit study. By calibration, the model is fitted to the data set on multiple aspects simultaneously. This is carried out by adjusting (preselected and sensitive) model parameters. The goal is to accurately fit the mathematical model - which by itself is a system of mass balances describing the plant performance - to the measured performance data. By doing this, the model input data (e.g. influent data, pump flows and control set points) are compared to and fitted on the reference data (e.g. measured effluent data and sludge production). If the measured data are correct and accurate, calibration is usually possible by adjusting only a few model parameters in a common range. In the event that the model does not give a satisfactory description of the reality, meaning that physical or biological model parameters need to be calibrated out of the common range to obtain a fit, four possible types of errors can be distinguished:

(1) *Programming errors in the model and/or software.* In this study an open source software was used which allows changes to be made in the models and calculation methods. In theory this could lead to calculation errors e.g. in the model stoichiometry or kinetics resulting in incorrect model results.
(2) *The plant design and/or flow scheme is modelled incorrectly*. Typically this is related to errors in the Process Flow Diagram (PFD), functioning of the hydraulics (e.g. unexpected occurrence of backflow or overflow) or incorrect process flow (control) settings. Errors of this type typically occur when the plant is not well documented (is older) or when local flow control systems are used which are not logged (e.g. in a central SCADA database).
(3) *The measured quality data is not correct.* This can be caused by inadequate sampling devices, non-representative sampling and inaccurate measurements due to inaccurate analytical procedures or inadequate sampling protocols.
(4) *There are unknown processes occurring in the plant that are not adequately described by the model.* Examples of such processes are occurrence of uncommon compounds in the influent, unexpected retention of activated sludge in the reactors, formation of scum and flotation layers, non-homogeneous mixing, undefined concentration gradients, non-homogeneous distribution of solids in tanks, settling problems, influent toxicity, etc.

Under conditions of normal use, software errors of the first category are unlikely to occur. When changes are introduced in the software calculation methods, it is recommended to do this carefully and always check the model results on consistency. Errors of the second and third category can be avoided by performing an adequate data check. Thus, by excluding (as much as possible) errors of the first three categories, it becomes possible to identify errors of the fourth category. In essence, the primary reason to perform an operational plant assessment is often to identify practical operational deficiencies of the fourth error type. Commonly, these types of 'errors' (better indicated as 'operational flaws') are identified in the model calibration phase, when fitting the model to the (validated) reference data leads to parameter settings out of the common range. This typically leads to a situation where the attention of the model study is redirected to the plant operation and finding the practical causes for the model misfit. By following a structural modelling approach, the data evaluation and model study helps to indicate operational flaws and shortcomings in the plant information and process description. These errors need to be addressed in order to obtain a fit of the model to the reality within a reasonable margin of error and to use the model for operational predictions and design purposes. Usually checks can be made based on plant inspections and additional measurements can be performed. However, this always has to be done in retrospect. When measuring in retrospect is not representative or satisfactory, e.g. when process conditions have changed too much, the information gap needs to be closed based on sound reasoning and expert judgment. To do this properly, supporting information collected in the inventory phase can be used to underpin the necessary estimations and assumptions in the design argumentation. Most often, information collected in the inventory phase is a valuable source of information for the rest of the plant study and should be collected with care, always with the particular purpose of modelling and trouble shooting in mind.

15.7 Part 1: Technical plant description

15.7.1 Facts and figures

WWTP Houtrust is located in the residential and coastal area of The Hague (Figure 15.1). The plant mainly receives sewage from domestic origin with a modest seasonal input of fish factory wastewater. The yearly average loading (Mixed Water Flow or Q24) is 330,000 Population Equivalents (design capacity 430.000 PE based on 136 g TOD per PE). Dry Weather Flow (DWF) is in average 57,000 m^3/d with a relative concentrated composition indicating that infiltration of foreign (surface or ground) water into the sewer system is limited. The sewer system has a relative large paved area (Figure 15.3). Average Rain Weather Flow (RWF) is 114,000 m^3/d and RWF peak flows are typical 3.5 DWF. With a moderate sea climate, the average temperature of operation is around 16.2°C with a minimum of approximately 11°C.

Table 15.1 Total reactor volumes of the activated sludge process of WWTP Houtrust (total of the two lanes; North + South). The volumes of both the 2008 refurbished design and the 2011 re-design are presented.

PFD code	Name of reactor	Number of tanks per lane	Unit	2008 refurbishment	2011 re-design
PS	Primary settling	3 in parallel	m^3	8,424	8,424
SEL	Selector tank	1 single	m^3	650	650
PDN	Pre-denitrification tank	1 single	m^3	1,310	1,310
ANA	Anaerobic tank	2 in series	m^3	6,000	3,000
ANOX	Anoxic tank	2 in parallel	m^3	4,600	9,000
AT	Aeration tank	4 in parallel	m^3	21,800	21,800
DA	De-aeration tank	4 in parallel	m^3	1,550	1,550
CL	Secondary clarifiers	8 in parallel	m^3	40,275	40,275
VAERO	Total aerated volume	-	m^3	21,800	21,800
VMLSS	Total activated sludge volume	-	m^3	35,910	37,310
VANOX	Total anoxic volume	-	m^3	6,560	10,960
VANA	Total anaerobic volume	-	m^3	6,000	3,000
	Anaerobic volume fraction of MLSS	-	%	17%	8%
	Anoxic volume fraction of MLSS	-	%	18%	29%
	Anaerobic contact time	-	day	1.0	0.5

WWTP Houtrust is largely constructed underground. A typical feature is the intermediate pumping station, which is used to elevate activated sludge from the anoxic zones towards the higher located aeration tanks. This means a considerable requirement of pumping energy while on the other hand the anoxic MLSS recycle is a free flowing recycle (however controlled) without energy requirement (Figure 15.7).

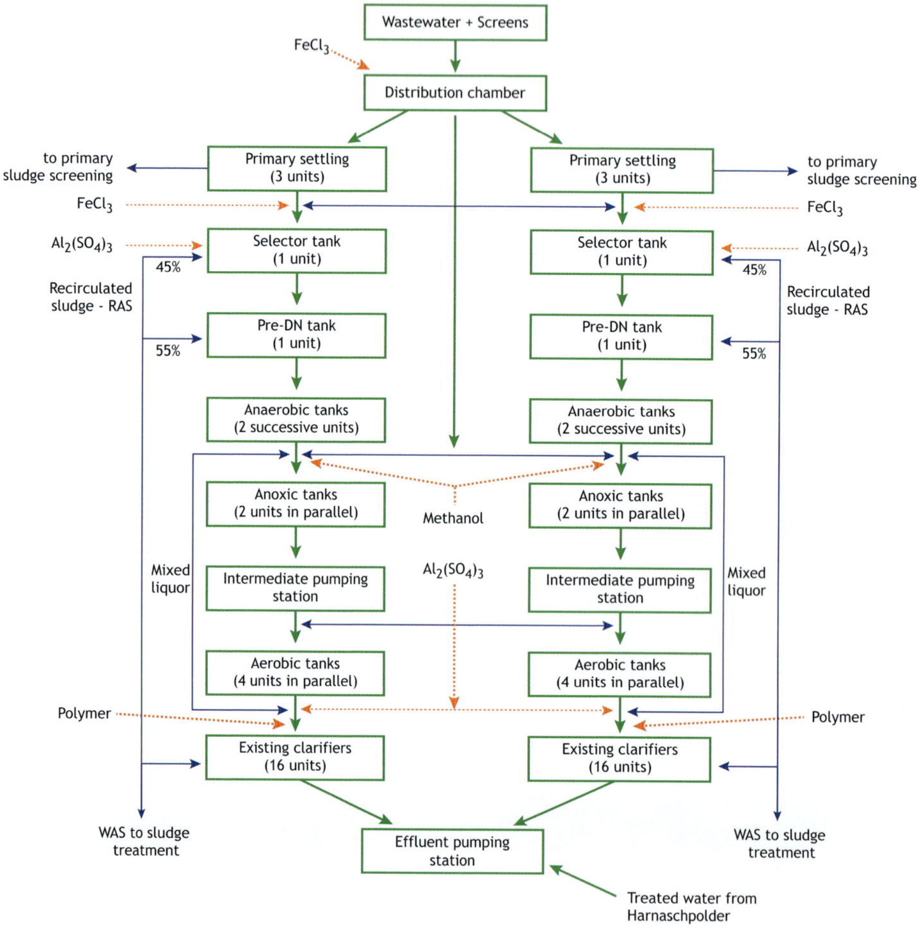

Figure 15.7 Simplified block process scheme of the WWTP Houtrust. The plant consists of two parallel lanes, North and South. The design is according to the Phoredox® concept. Iron dosage in the distribution chamber is currently not functioning.

The 2008 refurbishment of WWTP Houtrust was accomplished by redirecting flows within the existing reactor volumes. Thereby the additional activated sludge volume required for nutrient removal was found in 10 of the originally 16 primary settling tanks which became obsolete as a result of the lower influent flow in the new design. 4 former settling tanks were rebuilt as anaerobic zones; in total 6,000 m^3 anaerobic volume. 4 other settling tanks were rebuilt as anoxic zones. In these zones also the two flow dividers D18/D48 and the MLSS flow dividers were constructed. This reduced the total effective anoxic volume by 1,400 m^3 resulting in 4,600 m^3 anoxic volume. 2 primary settling tanks were rebuilt as selector and pre-denitrification tanks including a new distribution chamber. 6 primary settling tanks remained in operation. Table 15.2 shows the volumes of the 2008 design and adjustments on the 2011 re-design.

Table 15.2 Influent design capacity of WWTP Houtrust at 104% loading. The flows and units relate to the PDF in Figure 15.8.

Parameter	Description	Unit	PDF-Q4 Raw sewage	PDF-Q5 Dirty water	PDF-Q6 Inflow PS	PDF-PS Settling efficiency	PDF-Q7 Inflow AS tanks
Q	Flow	m^3/d	81,900	2,680	84,580	0.04	81,197
PE	Person equivalent	136 gTOD	448,117	52,362	500,479		373,483
TSS	Total suspended solids	kg/d	25,095	3,012	28,107	0.51	13,772
TCOD	Total COD	kg/d	41,600	4,992	46,592	0.32	31,683
TKN	Total Kjeldahl nitrogen	kg/d	4,233	466	4,699	0.11	4,182
TP	Total phosphorus	kg/d	686	185	872	0.11	776
TBOD	Total BOD	kg/d	15,600	2,262	17,862	0.29	12,682

In 2010 the biological reactor volumes were re-designed. Thereby the anoxic zones were enlarged at the cost of anaerobic volume. This was done by redirecting the MLSS recycle and incorporating one anaerobic tank in the anoxic recycle flow. This reduced the anaerobic zone by 3,000 m^3 and increased the anoxic zone by the same volume. The anaerobic hydraulic residence time was decreased from 1.0 to 0.5h. Model simulations showed that this was allowed and also this is in correspondence with the conclusions of Janssen (2001) that for Dutch wastewater practice 0.5 hours anaerobic HRT is sufficient for EBPR removal. To improve the mixing also the MLSS and D18/D48 flow dividers were removed. Thereby the anoxic volume was further enlarged by 1400 m^3 to 9,000 m^3 in total (Table 15.1) which further benefitted the available denitrification capacity.

Table 15.2 presents the contractual required treatment capacity of WWTP Houtrust (full loading including 4% flexibility/safety resulting in the contractual loading requirements of in total 104%). In the table dirty water (Q5) is composed of return water from the primary sludge thickening, centrate from the waste activated sludge thickening and centrate from dewatering. The P- and N-loads in dirty water mainly originate from the dewatering of digested sludge.

15.7.2 Description of the process flow system

The Process flow diagram (PFD) of Houtrust is presented in Figure 15.8. Creation of PFD is the first step in organizing and processing the collected plant data. It shows all relevant flows, flow dividers, valves, pumps, flow meters and reactors and is developed for the purpose of: (*i*) structured data collection and development of the measurement plan, (*ii*) data analysis and evaluation including mass balancing, error detection and statistical data reconciliation and (*iii*) process modelling and model construction (Meijer and Brdjanovic, 2012). The PDF is a single lane representation of the plant. The number of parallel flows and process units are represented by the flow dividers. For example, the process unit DEW in Figure 15.8 is representing two parallel operated sludge dewatering centrifuges. The feed of this unit is the sum of two parallel pumps Q34 which are individually registered in the SCADA database, and the flow divider indicates that the two flows are first mixed after being evenly divided to the two centrifuges.

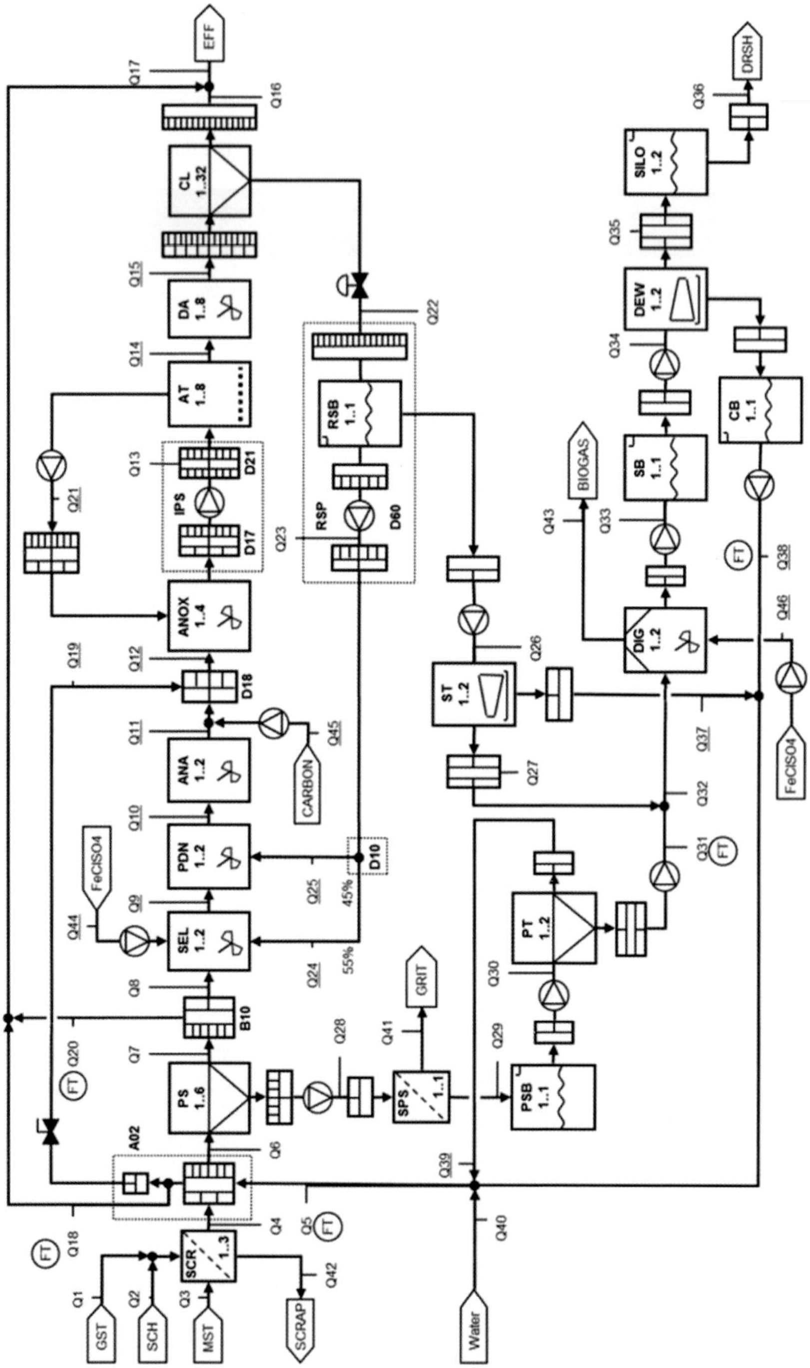

Figure 15.8 Process flow diagram of WWTP Houtrust including all combined flows, reactor units and flow dividers. Based on the diagram the system of mass balances was constructed and the measurement plan was developed.

Table 15.3 List of process units according to the process flow diagram in Figure 15.8. The flow divider B10 could be separated for the two lanes by a valve at operated in 2013. In the reference measurements performed in 2008-2009, two primary thickeners were operated, in 2013 only one. During the entire studied period only one digester was operated.

PFD	Nr.	Name	PFD	Nr.	Name
SCR	3	Influent screening	DA	8	De-aeration tank
AO2	1	Influent flow divider	CL	32	Secondary clarifiers
PS	6	Primary settling (in parallel)	RSB	1	Return sludge buffer tank
B10	2	Flow divider settled influent	ST	4	WAS thickener centrifuge
D10	2	RAS splitter	RSP / D60	6	Return sludge pumps
SEL	2	Selector zone	SPS	1	Screening primary sludge
PDN	2	Pre-denitrification zone	PSB	1	Primary sludge buffer
ANA	2	Anaerobic zone	PT	1-2	Primary thickening
ANOX	4	Anoxic zone (2 by 2 parallel)	DIG	1-2	Digester
D18	2	MLSS flow division	SB	1	Sludge buffer
IPS	2	Intermediate pumping station	DEW	4	Dewatering centrifuge
D17	2	MLSS flow divisions	SILO	2	Sludge storage silo
D21	2	Flow division	CB	1	Centrate buffer
AT	8	Aeration tank			

The process flow diagram was developed in the first stage of the plant evaluation study. In this phase it cannot yet be overseen which process units and flows are relevant for the study. Therefore the PFD is drawn up including all process units and flows. By also including all smaller flows and loads e.g. iron dosage, polymers for dewatering, process water etc., it is often possible to improve the results of the data evaluation.

Table 15.4 List of process flows according to the process flow diagram in Figure 15.8. Normally SCADA and/or database tags are included in the PFD list.

PFD	Nr.	Name	PFD	Nr.	Name
Q1	1	Influent sewer *Groenhovenstraat*	Q24	2	RAS to selector
Q2	1	Influent sewer *Scheveningen*	Q25	2	RAS to PDN
Q3	1	Influent sewer *Morsestraat*	Q26	4	Waste activated sludge (WAS) to CF
Q4	3	Total influent	Q27	2	WAS to digester
Q5	1	Dirty water	Q28	2	Primary sludge
Q6	6	Inflow primary clarifier	Q29	1	Primary sludge after screening
Q7	6	Pre-settled influent	Q30	2	Primary to thickener
Q8	2	Influent AS tanks	Q31	2	Primary to digester
Q9	2	Overflow selector	Q32	2	Feed digester
Q10	2	Overflow PDN	Q33	1	Digested sludge
Q11	2	Overflow anaerobic tanks	Q34	4	Inflow WAS dewatering centrifuges
Q12	4	Inflow anoxic tanks	Q35	4	Dewatered sludge
Q13	2	Intermediate PS to AT	Q36	1	Sludge evacuated to incineration
Q14	8	Aeration tank sensors	Q37	1	Centrate WAS thickening
Q15	8	Inflow clarifiers CL	Q38	1	Centrate dewatering
Q16	1	Effluent AS tanks	Q39	2	Overflow primary thickener
Q17	1	Effluent	Q40	2	Water
Q18	1	Bypass WWTP	Q41	1	GRIT
Q19	2	Bypass primary treatment	Q42	3	Screening
Q20	2	Bypass – emergency outlet	Q43	2	BIOGASS
Q21	8	Anoxic MLSS recycle	Q44	2	Iron dosage AS tanks
Q22	32	Underflow CL	Q45	2	CARBON
Q23	6	Return activated sludge (RAS)	Q46	2	$FeClSO_4$ digester

An accurate PFD is essential when assessing the plant data and performance. It contains critical information for the data evaluation and required to obtain reliable simulation results. The PFD is constructed based on technical drawings. These were checked in practice during several onsite inspections, thereby verifying all flow connections, valves, shutters and overflows. Based on the PFD, two tables are made; one with all process flows (numbered Q1, Q2, etc., Table 15.4) and one containing all process units (Table 15.3). The lists are crosschecked with tags in the SCADA database (tags not presented here). This is done to avoid misinterpretation with the technical staff, especially important as they were involved in the measuring campaign and data collection. Organisation of the collected plant data is done by combining the two lists in one matrix table (Table 15.5) resulting in a mathematical interpretation of the flow system. This matrix is used for data evaluation and reconciliation which is explained further on.

15.8 Part 2: Organizing plant data and development of a data model

15.8.1 Data improvement

Activated sludge modelling is a state-of-the-art technique and when performed correctly, gives the satisfactory mathematical description of the actual plant operation. The main goal of this study is to assess the effluent performance of WWTP Houtrust and to do it convincingly and decisive based on accurate simulation results. Accurate model predictions are only possible based on reliable input and plant data. This means that attention has to be given to the collection of accurate and representative performance data. In its raw form these measurements often contain inaccuracies and errors which are addressed by performing a data evaluation and crosscheck. According to this procedure the highest possible model accuracy is obtained.

For reliable effluent predictions, two aspects of the plant operation need to be known with particular high accuracy: the influent loading of the activated sludge process (in this study the pre-settled influent loading) and the production of waste activated sludge (WAS), which value is typically expressed by the Sludge Retention Time (SRT). This is caused by the formulation of the activated sludge model. Effluent concentrations are not simulated directly, but are calculated as the remainder of the modelled bioconversions and the produced biomass. Thus, the load from the influent which is not converted, remains in the effluent. As a result of this calculation, mathematically all model errors converge in the effluent result, which makes the prediction of effluent the least reliable aspect of the model. This is further emphasized by the usually low effluent concentrations (e.g. ammonium usually is measured in the range of 1.5-0.5 mg/L).

To minimize model errors and thereby improve the effluent prediction, a first requirement is to accurately check and establish all modelled flows. Secondly, the influent loading needs to be accurately established, for this load is the main source of substrate for the modelled bioconversions. The influent contains particulate non-biodegradables which directly result in the production of waste activated sludge. Therefore also the sludge production needs to be thoroughly checked on errors before applied in a model design study.

Most full-scale plant data contain measurement errors and misinterpretations (Barker and Dold, 1995; Meijer *et al.,* 2002; Puig *et al.,* 2008; Rieger *et al.,* 2011; Spindler *et al.,* 2014). Therefore data evaluation will be required for practically all model and/or design retrofit studies. In this study, state-of-the-art mathematical and statistical data reconciliation techniques are applied to check data used for modelling. The goal of this study is to demonstrate these methods and to demonstrate how this can be done in a feasible and practical manner.

Techniques for data reconciliation are available from process industry however, not widely applied in wastewater treatment. Statistical data reconciliation in wastewater treatment was introduced by Meijer *et al.* (2002) by using the combination of mass balance evaluation and statistical reconciliation adapted from the software "Macrobal". The general principle of checking plant data is to combine plant information from different sources in one overall mass balance study and to solve the system of interrelated linear equations simultaneously. Gross error detection and data reconciliation techniques for biochemical processes previously were developed by van der Heijden *et al.* (1994 a, b, c) and implemented in the free-domain software "Macrobal" by Hellinga *et al.* (1992). Macrobal was developed for balancing elements and conversion rates on a molecular level. Meijer *et al.* (2002) applied the Macrobal software to balance steady state wastewater treatment data.

The practical use of statistical data reconciliation on full-scale wastewater treatment was demonstrated by Meijer *et al.* 2002 and Puig *et al.* 2008. Their method was adopted by Meijer and Brdjanovic (2012) and incorporated in a practical modelling protocol which is put into practice in this study.

15.8.2 Development of the data model: The directed incidence matrix

To perform the data evaluation, first all data needs to be organized. This is done in what is called a data model, which has a central function in this study and is used to, (1) structure and organize the data in the inventory phase thereby helping to get an early overview of the (available) information, (2) construct a measurement plan, (3) evaluate the flows and mass balances over the plant, (4) serve as the input for the statistical data reconciliation and, (5) present all mass balance results in a single table overview, based on which the plant performance can be indicated. This table serves as the Key Performance Indication (KPI) of plant operation.

The ultimate goal of the data evaluation is to accurately solve the data model based on the corrected and reconciled plant data. For a large extent, the data model can be constructed in a simple spreadsheet. The data model is derived from the flow scheme in Figure 15.8. The PFD is translated into a matrix table (Table 15.5) which is a mathematical description of the flow network indicated as the *directed incidence matrix M* (Spindler *et al.*, 2014). Herewith, it is referred to matrix as the "data model" for it is a mathematical description of the Process Flow Diagram (thus a model) which presents the relation of all flows passing through the different units in one table overview and which is used to organize, present and check data and plan data acquisition.

The matrix in Table 15.5 is set up with the rows representing process flows (according to the flows in Table 15.4) and the columns representing mass balances over separate (or groups of) process units (according to the units listed in Table 15.3). As a definition of choice, outgoing flows from a process unit are indicated as negative values and ingoing flows as positive values. Empty value represents a zero value. To show the spatial distribution of the interrelated process units in Table 15.5, it is advised to number and arrange the flows and process units in the matrix starting from the influent and primary tanks and ending towards the dewatered sludge. Thus, beginning with the primary treatment, following the water line and ending with the sludge line and dirty water collection. Thereby in the table a diagonal is obtained which represents the gravitational flowing water through the wastewater treatment plant.

Table 15.5 *Directed incidence matrix* of the process indicated as "data model" being the mathematical interpretation of the flow system graphically represented by the PFD indicated. +1 indicates a feed and -1 an outflow. Horizontally the flows are listed and vertically the different interrelated mass balances of the individual process units. The diagonal in the table represents the flow through the process, from the influent to evacuated sludge.

PFD nr.	Short name	AO2	PS	B10	SEL	PDN	ANA	D18	ANOX	AT	DA	CL	Effluent	RSP D60	D10	ST	SPS	PSB	PT	Q32	DIG	SB	DEW	SILO	Dirty water
4	Total influent	1																							
5	Dirty water	1																							-1
6	Inflow primary clarifier	-1	1																						
7	Settled sewage		-1	1																					
8	Inflow activated sludge tanks			-1	1																				
9	Overflow selector				-1	1																			
10	Overflow PDN					-1	1																		
11	Overflow anarobic zone						-1	1																	
12	Inflow anoxic zone							-1	1																
13	Intermediate PS to aeration tanks								-1	1															
14	Aeration tank sensors									-1	1														
15	Inflow clarifiers										-1	1													
16	Effluent activated sludge tanks											-1	1												
17	Effluent												-1												
18	Bypass WWTP	-1											1												
19	Bypass primary treatment	-1						1																	
20	Bypass activated sludge tanks			-1									1												
21	Anoxic MLSS recycle								1	-1															
22	Underflow CL											-1		1											
23	RAS													-1	1										
24	RAS to selector				1										-1										
25	RAS to PDN					1									-1										
26	WAS to CF													-1		1									
27	WAS to digester															-1				1					
28	Primary sludge		-1														1								
29	Primary sludge after screening																-1	1							
30	Primary to thickener																	-1	1						
31	Primary to digester																		-1	1					
32	Feed digestor																			-1	1				
33	Digested sludge																				-1	1			
34	Inflow sludge dewatering centrifuges																					-1	1		
35	Dewatered sludge																						-1	1	
36	Sludge evacuated to incinerator																							-1	
37	Centrate WAS thickening															-1									1
38	Centrate WAS dewatering																						-1		1
39	Overflow primary thickener																		-1						1
40	Water																								
41	Grit																-1								
42	Strip																								
43	Biogass																				-1				
44	Iron dosing				1																				
45	Carbon							1																	
	Polymers thickening SS																								1
	Polymers dewatering																						1		
46	$FeClSO_4$ digester																				1				

15.8.3 Data evaluation planning

The critical measurements required to solve the system, can be relatively easily derived from Table 15.5. From the data model it becomes evident which measurements are minimally required to solve the system of mass balances and accordingly, the data model is the basis for the measurement plan. By combining multiple flow and mass balances in one system, as was done in Table 15.5, often it is possible to obtain a redundant mass balance system, meaning more measurements are available then minimally required to solve the system of equations. Redundancy makes it possible to crosscheck the measured data and detect inconsistency and possible errors in the (mass) flows. Redundancy also can be obtained by adding additional mass balances and/or measurements to the data model. Therefore the minimal (i.e. critical) measurement plan is extended by additional (redundant) measurements. As a result, the development of the data model and development of the measurement plan go hand in hand and becomes an iterative process. Taking measurements can be costly and time consuming and it is therefore advised to carefully plan and develop all steps before starting the measurements. Figure 15.9 shows the step-wise approach followed in this study. All this is done before commencing on the actual model calculations. The steps are explained further in this chapter.

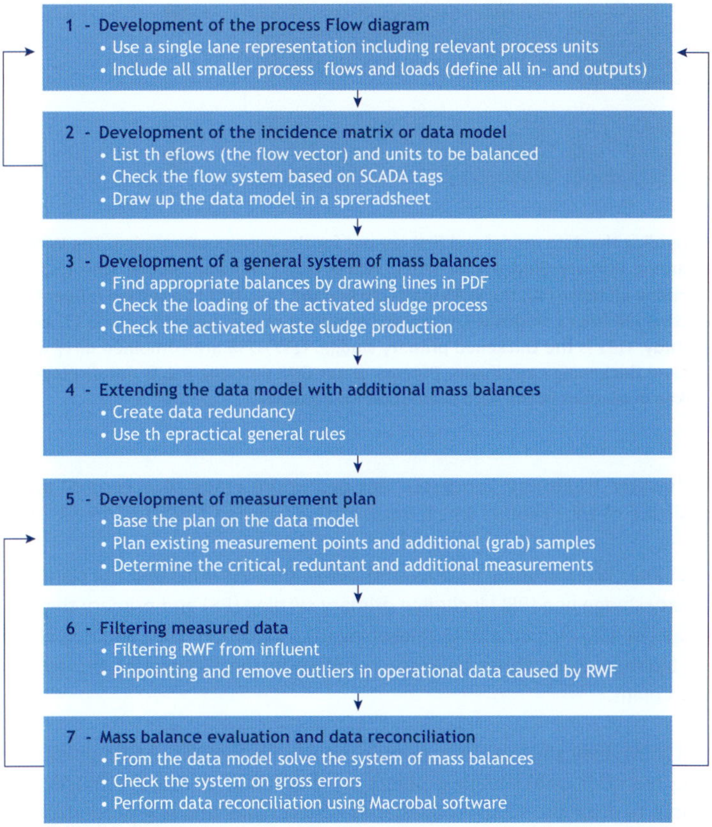

Figure 15.9 General and step-wise approach of the data collection, mass balance evaluation and statistical reconciliation. Each of the steps is discussed in this chapter. Feedback loops represent possible iteration steps when errors are detected in the PFD or in the (critical) measured data which cannot be repaired in the data reconciliation procedure.

15.18.4 Data model development

After establishing the flow balances, the influent loading of the activated sludge system (Q7) and the waste activated sludge production (Q26) needed to be established. It was explained in the introduction why these specific loads need to be known with the highest possible accuracy. Therefore, a measurement plan needed to be developed, which allows checking and improving the quality of this specific information and therefore has the necessary redundancy. This is achieved by applying the concept of overlapping balances, according to the method used by Puig *et al.* (2008) and explained by Rieger *et al.* (2010).

The first step was to define a set of main overlapping mass balance over the plant, including the two flows to be checked, Q7 and Q26 (Figure 15.10). These balances can be derived from the PFD (Figure 10.8). The balances all overlap with the waterline (WL) including the two critical measurement points (Q7 and Q26). Finally, the new set of grouped mass balances was also added to the data model in Table 15.5.

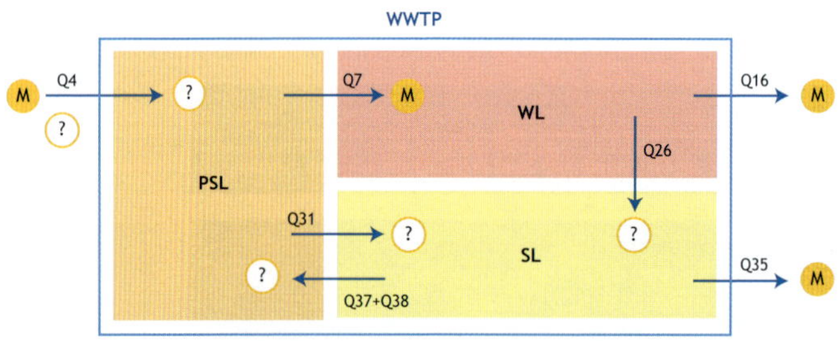

Figure 15.10 Extending the data model and first lay-out for the measurement plan. The WWTP is divided in three main sections: Primary Sludge Line (PSL), Water Line (WL) and Sludge Line (SL). The goal is to accurately determine all flows (Q), the WAS load Q26 and the raw (Q4) and settled influent loads (Q7). The M indicates existing (24-hour composite) measuring sampling points, question marks (?) are new sampling points for this study. Q31 is the thickened primary sludge. Q37+Q38 are combined dirty water flows from the dewatering and thickening centrifuges. Process water Q40 entering PSL is not displayed in this figure, however it is taken in account in the data evaluation for accuracy.

For most wastewater treatment processes this procedure leads to a general system of mass balances, consisting of four groups which interrelate according to Figure 15.10. These balances are:

- The water line (WL) including all activated sludge process units and the clarifier (CL);
- The primary sludge line (PSL) including primary settling (PS) and primary thickening (PT);
- The sludge line (SL) including WAS thickening (ST), digestion (DIG) and dewatering (DEW);
- The combination of all three balances resulting in the balance of the total wastewater treatment plant.

In Figure 15.10, M indicates existing sample points (automatic 24-hour composites). The question marks are additional sampling points for this study, in practice mostly grab samples. The choice where to draw the system boundaries – by actually drawing borders on a print out of the PFD – came from practical considerations such as; where are the most accessible sampling points?, which combination of units has the least variation over time and therefore delivers the most reliable balances?, etc. Instead of connecting the water line (WL) and the sludge line (SL) by the waste activated sludge Q26, one could also have chosen to include the activated sludge thickening unit to the water line, thereby identifying the thickened WAS flow Q37 and the centrate Q27 as the system boundary to the sludge line SL (see also the PFD in Figure 15.8). For practical reasons, Q26 was selected as the system boundary between WL and SL, because there two on-line TSS measurements were available. As the result of these practical considerations the measurement plan will be different for most situations.

A point of attention is that the choice of sampling points needs to be accurate. A misinterpretation can reduce the accuracy of the collected data set and in the worst case results in missing critical information making the measurements useless for modelling and plant assessment. For this reason a detailed and correct flow diagram is required (Figure 15.8).

Figure 15.10 is used to extend the data model in Table 15.5 by adding the new combined flow and mass balances. Figure 15.10 is translated in to a data model according to Table 15.6. This is carried out in the same way Table 15.5 is constructed except in Table 15.6 the columns represent groups of overlapping balances (WWTP, WL, PSL, SL) consisting of multiple combined and connected process units.

Table 15.6 Construction of the data model and measurement plan. The flow matrix is developed from Figure 15.10. Incoming flows are indicated in black, outgoing flows are negative and indicated in light blue. When the flow matrix is multiplied by the flow vector (in this case containing 11 flows), a system of flow balances is obtained. Thereof the balance residuals can be checked by vertical summation. Additional load balances (e.g. TP) can be added to the flow matrix (this is done by including the concentrations). Note that process water Q40 does not contain TP.

PDF nr.	Flow	WWTP in/out	WL in/out	PSL in/out	SL in/out	WWTP flow (m^3/d)	WL flow (m^3/d)	PSL flow (m^3/d)	SL flow (m^3/d)	WWTP TP (kg/d)	WL TP (kg/d)
4	Total influent	1		1		Q4		55,738		447	
16	Effluent	-1	-1			-55,737	-55,737			-122	-122
35	WAS	-1			-1	-49			-49	-325	
40	Process water	1		1		47		47			
38	Centrate from DEV			1	-1			287	-287		
7	Settled influent		1	-1			56,903	-56,903			557
26	WAS to THICK		-1		1		-1,166		1,166		-435
31	Primary sludge to DIG			-1	1			-212	212		
37	Centrate from WAS			1	1			1,042	-1,042		
						sum = 0	sum = 0	sum = 0	sum = 0	sum = 0	sum = 0

The data models in Table 15.6 is filled with measurements (flows and concentrations) and solved simultaneously. By multiplying the data model with the flow vector (in this example the flows Q4...Q37) a flow matrix is obtained. The flow vector consists of a combination of pump flows and gravitational flows, which are measured or unknown values. A minimal required amount of flow data is needed to solve the flow system. In the ideal situation, when all elements in the flow vector are measured and when these measurements are 100% accurate and also when the flow matrix (i.e. data model) itself is a correct representation of reality, vertical summation of the flows and loads in the data model results in zero balance residues (sum=errors=0). For example the flow balance over the water line in Table 15.6 is calculated by:

$$-Q16 + Q7 - Q26 = 0 \quad (15.1)$$

The general principle of detecting errors in the mass balance evaluation is that possible errors in the measurements and/or the matrix setup will appear in the calculated balance residues thus, when the flow measurements contain errors, the flow balance over the water will become:

$$-Q16 + Q7 - Q26 = error \quad (15.2)$$

The value of the balance residue/error should be in relation to the calculated standard deviation of this error. This is calculated based on error propagation rules. For calculation of the standard deviation σ_x the following relation is applied:

$$\sigma_X = \sqrt{\frac{\sum_i^n (x_i - X)^2}{n-1}} \tag{15.3}$$

In this equation n equals the total amount of measurements, x_i is a measurement value from the measurement series and X is the calculated average of the measurement series. For an average value $X \pm \sigma_x$ 68% of the measurements are expected in this interval. For an expectancy of 95% the interval $X \pm 2\sigma_x$ is applied. According to the rules of error propagation, when adding or subtracting average values according to: A + B = X or A – B = X (with standard deviations σ_A, σ_B), for X a new standard deviation is calculated according to $\sigma_A^2 + \sigma_B^2 = \sigma_X^2$. In a multiplication or division according to: A × B = X or A ÷ B = X (with standard deviations σ_A , σ_B), the new standard deviation of X is calculated from the relative standard deviations according to:

$$\left(\frac{\sigma_A}{A}\right)^2 + \left(\frac{\sigma_B}{B}\right)^2 = \left(\frac{\sigma_X}{X}\right)^2 \tag{15.4}$$

with relative standard deviations $\left(\frac{\sigma_A}{A}, \frac{\sigma_B}{B}\right)$

In these rules, for simplicity it is assumed there is no correlation between the two measured series A and B.

Typically, a first calculation check on the balance residues is done by solving/fitting the data model in a spreadsheet. When the system of balances cannot be solved/fitted based on the raw measurements, this means inconsistencies in the data are apparent and further analysis may be necessary. Rudimentary error analysis of the system can be done by means of a sensitivity analysis. Thereby realistic deviations in the flow vector (as well as in the concentration measurements in the matrix) are introduced with the goal to improve the fit of the balance system. For example, each of the measured values is adjusted in a predetermined range of e.g. ± 10% and thereby the resulting errors in the balance residues are evaluated. As a result of the applied variations, the balance residues of the system will become larger or smaller. Thereby the measured value(s) which affect the error(s) in the balance residues the most, i.e. requires the smallest alteration to correct the balance system, is/are most likely to contain the measurement error(s). Based on this information, it should be checked whether or not the identified origin of the error can be confirmed form operational practice.

15.8.5 Data representativeness improvement

The means to improve the quality of raw measured wastewater treatment data are often limited. Wastewater treatment processes are designed for dynamic and robust operation and not for precise measurement. Thereby measurements often are not fully representative even when the equipment is well-calibrated. To obtain better results in fitting the flow and mass balances over the plant, a technique to pre-process and filter influent and process data has been developed. Typically, data measured over longer periods are disturbed by rain events and process control out of the regular operation. Filtering techniques were applied to identify and remove these data from the dataset. Different filtering techniques were tested and it was concluded that much better results can be obtained in mass balancing when influent is filtered on rain events. After filtering mass balances give a better fil and a more consistent approximation of the average plant performance is obtained.

The reference data set was measured in the period 30/10/2008 to 18/12/2008. It was observed that during this period of six weeks, the loading and operation of the plant varied significantly as the result of several rain events. These events caused a considerable increase (by 33%) of the average influent flow (Mixed Weather Flow or MWF). Balancing raw MWF data showed inaccurate and therefore it is concluded that the data including Rain Weather Flow (RWF) are not representative for the evaluated period. Choosing shorter periods of 5-10 days, thereby excluding the rain events, resulted in better balanceable data however, also resulted in different influent loading conditions for the different selected periods. This indicates that by shortening the period over which the influent loading of the plant is evaluated, too few data are used to obtain a representative average influent loading. By taking the whole measured period and filtering out the rain events, much better balancing results are obtained. It was observed that the selection criterion for Dry Weather Flow days (DWF) is critical for the average results. Picking out rain events by hand, based only on the influent flow, did not result in satisfactory results. Therefore a filtering algorithm was developed based on the combination of three measured criteria for selection of DWF. The selected criteria are based on common process knowledge and are used in a filter including a tuning parameter according to Equation 15.5: RWF/DWF filter criterion:

$$IF\left(\frac{COD_t}{\overline{COD}} \times \frac{TKN_t}{\overline{TKN}} \times \frac{\vec{Q}}{Q_t} > 0.75\right) THEN\ (DWF)\ ELSE\ (RWF) \qquad (15.5)$$

In the filter, the normalized COD and TKN concentrations (the measured value at time = t divided by the calculated average) and the inversed normalized flow are multiplied and the result is compared to a dilution factor (the tuning parameter which is set to 0.75). In words, Equation 15.5 prescribes that when diluted COD and TKN influent concentrations are accompanied with an increased influent flow, rain is detected (RWF) and otherwise the measurements are indicated as dry weather flow (DWF). The filter setting (factor 0.75) is chosen such that the beginning of a rain event (the first flush) is included with DWF. This parameter setting is the result of several balancing attempts (not shown) and delivered the best fit, independent of the measured period that was selected. We tested the filtering technique on the whole year of influent data and also on several other data sets of other wastewater treatment plants, all giving similar improved results. It is therefore our conclusion that the selected filter setting (0.75) has a general validity for at least Dutch wastewater conditions (unpublished results). The filtered influent data are presented in Figure 15.11.

Table 15.7 and Figure 15.11 show that the mixed weather flow (MWF) is considerably higher than dry weather flow (74,078 compared to 55,840 m^3/d). Based on the raw mixed weather flow measurements, the system could not be balanced. By using filtered DWF, accurate balancing is possible (therefore see the final results in Table 15.10). It is observed that while the mixed and dry weather flows are quite different, the average mass loads are very similar. This indicates that filtering the raw influent data on rain events according to the proposed method does not alter the actual influent loading however, does deliver more representative and better balanced results. It is therefore concluded that the proposed filtering approach is justified.

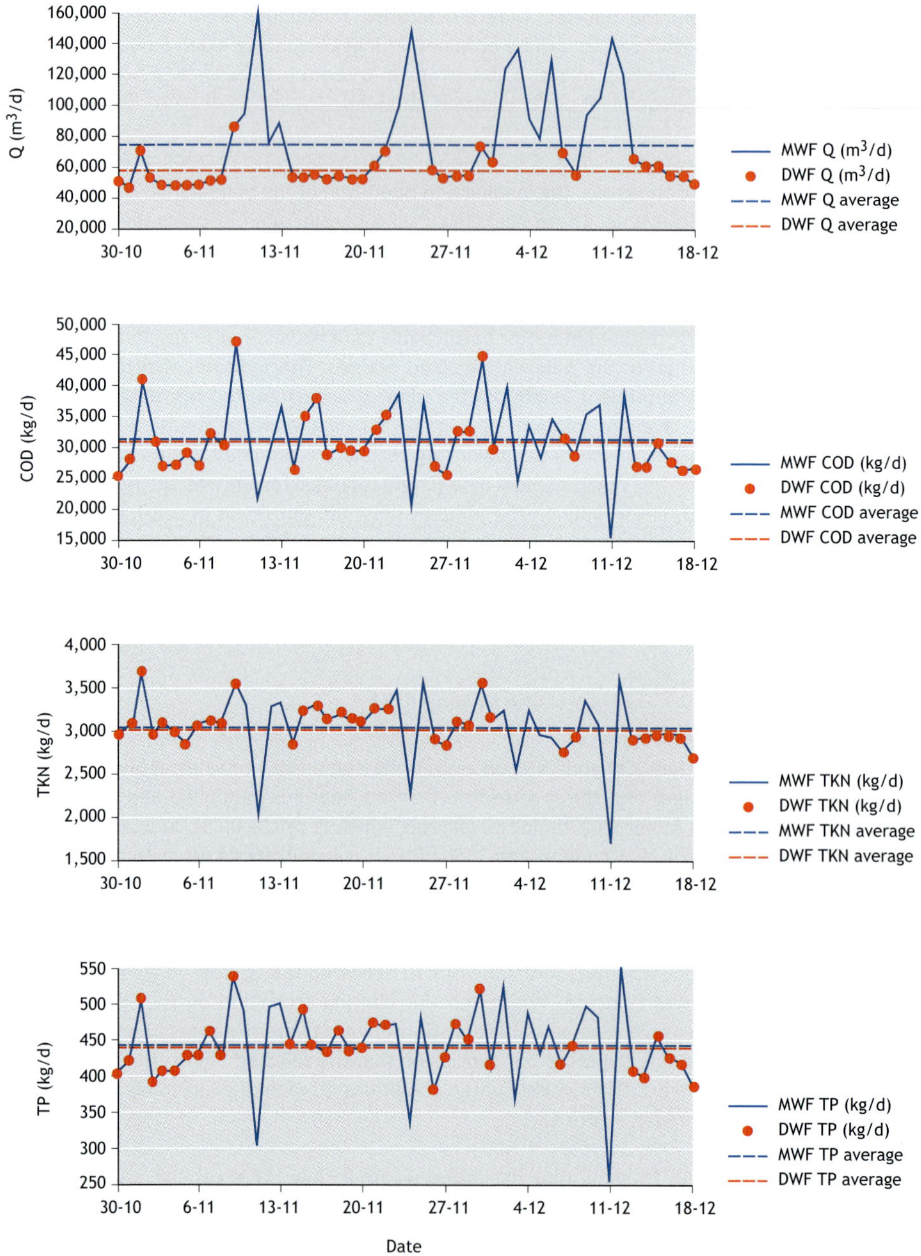

Figure 15.11 Results of filtering the influent data (Table 15.5) on dry weather flow conditions. The four graphs show the flow and COD, TKN and TP influent loads. While for the flow, filtering reduces the average, for the COD, TKN and TP loads dry weather and mixed weather result in very similar loads. The difference is therefore mainly expressed in the measured concentrations (Table 15.7). Filtering the data results in smaller standard deviation and makes it possible to evaluate longer periods of measured data, thereby obtaining a better representation for average conditions.

Table 10.7 Average wastewater composition and flow measured in the period 30/10 to 18/12/2008. RSD is relative standard deviation calculated from 30 measurement days. The weighed concentration is calculated from the average influent load. DWF and RWF are calculated from filtered influent data. Samples are taken using automated 24-hour flow proportional composite sampling equipment.

Parameter	Unit	Average ±	RSD%	Weighed ± RSD%	Parameter	Unit	Average ±	RSD%
Dry weather flow (DWF)					*Dry weather flow (DWF)*			
Q	m^3/d	55,840 ±	18%		Q	m^3/d	55,840 ±	18%
COD	mg/L	563.3 ±	11%	559.1 ± 25%	COD	kg/d	31,218 ±	17%
TKN	mg/L	57.3 ±	11%	56.3 ± 19%	TKN	kg/d	3,143 ±	6%
TP	mg/L	8.1 ±	10%	8.0 ± 20%	TP	kg/d	447 ±	8%
P.E.	PE/m^3	6.1 ±	10%	6.0 ± 23%	P.E.	PE/d	335,152 ±	14%
Mixed weather flow (Q24)					*Mixed weather flow (Q24)*			
Q	m^3/d	74,078 ±	41%		Q	m^3/d	74,078 ±	41%
COD	mg/L	468.6 ±	29%	418.2 ± 46%	COD	kg/d	30,982 ±	20%
TKN	mg/L	46.8 ±	31%	41.1 ± 43%	TKN	kg/d	3,046 ±	12%
TP	mg/L	6.7 ±	30%	5.9 ± 43%	TP	kg/d	441 ±	13%
P.E.	PE/m^3	5.0 ±	30%	4.5 ± 44%	P.E.	PE/d	330,148 ±	17%
Rain weather flow (RWF)					*Rain weather flow (RWF)*			
Q	m^3/d	113,909 ±	22%		Q	m3/d	113,909 ±	22%
COD	mg/L	299.8 ±	37%	279.2	COD	kg/d	31,802 ±	26%
TKN	mg/L	28.0 ±	34%	26.1	TKN	kg/d	2,972 ±	20%
TP	mg/L	4.2 ±	33%	3.9	TP	kg/d	443±	20%
P.E.	PE/m^3	3.1 ±	36%	2.9	P.E.	PE/d	333,717 ±	24%

From Table 10.7, it is observed that by filtering the raw MWF influent data, the calculated weighted average DWF concentrations (concentrations calculated from the average DWF loads, Table 15.7) become (more or less) identical to the measured average DWF concentrations (measured DWF concentrations, Table 15.7). Thus, after filtering, it becomes (more or less) indifferent which concentrations are used in the design evaluation for establishing the DWF loads; measured or weight averaged concentrations give very similar results in the calculated loads (Table 15.7). However, weight average concentrations have relative larger standard deviations. This is the direct result of the error propagation which occurs when calculating the weight average concentrations. From there it is concluded that through application of influent filtering, more accurate data are obtained with lower standard deviations and that as a result also the statistical data evaluation and reconciliation procedure results in higher accuracy.

Besides the influent data, also other process measurements can be improved via the filtering technique. Because RWF can exactly be pinpointed by the RWF/DWF filter, possible related disturbances (measured outliers) in other process data can be identified and removed from the dataset. In this study, it was possible to relate several outliers in e.g. the effluent quality and sludge dewatering performance to rain events. By excluding these outliers from the data set, the accuracy of the total data set improved and also the mass balancing results over the process units improved considerably. A side result from the filtering technique is that it becomes possible to calculate the load carried by rain water. From the difference between the DWF and RWF loads (Table 15.7) it is calculated that when the first flush is accounted to DWF, the load of rain water is close to zero. This is also the results of the filter setting. However, this assumption is justified when it is considered that the first flush mostly is of DWF sewage quality. The fact that this amount is transported to the plant with a high rate does not affect the plant evaluation when this evaluation is done over a longer period. A general warning coming to the forth from this analysis, is that unfiltered MWF concentration measurements cannot be used for design calculations, only the weighted average concentrations are valid. When mistaken, this can result in large design errors, clearly observed from the results in Table 15.7.

The proposed method of filtering influent data has considerable advantages in the modelling study, as follows:

- Filtering allows the use of more influent data measured over a longer time stretch. When more data are used, this results in a lower standard deviation and a higher accuracy.
- By discharging RWF data from the influent, lower standard deviations are obtained resulting in information with a higher accuracy.
- By using influent filtering, the selected measurement period becomes indifferent for the calculated influent loads. In this study it is shown that unfiltered influent data gives different results for different measured periods.
- The use of filtering allows more accurate detection of outliers in other process measurements besides the influent. This allows filtering out disturbance by rain, also in other parts of the process, thereby obtaining an overall improved data set which has a lower standard deviation, is more accurate and better balanceable.

From this study, it is concluded that as result applying filtering of influent and process data the accuracy of the data improved considerably. By removing rain disturbances, closing of the mass balances becomes easier and data reconciliation delivers even better results. It is concluded that the success and accuracy of this study largely relies on the developed and applied data filtering technique. The technique is generally applicable for modelling and plant assessment studies.

15.8.6 Data and parameter estimation based on activated sludge composition

Meijer and Brdjanovic (2012) propose to measure the aerobic sludge composition and use this information for model calibration. The results of the measurement are presented in Table 15.8. In the model the different particulate COD/VSS ratios are fitted and used to predict the VSS and TSS sludge production in the model (Table 15.16).These fractions are theoretical fractions and cannot be measured directly. Therefore the measured COD_X/VSS ratio is used as an estimate. Also the particulate nitrogen and phosphorus fractions of COD are fitted in the model from this measurement. To establish the accuracy of the measurement this is investigated.

Natural variation of the activated sludge composition fractions is expected to be low. For example, the COD_X/VSS composition of biomass has a common ratio of 1.42 gCOD/gVSS which by approximation also applies to activated sludge. This ratio comes from the natural composition (weight) of organic substance. Common ranges for activated sludge (measured in the end of the aeration tanks) are usually a little lower as the result of biodegradation and presented in the last column of Table 15.8. Two activated sludge compositions, measured in 2008 and 2013, are compared in Table 15.8.

It is concluded that during 5 years of operation and despite considerable design and operational changes in the WWTP Houtrust process, the composition is very similar for TSS, VSS, COD and N. Variations are observed between 2 and 7%. For phosphorus fractions, larger changes are observed. As the result of a considerable improved EBPR process in 2013, the P-content of the sludge increased in the 2013 measurement by 26%. The relative stable composition of the sludge can be explained by the fact that activated sludge largely contains of accumulated particulate materials originating from the sewage. This material will have a relative constant composition. Possible variations in the composition are likely to occur as the result of operational conditions (e.g. the SRT operation and dosage of Iron). When measuring the sludge composition in a modelling study, these operational conditions need to be stable. Thereby only a few sludge composition measurements are required for an accurate estimation. Sampling can be done accurately from aerated and well mixed tanks using grab samples.

Table 15.8 Measured activated sludge compositions in the aerated tanks in 2008 and 2013. The measurements are compared based on their relative change. The composition is calibrated in the model and typical composition ranges are presented in the last column.

Ratio	Unit	Measured 2008	Measured 2013	Relative Δ 2008-2013	Calibrated 2008	Calibrated 2013	Notes and typical ranges
VSS/TSS	g/g	0.759	0.742	-2%	0.765	0.742	For a BOD-removing process typically 0.82 to 0.79, for a BPR system typically 0.78 to 0.71
CODx/VSS	gCOD/g	1.371	1.428	4%	1.379	1.428	For organic material 1.42 gCOD/gVSS, for activated sludge the typical range is 1.32 - 1.40
CODx/TSS	gCOD/g	1.041	1.059	2%	1.062	1.060	Typically 1.05 to 0.95 depending on the VSS/TSS ratio
TPx/CODx	gP/gCOD	0.045	0.056	24%	0.045	0.056	Organic material 0.015 to 0.020 gP/gCOD, for a BPR process in the range of 0.035 to 0.055
TKNx/CODx	gN/gCOD	0.074	0.077	4%	0.073	0.077	Influent organic material is typically 0.030 gN/gCOD, for activated sludge this is in the range of 0.050 to 0.080
TKNx/TSS	gN/g	0.076	0.081	7%	0.077	0.082	Depending on VSS/TSS ratio
TPx/TSS	gP/g	0.047	0.059	26%	0.047	0.059	Depending on VSS/TSS ratio

15.8.7 Applicability of sludge composition measurements for data estimation

Based on the same principle explained in the previous section, Meijer and Brdjanovic (2012) propose to measure the aerobic sludge composition and use this information for estimating and checking data in different sections of the plant, thereby reducing the required measurement data for modelling. Based on this study the accuracy of this method is questioned. In this study it is investigated to what extent the aerobic sludge composition is representative for the sludge compositions in the return activated sludge (Q22, RAS), waste activated sludge (Q26, WAS), pre-denitrification tank (R3), anaerobic tank (R4) and thickened sludge (Q27). For each of these flows, the fractions are calculated based on the measured activated sludge composition in the aerobic tank and checked on available sludge composition measurements. Additionally, Macrobal software was used to evaluate the accuracy of the estimated sludge compositions, by including the estimated results in the data evaluation. The presumption was that the activated sludge in the different sections of the plant (especially the water line) is well mixed by the return sludge flow and anoxic recycle. It was expected, that by reasonable approximation, the measured aerobic activated sludge composition (Table 15.8) also would be representative for all other activated sludge flows, including the secondary clarifier and (thickened) waste sludge. However, the results of this study show that this assumption is only accurate for the return activated sludge (RAS) and waste activated sludge (WAS) and only under the condition that the process is low loaded and when no significant reactions in the settler are expected. Thereby, it is concluded that calculation is less accurate than direct measurement. Due to error propagation in the calculations, a large standard deviation is calculated for the estimated fractions. It is our conclusion that estimated data does not add much information to the balance system. Direct measurement of the sludge fractions in the different activated sludge flows is therefore preferred. Estimation of the composition of thickened waste sludge showed to be highly inaccurate and therefore not useful for accurate modelling. The composition of the activated sludge after centrifugation and adding polymers (Q27) was very different from the aerobic sludge composition. Most likely centrifuging affects certain particulate fractions more than others. Also dosage of dewatering polymers changed the composition. Depending on the type of polymer (often acryl amides), effect on TSS, VSS, COD and in minor extent also TN can be expected. This also means that the effect of polymers on the mass balances of thickening and dewatering centrifuges cannot be neglected.

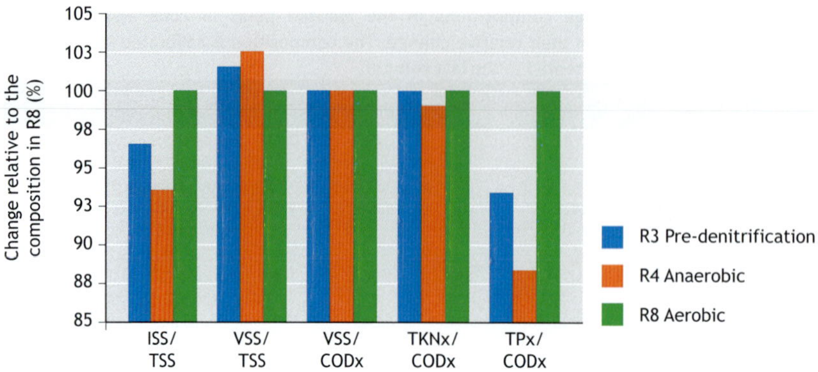

Figure 15.12 Comparison of the aerobic, anoxic and anaerobic sludge composition: Results are exemplary for this study and obtained from the 2013 plant evaluation study. The particulate VSS/COD and TKN/COD ratios are relative constant and can be estimated. ISS and TP related fractions varied over the plant and expected to change as the result of operation and cannot be estimated accurately.

Estimation of the sludge composition in the pre-anoxic and anaerobic tanks resulted in an acceptable estimation (within 2%) for VSS, particulate COD and particulate TKN. It is therefore concluded that these sludge fractions are more or less constant for all biological reactors. The ash fraction (ISS) and particulate TP showed a relative large change as the result of anaerobic release of poly-phosphate by Phosphorus Accumulating Organisms (PAO). This is also demonstrated based on model simulations (Figure 5.12); the figure shows the relative difference (in %) of the pre-anoxic sludge composition (R3) anaerobic sludge composition (R4) and aerobic sludge composition (thereby the aerobic composition is the reference set to 0 % change). Especially in EBPR systems, anaerobic release of poly-phosphate is expected, but also in other unaerated tanks holding activated sludge P-release can be expected which will reduce the ash fraction (ISS) of the anaerobic activated sludge by estimated 4-7%.

15.9 Part 3: Practical methods for creation of data redundancy

In operational practice, often not all flows (or loads) of a wastewater treatment system are measured. From the data model and based on the available measurements, it can be determined whether the flow system can be solved or is redundant. A system of conservation equations consisting of *n* unknown flows can be calculated if they are described by at least *n* independent mass balances. When more measurements are available than required to solve the system, the system is redundant. This allows the redundant measurements to be checked on errors. The order of redundancy of a measured variable increases with the number of independent conservation equations (mass balances) in which this variable can be expressed (Spindler *et al.,* 2014). Thereby the order of redundancy of the system determines the accuracy which can be obtained in the error detection and data reconciliation. A basic approach to create a system with redundancy is by extending the data model (Table 15.5) with additional mass balances. These mass balances are derived from the PFD (Figure 15.8). It is required to select mass balances which are independent. A practical method of creating redundancy is by selecting mass balances that are overlapping i.e. share one or more mass flows. A similar method was used by Puig *et al.* (2008) and Rieger *et al.* (2010). Redundancy is increased most effectively by including balances which are 'conserved' meaning that all mass loads are fully recoverable (thus measurable) in all flows entering and leaving the balanced sub-units. Typically this is the case when there are no conversions to the gas phase. Total Phosphorus (TP) is such a compound and under normal wastewater treatment conditions will not leave the system via the gas phase (e.g. via phosphine gas PH_3). In practice measuring the TP in wastewater treatment is a straight and cost effective method to create a data set with sufficient redundancy to check the critical data for design and operation (i.e. flow, biological load and sludge production data). Therefore measuring TP is useful, even when P removal itself is not of interest for the plant effluent performance. Including TP when modelling a plant increases the model accuracy. Therefore it is advised to use activated sludge models which are adapted for this purpose (these models are ASM2d, ASM3p, BioWin ASAD or one of these models combined with a dedicated model for EBPR, like the metabolic TUDP-model).

When it is possible to (accurately) measure the conversions taking place in a balanced unit, also non-conserved balances can be used to increase data redundancy. An example is the COD balance over the anaerobic digester. COD is non-conserved parameter; it is converted to methane (CH_4) biogas in digesters. However, in practice often both the biogas flow and the methane concentration are measured by on-line measurement devices. This means the COD balance over the digester can be closed and checked when all COD flows including the conversion to methane are measured.

In theory also iron (Fe_{tot}) is conserved and its measurements could be used for data reconciliation. Thereby dosage of iron for P removal needs to be taken in account. However, from this study it was concluded that the standard analytical procedures for measuring Fe_{tot} are not satisfactory and do not deliver reproducible results. The inaccuracy in the analytical measurements was accounted to difficulties in the analytical step of releasing all iron from the complex organic and inorganic matrix. As a result, the Fe_{tot} balances had large errors and were not usable fort data reconciliation. Similar, use of total inorganic suspended solids (ISS) measurements did not lead to satisfactory results. Two possible reasons were identified; (*i*) the (analytical) procedure to determine ISS is not accurate, and (*ii*) ISS is not a stable compound in wastewater treatment, but is formed and also converted in the different process units by a range of precipitation and re-dissolution reactions under which iron precipitation as the result of Fe dosage for P removal. Further research on these aspects is therefore required. Under certain conditions, very much depending on which process unit is balanced, Total Suspended Solids (TSS), Chemical Oxygen Demand (COD), Total Nitrogen (TN) and Total Kjeldahl Nitrogen (TKN) can be balanced by approximation. A common example is balancing TSS over the secondary settler. These types of mass balances are inherently inaccurate due to the constant conversions taking place however, if the data inaccuracy is high or when other means of creating redundancy in certain (uncertain) measurements is not possible, these balances can help to estimate these measurements and give direction towards a more certain system solution. In general it needs to be noted that by adding mass balances to the system, when calculating the loads, measurement errors in the concentrations as well as in the flows need to be taken in account. In this study we have done this by using the rules of error propagation and calculating the standard deviations for the balanced loads. As an approximation these also were used in the Macrobal data reconciliation procedure. Theoretically this approximation is not accurate and also Spindler *et al.* (2014) pointed out that in theory this problem reduces the accuracy and usefulness of mass balances in error detection and data reconciliation. It is however our conclusion that, as a result of the generally high uncertainty of wastewater treatment data, in an operational assessment every useful piece of additional information will add to the improvement of the results. By combining the flow and mass balances in one system it is attempted to reduce the large and often unavoidable errors and inaccuracy in wastewater treatment data to acceptable proportions, such that the data becomes useful for a meaningful plant assessment and design purposes. Some theoretical considerations on the proposed methods are discussed further on in this chapter.

15.9.1 Optimizing and reducing the requirement for data

A practical consideration in wastewater treatment is minimizing the amount of measurements for reasons of cost reduction; not only in relation to this study but for plant operation in general. When only taking in account the critical data for reliable assessment of the plant operation, these data by themselves often incorporate some information redundancy and therefore often allow to be checked (on some aspects) without requiring additional redundant measurements. However, the range and accuracy of a data check often can be improved considerably by adding only a few redundant measurements to the measurement system, as was the case in this study. Also it was possible to further optimize the (initial) data model and thereby reduce the amount of required measurements without reducing accuracy. This was done based on a list of practical rules for selecting the most effective balances. This is further discussed below. Optimization of the data model resulted in 30% less measurement without losing critical information, redundancy, range or accuracy of the data check. This shows that investing time in this step can be worthwhile.

15.9.2 Available methods for selecting balances

In general, the redundancy of a system increases with an increasing amount of independent mass balances. Spindler *et al.* (2014) demonstrated that from a flow system a large quantity of redundancy relations can be derived using a computer algorithm. Thereby also complex and less intuitive combinations of connected process units and balance equation are found. In theory these relations could be used to increase the redundancy and improve the reconciliation without adding additional measurements. Spindler points out that the automated method also derives meaningless and insensitive relations which do not add to the redundancy of the measurement system (Spindler *et al.* 2014). Meijer *et al.* (2002) and Puig *et al.* (2008) showed that based on common process knowledge, it is possible to construct a balance system with sufficient redundancy to check the critical data, thereby mainly focusing on the influent loading and waste sludge production. These results were obtained by intuitively selecting and testing redundant balances (in which all elements are measured) and balances which could be calculated from the measurements (directly or indirectly). Thereby it was helpful to literally draw up the balances in a printout of the PFD and including groups of process units according to the overlapping balance principle (Rieger *et al.*, 2010). In an attempt to make the intuitive method more structured, a list of general rules for selecting appropriate mass balances is drawn up in the next section. By following these rules, it should be possible to quickly produce a workable balance system, which can be further optimized and refined when needed. In this study this resulted in a system of balances with sufficient redundancy to check the critical data on gross errors and additionally, to perform the more advanced statistical data reconciliation incorporated in the Macrobal software.

15.9.3 Practical rules for selecting flow and mass balances

For an effective development of the data model practical rules are presented based on which the appropriate balances can be selected to be used in the data evaluation. These rules are based on our practical experience in data evaluation and several modelling studies under which published by Meijer *et al.* (2002) and Puig *et al.* (2008). As the each plant is different, a generic set of rules is presented here. By following these rules, it should be possible to obtain a data model which allows to be checked on errors and statistically improved using the Macrobal software.

Step 1: Tips for building a data model of flow balances

1. Start building the data model from flow balances.
2. Select flow balances that are *independent* thus cannot be derived from combinations of other balances. Dependent balances do not add information a system.
3. Select flow balances for which all elements are measured. These balances are redundant and allow the detection of measurement errors and data reconciliation. Redundancy also makes it possible to correct gross errors by calculation.
4. Select flow balances that can be directly calculated based on the available flow measurements. These balances are observable and by combination with other mass or flow balances can result in additional redundancy.
5. Select balances that overlap and thereby share one or more flows. These balances can increase the redundancy of the system with relative fewer measurements.
6. For accuracy also include small unknown flows like process water and overflows. If unknown, these flows can be estimated.

Step 2: Tips for extending a data model using mass balances

7. Select the mass balances in line with the flow balances.
8. Select mass balances for which it is possible to accurately measure all in and outgoing flows. Measuring overflows and particulate material (MLSS) in pipes and not well stirred tanks is often inaccurate. Preferably use 24-hour composite sampling. On-line sampling equipment can be used but should be validated by analytical samples. Grab sampling should be performed daily, at fixed time, to reduce the effect of the influent dynamics.
9. Select mass balances which can be calculated, thus in which only one element is not measured or cannot be accurately measured, for example, the mass balance over the gravitational primary sludge thickener where measurement of particulates in the overflow is inaccurate.
10. Balance measured compound that are conserved in the water phase (mass flow IN=OUT without elements leaving the system otherwise). These balances give the highest increase of redundancy of the system.
    - TP is conserved in the plant and therefore can be balanced for all units.
    - In theory, total iron should be conserved, however, in practice sampling and analytical measurement showed problematic and delivered unreliable results.
    - COD over the digester can be measured in the biogas and therefore is conserved.
11. Select mass balances for which the compound is practically conserved. This is true for element balances over sludge dewatering, thickening centrifuges and belt thickeners. It should be taken in consideration that by adding polymers to these types of process units also COD (and depending on the type of polymer also N) is added.
12. Select mass balances for which the compound is conserved under conditions of the balanced system.
    - For example, by approximation TSS and particulate COD and TKN can be balanced over the secondary clarifier. This is truth when the activated sludge processes is low loaded. However storage of sludge and dynamics of the return flow (rain) can make this balance inaccurate.
13. Select mass balances to approximate critical process flows which are not measured or cannot be verified elsewise.
    - NH_4 can be balanced over the anaerobic tank giving a crude approximation of the anaerobic recycle flow. This balance is inaccurate and a last resort, only to be used when no flow measurements are available. The error of the balance is caused by ammonification in the anaerobic tank and exchange of NH_3 with the gas phase.
    - NH_4 can be balanced over the anoxic tank giving crude approximation of the anaerobic recycle flow. The balance is a last resort, when no flow measurements are available,

while it is not accurate because the formation of ammonia from hydrolysis and the uptake as the result of anoxic growth.

14. Most small mass loads can be neglected or excluded for example, scum and grit. An exception is the polymer dosage which can carry a considerable COD load and is needed for accurate estimation of the waste sludge load.

Additional tips for improving the measurement results

- Select mass balances only for units were a significant concentration difference is expected in the main in- and outgoing flows. For example, in theory balancing COD or TSS over the anoxic tank is possible and could result in redundancy. In practice however, such balances do not add to the information.
- Avoid mass balances where large fluctuations of the measured compound can be expected in the main connecting flows. Process dynamics causes high variance of the measurement (resulting in a large standard deviation) which makes these measurements less useful for error detection and data reconciliation.
- Avoid mass balances over units in which considerable effects of storage can be expected. Examples of such balances are: the primary settler, gravitational sludge thickeners and secondary clarifier.
- Mass balancing primary clarifiers in general is unreliable as the result of storage of primary sludge combined with the effect of influent dynamics. A combined balance of the primary settler and thickener gives slightly better results however, also showed less accurate in practice.
- The balance over secondary clarifier often can be made under dry weather conditions and when on-line (high frequency) TSS sampling is used. Grab sampling of TSS in the return sludge flow is inaccurate.
- Only sample at stable dry weather conditions. Rain conditions have a large dynamic effect on the process. The occurrence of rain can be noticed in the measurements three days after the event and are not representative for average plant operation.
- The time span of the sampling period should not be too long, within one period of the SRT and preferably within two weeks of operation.
- No maintenance should be planned in the weeks preceding/during the sampling period.
- During the measuring period primary sludge, return sludge, and waste activated sludge should be operated as stable as possible. Also the plant should be operated under stable SRT conditions.
- Macrobal experiences problems balancing elements over identical process units which have a similar distribution for example, TSS and particulate COD balanced over the secondary clarifier. These balances however can be useful for gross error detection.

15.10 Part 4: Results of the data evaluation study

In the following sections the results of the 2010 data evaluation study is presented (Figure 15.4). The data model is developed in a series of iterative steps following Figure 15.9. Initially, two gross errors were identified in the flow data. This was concluded after calculating the system in Excel and evaluating the system based on a sensitivity analysis. These errors were confirmed from the results of the Macrobal evaluation. Macrobal is limited in the size of the input matrix. Therefore a compressed data model was developed containing the same redundancy (Table 15.9) and allowing both gross errors to be repaired by calculation. In the absence of more gross errors, reconciliation by Macrobal was possible thereby solving the complete flow system with zero balance residues (thus both Table 15.5 and 15.9) thereby improving the accuracy of the flow vector by reducing the standard deviations. This balanced data set is used as the reference data for the model simulation study.

Table 15.9 Final developed data model in this study. The model is initially set up in MS Excel and has an identical format as the Macrobal input table. The concentrations that need to be measured and are all included in the measurement plan. The data model is the result of intuitive choices and practical conditions and other data models could have been constructed.

PFD nr.	Flow	Water balance over process units (average DWF)											Mass balance over process units (average concentrations)								
		WWTP in/out	WL in/out	SL in/out	PS in/out	PT in/out	AS in/out	CL in/out	ST in/out	DIG in/out	DEW in/out	INT in/out	WWTP TP (mg/L)	WL TP (mg/L)	PS TP (mg/L)	SL TP (mg/L)	PS COD (mg/L)	SL COD (mg/L)	DIG COD (mg/L)	CL TP (mg/L)	ST TSS (mg/L)
4	Total influent	1			1								8		8		559				
17	Effluent	-1	-1					-1					-2	-2						-2	
36	Sludge to external incenera	-1		-1							-1		-6,618			-6,618		-167,771			
40	Water	1		1								1									
38	Centrate from dewatering										-1	1									
7	Settled influent		1		-1		1							10	-10		-381				
26	WAS to dewatering		-1	1				-1	1					-373		373		8,314		-373	7,965
28	Primary sludge			1	-1	1															
5	Dirty water			-1	1							-1			53	-53	1,574	-1,574			
34	Inflow sludge dewatering									-1	1										
31	Primary sludge to digesters					-1				1									43,041		
39	Overflow primary thickener					-1						1									
23	RAS						1	-1												-373	
15	Inflow clarifiers						-1	1												201	
27	WAS to digestor								-1	1									74,283		-71,384
37	Centrate WAS thickening								-1			1									-436
33	COD load inflow dewaterin																		-1		
28	TP load underflow PST														-1	1					
28	COD load underflow PST																-1	1			
43	Biogass																	-1,802	-1,802		

15.10.1 Construction of the data model and planning the measurements

Table 15.9 is developed according to the general approach in Figure 15.10. Four independent balances are identified: the total plant (WWTP), the water line (WL), sludge line (SL) and primary treatment (PT). Two combined balances are added to the system: the internal dirty water load (INT) and the combined activated sludge tanks (BIO). Additionally, five individual process units are added; the primary settler (PS), primary thickener (PT), digester (DIG), dewatering (DEW), and WAS thickening (ST). These are selected according to the list of rules presented previously resulting in the system presented (Table 15.9).

The required data redundancy is obtained by measuring all concentrations presented in Table 15.9 (TP in Q4, 16, 35, 7, 26, 28, 5, 23, 15, COD in Q4, 35, 7, 26, 31, 28, 5, 34, 27, 43 and TSS in Q26, 37, 27). All COD and TSS measurements in Table 15.9 are critical and required to perform the model study and plant assessment. Of the TP measurements, raw influent (Q4), pre-settled influent (Q7), effluent (Q16) and dirty water (Q5) are critical for the assessment. In the evaluated system, 5 redundant measurements (TP in Q35, 26, 28, 23 and Q15) are used to effectively perform the presented data check and improve the overall accuracy of the study. Hence, the (financial) investment in additional measurements for application of data reconciliation is relative small; for 5 days sampling only 25 additional analytical grab measurement are required on a total of 455 measurements meaning 5% more measurements thereby making data reconciliation possible. It is therefore concluded that obtaining accurate results can be done at relative low additional costs. In the total N-tot project, the costs for data evaluation are approximately 30-40% of the total investment. This involves all costs related to acquiring the data including analytical measurements, equipment, manpower to collect and analyse the measurements and the time invested in the desk study during to process the information towards final application in modelling.

15.10.2 Error detection and data reconciliation

After performing the measurements, filtering the raw data and calculating the average results, the data model in Table 15.9 was filled and calculated in a spreadsheet. Thereby gross errors in the raw measurements became apparent in the balance residuals. A sensitivity analysis indicated two possible gross errors; the dirty water flow (Q5) and the primary sludge flow (Q28). Both errors were confirmed by statistical evaluation of the system using Macrobal. The software indicated a fail of the statistical test in its attempt to close the system within the designated boundary of accuracy set to 96% based on the statistic h-test. From a visual plant inspection both errors were confirmed. The dirty water flow meter (Doppler type) was not working properly as the result of a partly filled transport pipe. The primary sludge flow registered in SCADA was a calculated value based on the pump operating hours (on/off). A theoretical pump curve was used to calculate the flow and the pumps had known problems of wear. It was therefore concluded that the measurements of Q5 and Q28 were most likely incorrect and therefore were discarded. The redundancy of the system allowed both errors to be repaired by calculation after which no other gross errors were indicated and reconciliation was possible; the system passed the statistic h-test on a predestined confidence level of 96% (by 0.46 with a critical value <13.2). Thus, by rejecting the dirty water flow and primary sludge flow, the measurement set was accepted and the reconciliation was performed according to Table15.10 (representing the Macrobal output). Table 15.11 shows the result of the balanced system with zero calculated residues. This table serves as the basis for the model calibration which is described further on.

Table 15.10 Results of the data reconciliation: The statistical reconciliation was performed using Macrobal. The flows are according to the PFD in Figure 15.8. In addition three loads are calculated and confirmed by reconciliation. Two gross errors were corrected and the accuracy of the data set was improved by reduction of the Relative Standard Deviation (RSD) of the measurements of which the percentile correction relative to the raw measurements is presented in the table. The colour band gives indication of the accuracy of the results. Gross errors are red, balanced data green, and calculated data from the balanced system yellow.

PFD nr.	Mass & water flows	Parameter	Unit	Raw value	±	%RSD	Gross error correction	Δ%	Improved data	Correction factor	Correction %	Origin	Error evaluation
4	Total influent	Q	m³/d	55,840	±	18%	1.0000	0.0%	55738	0.998	-0.2%	Bal	Balanced in reconciliation
17	Effluent	Q	m³/d	6,492	±	7%	0.9050	-9.5%	5875	0.905	-9.5%	Meas	Corrected gross error
36	Sludge to external inceneration	Q	m³/d	56,984	±	18%	1.0000	0.0%	56902	1.000	0	Calc	Calculated
40	Water	Q	m³/d	118,881	±	17%	1.0000	0.0%	119899	1.000	0	Calc	Calculated
38	Centrate from dewatering	Q	m³/d	55,840	±	18%	1.0000	0.0%	55737	1.000	0	Calc	Calculated
7	Settled influent	Q	m³/d	61,757	±	16%	1.0000	0.0%	62994	1.020	2.0%	Bal	Balanced in reconciliation
26	WAS to dewatering	Q	m³/d	1,189	±	5%	1.0000	0.0%	1166	0.981	9.4%	Bal	Balanced in reconciliation
28	Primary sludge	Q	m³/d	123	±	16%	1.0000	0.0%	124	1.003	0.3%	Bal	Balanced in reconciliation
5	Dirty water	Q	m³/d	4,307	±	9%	1.0000	0.0%	4710	1.094	9.4%	Meas	Corrected gross error
34	Inflow sludge dewatering	Load	kg/d	203	±	59%	1.0000	0.0%	203	1.000	0	Calc	Calculated
31	Primary sludge to digesters	Load	kg/d	18,729	±	27%	1.0000	0.0%	18703	0.999	0	Calc	Calculated
39	Overflow primary thickener	Q	m³/d	223	±	33%	1.0000	0.0%	212	0.951	-4.9%	Bal	Balanced in reconciliation
23	RAS	Load	kg/d	7,816	±	71%	1.0000	0.0%	7401	0.947	-5.3%	Calc	Confirmd by reconciliation
15	Inflow clarifiers	Q	m³/d	314	±	71%	1.0000	0.0%	336	1.068	6.8%	Bal	Balanced in reconciliation
27	WAS to digestor	Q	m³/d	45	±	120%	1.0000	0.0%	49	1.094	9.4%	Bal	Balanced in reconciliation
37	Centrate WAS thickening	Q	m³/d	1,066	±	5%	1.0000	0.0%	1042	0.978	-2.2%	Calc	Confirmd by reconciliation
33	COD load inflow dewatering	Q	m³/d	256	±	70%	1.0000	0.0%	287	1.000	0	Calc	Calculated
28	TP load underflow primary settling	Q	m³/d	4,508	±	10%	1.0000	0.0%	4498	1.000	0	Calc	Calculated
28	COD load underflow primary settling	Q	m³/d	47	±	56%	1.0000	0.0%	47	1.000	0.0%	Bal	Balanced in reconciliation
43	Biogass	Q	m³/d	6,077	±	8%	1.0000	0.0%	6063	0.998	-0.2%	Bal	Balanced in reconciliation

Table 15.11 Result of the reconciled data calculated from Table 15.9 using the reconciled flow vector presented in Table 15.10. This table format is used for presenting balanced data in relation to the overall flow system and thereby serves as a key performance indication of the combined process units in the wastewater treatment and effectively used in plant performance assessment.

PFD nr.	Flow	Water balance over process units (average DWF)											Mass balance over process units (average concentrations)								
		WWTP Flow (m^3/d)	WL Flow (m^3/d)	SL Flow (m^3/d)	PS Flow (m^3/d)	PT Flow (m^3/d)	AS Flow (m^3/d)	CL Flow (m^3/d)	ST Flow (m^3/d)	DIG Flow (m^3/d)	DEW Flow (m^3/d)	INT Flow (m^3/d)	WWTP TP (kg/d)	WL TP (kg/d)	PS TP (kg/d)	SL TP (kg/d)	PS COD (kg/d)	SL COD (kg/d)	DIG COD (kg/d)	CL TP (kg/d)	ST TSS (kg/d)
4	Total influent	55,738			55,738								447		447		31,161				
17	Effluent	-55,737	-55,737					-55,737					-122	-122						-122	
36	Sludge to external incenerа	-49		-49							-49		-325			-325		-8,230			
40	Water	47		47								47									
38	Centrate from dewatering										-287	287									
7	Settled influent		56,903		-56,903		56,903							557	-557		-21,702				
26	WAS to dewatering		-1,166	1,166				-1,166	1,166					-435		435		9,696		-435	9,289
28	Primary sludge			4,710	-4,710	4,710															
5	Dirty water			-5,875	5,875							-5,875			313	-313	9,244	-9,244			
34	Inflow sludge dewatering									-336	336										
31	Primary sludge to digesters					-212				212									9,132		
39	Overflow primary thickener					-4,498						4,498									
23	RAS						62,994	-62,994												-23,501	
15	Inflow clarifiers						-119,897	119,897												24,057	
27	WAS to digestor								-124	124									9,193		-8,834
37	Centrate WAS thickening								-1,042			1,042									-454
33	COD load inflow dewaterin																		-7,401		
28	TP load underflow PST														-203	203					
28	COD load underflow PST																-18,703	18,703			
43	Biogass																	-10,925	-10,925		

15.11 Part 5: Model and calibration results

The Houtrust plant model is constructed according to the set-up in Figure 15.13. Before starting with simulation (starting with model calibration), the checked input data are implemented in the model and the model is tested on constancy. The model influent is primary settled quality. The primary settling tank (Figure 15.13, R1 PS) is only a mixing tank without reactions. Figure 15.13 shows the incorporation of a (simplified) digester model with centrate production. During the model calibration the digester model was not connected and the centrate load is accounted in the settled influent quality, which is measured in the overflow of the primary settlers. In Table 15.12, the theoretical maximum N and P centrate loads are calculated without taking in account the struvite formation. When the model is extrapolated and used for design evaluation and optimization, the digester model is connected to the influent and used to simulate a relative increase or decrease of N and P in the centrate as the result changing sludge composition, This was expected as the result from e.g. a dosage of external carbon, iron or an improved EBPR performance.

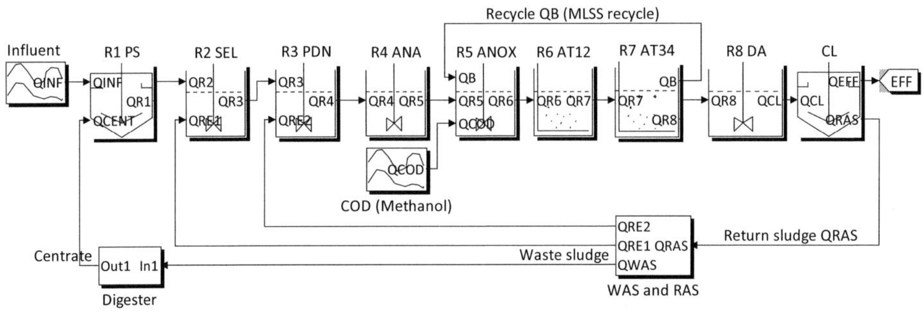

Figure 15.13 Model set-up of WWTP Houtrust. The digester is not connected. The primary settling tank R1 PS is only a mixing volume without reactions or settling. The model influent is settled sewage flow.

The performance of WWTP Houtrust is evaluated using the IAWQ activated sludge model (ASM). The modelling platform is SIMBA version 4.2 (Alex *et al.,* 1997). The applied activated sludge model is ASM2d (Henze *et al.,* 1999) combined with the TUDP-model (Meijer, 2004). This model is validated under lab- and full-scale conditions and in full-scale practice in the Netherlands. The model of Houtrust is calibrated in steady state thereby using the step wise calibration method proposed by Meijer *et al.* (2002) and Meijer and Brdjanovic (2012). In the static calibration procedure, the validated mass balances are reproduced/fitted in the model.

In Table 15.12, the validated influent and modelled influent loading of Houtrust is expressed in typical design values and compared for accuracy. Small deviations of the model are the unavoidable result of rounding errors in model input calculations. In the table, the modelled centrate is an estimate fraction from the settled influent and as such not used in the activated sludge model calculations.

Table 15.12 Results the design retrofit for the (settled) influent loading. Balanced and validated measured data are compared to the results of the calibrated model. The last column gives the absolute deviation of the simulation compared to the measurement. Small rounding errors are made in the model input. Centrate is calculated as the theoretically maximum from waste activated sludge and is not included in the calculations.

Settled influent load and external loading	Unit	Measured	Model	Deviation
Fraction of the maximum design loading (100%)	%	78%	78%	0%
PE loading	136 g TOD/PE.d	264,869	266,011	0%
TCOD	kgCOD/d	21,704	21,716	0%
External carbon source	kgCOD/d	1,905	1,899	0%
TKN	kgN/d	3,133	3,165	1%
Estimated fraction of NH_4 in centrate	kgN/d	254	294	15%
Estimated fraction of NH_4in centrate	%	0.08	0.09	14%
TP	kgP/d	557	561	1%
Estimated fraction of P in centrate	kgP/d	138	254	84%
Estimated fraction of P in centrate	%	0.25	0.45	83%
Influent BOD_5	kgBOD/d	10,169	10,168	0%
Influent BOD_5	kgBOD/d	12,074	12,068	0%
Biodegradable COD (COD_{BD})	kgCOD/d	11,890	11,861	0%
Influent ratio BOD_5/N	kgBOD/kgTKN	3.2	3.2	-1%
Influent ratio TCOD/N	kgCOD/kgTKN	6.9	6.9	-1%
Influent ratio TCOD/P	kgCOD/kgTP	39.0	38.7	-1%
Sludge loading	kgBOD_5/kgTSS.d	0.07	0.07	0%
Sludge loading	kgOD/kgTSS.d	0.14	0.14	0%
Sludge Loading	Kg TKN/kg TSS/d	0.02	0.02	1%

15.11.1 Model calibration

Model calibration was carried out in two trials. To improve the model fit, the model set-up was detailed (mainly in the aeration set-up) and some modelled assumptions were re-evaluated. In this chapter only the final results are presented. The steady state calibration/fit was performed according to Meijer *et al.* (2002) and Meijer and Brdjanovic *et al.* (2012). Mainly kinetic parameters were adjusted, based on their sensitivity. Stoichiometry was only adjusted when this was supported by measurements, e.g. the activated sludge fractions. To fit the model, as few as possible sensitive parameters were adjusted. When adjustments needed to be made outside of the acceptable range, this was an indication of process limitations (assuming the mass balances and model assumptions are correct). The result of the calibration is a model which fits the measured reference period. From the calibration results valuable insight in the plant (under)performance was obtained. The model was fitted accurately to the validated data by adjusting only 6 of in total 65 model parameters, whereof 3 needed to be fitted outside of an acceptable range. The required parameter adjustments indicate that there is a limitation in the biological process related to denitrification and denitrifying phosphate uptake (EBPR). Because balanced data were used, all flows and mass balances (COD, TN and TP) in the model can be accurately fitted simultaneously (Table 15.19). This is a strong indication that the model is accurate and there are no (hidden) errors in the model set-up, plant description and/or flow system. The fact that this can be achieved is the direct result of performing the data evaluation procedure and it makes very clear what the advantages of following the proposed methods in a modelling study are.

Some model uncertainty remains where the model is fitted using non-validated data, e.g. the plant internal (on-line) ammonium and nitrate measurements and the MLSS recycle flow Q21. Thereby the model was used to correct these measurements. This is allowed within the boundaries of the sound framework of well-established mass balances. In the calibration procedure, the internal MLSS recycle flow is adjusted to get a good fit of the on-line ammonium measurements in the anoxic tank. Therefore in the model, the internal MLSS recycle was corrected by approximately 15%.

Another general model uncertainty and often sensitive aspect is oxygen modelling. Two of the calibrated parameters were sensitive towards the oxygen modelling approach: the Monod kinetics parameters K_{O2} and K_{NH4} (Table 15.16). These parameters were adjusted to correct for errors introduced in the model when simulating the oxygen control system using a static approach. One of the model refinements was placing, parallel to the aeration zones, unaerated zones to simulate oxygen gradients in the tanks. As a result, denitrification occurring in the activated sludge yanks was better simulated, while the parameter adjustments of both K_{O2} and K_{NH4} could be maintained within an acceptable range. In the following section, step by step the calibration procedure is explained.

Table15.13 Results of the design retrofit for the SRT and waste sludge production. Balanced and validated measured data are compared to the results of the calibrated model. The last column gives the absolute deviation of the simulation compared to the measurement. The sludge production is simulated within 3% accuracy. TSS in the effluent is not modelled but used as an input parameter.

Parameter	Unit	Measured	Model	Deviation
Temperature	°C	17.3	17.3	0%
MLSS in aeration tanks	$gTSS/m^3$	4,284	4,283	0%
SRT	d	15.9	16.1	1%
Aerobic fraction of HRT (aeration time)	%	0.7	0.7	1%
SRT aerobic (actual aeration)	d	6.3	6.5	2%
SRT aerobic max (full aeration)	d	9.7	9.8	1%
Sludge in effluent	kgCOD/d	307	343	12%
Waste activated sludge (WAS)	kgCOD/d	9,702	9,926	2%
Total sludge production (effluent + WAS)	kgCOD/d	10,009	10,269	3%
Total sludge inorganic content (ISS)	g/gTSS	0.24	0.24	-2%
Iron dosage	kgFe/d	224	224	0%
Production of $FePO_4$	$kgFePO_4/d$	472	472	0%
Production of $Fe_{(III)}OH$	kgFeOH/d	79	79	0%
Total chemical sludge	kgTSS/g	551	551	0%
Sludge loss to effluent	kgTSS/d	361	325	-10%
Waste activated sludge (WAS)	kgTSS/d	9,289	9,085	-2%
Total sludge production	kgTSS/d	9,650	9,410	-2%

Step 1: Calibrating the TP balance and SRT

Before the model is accurately calibrated, TSS was not available to fit the SRT. Initially the correct SRT can be obtained by fitting the TP balance in the model. This resulted in the appropriate waste sludge flow and a SRT of 15.9 ± 0.4 days as calculated from the validation data. The TP balance however, is also depending on the EBPR process. As long as this process is not yet calibrated correctly, an artificial controlled iron dosage is used to fix TP in the effluent and waste sludge. After EBPR is calibrated, this artificial dosage is replaced with the measured iron dosage. Because several model processes influence each other, the calibration steps are repeated one or two times to get an accurate fit.

Step 2: Calibrating the COD mass balance

In this calibration step, the COD balance was fitted in the model. Therefore first the effluent COD is fitted to the measured effluent concentration by adjusting the loss of solids of the model clarifier to the effluent by setting this value to 0.073% (as a percentage of the RAS concentration). The COD waste sludge production and COD composition of the activated and waste sludge are fitted to the measurements by adjusting the influent $X_I/(X_I+X_S)$ fraction. To reduce the need for multiple model iterations, a model controller is used which automatically adjusts the influent composition with each parameter adjustment. This resulted in an influent inert particulate COD ratio of 48.1% (fraction particulate influent X_I of the total particulate COD $X_I/(X_I+X_S)$). Influent X_I represents the COD biodegradability in the process, which varies depending on the operated process conditions and influent characteristics. A common value of X_I is between 40-60%.

Step 3: Calibrating the MLSS nitrogen content

By adjusting the nitrogen fractions of the activated sludge according to the measurements, the nitrogen content of the sludge was fixed. This results in the fractions i_{NSF} of 0.020, i_{NSI} of 0.050, i_{NBM} of 0.076, i_{NXI} of 0.076 and i_{NXS} of 0.076 (Table 15.16). Model shortcomings in ASM2d determine that both the influent composition and activated sludge composition need to be calibrated with the same nitrogen COD fractions. This complicates the model calibration. Several iterations are applied to find appropriate fractions which accommodate the influent and activated sludge composition simultaneously. Due to the nature of solids separation in the clarifier, particulate nitrogen (and phosphorus) in the effluent was not well predicted by the model. Therefore the ASM2d model results were mainly evaluated based on the effluent ammonium and nitrate (and ortho-phosphate) concentrations.

Step 4: Calibrating nitrification

After fitting both the SRT and the MLSS composition in the activated sludge tanks, nitrification was calibrated. Nitrification is the direct result of the aeration input, oxygen modelling and aeration control. In the model the airflow is fitted to the measured input (390,200 Nm^3/d, Table 15.14).

Table 15.14 Results of the design retrofit of the aeration design. Balanced and validated measured data are compared to the results of the calibrated model. The last column gives the absolute deviation of the simulation compared to the measurement. The aeration is fitted within 5% accuracy. The modelled oxygen set-point in the aerated fraction of the aerobic tank is 2 mgO_2/L. Including the anoxic fraction of the aerobic tanks the average oxygen concentration is 1.7 mgO_2/L.

Parameter	Unit	Measured	Model	Deviation
Set point O_2 in the aerated tanks	gO_2/m^3	1.7	2.0	19%
Oxygen deficit	-	1.2	1.3	4%
OC dirty water	kgO_2/d	21,008	21,954	5%
Alpha factor	-	0.78	0.78	0%
OC clear water	kgO_2/d	26,933	28,146	5%
Air input	Nm^3/d	390,198	408,679	5%
Aeration depth	M	4.2	4.2	0%
Aeration efficiency	$gO_2/Nm^3.m$	16.4	16.4	0%

Air is distributed in the 4 successive aerobic tanks according to the actual aeration diffuser distribution (aerobic compartments 1, 2, 3 and 4 equipped with 190, 150, 124 and 95 diffusers, respectively). With local DO set-point controllers, effluent ammonium was controlled at the measured effluent set-point of 0.35 $mgN\text{-}NH_4/L$ by controlling the 4 modelled aerated tank reactors R6.1, R6.2, R7.1 and R7.2. The average oxygen concentration in the aerobic tanks was measured and it was 1.7 mgO_2/L. This included approximately 35% of anoxic time (volume). Without anoxic time, the average oxygen concentration in the tanks was estimated at 2.0 mg/L. In the model, the dynamic oxygen control is approximated by a static model by introducing 35% anoxic volume as a part of the total aerobic volume and splitting the flows accordingly. This set-up allows simultaneous nitrification and denitrification to occur in aeration tanks. Eventually, aeration was modelled by adjusting the parameter K_{NH4} to 1.0 $mgN\text{-}NH_4/L$ (Table 15.16) resulting in an acceptable simultaneous fit (*i*) of the air input (fitted result 408,679 Nm^3/d, Table 15.14), (*ii*) the actual measured DO level in the different aerated tanks (approximately 2.0 mgO_2/L during aeration and 1.7 mgO_2/L on average) and, (*iii*) the effluent ammonium concentration simulated at 0.3 $mgN\text{-}NH_4/L$ (Table 15.17). The modelled aeration results are presented in the Figure 15.14.

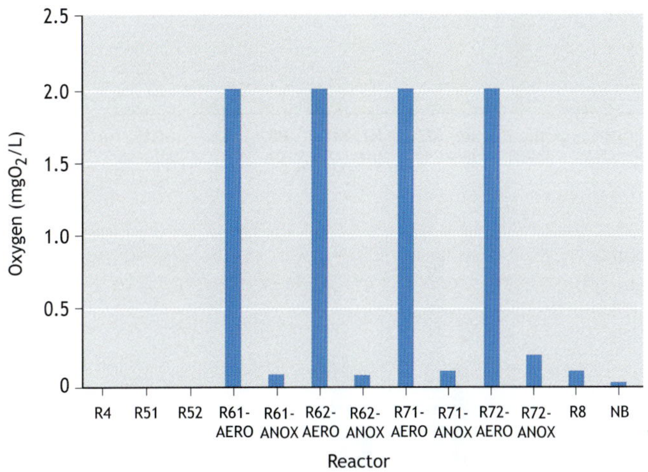

Figure15.14 Simulated oxygen profile including stimulation of simultaneous nitrification and denitrification (SND) based on the actual anoxic residence time in the aerobic tanks during oxygen control. The air input is fitted to the measured flow of 390.200 Nm^3/d. During aerated periods, DO is controlled in each tank at 2.0 mgO_2/L. Air is distributed over the aeration units according to the actual diffuser distribution. SND is introduced by determining 35% parallel anoxic volume, equal to the total measured anoxic time and distributing the flow over the aerobic/anoxic tanks with the same fraction. This resulted in an average measured anoxic DO concentration of 1.7 mgO_2/L after mixing the flows coming from the anoxic and aerobic volumes.

Step 5 Calibrating the internal MLSS recycle

After calibrating the overall nitrification by adjusting the aeration, the measured ammonium profile over the reactors was calibrated. To simulate the measured ammonium concentration in the anoxic tank R5, the anoxic MLSS recycle (Q21) was reduced. The on-line ammonium measurements in the anoxic tanks were re-evaluated and corrected based on analytical measurements (12.3 $mgNH_4/L$). In the model, this concentration was fitted with an anoxic MLSS recycle of 87,850 m^3/d, being 85% of the designed flow capacity and within the accuracy range of the measurement (±15%). The MLSS recycle is estimated in the model, rather than based on mass balancing. For the model it is more accurate, taking in account the effect of ammonium release due to ammonification in the anaerobic and anoxic tanks.

Step 6: Calibrating pre-denitrification

In the final calibrated model, 65% of denitrification occurs in the pre-denitrification reactors (Figure 15.15): the selector (SEL), pre-denitrification tank (PDN), anaerobic tank (ANA) and anoxic tank (ANOX).

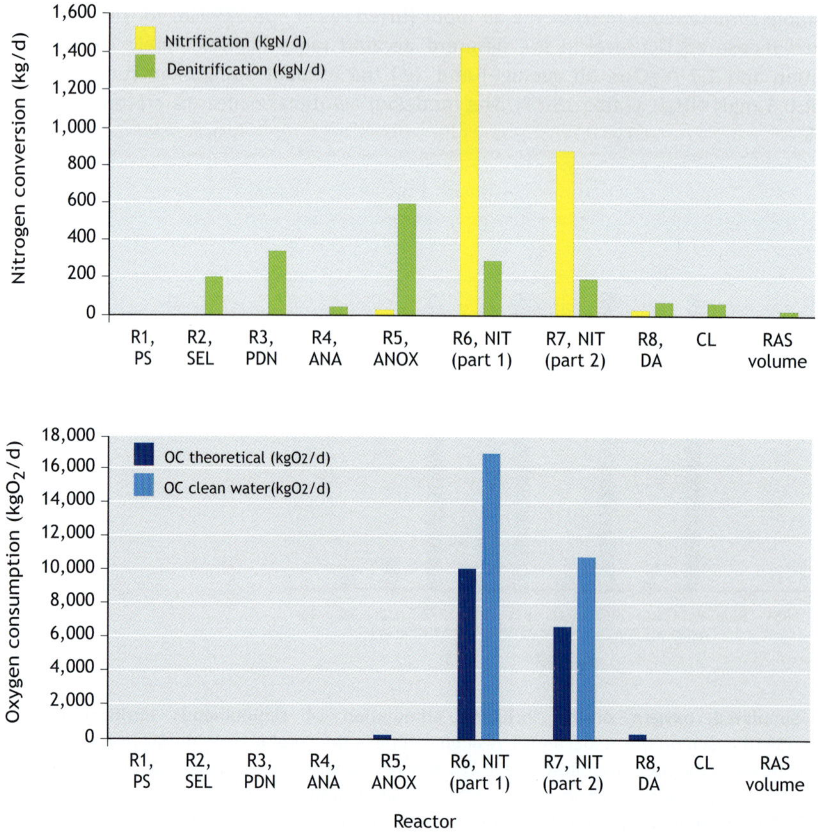

Figure 15.15 Calibration round 2: simulated internal Nitrogen conversions and oxygen uptake. The mass (kg) denitrified in the selector, PDN, ANA and ANOX are limited by the recycle of nitrate. 27% of denitrification is SND, 8% post denitrified and 65% pre-denitrified. The model shows that recycle of oxygen in the process is not significant.

The model shows that (pre-)denitrification is limited by nitrate recycled by the (anoxic) RAS and MLSS recycle flows. As a result, denitrification and effluent nitrate are sensitive to the anoxic recycle rates. The RAS flow (Q23) was fixed based on the mass balance evaluation on 64,160 m^3/d and therefore, was not changed during calibration. The MLSS recycle (Q21) was fitted based on the measured ammonium concentration in the anoxic tank (Q13, Table 15.16, Figure 15.16 features corrected value of 12.3 mgN-NH_4/L). Thereby, both recycles are fixed (Q21 and Q23). When the plant was simulated using these internal recycle rates, considerable more (pre)-denitrification was predicted then measured. This indicates limitation of denitrification in the full-scale plant. To fit pre-denitrification in the model, the PAO denitrification capacity needed to be reduced from 80% (default setting) to 13.5% (ηP_{NO3}, Yable 15.16). This adjustment is out of the acceptable range and strongly indicates that in the process, denitrification and/or anoxic PAO activity was limited. This calibration was however required, to simulate the nitrate profile and the measured phosphate effluent concentration (1.8 $mgPO_4/L$). The out of range parameter adjustments indicated clearly that in the plant other processes occur, which strongly limit the anoxic EBPR activity, more than expected based on the model. This issue is discussed further on.

Step 7: Calibrating post-denitrification

The model shows that 8% of denitrification occurs as post-denitrification in the de-aeration tank, the settler sludge blanket and return sludge (Figure 15.15). Post-denitrification is estimated from the measured concentration difference between nitrate in the aeration tank and the effluent (respectively 10.1 and 9.8 $mgNO_3/L$). Post-denitrification is calibrated by adjusting the activated sludge volume of the clarifier sludge blanket. A fit was obtained by introducing an anoxic sludge blanket volume of 805 m^3, resulting in the nitrate profile as presented in Figure 15.16. With a total clarifier volume of 40,276 m^3, the modelled sludge blanket is only 2% and not significant.

Table 15.15 Results of the design retrofit for the distribution of sludge activity over the bioreactors. Balanced and validated measured data are compared to the results of the calibrated model. The last column gives the absolute deviation of the simulation compared to the measurement. Accurate reproduction in the model is obtained.

Parameter	Unit	Measured	Model	Deviation
Effective aerobic fraction of the activated sludge	%	40%	40%	1%
Effective SND fraction of the activated sludge	%	25%	25%	-2%
Pure anoxic fraction of the activated sludge	%	18%	18%	0%
Anaerobic fraction of the activated sludge	%	17%	17%	0%
Anoxic COD utilization	%	42%	43%	3%
Aerobic COD utilization	%	58%	57%	-2%
Nitrification efficiency	%	73%	74%	1%
Total denitrification efficiency	%	76%	76%	1%

Step 8: Calibrating SND

After calibrating pre-denitrification and post-denitrification, 27% of denitrification remains to be denitrified under partial aerated (SND) conditions (Figure 15.15). The amount is determined from the TN mass balance, which indicates in total 1,798 $kgN\text{-}NO_3/d$ is denitrified (Table 15.19) and subtracting post and pre-denitrification. SND was simulated correctly (average in time) by introducing 35% of anoxic volume for each aerobic compartment (R6.1 to R7.2) thereby correctly fitting the TN balance. The primary parameter to calibrate SND is K_{O2} thereby remained on the model default setting of 0.2 mgO_2/L (Table 15.16).

Step 9: Calibrating EBPR

Based on the default parameter settings for PAO, the model predicts excellent EBPR activity resulting in a low effluent phosphorus concentration (<0.2 $mgPO_4/L$) while the measured effluent PO_4 concentration is much higher (1.8 $mgPO_4/L$, Table 15.17). To fit the measured effluent value, EBPR needs to be suppressed in the model. This was done by reducing the poly-phosphate formation rate k_{PP} (0.1 $gPP/gCOD_{PAO}.d$, Table 15.16) by 78% to 0.020. This adjustment is outside the acceptable range and strongly indicates that EBPR in the process was inhibited. There are different possible practical causes for such an inhibition. The observed underperformance most likely is caused by biological limitations due to irregular process conditions e.g. insufficient internal mixing, sludge settling in the reactors, irregular distribution of flow through the plant or other unknown reasons. Because the applied process model is a simplification of reality, it is not equipped to (directly) predict such inhibitions. Indirectly, the calibration and sensitivity analysis indicate three main causes of the observed limitations; (1) the anoxic pre-denitrification activity is limiting and/or, (2) the phosphorus uptake in the aeration tanks is limited and/or, (3) phosphorus is released in the aeration, de-aeration and/or secondary clarification tanks. These were decisive leads which eventually resulted in several process modifications thereby resolving the limitations in the Houtrust WWTP.

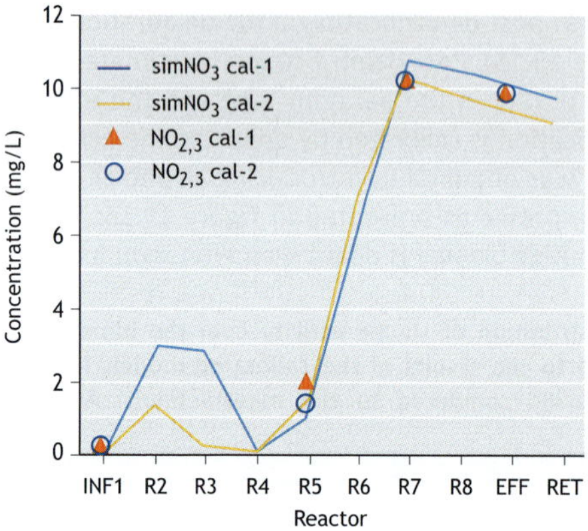

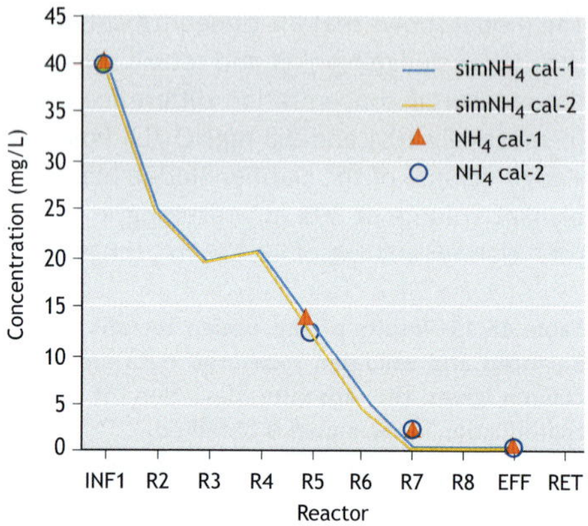

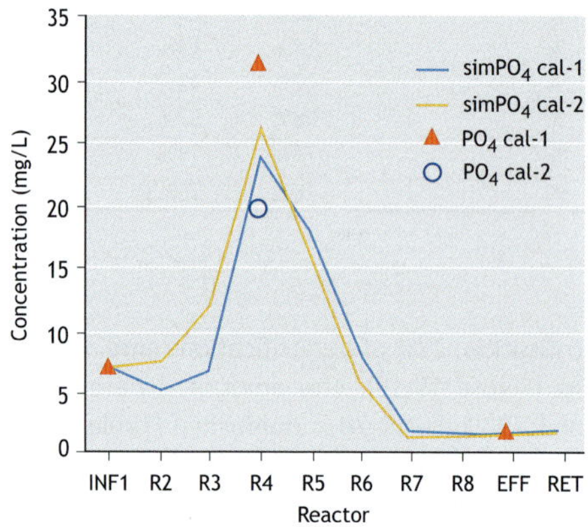

Figure 15.16 Simulation of the initial and final calibration (bold). The lines are simulated nitrate, ammonium and phosphate concentration profiles and dots the analytical measurements. The reactor compartments are according to Figure 15.13. In the final calibration ammonium in the anoxic tank R5 was corrected. A high variation of phosphate in the anaerobic reactor R4 was measured of which the minimum and maximum values are plotted.

It can be concluded that it was possible to accurately fit the validated mass balances in the model, which indicated that the model is consistent and accurate (Table 15.19). Most of the calibrated parameters are supported by validated measurements (Table 15.16). The non-validated MLSS recycle flow is adjusted within a reasonable range to fit the model to the on-line nitrate and ammonium measurements, which were checked in practice. Nitrification is calibrated within the default range, indicating that the removal of ammonium is not biologically limited. Two parameters related to denitrifying biological phosphorus removal are calibrated out of range. To simulate the measured denitrification, the denitrifying activity of PAO was reduced out of the normal range. In addition, the poly-phosphate uptake capacity of PAO was reduced out of range (Table 15.16). The model predicts better EBPR performance than is measured in practice. This is a strong indication that the treatment process is limited in denitrifying EBPR capacity. Thereby the calibration brought to the forth conclusive information on the process performance and gave direction on how to improve the plant operation.

Table 15.16 Calibration parameters: The selected list contains sensitive model parameters which are adjusted to accurately fit the model to the validated data and mass balances (results in Table 15.19). Green: parameter values are calculated from validated data or calibrated in the default range. Yellow: adjustment of a known inaccurate measurement to fit the validated data. Red: adjusted out of the acceptable range to fit the model. The calibration results clearly indicate problems around denitrifying EBPR capacity in the treatment process.

Parameter	Adjusted	Default	Unit	Parameter description	Method
$X_I/(X_I+X_S)$	0.481	0.85-0.30	gCOD/gCOD	Fraction inert particulate COD in influent	Fitted on sludge production and TSS/COD ratio
$S_I/(S_I+S_F)$	0.195	n.a.	gCOD/gCOD	Fraction inert soluble COD in influent	Fitted on soluble COD in effluent measurement
K_{NH4}	1.000	1-0.5	gN/m^3	Saturation coefficient for ammonium (substrate), autotrophic growth	Fitted on effluent ammonium and measured aeration input
K_{O2}	0.200	0.4-0.2	gO_2/m^3	Monod Saturation/inhibition coefficient for oxygen	SND fitted on TN balance
η_{HNO3}	0.800	0.8-0.6	-	Reduction factor for denitrification, heterotrophic growth	Pre-denitrification fitted on TN balance
η_{PNO3}	0.16	0.8-0.6	-	Reduction factor for denitrification, PAO	Pre-denitrification fitted on TN balance
g_{PP}	1.00	1.0-0.22	-	Saturation reduction factor for PP formation	Reduction factor not applied in full scale systems
k_{PP}	0.020	0.1	$gP/gCOD_{PAO}.d$	Poly-P formation rate PAO	EBPR fitted on effluent phosphate
Recycle Q21	87849	103,352	m^3/d	Anoxic MLSS recirculation	Fitted on anoxic ammonium concentration
Sludge blanket	805	n.a.	m^3	Biological activity in clarifier	Post-denitrification fitted on difference NO_3 aeration tank and effluent
iN_{SF}	0.020	0.03	gN/gCOD	N-fraction in fermentable substrate S_F	Fitted on influent NH_4/TKN ratio
iN_{SI}	0.050	0.01	gN/gCOD	N-fraction in Inert soluble S_I	Fitted on effluent TKN concentration
iN_{BM}	0.076	0.07	gN/gCOD	N-fraction in biomass X_{BM}	Fitted on aerobic MLSS composition
iN_{XI}	0.076	0.03	gN/gCOD	N-fraction in Inert particulate X_I	Fitted on aerobic MLSS composition
iN_{XS}	0.076	0.03	gN/gCOD	N-fraction in Particulate substrate X_S	Fitted on aerobic MLSS composition
iP_{SF}	0.008	0.01	gP/gCOD	P-fraction in fermentable substrate S_F	Fitted on influent PO_4/TP ratio
iP_{SI}	0.011	0	gP/gCOD	P-fraction in Inert soluble S_I	Fitted on effluent TP concentration
iP_{BM}	0.011	0.02	gP/gCOD	P-fraction in biomass X_{BM}	Assumed/default
iP_{XI}	0.011	0.01	gP/gCOD	P-fraction in Inert particulate X_I	Assumed/default
iP_{XS}	0.011	0.01	gP/gCOD	P-fraction in Particulate substrate X_S	Assumed/default
EFF_{SS}	0.073	n.a.	% of RAS	Loss of solids clarifier to effluent	Fitted on TSS effluent

By calibrating the denitrification and EBPR performance, thereby including the measured inhibitions in the model, an accurate fit of the validated data and performance is obtained (these results are presented in several tables, eg. 15.17 and 15.19). The calibrated model describes the validated measurements within 5% accuracy. This accuracy allows the model to be used for extrapolation to full loading conditions and eventually re-design of the activated sludge system according to the new volume arrangement according to Table 15.1.

Table 15.17 Effluent results: Balanced and validated measured data is compared with the results of the calibrated model. The last column gives the absolute deviation of the simulation compared to the measurement. Nitrate and ortho-phosphate are accurately fitted. As explained previously, the effluent prediction is the least accurate aspect of the model as the result of the method of calculation and the fact that the predicted values are very small numbers. Particulate effluent fractions including COD biodegradable, TP and TN are less accurately predicted because the exact processes in the secondary settler are not well described. TKN is less accurate due to over-simplification of the ASM2d model regarding the use of one set of nitrogen fractions to fit both the influent and activated sludge fractions.

Parameter	Unit	Measured	Model	Deviation
NO_3	mgN-NO_3/L	9.8	9.8	0%
NH_4	mgN-NH_4/L	0.4	0.3	-31%
TKN	mgN/L	2.6	2.4	-10%
TN	mgN/L	12.4	12.1	-2%
PO_4	mgP-PO_4/L	1.8	1.8	-1%
TP	mgP/L	2.2	2.4	9%
$COD_{filtered}$	mgCOD/L	30.7	32.1	5%
$COD_{biodegradable}$	mgCOD/L	5.7	0.3	-95%
TSS	mgCOD/L	6.5	6.2	-5%
TCOD	mgCOD/L	36.2	38.3	6%

Table 15.18 Removal efficiency (effluent vs. influent): Accurate model prediction is possible.

Item	Unit	Measured	Model	Deviation
TSS	%	94%	96%	2%
COD	%	91%	90%	-1%
TN	%	78%	79%	1%
TKN	%	95%	96%	1%
TP	%	78%	76%	-2%

Table 15.19 Mass balance results of the flows over the total plant: Balanced and validated measured data is compared with the results of the calibrated model. The last column gives the absolute deviation of the simulation compared to the measurement. The mass balances can be closed within 3% accuracy.

Item	Unit	Measured	Model	Deviation
COD to CO_2	kgCOD/d	11,890	11,861	0%
COD to biomass growth	kgCOD/d	9,702	9,619	-1%
COD oxidized	kgCOD/d	6,927	6,739	-3%
COD denitrified	kgCOD/d	4,963	5,122	3%
Nitrified N Load	kgN/d	2,280	2,336	2%
Denitrified N load	kgN/d	1,729	1,785	3%
Nitrogen removed via WAS	kgN/d	708	697	-2%
Oxygen consumption for nitrification	kgO_2/d	10,419	10,675	2%
Oxygen saved by denitrification	kgO_2/d	4,963	5,122	3%
Oxygen consumption COD conversion	kgO_2/d	6,927	6,739	-3%
Total theoretical oxygen demand (ThOD)	kgO_2/d	17,363	17,415	0%

15.11.2 Determining calibration accuracy by parameter sensitivity analysis

In the final model calibration, several parameters were adjusted (Table 15.16). The larger part of these parameters is directly derived from actual plant measurements and therefore is considered as a model input (e.g. the sludge fractions regarding the MLSS sludge composition, Table 15.8). These parameters are coloured green and indicated reliable and validated parameter estimates. Five parameters in Table 15.16 are indicated as critical for the model results and/or highly sensitive for the model results (yellow and red). These are K_{NH4}, K_{O2}, η (both reduction factors are tested at the same time), k_{PP} and the operation parameter internal anoxic MLSS flow Q21. Thereby the parameters in red are calibrated using out of range values. To quantify the model uncertainty, these sensitive parameters are tested in a sensitivity analysis. The reference simulation for the sensitivity analysis is the calibrated model (simulation indicated in following figures as 'calibration'). The sensitivity is tested by returning each of the 5 selected parameters to ASM2d and TUDP model default setting and simulating the model. Additionally, the model is tested by running all parameters on default (simulation indicated 'default'). Based on default parameter settings also the effect of the aeration set-up is tested by comparing the results with a theoretically optimal cascade oxygen controlled process (indicated 'default+O_2-control'). The difference between the two aeration simulations gives a qualitative expression of the effect of the used oxygen control set-up in the model (compare 'default' and 'default+O_2-control').

The sensitivity is calculated for the nitrate, ammonium and phosphate concentration profiles over the plant (Figure 15.17, 15.18 and 15.19). The results show how the selected calibration parameters affect the model results. Changing the parameters in the sensitivity tests, results in slight changes in the modelled sludge composition and SRT. Direct comparison to the calibration is therefore not accurate. However, the sensitivity 'band' gives a general indication of the model uncertainty and the sensitivity results may be compared relative to each other.

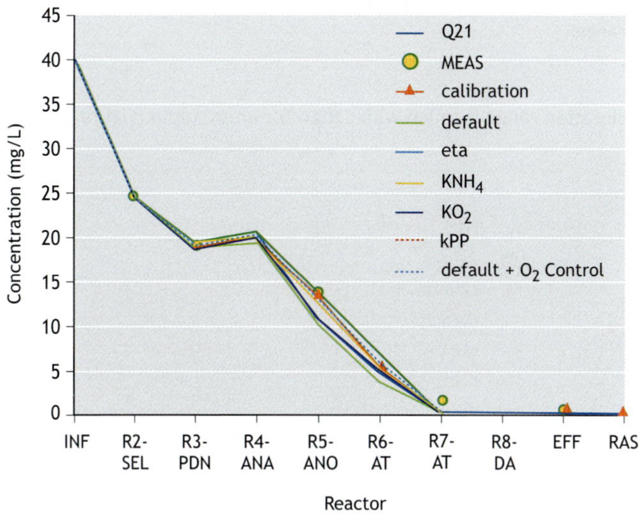

Figure 15.17 Parameter sensitivity calculation for effluent ammonium (NH_4): The calibrated model is the reference simulation. For each parameter the values are changed to the IAWQ/TUDP model default (Table 15.16)

For ammonium (nitrification) the model shows a relative narrow band indicating a low sensitivity. Regarding ammonium, the highest sensitivity is simulated for the internal MLSS recycle in R5. For the effluent K_{NH4} is the most sensitive. For each of the sensitivity tests, little effect is observed on the ammonium concentration in the last aeration tank R7. This measurement was not validated in the data check and could not be simulated accurately, indicating the measurement is incorrect.

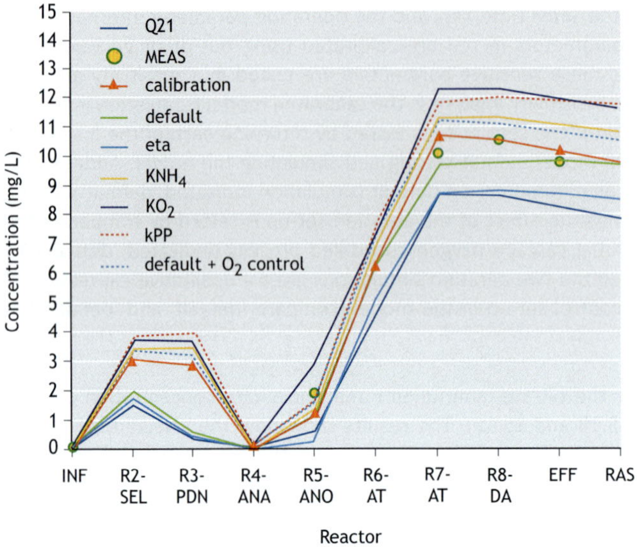

Figure 15.18 Parameter sensitivity calculation for effluent nitrate (NO_3): The calibrated model is the reference simulation. For each parameter the values are changed to the IAWQ/TUDP model default (Table 15.16). This results in a calibration range.

For nitrate (Figure 15.18) the calibration range is much wider than for ammonium (Figure 15.17).

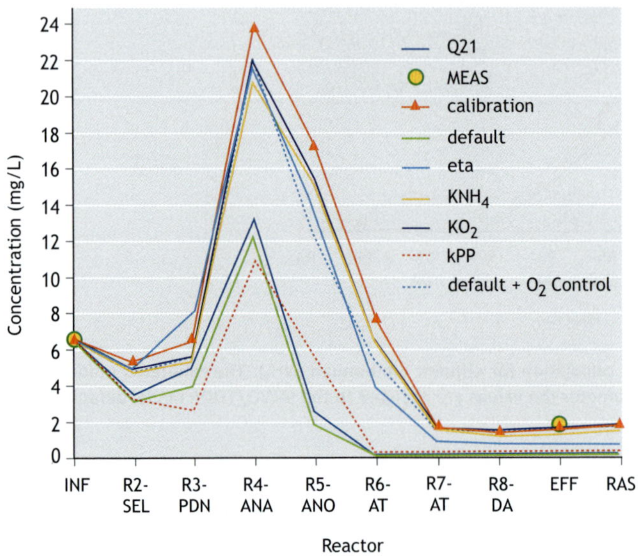

Figure 15.19 Parameter sensitivity calculation for effluent phosphate (PO_4): The calibrated model is the reference simulation. For each parameter the values are changed to the IAWQ/TUDP model default (Table 15.16).

This shows how denitrification is more affected by calibration. The figure shows that default IAWQ model parameter values already predict good results for denitrification (Figure 15.17, default simulation). Adjustment of denitrification was however required to predict the EBPR process, including the high effluent phosphate concentration (1.8 mgP-PO_4/L). This is an indication that not conventional denitrification, but most likely denitrifying P-removal is limited in the process.

The overall denitrified load is fitted by adjusting K_{O2} (Table 15.16). This parameter is most sensitive for simultaneous nitrification and denitrification (SND) occurring in the aeration tanks (R6 and R7) and therefore selected to fit SND in the model. The nitrate concentration in the anoxic tank R5 is most sensitive for the anoxic MLSS recycle flow. The recycle has a reverse effect on the modelled concentrations in R5; increasing the recycle flow increases nitrate and at the same time, dilutes the ammonium concentration in R5. The high sensitivity of the MLSS recycle indicates that the corrected flow value most likely is justified. From the relative low recycled nitrate load it is concluded that in the actual process the anoxic MLSS recycle flow (Q21) most likely is limiting for denitrification.

When default TUDP model settings are applied in the model, EBPR shows excellent phosphate removal. To decrease the EBPR efficiency in the model, k_{PP} was the most sensitive parameter (Figure 15.19). To simulate the effluent phosphate concentration of 1.8 mgP-PO_4/L, the value for k_{PP} was changed out of the acceptable range and reduced to 20% of the default value. Again, this is an indication of a process limitation in the (denitrifying) phosphate uptake.

15.11.3 Dynamic temperature modelling

The model was calibrated using an average temperature over the measured reference period (17.3°C). To test the sensitivity of the model effluent results on the yearly temperature variation, the calibrated model is validated using a measured temperature profile over a period of three years (starting 2005-2007). The dynamic temperature curve is adjusted in such a way that the simulation starts at the calibrated steady state value of 17.3°C.

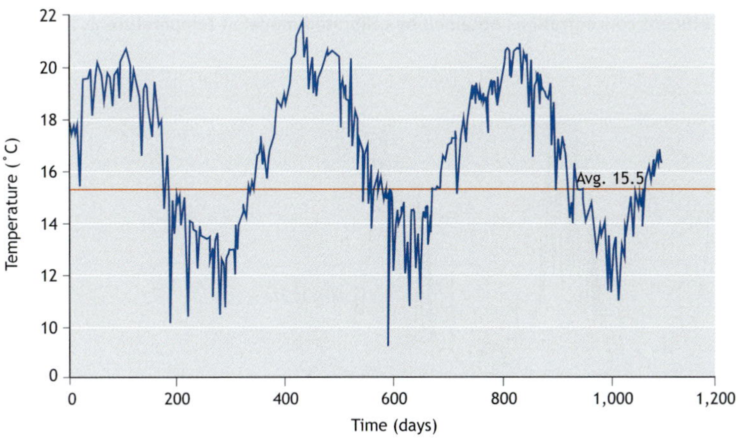

Figure 15.20 Measured temperature profile of the influent of WWTP Houtrust (2005-2007) used for dynamic temperature simulations.

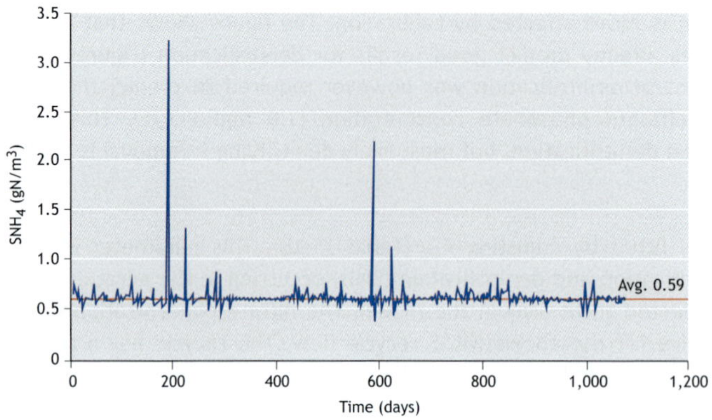

Figure 15.21 Ammonium effluent concentrations obtained by calibrated model at temperature as in Figure 15.20.

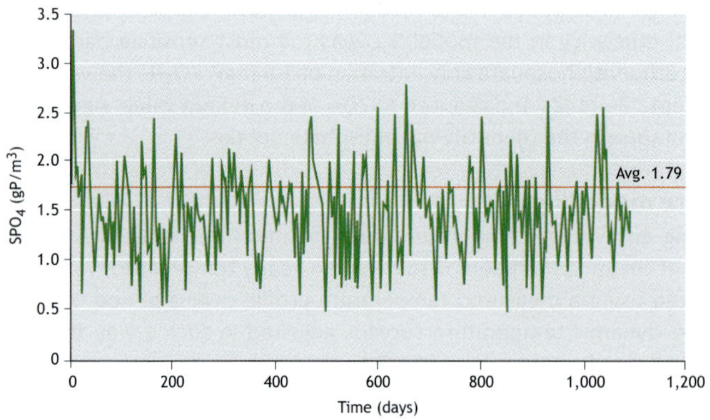

Figure 15.22 Phosphate effluent concentrations obtained by calibrated model at temperature as in Figure 15.20.

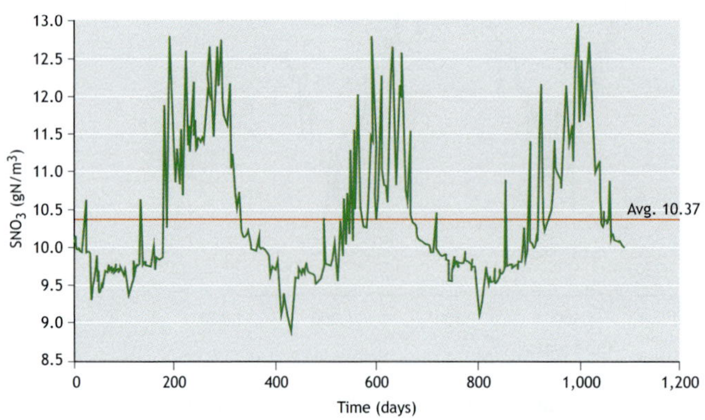

Figure 15.24 Nitrate effluent concentrations obtained by calibrated model at temperature as in Figure 15.20.

Figures 15.21-23 show that to simulate the average effluent concentration there is no significant difference in using a static design temperature (17.3°C) or a dynamic temperature profile (with identical average value). For ammonium, phosphorus and nitrate very similar average concentrations are calculated. It is therefore justified to calibrate the model using an average temperature input. The dynamic temperature simulation gives more realistic effluent result than a series of static temperature simulations. Because temperature develops gradually, the activated sludge composition is allowed to adapt to new temperature conditions. The dynamic temperature simulations show that sludge adaptation needs to be taken in consideration when predicting possible wash-out of nitrifying bacteria at the minimum design temperature. Figure 15.21 shows that with a dynamic temperature profile the temperature can drop to the minimum value for shorts periods of time (around day 200 and day 600) without losing nitrification capacity. During this time, nitrification slows down resulting in an increased effluent concentration however, nitrification wash-out does not occur immediately. When the same temperature would be simulated using a static approach, complete wash-out would be established. This analysis shows that a static approach is not realistic and will result in an overestimation of the minimal required aerobic SRT and/or volume. More reliable results are obtained when evaluating the minimum design temperature using a dynamic temperature profile.

The temperature sensitivity range, plotted in Figure 15.24, is obtained from the dynamic temperature simulations in Figures 15.21-23. The figure shows the expected range for the concentrations of nitrate, ammonium and phosphate concentrations in the different reactors as a function of the maximum expected temperature variation. The simulated bandwidth of the design temperature is considerable higher than the band found from the sensitivity test regarding the calibration parameters. This is a good indication that the calibration is accurate and performed correctly.

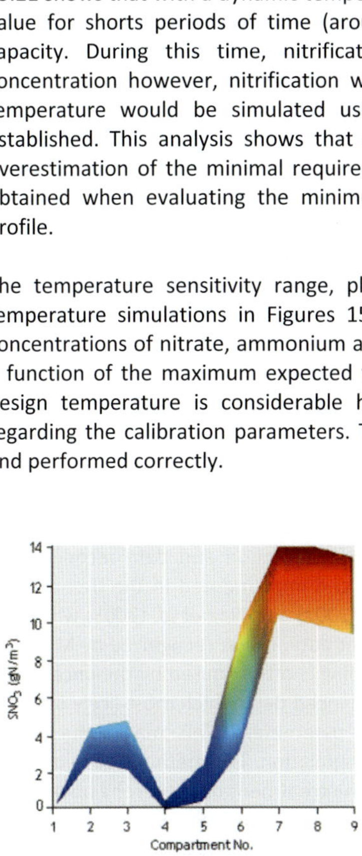

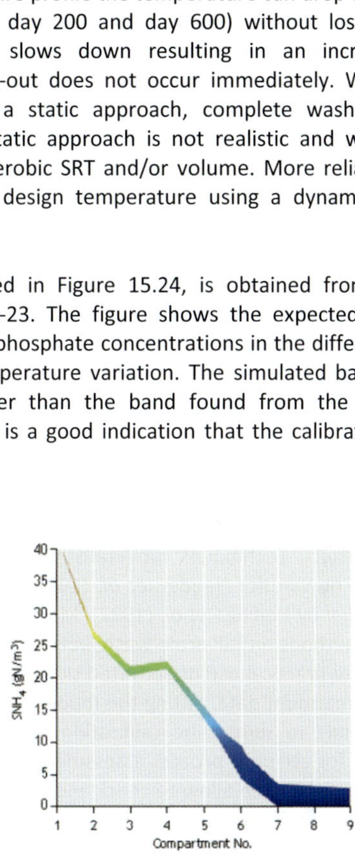

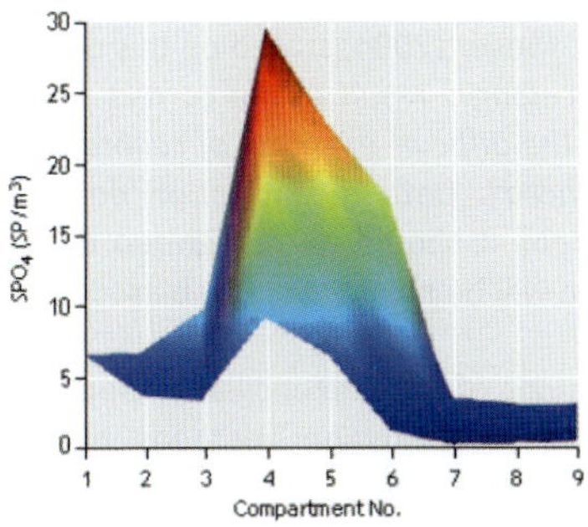

Figure 15.24 Model sensitivity towards the yearly temperature variation: Shown is the concentration range over the tank reactors calculated for nitrate, ammonium and phosphate. (1=INF, 2-SEL, 3=PDN, 4=ANA, 5=ANO, 6=AT, 7=AT, 8=DA, 9=EFF).

It can be concluded that overall the model is not particularly sensitive towards the applied parameter changes needed to calibrate the model. With default model parameter settings the model already gives reasonable results for nitrification and denitrification, except for EBPR. Satisfactory calibration was possible by changing only 3 preselected model parameters out of more than 65 model parameters in total. Most parameter adjustments are based on validated measurements (e.g. sludge composition measurements). Two parameter changes related to PAO are far out of the normal range of calibration. The TUDP model predicts far better phosphorus removal than observed in practice. Therefore the main effort of the calibration is to predict EBPR, and thereby maintaining the currently appropriate nitrification and denitrification rates.

The sensitivity analysis of the calibration parameters results in a band of accuracy for the effluent prediction (Figures 15.17-19). This bandwidth is in the range of the operational uncertainty including day and night operation, rain events and temperature variations. This is also demonstrated by the sensitivity of the simulated effluent quality as the result of yearly temperature variation further on (compare the band of accuracy in Figures 15.17-19 and 15.24). The accurate and validated data leaves less freedom to free adjustment of model parameters. This narrows the sensitivity band and increases the accuracy of the model (effluent) predictions. The high data accuracy directs the calibration towards those processes that actually require calibration. This makes it possible to detect process limitations in the early calibration phase of the model study. It is concluded that the accuracy of the model is improved by using the proposed methodology and thereby also the effective use of the model is improved for extrapolation and design purposes.

15.12 Discussion on data accuracy in wastewater treatment

This study demonstrates how evaluation and reconciliation of raw and often inconsistent wastewater treatment data, produces a consistent data set which is appropriate for modelling and represents a likely state of operation. Statistical improvement of data by reconciliation is only possible after removing systematic (gross) errors from the data, since such errors bias the reconciliation results. Techniques for (statistic) gross error detection and data reconciliation were developed by van der Heijden *et al.* (1994 a, b, c) and implemented in the free-domain software *Macrobal* by Hellinga *et al.* (1992). Meijer *et al.* (2002) and Puig *et al.* (2008) were first to apply these methods on full-scale wastewater treatment data. It was recognized by Meijer *et al.* (2002) that the practical use of activated sludge modelling is often limited by the uncertainty of wastewater treatment data. Fitting the model on unbalanced data which most likely includes errors, often leads to unrealistic calibration of model parameters. This makes the model practical useless for accurate process assessment and extrapolation. The methods developed by Hellinga and van de Heijden allow to check and improve the treatment data thereby obtaining a balanced data set which can be modelled accurately. Meijer and Puig were aware of the theoretical bottlenecks and limitations of applying reconciliation on wastewater treatment data however, also were convinced by the benefits the method offered in the development of a structured and reproducible method for model calibration. Meijer (2004) stated on this dilemma the following: *"Statistical data reconciliation does not necessarily result in the correct solution (read state estimation), but does relive the modeller from the difficult dilemma of calibrating an activated sludge model on data which cannot be balanced".* Data reconciliation delivers a balanced data set which allows accurate modelling. Thereby it should be recognized that the estimated state estimation resulting from the reconciliation will be mo\st likely be biased.

In 2014, Spindler wrote the following regarding the limitations of data reconciliation applied on wastewater treatment data: *"There are effective data reconciliation methods available developed for (chemical) process engineering. However, the influent being the main disturbance to the process of wastewater treatment causes a considerable difference to most industrial processes where process dynamics stay within closer boundaries and/or have lower frequency. Another important difference to most industrial processes is the low concentrations and significant heterogeneity (dissolved/suspended) of the relevant compounds. The various sources of measurement random errors (sampling, matrix, range of expected values, dynamic flows and concentrations) add up to comparatively larger uncertainty and make it difficult to determine the measurement variances".* Spindler concludes that the (largely uncontrolled) dynamics of wastewater treatment makes it difficult to distinguish random measurement errors from the 'normal' variance caused by the (relative large) process dynamics. They also conclude that measuring wastewater treatment is cumbersome and that measurement errors often are the sum of many factors. Thereby it is concluded that a high variance of the measurements (expressed by the standard deviation) reduces the accuracy in which the process state can be estimated and also reduces the possibility to detect gross and systematic errors by mass balance evaluation. They point out that the problem of distinguishing between measurement variance and process dynamics in theoretically limits the use of statistical data reconciliation in wastewater treatment.

In addition of the observations of Spindler we have observed that also the representativeness of measurements in full-scale wastewater treatment is generally poor. This means that even by accurate measurement, the process state most likely is misinterpreted. The large-scale treatment processes are designed for robust operation and not for precise measurement. In practice most plants cope with non-ideal mixing conditions in rectors and non-homogeneous distribution of particulates in flows and pipes. As a result measurements are most likely performed in one or another (slow or fast changing) dynamic gradient which does not represent the average nor actual state of the measured system. Fast changing gradients are typically measured in aerated reactors for oxygen, ammonium and nitrate. Slower and plant wide gradients are caused by non-homogeneous distribution of suspended solids as the result of insufficient flow and/or poor mixing in the reactors which can result in low frequency concentration variations in the activated sludge tanks and settlers. This non-ideal behaviour may appear randomly or cause systematic errors in the measurements. Thus, it is our conclusion that accurately measuring the state of a treatment plant is practically impossible and that estimation of the process state therefore is always required, thereby including non-representative plant measurements. As result of this relative high uncertainty in general, only larger gross errors will be detectable from treatment data and can be removed. Consequently, estimation of the average state of operation of a wastewater treatment, either by accurate measurement, data reconciliation or other means, most probably will be biased. In the absence of more accurate alternatives, data reconciliation still delivers the most probable state estimation. Thereby, the method gives the means to do this objectively, thus without the influence of the modeller, who may be biased in his or her assumptions. Also the practical benefits of the reconciliation technique in obtaining a balanced data set for application in activated sludge modelling are evident.

Taking in account this knowledge, in this study a range of practical measures were taken in the process of collecting plant data to improve the robustness of the estimation:

- Taking in account the (most likely) biased estimation of the state of the wastewater treatment process, this study focused on accurately measuring the waste activated sludge production (WAS). Therefore two on-line TSS measurements were used measuring the waste sludge concentration.
- Also the data model and reconciliation were developed with the goal to obtain maximum redundancy on the waste sludge production. By accurately establishing the

waste sludge production the bias in the estimation of the operational state of the plant could be redirected towards less critical measurements for modelling and plant assessment.

- Two weeks prior to the measurement campaign, the operation of the plant was remained as stable as possible; no maintenance was conducted during this period and all control settings remained unchanged (wherever this was allowed). The plant was operated under constant SRT and MLSS controlled conditions.
- A known method to reduce the variance of dynamic measurements is filtering the data. Thereby partial influence of process dynamic is removed from the data.
- Additional measures were taken in the grab sampling period; DWF conditions were defined as a DWF day proceeded by at least three dry days thereby excluding possible after-effects of rain.
- The sludge production measurements are filtered based on dry and wet weather conditions over a period of two weeks prior to the grab sampling period.
- By focusing on average DWF operation over a period of 1-2 weeks, being less than the operated sludge retention time, changes in the activated sludge composition are avoided and also effects of changing temperature could be excluded.
- Based on our expert knowledge and practical experience, certain measurements were not used or only were measured as an indication (for example grab measurements of particulates in the return sludge flow, centrate, and overflow of the primary thickening).

By taking in account these practical measures the data quality in this study improved considerably. The results of this study show that balancing is possible without exceptional adaptations of the measured data in the reconciliation procedure and all within a reasonable range of the standard deviation. This study clearly shows that data reconciliation is a practical useful and very effective tool in obtaining quality data resulting in accurate modelling and effective plant assessment which otherwise would not have been possible.

15.13 Final remarks

This study demonstrates how effective assessment of full-scale wastewater treatment performance is possible using process modelling. High accuracy can be obtained in the assessment by applying data filtering and statistical improvement techniques. Valuable process information is obtained from the modelling study, which up to this date is used to maintain the process at high efficiency. The goals of the N-tot study and upgrading study are fully achieved and have resulted in considerable improvement of the plant operation since. The study convincingly demonstrated that the redesigned plant is capable of meeting the contractual performances. By measurements it is established that nitrogen removal improved by 33% and phosphorus removal by 50%. Such substantial improvement of the plant removal efficiency was obtained by introducing relative small changes in the flow system of WWTP Houtrust, all achieved within existing volumes. Without the application of activated sludge modelling such a result would not have been possible and most likely the conventional approach would have resulted in a more traditional optimization by a volume extension with considerable higher cost. It is demonstrated how a reliable model is developed, capable of accurate predictions through the use of mass balance evaluation and statistical data reconciliation. The study shows that after performing a proper data evaluation, modelling become easier, and reliable and a very effective tool for operational troubleshooting, plant design and performance improvement. By following the modelling protocol developed by Meijer and Brdjanovic (2012), useful results of the study can be already obtained in an early project phase, which allow for cost-effective actions eventually resulting in a truly successful upgrade of WWTP Houtrust.

References

Barker, P. S. and Dold, P. L. (1995) Cod and Nitrogen Mass Balances in Activated-Sludge Systems. Water Research, 29(2), 633–643.

Hellinga C. (1992) Macrobal 2.02. Delft University of Technology, the Netherlands. M.C.M.vanLoosdrecht@TNW.TUDelft.NL. http://www.bt.tudelft.nl/miscell

Henze, M., Gujer, W., Mino, T., van Loosdrecht, M. (2000) Activate sludge models ASM1, ASM2, ASM2d ASM3, Scientific and Technical Report No. 9, IWA Publishing, London, UK.

Janssen P.M.J., (2001) Neerslag 2001-1. Praktijkervaringen met biologische P-verwijdering.

Meijer S.C.F., van der Spoel H., Susanti S., Heijnen J.J. and van Loosdrecht M.C.M. (2002) Error diagnostics and data reconciliation for activated sludge modelling using mass balances. Water Sci. Tech. 25(6), 145-156.

Meijer S.C.F. (2004) Theoretical and practical aspects of modelling activated sludge processes. TU Delft, ISBN 90-9018027-3, 218 p.

Meijer and Brdjanovic (2012) A Practical Guide to Activated Sludge Modeling. UNESCO-IHE, ISBN 9789073445260, 291 p.

Puig, S., van Loosdrecht, M. C. M., Colprim, J., and Meijer, S. C. . (2008) Data 512 evaluation of full-scale wastewater treatment plants by mass balance. Water Research, 42, 4645–4655.

Rieger, L., Takács, I., Villez, K., Siegrist, H., Lessard, P., Vanrolleghem, P. A., and Comeau, Y. (2010) Data Reconciliation for Wastewater Treatment Plant Simulation Studies—Planning for High-Quality Data and Typical Sources of Errors. Water Environment Research, 82(5), 426–433.

Spindler, A. and Vanrolleghem, P. A. (2012) Dynamic mass balancing for wastewater treatment data quality control using CUSUM charts. Water science and technology: a journal of the International Association on Water Pollution Research, 65(12), 2148–2153.

Spindler, A. (2014) Structural Redundancy of Wastewater Treatment Data. An Algorithm Derived from Data Reconciliation Methods. Water Research (to be published) Institute of Water Quality and Resource Management, Vienna University of Technology, Karlsplatz 13/226-1, 1040 Wien, Austria.

STOWA (1993) Handboek chemische P-verwijdering, STOWA 1993-06

STOWA (2001) Handboek biologische fosfaatverwijdering, STOWA 2001-15

Van der Heijden R.T.J.M., Heijnen J.J., Hellinga C., Romein B., Luyben K.Ch.A.M. (1994a) Linear constraint relations in biochemical reaction systems: I. Classification of the calculability and the balanceability of conversion rates. Biotechnol. Bioeng. 43, 3-10.

Van der Heijden R.T.J.M., Romein B., Heijnen J.J., Hellinga C., Luyben K.Ch.A.M. (1994b) Linear constraint relations in biochemical reaction systems: II. Diagnosis and estimation of gross errors. Biotechnol. Bioeng. 43, 11-20.

Van der Heijden R.T.J.M., Heijnen J.J., Hellinga C., Romein B., Luyben K.Ch.A.M. (1994c) Linear constraint relations in biochemical reaction systems: III. Sequential application of data reconciliation for sensitive detection of systematic errors. Biotechnol. Bioeng. 43, 781-791.

Figure 15.25 Panoramic view of WWTP Houtrust (Photo: Delfluent Services BV).

Figure 15.26 Plant laboratory at WWTP Houtrust (Photo: Delfluent Services BV).

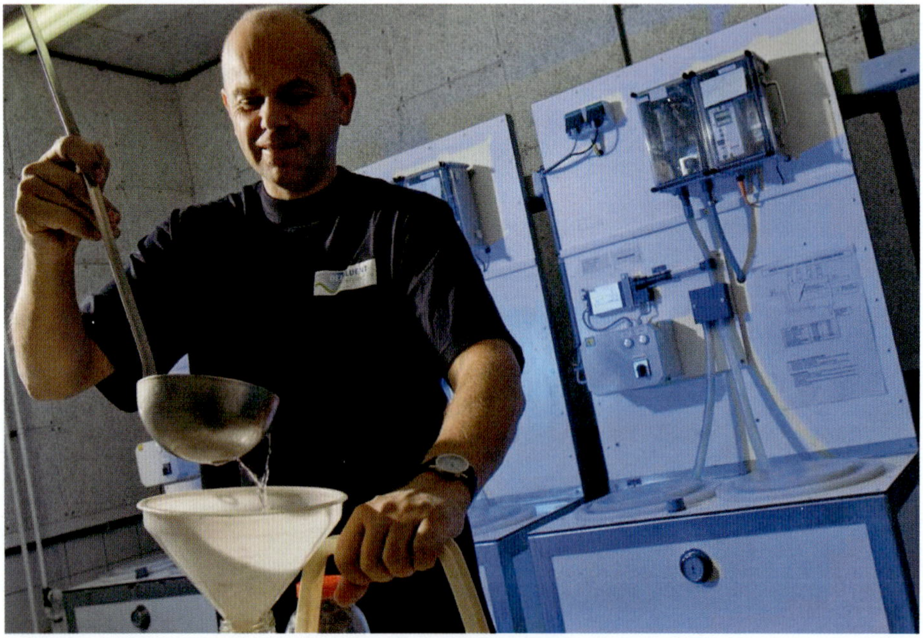

Figure 15.27 Effluent sampling at WWTP Houtrust (Photo: Delfluent Services BV).

Chapter 16: WWTP UPM, Uruguay

Modelling pulp mill wastewater treatment

Bentancur S., Lopez-Vazquez C.M., Garcia H.H.A., Travers D., Brdjanovic D.

This chapter is based on the MSc thesis "Evaluation and modelling of a pulp mill wastewater treatment plant in Fray Bentos, Uruguay" by Bentancur S. (2014). UNESCO-IHE Institute for Water Education. Delft, the Netherlands.

16.1 Introduction

16.1.1. Background

Pulp and paper mill industrial processes have high energy consumption and high requirements for fresh water and chemicals. Regarding the fresh water consumption, the pulp and paper industry ranks third in the world, just after the primary metal and the chemical industries (Thompson *et al.*, 2001). Consequently, large effluent volumes are generated by pulp and paper mill industries.

During the last decade, several pulp mills have started their operations in South America and specifically in Uruguay. Some of the main drivers to this trend include the wide and attractive natural reserves present in the region, as well as the good economic perspectives predicted for South America. Uruguay is emerging as a new target country besides Brazil and Chile (BRACELPA, 2011). Particular attention has been given to the treatment of the industrial effluents produced by this industrial sector. Consequently, these industries have incorporated state-of-the-art technologies to treat their effluent discharges to minimize potential environmental impacts on the receiving water bodies.

16.1.2. Description of the Metsä-Botnia pulp mill plant in Uruguay

The present study was carried out at a pulp mill facility located in Uruguay, South America. The pulp mill was designed and built by the Finnish company, the second largest pulp producer in Europe, which operates 5 additional plants in Finland. Metsä-Botnia owns 60% of the total forestry industrial companies in Uruguay. The pulp mill is part of an international group, with presence in 67 countries, and with production plants in 17 countries. The company has three business groups: energy and pulp, paper, and engineered materials.

The plant started its operation in November 2007. It is located 5.2 km away from the city of Fray Bentos (Río Negro Department), next to the Uruguay River (Figure 16.1), and 1.2 km from the International bridge "Libertador General José de San Martin", which connects Uruguay and Argentina. The pulp mill, with a production capacity expressed in air dry ton (ADT) per year of 1 100 000, is one of the most modern in the world. The production process was designed incorporating the best available technologies (BAT) for chemical pulp manufacturing. The fully loaded factory can process up to 3 million m^3 of eucalyptus per year (mainly *Eucalyptus Grandis*)

(Saarela *et al.,* 2008). The location of the processing plant next to the Uruguay River is strategic since the river acts as a reliable source of water supply and as an easily accessible mean to transport the final products to their final destinations. The total area of the plant is approximately 550 ha, but currently no more than 80 ha are used. The facilities are divided in three main areas: (i) wood handling, (ii) fiberline (cooking, pulp washing, oxygen delignification, screening and bleaching) and (iii) recovery line (chemical recovery and electric power generation) (Saarela *et al.,* 2008).

Figure 16.1 Pulp mill in Fray Bentos, Uruguay (Portal de Noticias TIEMPO, 2013).

The wastewater treatment plant of the pulp mill has a treatment capacity of 25 m^3/ton ADT, equivalent to approximately 73,000 m^3/day. Currently, the pulp mill treats its effluent discharges biologically in an activated sludge (AS) system that has a stable, reliable, and robust operation.

Commonly, most of the activated sludge processes dealing with pulp and paper mill effluents need to dose nutrients to the biological treatment system since the concentrations of nitrogen and phosphorus in the receiving stream to the wastewater treatment plant are usually nutrients deficient. However, at the pulp mill in Uruguay, the situation is the opposite. Mostly, *Eucalyptus Grandis* is used for the pulp production. *Eucalyptus Grandis* is a hard wood with a different phosphorus composition (in terms of both concentration and chemical structure) compared to other hard woods utilized in other parts of the world. Thus, the wastewater generated contains a higher phosphorus concentration than other pulp mill wastewaters. The concentration of phosphorus in the receiving wastewater is high enough for the correct operation of the biological system, yet low phosphorus concentrations are observed in the treated effluent. Therefore, the phosphorus observed in the treated effluent needs to be treated to comply with the discharge standard of 5 mgTP/L at any given sample, and of 74 kgTP/day as a monthly load average. The total phosphorus (TP) effluent concentration usually meets the discharge standards, and very rarely comes closer to the limit. Moreover, the wastewater treatment plant (WWTP) has a high removal efficiency meeting all other discharge standards set by the local environmental agency at 60 $mgBOD_5/L$ and 150 mgTSS/L. For instance, from May to August 2013, the following parameters were reported by the pulp mill: 285 mgCOD/L, 7 $mgBOD_5/L$, 10 mgTSS/L, 2.9 mgTN/L and 0.49 mgTP/L. However, and in spite of the high removal efficiencies exhibited by the WWTP (and the acceptable effluent concentrations of all the parameters of concern), there is a strong interest to explore different operational conditions and plant configurations to maximize the phosphorus removal efficiency (and decrease even more the phosphorus concentration in the effluent) in a cost-effective and reliable manner.

The main objective of the present research was to apply mathematical modelling to describe and assess the performance of the WWTP. In addition, a model-based assessment of different operational and configuration alternatives to enhance the plant performance and efficiency was carried out. To achieve the aforementioned objective, the STOWA protocol for dynamic modelling of activated sludge plants was applied (Hulsbeek *et al.,* 2002). Using the calibrated model, different scenarios were evaluated not only to maximize the phosphorus removal performance of the industrial WWTP but also to evaluate the potential recovery of nutrients and energy (through biogas production).

16.1.3 Pulp mill manufacturing processes

The evaluated pulp mill uses the chemical Kraft process, which is the dominant pulping process in the pulp and paper industry worldwide. Its robustness, the ability of the process to handle almost all species of soft and hard wood, and the favourable economics (due to a high chemical recovery efficiency of up to 97%), provide the Kraft process a considerable advantage over other pulping processes (Tran and Vakkilainnen, 2008). The pulp is bleached without elementary chlorine free (ECF) according to the recommendations of the Integrated Pollution Prevention and Control directive (IPPC) and the BAT Reference (BREF) that defines the BAT documents (Saarela *et al.,* 2008).

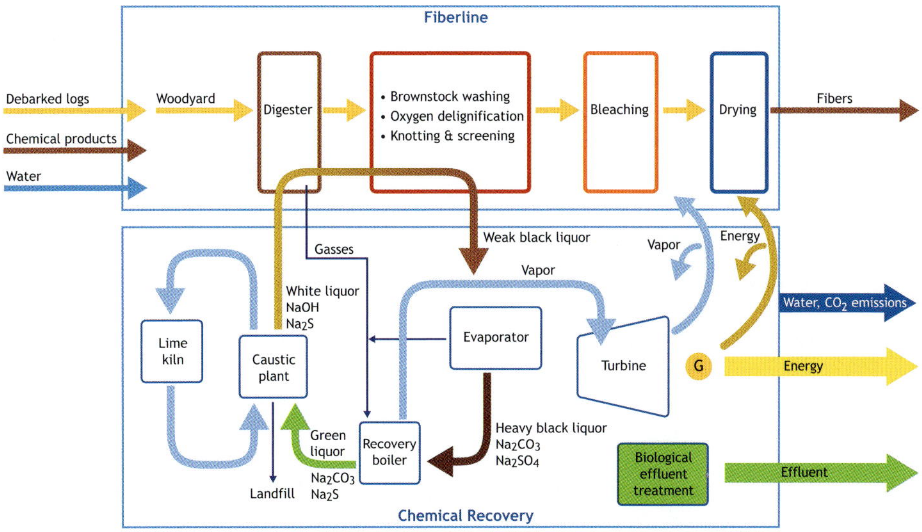

Figure 16.2 The Kraft pulping process at the evaluated pulp mill (Faroppa and Annala, 2004).

The completed fibre processing line and the chemical recovery process are presented in Figure 16.2. Once the wood arrives to the plant (as debarked logs), it goes to the woodyard and is stored according to the different species of eucalyptus. Thereafter, it is sent to two parallel chipping lines. In the next processing step, the Kraft pulping process starts and the wood chips are digested at high temperature and pressure in a water solution composed of sodium sulphide (Na_2S) and sodium hydroxide ($NaOH$) also known as the white liquor. The white liquor chemically dissolves the lignin that binds the cellulose fibres together. The white liquor is combined with the used pulping chemicals producing a weak black liquor that is separated from the pulp by washing. The black liquor is sent to the Kraft recovery system (where the inorganic pulping chemicals are recovered for reuse), while the dissolved organics are used as a fuel to produce steam and power. Per every ton of pulp produced, the Kraft pulping process generates about 10 tons of weak black liquor that needs to be processed through the chemical recovery process (Tran and Vakkilainnen, 2008). Once the pulp is cooked and washed, it passes through the oxygen delignification, knotting and screening processes. The separated and washed knots are returned to the chip feeding system to be re-cooked, while the final reject comprising mostly impurities (like sand) is removed from the system as a solid waste (Saarela *et al.,* 2008). One of the most critical processes is the pulp bleaching. In this process, the required pulp quality in terms of brightness and cleanliness should be reached. The evaluated pulp mill uses bleaching without elementary chlorine. The elementary chlorine free (ECF) process is applied, which uses chlorine dioxide. It allows to handle the high level of chlorinated organic compounds and to prevent the formation of the highly toxic polychlorinated aromatic compounds (Saarela *et al.,* 2008). The bleaching plant is the only open process. The evaluated pulp mill typically discharges approximately 20-25 m^3/ADT of bleaching filtrates. This effluent goes to the wastewater treatment plant. After that, the bleaching pulp is dried up in two parallel drying machines. With regard to the chemical recovery process, Tran and Vakkilainnen (2008) explained that the process has three main functions: (i) minimizing the environmental impact of waste material (black liquor) from the pulping process; (ii) recycling pulping chemicals such as $NaOH$ and Na_2S; and (iii) co-generating steam and power. The Kraft recovery process is shown in Figure 16.3.

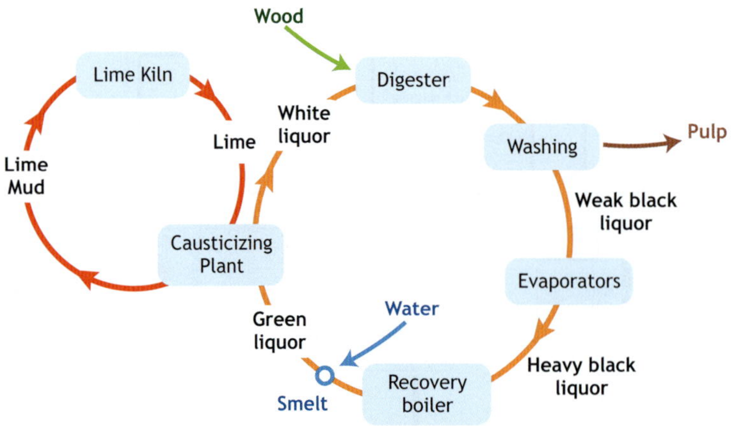

Figure 16.3 The Kraft Recovery Process (Tran and Vakkilainnen, 2008).

The weak black liquor from the fiberline is taken to the evaporation plant, where the solids concentration on the black liquor from the pulp washing is increased. The nominal evaporation capacity is 1,100 m^3 water/h. The black liquor is concentrated up to 80% solids content to be fired in the recovery boiler, which has a combustion capacity of 4,450 ton solids/d. The vapor pressure and temperature are 94 bar and 490°C, respectively, with a vapor generation capacity of 182 kg/s. The concentrated black liquor is sprayed into the lower part of the recovery boiler, where it is burned in an oxygen deficient environment so that Na_2S is formed (Tran and Vakkilainnen, 2008). Inorganic sodium and sulphur are recovered as a molten smelt which consists mostly of Na_2S and sodium carbonate (Na_2CO_3) (Tran and Vakkilainnen, 2008). The molten smelt enters a dissolving tank where it is dissolved in water to form a green liquor. The green liquor is then sent to the causticizing plant, where it reacts with lime (CaO) to convert the Na_2CO_3 into NaOH (Tran and Vakkilainnen, 2008). The conversion is measured by the causticizing efficiency, typically between 80 to 83%. The Na_2S passes through the causticizing step unchanged. The causticized green liquor is known as the white liquor containing mostly NaOH and Na_2S. It is returned to the digester for reuse in the pulping process. The precipitated $CaCO_3$ (lime mud) from the causticizing reaction is washed, and sent to a lime kiln where it is heated at high temperatures to regenerate CaO for reuse (Tran and Vakkilainnen, 2008). The evaluated pulp mill has a power generation capacity of 120 MW. Electric power is generated by two 80 MW turbines. One turbine is an extraction back up pressure turbogenerator and the other an extraction back pressure turbogenerator with condensing tail (Saarela *et al.,* 2008). That is, the pulp mill recovers not only the chemicals that are being used in the production process, but also energy from processing the black liquor. This is an example of the good practices currently encouraged considering the increasingly stricter environmental regulations.

16.1.4 Wastewater treatment plant

According to Saarela *et al.,* (2008), the evaluated pulp mill wastewater treatment plant was designed for an influent flow of 25 m^3/ADT (approximately 0.8 m^3/s). The plant has a mechanical pre-treatment followed by a biological activated sludge treatment for the removal of organic matter, total suspended solids (TSS), and adsorbable halogenated (AOX) compounds. The layout of the evaluated pulp mill WWTP is presented in Figure 16.4.

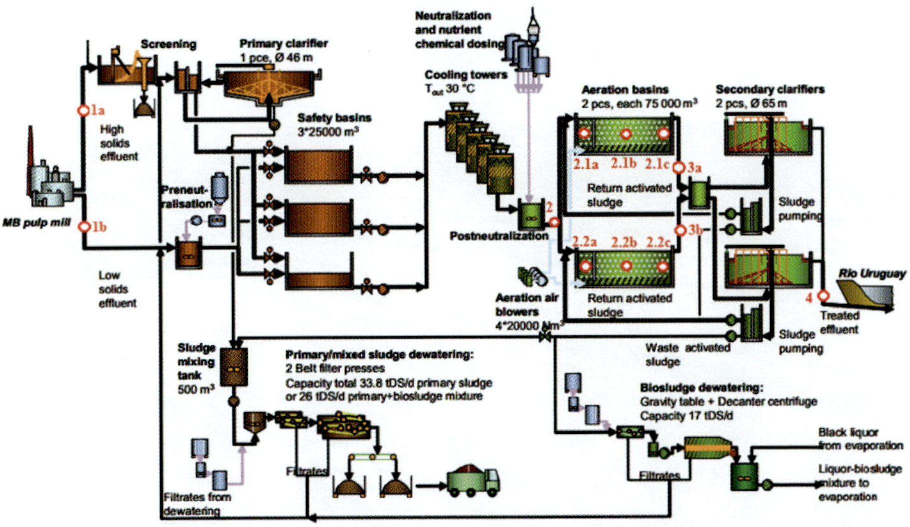

Figure 16.4 Process flow diagram of the evaluated pulp mill WWTP. The circles and numbers indicate the location of the sampling points used for the execution of the sampling campaign.

The industrial wastewater treated at the WWTP consists of two streams: a high solids concentration stream, and a low solids concentration stream. Firstly, the high solids stream is composed of wastewater generated from the white liquor plant, evaporation plant, recovery boiler, drying process, wood handling, water treatment, and fiberline. This stream is approximately one third of the entire influent to the evaluated pulp mill WWTP. As shown in Figure 16.4, the incoming stream is sent through a coarse screening to remove larger solids and later on through a primary clarifier (1 unit, ∅=46 m) to remove settleable solids (primarily fibres). Then, this stream is sent to the 3 safety and equalization basins (of 25,000 m^3 each one). The low solids stream represents about two thirds of the influent flow to the plant. The low solids stream is mostly generated in the acid and alkaline bleaching processes and to a lesser extent at the chemical and white liquor plants. The low solids stream is discharged directly in the pre-neutralization tank (to adjust the pH). Thereafter, the neutralized stream is stored in the safety and equalization basins where it gets mixed with the high solids stream (Figure 16.4).

Once the two flows are mixed in the safety and equalization basins (where the variations in flows and concentrations are balanced) the equalized volumes are directed to the cooling towers to decrease the wastewater temperature from 55°C to approximately 30°C to prevent the inhibition of the biological treatment (Figure 16.4). Thereafter, the effluent is post-neutralized for pH adjustment, urea is added, and the effluent flow passes through the biological activated sludge treatment consisting of two aerobic reactors in parallel (each one of 75,000 m^3) (Figure 16.5). The oxygen for both aeration basins is supplied by four air blowers (with an installed capacity of 20,000 Nm^3 each). The treated effluent from the aerobic reactors is sent to the secondary clarifiers (two units in parallel, each one with a diameter of 65 m). Finally, the treated effluent is discharged into the Uruguay River (Figure 16.6).

Figure 16.5 Secondary treatment at the evaluated pulp mill WWTP: Conventional aerobic activated sludge system (Photo Bentancur S.).

Figure 16.6 Secondary treatment at the pulp mill WWTP of study: Effluent to Uruguay River (Photo Bentancur S.).

After the secondary clarifier, the settled sludge is separated into the return of activated sludge (RAS) and waste activated sludge (WAS). The RAS flow is returned to the aeration basin, whereas the WAS flow is directed either to the sludge mixing tank (which stores the primary sludge) or to the bio-sludge dewatering process (Figure 16.4). The primary sludge treatment line begins at the sludge mixing tank (with a storage capacity of 500 m^3). Then, the sludge is treated in a primary dewatering unit with two belt filter presses with a total capacity of 33.8 total dry solids (TDS)/d. After dewatering, the filtrates are returned to the screening and the solids are transported to the final disposal site (forest plantations). In the same manner, the WAS that is directly sent to the bio-sludge dewatering, treated in the gravity table and decanter centrifuge with a capacity of 17 tDS/d. Likewise, the filtrates from the primary sludge dewatering process are sent back to the screening process and the liquor-bio-sludge mixture is sent to evaporation.

16.2 Materials and methods

The methodology applied in the present research was based on a practical protocol for dynamic modelling of activated sludge systems (Hulsbeek *et al.,* 2002) and the methodology described by Meijer and Brdjanovic (2012).

16.2.1 Process description

During this phase, a complete inventory of data and processes information was executed. The inventory included: the evaluated pulp mill low and high solids effluent composition, information regarding primary treatment, activated sludge process components and internal recycle flows in the water line, sludge treatment and effluent quality. A process flow diagram was drawn with the aim to show the WWTP treatments technologies, components and model boundaries (Figure 16.4).

16.2.2 Data collection and evaluation

In order to obtain a general overview of the WWTP, data regarding the components of the treatment systems, operation, and performance were provided for the period May to August 2013. The historical data was used to make a preliminary model of the plant. In addition, a sampling campaign was designed and implemented to complete the required information. The sampling program was carried out from October 21 to 28th, 2013. For this purpose, 24 h representative composite samples were collected during five consecutive days to characterize the influent and effluent. In Figure 16.4, the location of the sampling points can be observed. Their locations are: the high and low solids content influents before screening (sampling point 1a) and before the pre-neutralization (point 1b); after the cooling towers just before the aeration basins (sampling point 2); along the aeration basins (points 2.1a, 2.1b, and 2.1c); after the aeration basins (3a and 3b); and, at the treated effluent from the WWTP to the Uruguay River (sampling point 4). An overview of the monitored points, location, and routine parameters are presented in Table 16.1.

Table 16.1 Location of the sampling points for the execution of the sampling campaign at the evaluated pulp mill WWTP, including the analytical parameters of interest.

No	Location	pH	T	DO	TSS	COD	TP	TN	BOD_5	NO_3^-	NO_2^-	AOX
1a	Before screening	■	■		■	■	■	■				
1b	Before pre-neutralization	■	■		■	■	■	■				
2	After cooling towers	■	■		■	■	■	■	■			
2.1a	Aeration 1-Selector			■								
2.1b	Aeration 1-Zone 2			■								
2.1c	Aeration 1-Zone 3			■								
2.2a	Aeration 2-Selector			■								
2.2b	Aeration 2-Zone 2			■								
2.2c	Aeration 2-Zone 3			■								
3a	After Aeration 1	■			■							
3b	After Aeration 2	■			■							
4	Effluent	■	■		■	■	■	■	■	■	■	■

Parameters such as total COD, TN, TP, TSS and inorganic suspended solids (ISS) were analyzed at the laboratory facilities at the evaluated pulp mill. The rest of the parameters that could not be analyzed at the site (including NO_2^-, NO_3^-, NH_4^+, BOD_{10} and PO_4^{3-}) and therefore were analyzed at the Technological Laboratory of Uruguay (LATU). Total influent COD ($COD_{,inf,tot}$) was fractionated into soluble COD ($COD_{,inf,sol}$) and particulate COD ($COD_{,inf,part}$). For such a purpose, samples were filtered through a 0.45 μm pore size filter to determine the fraction of soluble COD. The soluble unbiodegradable influent COD was estimated to be equal to the COD in the effluent. That assumption is possible since the concentration of TSS in the effluent is low. In order to estimate the biodegradable COD (BCOD), the influent BOD_{10} was measured and used to estimate the total BOD based on an estimation of the BOD degradation curve (Hulsbeek *et al.*, 2002). The constant k_{BOD} and the correction factor f_{BOD} used in the calculation were assumed to be similar to typical values from municipal wastewater mostly because these data cannot be found for industrial wastewaters. With the aim of evaluating the historical data obtained from the WWTP, different mass balances were executed for: the flows, suspended solids, COD, TN and TP. In addition, a hydraulic model of the plant was developed and using the data collected during the sampling campaign, a complete wastewater characterization of the influent and effluent was carried out.

16.2.3 Steady-state model and aerobic batch tests

Using the data collected, a steady-state model was applied (Ekama and Wentzel, 2008) to estimate the aerobic degradation of organic material and the nutrient (P and N) requirements. In parallel, an aerobic batch test was executed to measure the nutrient requirements of the activated sludge biomass. For this purpose, a reactor set-up was assembled at the pulp mill chemical laboratory facilities (Figure 16.7). Mixed liquor activated sludge collected at the end of the aeration basin was aerated for 6 h prior to the execution of the test. Thereafter, the sludge was added into the beaker. The pH was adjusted to 7.5 with HCl 0.1 M and NaOH 0.1 M solutions. Allyl-N-thiourea was added as nitrification inhibitor. In addition, 10.3 mg of urea and 53.0 mg of sodium tri-phosphate were added as N and P sources, respectively. At minute zero, the influent was added into the batch reactor, and several samples were taken from minute 5 to minute 120. Also, samples were collected after 24 and 48 h.

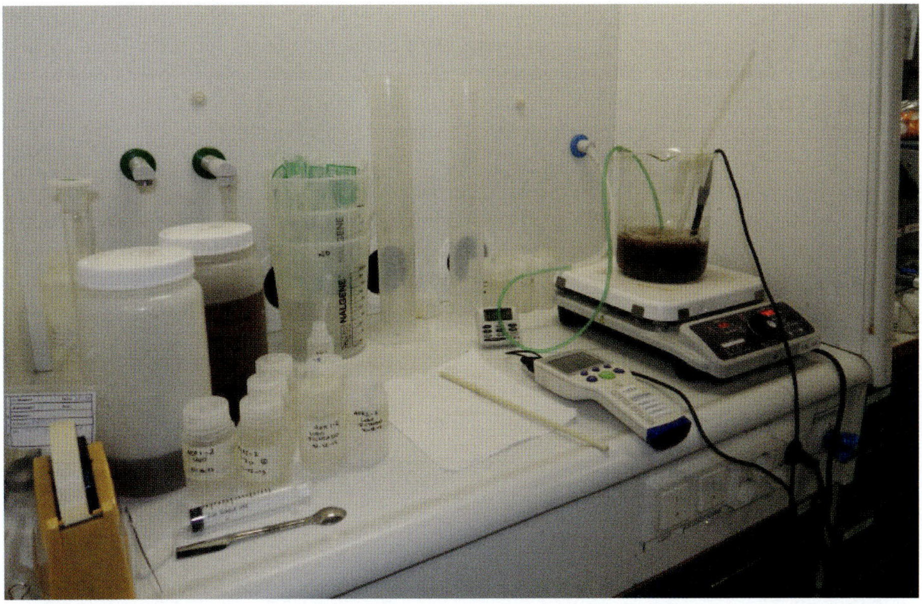

Figure 16.7 Aerobic batch set up assembled at the pulp mill laboratory facilities (Photo Bentancur S.).

16.2.4 Model calibration and validation

The model was calibrated following a step-wise approach by adjusting key wastewater fractions and kinetic and stoichiometric model parameters. The calibration was carried out until the model provided a satisfactory description of the parameters routinely measured at the WWTP. For model validation, two sets of data were used. The first set of data corresponded to winter conditions (from May to August 2013) and the second to summer conditions (from September to December 2013).

16.2.5 Scenarios assessment

Once the model was calibrated and validated, different simulations were executed to (i) evaluate the potential to achieve an improved effluent quality with minimum plant modifications, (ii) assess the capacity and robustness of the plant under different operational conditions, and (iii) recover energy and nutrients (as biogas and struvite, respectively). Thus, the following scenarios were studied:

1. Implementation of chemical phosphorus removal
2. Implementation of enhanced biological phosphorus removal
3. Assessment of higher influent flows
4. Discharge and co-treatment of municipal wastewater
5. Effects of shorter SRT on plant performance
6. Operation of the plant with only one treatment line
7. Implementation of an anaerobic digester for sludge waste treatment
8. Nutrients recovery in the anaerobic digester supernatant

16.2.5.1 Implementation of chemical phosphorus removal

The potential implementation of chemical phosphorus removal was assessed. With the objective of determining the optimum coagulant dose for chemical P-removal, different jar test experiments were carried out. The experiments were divided in two as shown in Table 16.2. In the first set of experiments, different coagulant doses were added at the following mol-to-mol ratios: 1, 2.5, 5 and 10. In the second set, the coagulant dose was determined based on the iron destabilization diagram for sweep coagulation (Lawler and Benjamin, 2006).

The experiments were executed on mixed liquor activated sludge (collected at the end of the aerobic reactor in sampling points 3a and 3b) and pulp mill WWTP effluent (after secondary clarifier, sampling point 4). Tests were conducted at the default wastewater pH (7.2-7.9) as well as at adjusted pH values (pH 5.5 and 6.0) in accordance to reported literature data. A cost analysis was performed to estimate the annual costs for chemicals. For this purpose, the total coagulant needs were quantified considering the coagulant characteristics (a commercial 40% ferric chloride solution) and its average commercial price of €100 per ton (Paul et al., 2001).

Table 16.1 Jar test experiments carried out with mixed liquor activated sludge and WWTP effluent for chemical phosphorus removal.

E1A	E1B	E2A	E2B
Mixed liquor AS		Mixed liquor AS	
Default pH: 7.4	Adjusted pH: 5.5	Default pH: 7.2	Adjusted pH: 6.0
E1C	E1D	E2C	E2D
Effluent		Effluent	
Default pH: 7.7	Adjusted pH: 5.5	Default pH: 7.9	Adjusted pH: 6.0
Experiment 1 Chemical precipitation		Experiment 2 Sweep coagulation	
Beaker	Chemical dosage ($mgFeCl_3x6H_2O$)	Beaker	Chemical dosage ($mgFeCl_3x6H_2O$)
1	1.5	1	20
2	3.8	2	100
3	7.5	3	180
4	15.1	4	270

16.2.5.2 Implementation of enhanced biological phosphorus removal

After the model was calibrated and validated, this upgrading scenario was assessed using the calibrated model in BioWin. The volume of a selector located in the aerobic tank (corresponding to 25% of the total volume ~35,000 m^3) was converted to an un-aerated zone (anaerobic).

16.2.5.3 Assessment of higher influent flows

The impact of higher influent flows generated at the evaluated pulp mill plant, due to a potential increase in the production capacity, was evaluated. The increases of the evaluated influent flows varied from 20 to 100% the average influent flows treated at the WWTP.

16.2.5.4 Discharge and treatment of municipal wastewater

This scenario was evaluated to determine whether the pulp mill WWTP could have the capacity to co-treat the municipal wastewater from the city of Fray Bentos, in addition to the industrial effluents from the pulp mill production processes. For this purpose, 70% of the wastewater generated by the 24.406 person equivalent (PE) of the City of Fray Bentos was considered. Representative municipal wastewater characteristics and concentrations were taken from typical raw municipal wastewater compositions as reported by Meijer and Brdjanovic (2012).

16.2.5.5 Effects of shorter SRT on plant performance

Shorter SRT can lead to higher fractions of active biomass and less generation and accumulation of inert materials. This could lead to a higher uptake of nutrients for biomass growth purposes. In addition, a less stabilized sludge will be produced which, if anaerobically treated, may lead to higher biogas production. Thus, using the calibrated model, the effects of shortening the SRT were simulated and assessed. The SRT at the evaluated pulp mill WWTP was sequentially shortened from 32 to 25, 20, 15 and 10 days by increasing the WAS withdrawn per day to assess its impact on the plant performance and effluent quality.

16.2.5.6 Operation of the plant with only one treatment line

This scenario was carried out to assess the response of the WWTP during a maintenance period when only one treatment line could be in operation.

16.2.5.7 Implementation of an anaerobic digester for sludge waste treatment

The main objective of executing this upgrading scenario was to analyze the implementation of an anaerobic digester to stabilize the sludge from the primary and secondary clarifiers. The biogas production, energy generation, and energy savings were studied to estimate the potential benefits of the implementation of an anaerobic digester.

16.2.5.8 Nutrient recovery in the anaerobic digester supernatant

As presented in previous scenarios, the implementation of an anaerobic digester for sludge waste stabilization may be a promising alternative to reduce the operational costs. However, that potential upgraded scenario will lead to a high generation of soluble nutrients (in the form of orthophosphate and ammonia) rather attractive for recovery. Thus, in this scenario, the potential recovery of those nutrients by struvite formation and precipitation was studied. The precipitation was assessed after simulating the addition of magnesium hydroxide and estimating the precipitation of struvite. Usually, struvite formation tends to occur inside the digesters (or in the supernatant); but even more favourable conditions for struvite precipitation exist in the anaerobic sludge digesters of WWTP performing enhanced biological phosphorus removal (Takács, 2008).

16.3 Results and discussion

16.3.1 Data collection and evaluation

Based on the data collected at the evaluated pulp mill WWTP from May to August 2013, the influent and effluent parameters were assessed. The influent flow fluctuated considerably, but they show relatively stable moving average values for the 5 and 10- day reporting periods. The average influent flow for the reporting period was 68,488 ± 9,762 m^3/d, with minimum and maximum values of 51,567 ± 2,335 and 74,726 ± 564 m^3/d, respectively. Originally, the influent flow to the biological treatment was unknown. Therefore, it was estimated considering the effluent flow to the Uruguay River, and the waste activated sludge (WAS) flow. Mass balances were executed to verify the reliability and completeness of data. For instance, to determine the flow of the return of activated sludge (RAS), a solids mass balance was performed. Thus, the sludge recycle ratio was estimated to be around 112% with regard to the influent flow (Qin). During a visit to the WWTP, it was confirmed that the average value fluctuated between 110 and 120%. The total phosphorus (TP) concentration in the sludge was also determined by performing a TP mass balance. The SRT of the evaluated pulp mill WWTP was estimated at approximately 32 days. Apparently, this value is extremely high considering that the biological system performs only organic matter removal. However, the system appears to have been designed as an extended aeration process to minimize the sludge production.

Table 16.3 shows a comparison between the characteristics of the influent wastewater of the evaluated pulp mill WWTP (based on data collected from May to August 2013) and typical wastewater characteristics from a raw municipal wastewater and other pulp mills WWTP.

Table 16.3 Comparison between the typical wastewater characteristics of the influent wastewater to the evaluated pulp mill WWTP, typical raw municipal wastewater, and other pulp mills WWTP.

	Raw Municipal Sewage Meijer and Brdjanovic (2012)	**Pinus Radiata Kraft Pulp Mill** (Chile) Diez et al. (2002)	**Stora Enso Fine Paper Oulu Kraft Pulp Mill** (Finland) Keskitalo and Leiviska (2010)	**Bleached Kraft Pulp Mill** (Uruguay) May to August 2013
COD total	750	1,208	1,167	1,588
COD soluble	300	-	-	1,243
BOD 5	350	319	255	911
VFA (as acetate)	30	-		217
N total	60	6.1	6.6	6.5
Ammonia-N	45	-	0.2	0.2
P total	15	1.1	1.7	3.7
Ortho-P	10	-	2.4	2.2
TSS	400	147	-	220
VSS	320	-	-	175

The influent wastewater of the pulp mill WWTP evaluated in this study exhibited higher concentrations of organic matter (including the biodegradable organic fraction) than a typical raw municipal wastewater. The COD/BOD ratio was lower than 2, indicating a considerable presence of biodegradable material in the influent. When comparing the nutrient composition, the differences are even higher. Typical total nitrogen (TN) and TP concentrations in municipal wastewater are 60 mgN/L and 15 mgP/L, respectively, whereas the pulp mill wastewater generated at the plant of study has TN and TP concentrations of approximately 6.5 mgN/L and 3.7 mg/L, respectively.

When comparing the pulp and paper mill wastewaters, the pulp mill wastewater from the plant in Uruguay has similar characteristics (in terms of total COD, TN, TSS and ammonia) to other pulp and paper mill wastewaters in the world. But, the BOD_5 and TP are considerably higher (and even higher than those observed by the same company in Finland). This may be attributed to the high P content of the wood used for the production of cellulose in Uruguay. It must be underlined that at the evaluated pulp mill WWTP, 1,500 kg of urea are dosed on a daily basis to the influent wastewater to cover the nitrogen requirements for biomass growth.

The pulp mill WWTP in Uruguay has remarkable removal efficiencies (Table 16.4). The processing plant applies the best available technologies, and the WWTP exhibits a good treatment performance. Consequently, the water quality of the effluent meets the local standards established by the local Environmental Regulatory Authority, DINAMA (Decree 253/79). In addition, the nutrient concentrations observed in the treated effluent are lower than other reported treated effluents from similar plants in Europe.

Table 16.4 Evaluated pulp mill WWTP removal efficiencies observed from May to December 2013 with an average flow of 65,639 m^3/d.

Sampling point/Parameter	Total COD	BOD_5	Total N	Total P	TSS
Influent to AS system	1,499 mg/L	811mg/L	16.3 mg/L		250 mg/L
Effluent to Uruguay River	284 mg/L	6.7 mg/L	2.3 mg/L	0.6 mg/L	8.0 mg/L
Removal efficiency	81%	99%	86%	80%	97%
DINAMA regulations (discharge limit)	-	60.0 mg/L	-	5.0[a]	150.0

[a] Currently this parameter is limited to a monthly average load of 74 kg/d.

16.3.2 Evaluated Pulp mill WWTP process plant layout

The pulp mill WWTP has two parallel treatment lines. Each line includes one aeration tank, and one secondary clarifier. The aeration tanks have a length of 190 m, width of 42 m, and height of 9.5 m. The diameter of each secondary clarifier is 65 m. Each treatment line has one recirculation of activated sludge (RAS). The average recirculation rate observed during the sampling campaign was 115% with respect to the influent flow. However, it can vary between 110 and 150%. Besides, there are two WAS extraction lines that direct the sludge to the side-stream bio-sludge treatment line at a flow ratio of approximately 5%. The aeration tanks do not have compartments. The oxygen level is not uniform along the length of the tanks. There are three different zones, namely: selector, zone 2 and zone 3 (Figure 16.8). In the aeration tank 1, the selector works at an average DO concentration of 1.07 mgO_2/L, zone 2 at 0.77 mgO_2/L, and zone 3 at 2.57 mgO_2/L. In the aeration tank 2, the selector works at an average DO concentration of 0.77 mO_2/L, zone 2 at 1.99 mgO_2/L and zone 3 at 3.38 mgO_2/L. The different DO concentrations observed are mainly caused by the organic matter degradation along the reactor length (with a higher DO consumption in the selector and a lower in zone 3). Nevertheless, the submerged air bubble diffusers provide satisfactory mixing conditions.

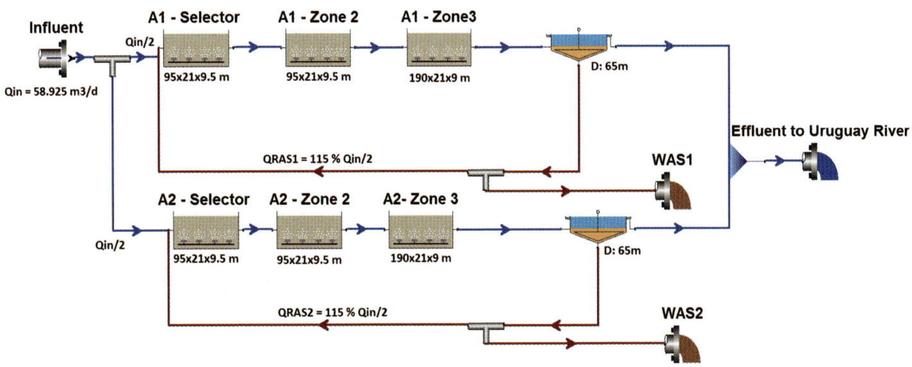

Figure 16.8 Simplified hydraulic flow diagram of the pulp mill WWTP.

16.3.3 Evaluated pulp mill wastewater characteristics and fractionation

The influent wastewater characteristics of the pulp mill WWTP observed during the sampling period are shown in Table 16.5; whereas, the wastewater fractionation is shown in Table 16.6.

Table 16.5 Influent wastewater characteristics for the modelling period at the pulp mill WWTP of study.

Parameter	Symbol	Influent	Effluent	Unit
Total COD	COD_{total}	1,588	268	$gCOD/m^3$
COD soluble[1)]	COD_{filt}	1,243	260	$gCOD/m^3$
BOD_{10}	BOD_{10}	911	6.9	$gBOD/m^3$
Influent BOD_{10} soluble[1)]	$BOD_{10, filt}$	749		$gBOD/m^3$
Influent Volatile fatty acids	VFA	217		$gHAc/m^3$
Total Nitrogen[2)]	TN	15.8	1.3	gN/m^3
Total Kjehldahl Nitrogen	TKN	15.8	1.2	gN/m^3
Ammonium	NH_4	0.23	0.04	gN/m^3
Nitrite	NO_2	ND	ND	gN/m^3
Nitrate	NO_3	0.03	0.1	gN/m^3
Total Phosphorus	TP	3.67	0.58	gP/m^3
Total Phosphorus soluble[1)]	TP_{filt}	2.35	0.54	gP/m^3
Orto-phosphate	PO_4	2.15	0.47	gP/m^3
Total Suspended Solid	TSS	220.2	4.1	$gTSS/m^3$
Inorganic Suspended Solid	ISS	45.4	[3)]	$gISS/m^3$
Volatile Suspended Solid	VSS	174.8		$gVSS/m^3$
Temperature	T	30		^oC

[1)] 0.45 μm membrane filtered.
[2)] TN after urea addition.
[3)] Not determined.

Table 16.6 Influent wastewater concentrations of the pulp mill WWTP for the modelling period.

Description	Symbol	Value	Unit
Soluble compounds			
Oxygen (negative COD)	S_{O2}	0.0	$gO2/m^3$
Readily biodegradable organics	S_F	803	$gCOD/m^3$
Volatile fatty acids	S_A	217	$gCOD/m^3$
Ammonium and ammonia nitrogen	S_{NH4}	0.2	gN/m^3
Nitrate and Nitrite nitrogen	S_{NO3}	0.0	gN/m^3
Nitrogen	S_{N2}	0.0	gN/m^3
Inorganic soluble phosphorus	S_{PO4}	2.1	gP/m3
Soluble inert organic matter	S_I	223	$gCOD/m^3$
Alkalinity	S_{ALK}	6.0	moleHCO/L
Solid compounds			
Particulate unbiodegradable organic matter	X_I	247	$gCOD/m^3$
Slowly biodegradable substrate	X_S	98	$gCOD/m^3$
Active heterotrophic biomass	X_H	0.0	$gCOD/m^3$
Phosphate accumulating organisms	X_{PAO}	0.0	$gCOD/m^3$
Poly-phosphate	X_{PP}	0.0	gP/m^3
Poly-hydroxy-alkanoates	X_{PHA}	0.0	$gCOD/m^3$
Glycogen	X_{GLY}	0.0	$gCOD/m^3$
Active autotrophic biomass	X_{AUT}	0.0	$gCOD/m^3$

Table 16.7 shows the evaluated pulp mill influent COD fractions compared to both municipal influent COD fractions, and to other pulp and paper mill influent COD fractions.

Table 16.7 Summary of the evaluated pulp mill WWTP COD influent fractions and comparison against other municipal and pulp and paper mill industrial wastewaters.

Influent fractions	Pulp and paper mill (Canada) Barañao and Hall (2004)	Stora Enso Fine Paper Oulu pulp mill (Finland) Keskitalo and Leiviska (2010)	Municipal wastewater (ASM3 default) (Gujer et al., 1999)	Pulp mill (Uruguay) STOWA protocol (Hulsbeek et al., 2002)
S_S	0.49	0.31	0.43	0.64
S_I	0.14	0.26	0.13	0.14
X_S	0.30	0.35	0.33	0.06
X_I	0.07	0.08	0.11	0.16

The fractions differ with respect to municipal wastewater fractions. The evaluated pulp mill effluent exhibited a higher fraction of readily biodegradable COD (Ss, 0.64) than a typical municipal wastewater (0.43). The slowly biodegradable substrate (Xs) together with the particulate inert organic material (X_I) (that is, the particulate fraction) accounted for 22% at the evaluated pulp mill wastewater influent COD. However, the particulate fraction of a typical municipal wastewater is much higher (approximately 44% of the influent COD). The RBCOD (S_S) observed at the evaluated pulp mill wastewater (0.64) was higher than the RBCOD observed for the pulp and paper mill wastewater both in Canada (0.49), and in Stora Enso Fine Paper Oulu pulp mill in Finland (0.31). Both the slowly biodegradable substrates (X_S) and the particulate inert organic material (X_I) were lower (0.06 and 0.16, respectively) than the particulate fractions from other pulp mills (from 0.30 to 0.35 and 0.07 to 0.08, respectively).

16.3.4 Steady-state model and aerobic batch tests

During the steady-state calculations, the wastewater characterization and fractionation obtained in the previous section were used. The two most important system characteristics such as temperature (30°C) and SRT (32 d) were used as a starting point. Other stoichiometric and kinetic parameters were considered in accordance to Henze *et al.* (2008). Results drawn from the steady-state model indicated that the influent TP (3.7 mgP/L) and TN (15.8 mgN/L) concentrations at the evaluated pulp mill WWTP were not sufficient to cover the biomass growth requirements, suggesting that the plant may suffer of nutrient limitations. However, in practice, the evaluated pulp mill WWTP operates very efficiently regarding COD, BOD, TSS, TN and TP parameters.

To assess the actual nutrient requirements, an aerobic batch test experiment was executed and the nutrient requirements were estimated based on the organic matter removal (on a COD basis) and the phosphorus and nitrogen consumption. The phosphorus removed by the sludge for biomass growth requirements (f_p) was estimated to be approximately 0.012 mgP/mgVSS. This value is approximately half of that reported value for municipal wastewaters (0.025 mgP/mgVSS). Applying a similar approach, the biomass nitrogen requirements were estimated (f_N). The f_N calculated was approximately 0.037 mgN/mgVSS, which is lower than the value usually observed in municipal wastewater treatment plants (0.1 mgN/mgVSS). Moreover, during the calibration phase of the model, f_P and f_N were further adjusted to 0.004 mgP/mgVSS and 0.025 mgN/mgVSS, respectively, to achieve a satisfactory description of the pulp mill effluent concentrations. The model suggested to applying an even lower assimilation of nutrients than that obtained in the laboratory batch experiments. To verify such low nutrient assimilation, a mass balance (including the analytical determination of N and P on the primary and secondary sludge) was conducted. The mass balance results supported and confirmed such low nutrient assimilation requirements.

16.3.5 Model calibration

The evaluated pulp mill WWTP layout was introduced in BioWin as shown in Figure 16.8. A preliminary simulation was carried out with the influent data from the wastewater characterization and fractionations obtained from the sampling program (Tables 16.5 and 16.6). Initially, the default values of BioWin were used for the kinetic and stoichiometric parameters. Thereafter, a step-wise calibration procedure was performed in accordance to Hulsbeek *et al.* (2002) and Meijer and Brdjanovic (2012):

1) The COD biodegradable and unbiodegradable particulate fractions (X_S and X_I, respectively) were adjusted to 0.18 and 0.03, respectively, to match the MLSS concentration of 5.3 gTSS/L of the plant.
2) The aerobic decay rates of OHO and AOB were decreased to 0.24 d^{-1} and 0.04 d^{-1}, respectively, to describe the MLVSS and ammonia concentrations in the aerobic reactors.
3) The nitrate concentration was calibrated by adjusting the half-saturation constant of dissolved oxygen of OHO (K_{OH}) to 0.35.
4) The nutrient content in the biomass, endogenous residues and other particulate compounds were decreased to 0.004 (for phosphorus compounds) and 0.025 (for nitrogen) to describe not only the low nutrient assimilation observed in the sludge of the pulp mill but also the low effluent ammonia and orthophosphate concentrations.

Table 16.8 displays a summary of the values adjusted during the model calibration.

Table 16.8 Summary of parameters adjusted during the calibration of the model.

Wastewater characteristics	Symbol	Unit	Calibrated value	Biowin default value	Wastewater characterization value
Readily biodegradable COD	F_{bs}	$gCOD/gCOD_{tot}$	0.64	0.27	0.64
Acetate	F_{ac}	gCOD/gRBCOD	0.15	0.15	0.21
Non-colloidal slowly biodegradable COD	F_{xsp}	gCOD/gSBCOD	0.50	0.50	0.50
Unbiodegradable soluble COD	F_{us}	$gCOD/gCOD_{tot}$	0.14	0.08	0.14
Unbiodegradable particulate	F_{up}	$gCOD/gCOD_{tot}$	0.03	0.08	0.16
Ammonia	F_{na}	gNH_3-N/gTKN	0.02	0.75	0.02
Particulate organic nitrogen	F_{nox}	gN/g OrgN	0.23	0.25	0.11
Soluble unbiodegradable TKN	F_{nus}	gN/gTKN	0.05	0.02	0.04
N/COD for unbiodegradable particulate COD	F_{upN}	gN/gCOD	0.04	0.04	0.02
Phosphate	F_{po4}	gPO_4-P/gTP	0.58	0.75	0.58
P/COD for unbiodegradable particulate COD	F_{upP}	gP/gCOD	0.01	0.01	0.00
Kinetic and stoichiometric parameters	**Symbol**	**Units**	**Calibrated value**	**Biowin default value**	**Arrhenius**
AOB: Aerobic decay rate	b_A	1/d	0.04	0.17	1.029
OHOs: Aerobic decay	b_H	1/d	0.24	0.62	1.029
Switches: Aerobic den. DO half saturation	$k_{O,H}$	mgO_2/L	0.35	0.05	-
Stoichiometric parameters					
Common: N in endogenous residue		mgN/mgCOD	0.025	0.07	
Common: P in endogenous residue		mgP/mgCOD	0.004	0.022	
Common: Endogenous residue COD/VSS ratio		mgCOD/mgVSS	1.48	1.42	
Common: Particulate substrate COD/VSS ratio		mgCOD/mgVSS	1.48	1.6	
Common: Particulate inert COD/VSS ratio		mgCOD/mgVSS	1.48	1.6	
OHOs: N in biomass		mgN/mgCOD	0.025	0.07	
OHOs: P in biomass		mgP/mgCOD	0.004	0.022	
OHOs: Fraction to endogenous residue	f_H	-	0.2	0.08	

After the calibration phase, the model predictions regarding the effluent COD, TSS, TN, TP, ammonia and nitrate were satisfactorily described as displayed in Table 16.9. Overall and besides the wastewater fractionation, only three kinetic parameters were adjusted (the aerobic decay rates of OHO and AOB and the half-saturation coefficient for dissolved oxygen of OHO). The main differences between the pulp mill model and the default values from BioWin for conventional municipal wastewaters are the extremely low stoichiometric parameters of the N and P fractions stored in the biomass. The default values used by BioWin are 0.07 mgN/mgCOD and 0.022 mgP/mgCOD, respectively, whereas the calibrated values are 0.025 mgN/mgCOD and 0.004 mgP/mgCOD. This suggests that the extremely low N/COD and P/COD ratios observed in the pulp mill could have possibly lead to the development and proliferation of microbial populations with low nutrient requirements.

Table 16.9 Results from the model validation for the two periods of study: winter and summer 2013.

Parameter	Unit	Influent	Effluent (Model prediction)	Effluent (Average measured values)	Model deviation
Calibration					
Collected data 21/10 to 28/10/2013					
Model calibration conducted with data obtained during the sampling campaign carried out in this research at pH 6.5 and 31°C					
Flow	m^3/d	58,925	56,509	56,506 ± 2,736	0%
Total COD	mgCOD/L	1,588	265	268 ± 11.0	1%
Total N	mgN/L	15.8	1.31	1.28 ± 0.18	2%
Total P	mgP/L	3.67	0.62	0.58 ± 0.06	7%
Nitrate N	mgN/L	0.03	0.00	0.10 ± 0.07	100% [a)]
TSS	mgTSS/L	220	8.0	4.1 ± 3.7	96% [b)]
Validation					
Collected data 01/05 to 31/08/2013					
Model validation conducted with data obtained from the regular monitoring of the plant at pH 6.5 and 31.7°C					
Flow	m^3/d	68,255	65,457	65,384 ± 6,044	-451
Total COD	mgCOD/L	1,430	2,38	283 ± 45.5	43.9
Total N	mgN/L	16.1	1.34	2.88 ± 1.65	1.5
Total P	mgP/L	2.74	0.68	0.49 ± 0.12	-0.3
Nitrate N	mgN/L	0.03	0.08	0.03 ± 0.05	-0.1
TSS	mgTSS/L	235.0	7.3	9.6 ± 6.3	1.4
Validation					
Collected data 01/09 to 20/12/2013					
Model validation conducted with data obtained from the regular monitoring of the plant at pH 6.5 and 31.6°C					
Flow	m^3/d	63,022	60,438	60,492 ± 6,241	-110
Total COD	mgCOD/L	1,568	261	286 ± 37.2	24.1
Total N	mgN/L	16.6	1.33	1.67 ± 0.38	0.3
Total P	mgP/L	3.17	0.59	0.61 ± 0.26	0.04
Nitrate N	mgN/L	0.03	0.01	0.03 ± 0.07	0.02
TSS	mgTSS/L	263.5	7.9	6.4 ± 2.9	-1.94

[a)] The nitrate concentrations predicted by the model differ slightly compared to the measured values. This difference was assumed acceptable considering the limitations of the model and the high sensitivity of this parameter due to its relatively very low values.

[b)] **The TSS concentrations predicted by the model differ from the measured values. This difference was assumed acceptable because of the high sensitivity of this parameter due to its relatively low values.**

16.3.6 Model validation

The predictions of the model were validated by comparing the obtained results with data from two different operational periods of the pulp mill WWTP (Table 16.9): May-August 2013 and September-December 2013. Overall, the model was able to provide a satisfactory description of the plant performance as observed by the relatively good agreement between the model predictions and the average measured values. Apparently, certain parameters, such as TSS and nitrate, differed. But, it must be considered that these concentrations are extremely low (and sometimes close to zero), that even a minimal increase of 0.1 mg/L can lead to an apparently high deviation between the predicted and measured value. Thus, such low differences were considered acceptable and it was assumed that the model was adequately validated.

16.3.7 Assessment of plant improvement and upgrading scenarios

Using BioWin, eight different configurations and operational alternatives were assessed to evaluate the robustness of the plant and potential for improvement. The assessed scenarios were evaluated from a theoretical perspective. Even though the conclusions drawn below were based on a calibrated/validated model describing the current operation of the WWTP at the evaluated pulp mill, additional experimental *in situ* validation would be desirable to fully confirm the applicability of the proposed alternatives. For instance, it would be necessary to demonstrate the appropriateness of every single alternative described below by conducting experimental pilot scale research at the pulp mill facility.

16.3.7.1 Implementation of chemical phosphorus removal

A jar test evaluation was carried out at the evaluated pulp mill laboratory facilities to define an adequate coagulant dose. Table 16.10 displays the results of the jar test experiments. A low soluble phosphorus concentration of approximately 0.26 mgP/L can be achieved with 100 mg$FeCl_3 \cdot 6H_2O$/L (experiment E2A, beaker No. 2). An even lower concentration can be reached if 270 mg$FeCl_3 \cdot 6H_2O$ are dosed (experiment E2A, beaker No.4), resulting in a phosphorus concentration of 0.11 mgP/L. However, the coagulant requirements are considerably higher (270 mg/L *versus* 100 mg/L) for a relatively marginal removal improvement (from 0.26 to 0.11 mgP/L). Thus, the addition of 100 mg$FeCl_3 \cdot 6H_2O$/L (experiment E2A, beaker No. 2) was selected as the best alternative. This also suggested that the coagulant addition could take place in the end or after the aerobic tank, following a sweep coagulation mechanism, at the regular pH observed at the plant (pH 7.2) without requiring any pH adjustment. The coagulant concentration and addition point were selected and implemented in Biowin (Figure 16.9), leading to a similar outcome of about 0.1 mgP/L in the effluent of the plant.

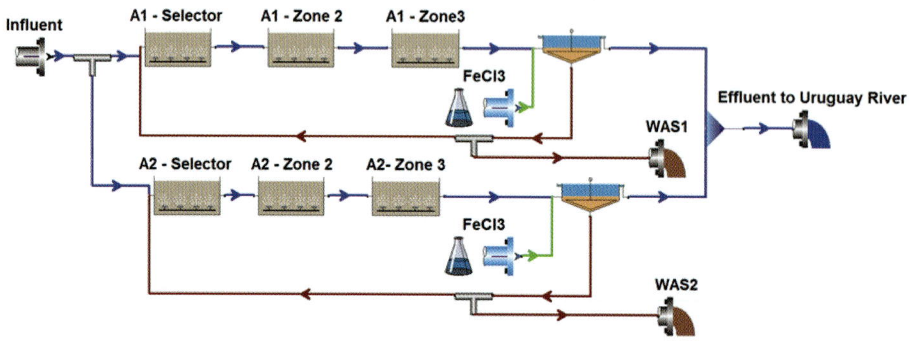

Figure 16.9 Modified layout of WWTP with chemical P removal.

Assuming an average current effluent TP concentration of 0.62 mgP/L (as observed during the sampling campaign) and an effluent flow of 58,925 m^3/d, the estimated phosphorus load is 36.5 kgP/d. Then, if chemical phosphorus removal is applied and an effluent TP concentration of 0.1 mgP/L can be achieved, the effluent phosphorus load will be reduced to 6.5 kgP/d, which is far even lower than the discharge limit set by the local environmental agency of 74 kgP/d. However, the volumes of excess sludge produced per day are estimated to increase from 321 tonTSS/d to 326 tonTSS/d. Or, an increase of 9.8 tonTSS/d if the two treatment lines are considered as parallel lines.

Table 16.10 Results of the jar-tests executed on activated sludge and mixed liquor from the pulp mill wastewater plant of study to assess the chemical phosphorus removal potential under different pH and coagulant doses.

	E1A				**E1B**			
Initial Mixed liquor AS:	Mixed liquor AS from end of aeration basins							
$P_{soluble}$=0.88 mgP/L	Default pH 7.4				pH adjusted to 5.5			
	Beaker 1	Beaker 2	Beaker 3	Beaker 4	Beaker 1	Beaker 2	Beaker 3	Beaker 4
P total	-	-	-	-	-	-	-	-
P soluble (0.2 μm pore size filter)	0.85	0.85	0.78	0.7	1.49	1.5	1.46	1.53
	E1C				**E1D**			
Initial effluent:	Effluent from secondary clarifier							
P_{total}=0.72 mgP/L $P_{soluble}$=0.68 mgP/L	Default pH 7.7				pH adjusted to 5.5			
	Beaker 1	Beaker 2	Beaker 3	Beaker 4	Beaker 1	Beaker 2	Beaker 3	Beaker 4
P total	0.76	0.72	0.74	0.7	0.69	0.67	0.68	0.69
P soluble (0.2 μm pore size filter)	0.62	0.61	0.64	0.51	0.63	0.63	0.65	0.61
	E2A				**E2B**			
Initial Mixed liquor AS:	Mixed liquor AS from end of aeration basins							
$P_{soluble}$=0.88 mgP/L	Default pH 7.2				pH adjusted to 6			
	Beaker 1	Beaker 2	Beaker 3	Beaker 4	Beaker 1	Beaker 2	Beaker 3	Beaker 4
P total	-	-	-	-	-	-	-	-
P soluble (0.2 μm pore size filter)	0.7	0.26	**0.23**	**0.11**	0.96	0.35	**0.16**	**0.17**
	E2C				**E2D**			
Initial effluent:	Effluent from secondary clarifier							
P_{total}=0.72 mgP/L $P_{soluble}$=0.68 mgP/L	Default pH 7.9				pH adjusted to 6			
	Beaker 1	Beaker 2	Beaker 3	Beaker 4	Beaker 1	Beaker 2	Beaker 3	Beaker 4
P total	0.66	0.68	0.66	0.66	0.67	0.69	0.68	0.67
P soluble (0.2 μm pore size filter)	0.40	**0.23**	**0.18**	**0.16**	0.44	**0.22**	**0.19**	**0.12**

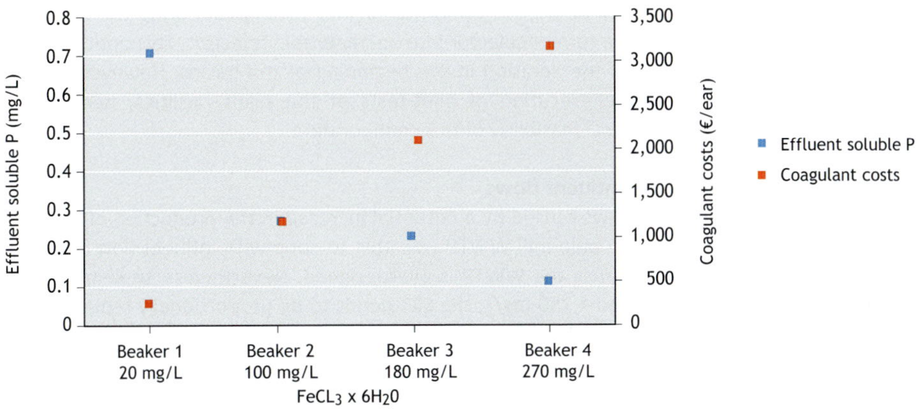

Figure 16.10 Soluble phosphorus concentrations in the effluent of the plant as a function of the coagulant dose and costs of coagulant per year.

Based on the results from the jar test experiments (Table 16.10, E2A) and the operational conditions at the plant, the actual coagulant dose required and their associated costs were estimated. Figure 16.10 shows the effluent P soluble concentrations as a function of the coagulant dose and costs of coagulant per year. As observed, the addition of 100 mg $FeCl_3$/L appears to be a satisfactory option to decrease and secure a low P concentration in the plant at an estimated annual cost of € 1.159.459.

16.3.7.2 Implementation of enhanced biological phosphorus removal (EBPR)

By converting the former selector (35,000 m^3) located at the beginning of the aeration basin to an unaerated zone, EBPR can be achieved at the evaluated pulp mill WWTP. The modelling results predicted that the TP concentration in the effluent can be decreased from 0.62 to 0.26 mgP/L (Figure 16.11). For EBPR to develop, anaerobic conditions must be created and a different group of heterotrophic microorganisms (known as the poly-phosphate accumulating organisms, PAOs) shall proliferate to enhance the biological P-uptake.

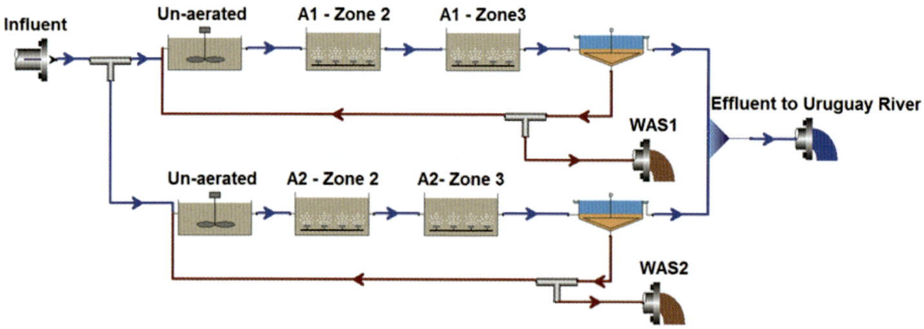

Figure 16.11 Modified layout of WWTP with un-aerated reactor zone.

The proposed configuration may be a good option to further P removal during the biological treatment by only converting the former selector into an anaerobic selector. This option can be easily achieved by switching off the aeration in the beginning of the basins. However, further research activities, such as the execution of pilot-tests at the plant facilities, need to be conducted to validate the technical feasibility of this configuration.

16.3.7.3 Assessment of higher influent flows

The impact of higher influent flows caused by a potential increase in the production of the pulp mill industry was evaluated. The pulp mill WWTP was able to cope with influent flows from 20 up to 100% higher, mostly, because the WWTP is overdesigned. Nevertheless, to keep a MLSS concentration between 4,215 and 4,750 mg/L, the SRT needs to be proportionally reduced from 32 to 26, 23, 20, 18 and 16 days, as the influent flow increases from 0 to 20, 40, 60, 80 and 100%.

However, even though the WWTP would still meet the discharge limits, the effluent TP concentration would increase from 0.62 mgP/L to 0.87 mgP/L. This option needs to be carefully addressed and perhaps combined with another alternative (such as chemical P-removal and EBPR implementation) to secure the compliance of the discharge standards if higher flows need to be treated.

16.3.7.4 Discharge and co-treatment of municipal wastewater

If the municipal wastewater from the city of Fray Bentos (with estimated flow of approximately 3,417 m^3/d) would be co-treated at the evaluated pulp mill WWTP, the modelling studies indicate that the total effluent P concentrations would considerably increase (up to approximately 0.8 mgTP/L) (Table 5.11). However, if EBPR is implemented (by introducing an un-aerated or anaerobic zone, as previously described) the evaluated pulp mill WWTP may have the capacity to satisfactorily treat both effluents while still achieving a final total P concentration in the treated effluent of approximately 0.11 mgP/L.

Table 16.11 Effects of the discharge and co-treatment of municipal wastewater from the City of Fray Bentos at the evaluated pulp mill WWTP.

Parameter	Unit	Pulp mill influent	Municipal influent	Total influent	Effluent	Aerobic reactor
Flow	m^3/d	58,925	3,416	62,341	59,926	
COD	mgCOD/L	1,588	750	1542	260	
TSS	mgTSS/L	167			7.6	
TN	mgTN/L	15.8	60	18.2	1.9	
TP	mgTP/L	3.7	15	4.3	0.1	
MLVSS	mgVSS/L					4,236
MLSS	mgTSS/L					5,269
SRT	days					31

16.3.7.5 Effects of shorter SRT on plant performance

According to model predictions, decreasing the SRT from 32 d to 25, 20, 15 and 10 d would lead to an increase in the effluent phosphorus concentrations from 0.6 mgP/L to 0.7, 0.9, 1.0 and 1.2mgP/L, respectively. A decrease in the SRT may decrease the concentration of total suspended solids in the reactor. Therefore, less biomass would be present and the adsorption of colloidal phosphorus to particulate organic phosphorus would also decrease. This apparently leads to the increase in the effluent TP concentration. This alternative cannot be implemented alone but, if needed, it will need to be combined with another operating strategy such as the implementation of chemical P precipitation or EBPR. Thus, decreasing the SRT may contribute to reduce the aeration costs but a higher sludge generation will be observed.

16.3.7.6 Operation of the plant with only one treatment line

If due to maintenance purposes the plant needs to operate with only one wastewater treatment line (Figure 16.12), the model-based results indicate that this will increase the effluent TP concentration up to 0.8 mgP/L. This will be close to the maximum permissible discharge limit. Thus, in principle, this practice could be applied, but it is not recommendable for long term periods, only for maintenance purposes and in case of contingencies.

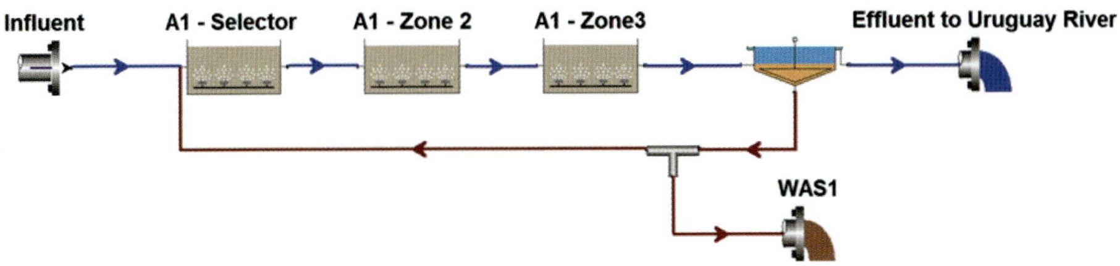

Figure 16.12 Modified pulp mill WWTP layout with only one treatment line in operation.

16.3.7.7 Implementation of an anaerobic digester for sludge waste treatment

The incorporation of an anaerobic digester will allow the WWTP to produce energy due to the potential biogas generation. However, the addition of the anaerobic digester would also promote the release of P in the anaerobic digester. This can lead to an increase in the effluent TP concentration due to the internal recirculation flows from the sludge handling facilities to the main treatment lines. Nevertheless, the incorporation of an anaerobic digester, together with the introduction of an unaerated zone (for EBPR implementation as previously discussed), may alleviate that increase in the effluent TP concentration. In addition, by shortening the SRT, and increasing the production of a less stabilized sludge, higher biogas productions could be reached. Figure 16.13 shows the modified layout of the pulp WWTP after the introduction of an anaerobic digester.

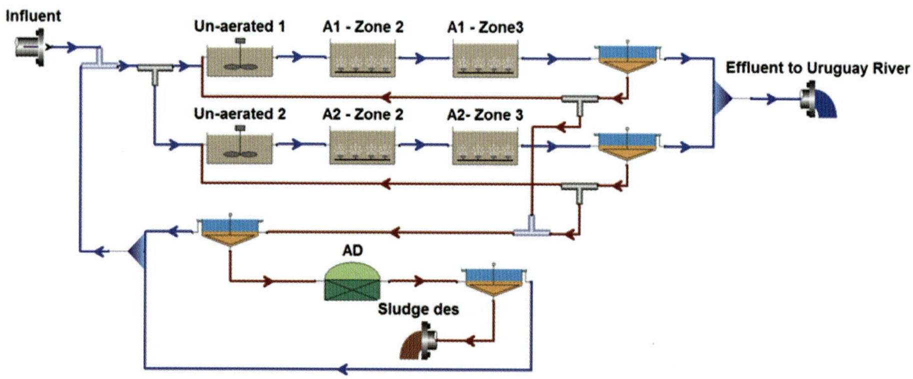

Figure 16.13 Pulp mill WWTP modified layout after the introduction of an anaerobic sludge digester.

The anaerobic sludge digester considered in this study was designed with a retention time of 21 d. The potential methane generated at the plant by the anaerobic digester was estimated as a function of the SRT applied in the mainstream activated sludge treatment line, which could be shortened to maximize the generation of sludge as shown in Figure 16.14.

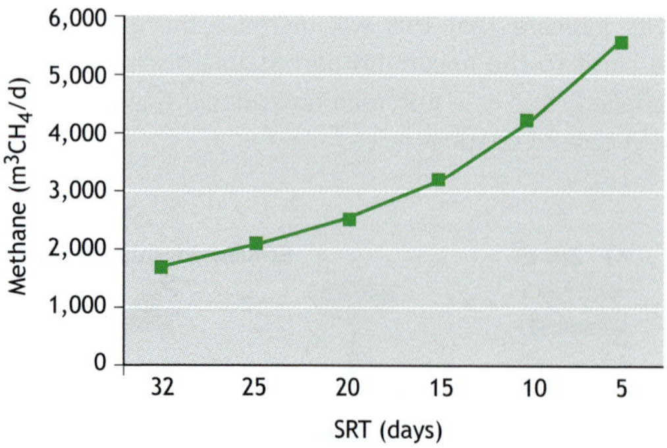

Figure 16.14 Potential methane production at the evaluated pulp mill as a function of the SRT after the potential implementation of an anaerobic digester.

If the SRT is shortened to 5 days, this may lead to an increase of 300% in the methane production. This suggests that the methane production would increase from 1,736 m^3/d (at 32 d SRT) to 5,568 m^3/d (at 5 d SRT). According to the model simulations, the combined implementation of EBPR and the introduction of the anaerobic digester will not affect the effluent quality. The effluent concentrations will remain practically unchanged. For instance, the effluent COD and TP concentrations would remain around 260-280 mg/L and 0.1 mg/L, respectively, regardless the applied SRT.

The energy production by the methane produced in the anaerobic digester, the energy consumption (only considering aeration requirements) at the activated sludge process, and the potential net energy savings at the evaluated pulp mill WWTP were estimated after the implementation of the previously discussed scenario. The y-axis at the left on Figure 16.15 shows the energy production (red line), energy consumption (blue line), and net energy consumption (green line) as a function of the SRT. The y-axis at the right shows the net energy savings expressed in percentage as a function of the SRT. 16The potential net energy savings may increase from 24% to even 88% of the aeration requirements when implementing the proposed scenario decreasing the SRT from 32 to 5 d. Overall, this alternative looks rather promising and deserves to be thoroughly analyzed due to the potentially remarkable benefits. However, further research activities such as the execution of pilot-tests at the plant facilities need to be conducted to validate the technical feasibility of this configuration.

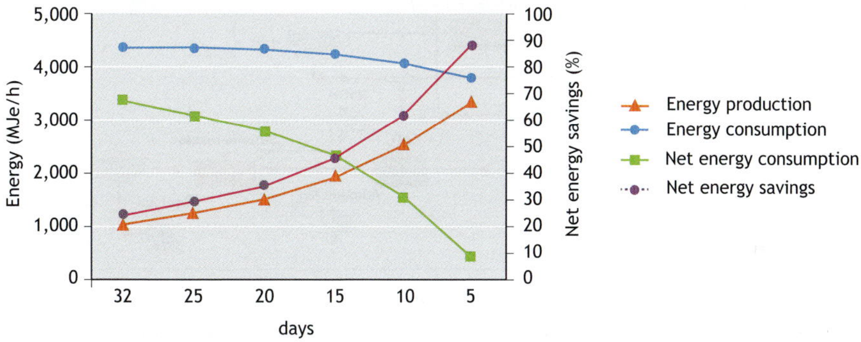

Figure 16.15 Potential energy savings at the pulp mill as a function of the applied SRT after the potential implementation of an anaerobic digester.

16.3.7.8 Nutrients recovery in the supernatant of the anaerobic digester

Further improvements can be incorporated to the previously described scenario. The nutrients (nitrogen and phosphorus) that would be present in the supernatant of the anaerobic sludge digester may be recovered through the addition of magnesium for their precipitation as magnesium-ammonium-phosphate (MAP), also known as struvite. The modified layout indicating the required changes and additions to the WWTP to achieve that goal are shown in Figure 16.16. Table 16.12 shows the design parameters for the precipitation of struvite in the reactor including the chemical addition dose and the effluent composition.

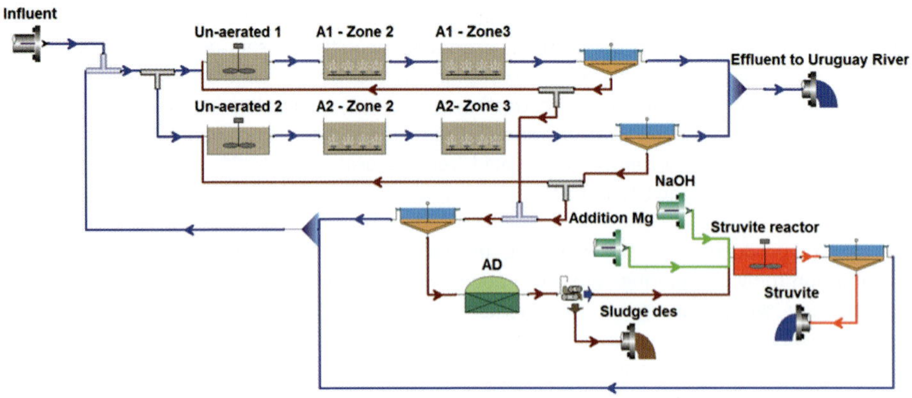

Figure 16.16 Modified layout of the pulp mill WWTP with a side-stream struvite recovery reactor.

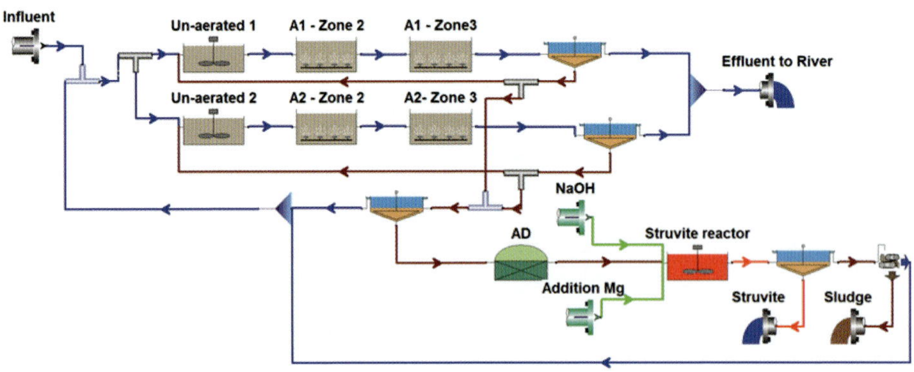

Figure 16.17 Modified layout of pulp mill WWTP in accordance to current practices implemented in Amsterdam West WWTP for phosphate recovery.

Table 16.12 Design and operating parameters for the recovery of nutrients as struvite in the sludge treatment line.

Parameter	Influent	Effluent	Reactor	Struvite reactor	Unit
Flow	58,925	58,604			m^3/d
COD	1,588	266			mgCOD/L
TSS	167	6.4			mgTSS/L
TN	15.8	1.5			mgTN/L
TP	3.7	0.2			mgTP/L
MLVSS			3,451		mgVSS/L
MLSS			4,244		mgTSS/L
Mg flow				10	m^3/d
Mg concentration				6,670/8,000	mgMg/L
Struvite precipitation				1,430/1,611	mgISS/L
Struvite production				346/432	kg/d

The proposed configuration may lead to the production of up to 346 kg of struvite per day. Recent research carried out at the Amsterdam West WWTP has shown that by precipitating struvite in the digested sludge (before dewatering) the performance of the sludge dewatering centrifuge can increase (van Nieuwenhuijzen *et al.,* 2009; Bergmans, 2011). The amounts of struvite recovered may be higher; eventually, reaching a concentration of approximately 1,611 mg/L corresponding to a load of 432.5 kgISS/d as shown in Table 16.12. The plant layout describing this alternative is described in Figure 16.17.

Though this case-study was carried out using the general activated sludge and anaerobic digestion model (ASDM) from Biowin, it is one more example of the needs to improve the coupling of ASM models (Henze *et al.*, 2000) with ADM1 (Batstone *et al.*, 2002), particularly because of the anaerobic digestion fate of EBPR sludge (Ikumi *et al.*, 2014). This can contribute towards the development of an integrated resource recovery strategy, while complying with the corresponding discharge limits. Further research, like the execution of pilot-tests at the plant facilities, the anaerobic digestion of the (EBPR) sludge and the nutrient recovery, needs to be conducted to validate the technical feasibility of this configuration. In this regard, the model developed by Ikumi *et al.* (2011) could be tested and compared against the outcomes obtained from Biowin.

16.4. Conclusions

The pulp mill WWTP evaluated in this study operates efficiently and therefore meets the effluent water quality standards established by the local environmental regulations (DINAMA, Decree 253/79). The model-based assessment indicates that its robustness may even allow the pulp mill WWTP to treat higher flows and organic loads while still meeting the discharge standards. Interestingly, the plant exhibited extremely low nutrient requirements (f_P of 0.012 mgP/mgVSS and f_N of 0.037 mgN/mgVSS) compared to the average requirements reported for municipal wastewater (fp: 0.025 mgP/mgVSS and fn: 0.10 mgN/mgVSS). This may reflect the selection of microbial populations with low nutrient requirements due to their relatively low availability in the influent. Although the ASDM model from BioWin (developed by EnviroSim) was specifically designed for modelling municipal wastewater treatment plants, it was capable of predicting the performance of the pulp mill WWTP. Only minor adjustments were needed that included the adjustment of certain kinetic and stoichiometric parameters like the OHOs and AOB aerobic decay rates and the N and P fractions in the biomass. Based on both the model predictions and laboratory experiments, the effluent TP concentrations can be further decreased by incorporating chemical sweep coagulation. If EBPR is implemented in the system, through the introduction of an unaerated section equivalent to 25% of the total volume, the total P concentration in the treated effluent may be further decreased. This could allow the pulp mill WWTP to co-treat also the municipal wastewater from the city of Fray Bentos (together with the pulp mill wastewater). The performance of the WWTP may be probably further improved. The effluent TP concentration could probably decrease and a high biogas production and energy savings may be simultaneously achieved by introducing the following modifications/additions: (i) incorporating an anaerobic digester; (ii) converting the actual selector zone of the aeration basin into an unaerated zone; and (iii) decreasing the SRT. Further research, like the execution of pilot-tests at the plant facilities, the anaerobic digestion of the (EBPR) sludge and the nutrient recovery, needs to be conducted to validate the technical feasibility of the new proposed configurations.

Acknowledgements

The authors would like to express their gratitude to the staff of the pulp mill in Uruguay where this research was conducted. Moreover, the authors would like to acknowledge the financial support provided by both the Agencia Nacional de Investigación e Innovación (ANII) in Uruguay and the Dutch government through awarding an MSc scholarship to the first author of this work. In addition, the authors would like to acknowledge the financial contribution provided by the Universidad Católica del Uruguay (UCUDAL) to carry out almost all the analytical work involved in this research, and to cover some transportation and accommodation costs.

References

Barañao P.A, Hall E.R. (2004) Modelling carbon oxidation in pulp mill activated sludge systems: Calibration of Activated Sludge Model No 3. Wat Sci Tech, 50(3): 1-10.

Bergmans B. (2011) Struvite recovery from digested sludge at WWTP West. MSc Thesis. Delft University of Technology. Delft, The Netherlands.

Bastone D.J.. Keller J., Angelidaki I., Kalyuzhyi S.V., Pavlostathis S.G., Rozzi A., Sanders W.T.M., Siegrist H., Vavillin V.A. (2002) The IWA Anaerobic digestion model No. 1 (ADM1). Wat Sci Tech, 45(10):65-73.

BRACELPA (2011) http://www.bracelpa.org.br/eng/estatisticas/pdf/booklet/booklet.pdf (accessed August 2013).

Ekama G.A., Wentzel M.C. (2008) Organic Material Removal. In: Biological Wastewater Treatment: Principles, Modelling and Design. eds. Henze M., van Loosdrecht M.C.M., Ekama G.A., Brdjanovic D. IWA Publishing. London, U.K.

Faroppa C., Annala K (2004) UPM Fray Bentos. Informe Ambiental - Resumen. (*In Spanish*). http://www.upm.com/uy/planta/medio-ambiente/estudios/estudios-por-la-empresa/Documents/Evaluaci%C3%B3n%20de%20Impacto%20Ambiental%20-%20UPM%20Fray%20Bentos.pdf (accessed July 2013).

Gujer W., Henze M., Mino T. and van Loosdrecht M.C.M. (1999) Activated Sludge Model No. 3. Water Sci. Technol., 3, 183-193.

Henze M., Gujer W., Mino T., van Loosdrecht M.C.M. (2000) Activated sludge models ASM1, ASM2, ASM2d and ASM3. IWA publishing. London, U.K.

Henze M., van Loosdrecht M.C.M., Ekama G.A., Brdjanovic D (2008) Biological Wastewater Treatment: Principles, Modelling and Design. IWA Publishing. London, U.K. ISBN: 1843391880.

Ikumi D.S., Brouckaert C.J. Ekama G.A. (2011) A 3-phase anaerobic digestion model. In: Proceedings of the 8th IWA International Symposium on Systems Analysis and Integrated Assessment in Water Management (Watermatex2011). San Sebastian, Spain, June 20-22th, 2011.

Ikumi DS, Harding TH, Vogts M, Lakay MT, Mafungwa H, Brouckaert CJ, Ekama GA (2014) Mass balances modelling over wastewater treatment plants III. Final Report. Water Research Comission. South Africa.

Hulsbeek J.J.W., Kruit J., Roeleveld P.J. and van Loosdrecht M.C.M. (2002) A practical protocol for dynamic modelling of activated sludge systems. Water Sci. Tech., 45, 127-136.

Keskitalo J, Leiviskä K (2012) Evaluating the Potential for Process Control in Pulp Mill Wastewater Treatment Plant by Simulation. Mathematical Modelling. Oulu: University of Oulu.

Keskitalo J, Leiviskä K (2010) Mechanistic modelling of pulp and paper mill wastewater treatment plants. Report A N°41, Oulu: University of Oulu, Control Engineering Laboratory.

Lawler D.F., Benjamin M.M. (2006) Particle treatment processes: Common elements. By: Desmond F. Lawler and Mark M. Benjamin, Chapter 1, pp 1-82.

Meijer S.C.F. and Brdjanovic D. (2012) A Practical Guide to Activated Sludge Modelling. UNESCO-IHE Lecture Notes, 280 p. ISBN: 9073445264.

Paul E., Laval M.L., Sperandio M. (2001) Excess sludge production and costs due to phosphorus removal. Environ Technol, 22(11): 1363-1371.

Portal de noticias TIEMPO. July 9th, 2013. http://www.tiempo.com.uy/forestales/1833-la-verdad-sobre-upm (accessed May 1st, 2014).

Saarela S., Garcia E., Eluen I., Fernandez V., Kovasin K. (2008) "Celso Foelkel". The first year of operation of the BOTNIA Fray Bentos pulp mill in Uruguay. 2008. http://www.celso-foelkel.com.br/artigos/outros/Botnia-Saarela-First%20year%20operation.pdf (accessed July 2013).

Takács I. (2008) Experiments in Activated Sludge Modelling. PhD Thesis, Ghent University. Belgium.

Thompson G., Swain J., Kay M., Forster C.F. (2001) The treatment of pulp and paper mill effluent: a review." Biores Technol 77(3): 275-286.

van Nieuwenhuijzen A.F, Havekes M., Reitsma B.A,, de Jong P. (2009) Wastewater treatment plant Amsterdam West: new, large, high-tech and sustainable. Water Practice Tech, 4(1), doi:10.2166/wpt.2009.006.

Chapter 17

The past, present and future of wastewater treatment modeling

Lopez-Vazquez C.M., Hooijmans C.M., Meijer S.C.F., Brdjanovic D. and van Loosdrecht M.C.M.

17.1 Introduction

Activated sludge systems have been applied for 100 years. In the course of the years, researchers have developed various models to describe the activated sludge processes. The main aim has been to get a better understanding of the conditions that favour the conversions of carbon, nitrogen and phosphorus present in wastewater, and associated oxygen consumption and sludge production. The current chapter presents reflection on the historical developments, state-of-the-art of activated sludge modelling and future trends supported by a literature inventory and a statistical analysis of publications dealing with the activated sludge models and modelling.

17.2 Historical overview

Before the 80's, several research groups worked independently from each other on developing models of activated sludge. Each group developed and applied their own approach and notation, first in steady-state models, and later on, in dynamic models. In the early 80's, Poul Harremoës, President of IAWPRC (The International Association of Water Pollution, Research and Control, later IAWQ, International Association of Water Quality; nowadays, IWA, International Water Association) initiated the idea to combine the most relevant and applied models and to work together at an international level to accelerate development of a common, unified model. As a consequence, in 1982, the "Task group on mathematical modelling for design and operation of biological wastewater treatment" was established with Gerrit Marais (University of Cape Town), Leslie Grady (Clemson University), Willy Gujer (EAWAG), Tomonori Matsuo (Tokyo University) and Mogens Henze (Technical University of Denmark) as chairman. This joint activity resulted in the development of the first dynamic Activated Sludge Model, called in short ASM1 (Henze *et al.,* 1987). The ASM1 model is a structured model based on Monod kinetics that predicts the processes of biological (bacteriological) reactions. The ASM1 models COD and N removal, oxygen consumption and sludge production. Wastewater is characterized in terms of 7 dissolved and 6 particulate components that are used to describe 2 biomass groups, 7 fractions of COD (organic material) and 4 fractions of nitrogen (Henze *et al.*, 1987; Gujer and Henze, 1991). Dissolved oxygen concentration and alkalinity are also included as part of the wastewater characteristics. From the 8 processes of the model, 3 are related to the growth of heterotrophic and autotrophic organisms, 2 describe the biomass decay (death-regeneration theory, Dold *et al.,* 1980), and 3 related to hydrolysis. The model is presented in a matrix format, also known as the Petersen matrix or Gujer matrix (Petersen, 1965; Takács *et al.*, 2007). This matrix contains stoichiometric coefficients and a kinetic vector. All state variables involved in a process are displayed in columns, and all processes where a state variable is involved are presented in the rows of the matrix. Already in use in chemical modelling (Petersen, 1965), this representation helped to present the model in a condensed form. It facilitated its publication, interpretation and comparison not only between models, but also between processes and compounds. Certain major limitations of ASM1 are, for example, that it only describes heterotrophic and autotrophic reactions under aerobic and anoxic conditions (in which, for instance, ordinary heterotrophs consume carbonaceous substrates and autotrophic nitrifying organisms oxidize ammonia to nitrate), but it does not include enhanced biological phosphorus removal (EBPR) processes (Gujer and Henze, 1991). Despite the fact that to a great extent knowledge of EBPR processes was already available when ASM1 was developed (van Loosdrecht *et al.*, 1997), EBPR was not included in ASM1 since most of the wastewater treatment plants at that time did not incorporate biologically enhanced (or chemical) phosphorus removal (Fenu *et al.*, 2010).

Throughout the years, several research groups started to work on the description of EBPR for its incorporation in the dynamic activated sludge models, mostly based on directly measurable soluble compounds. From the mid 80's to the mid 90's, the EBPR process grew in popularity and the understanding of the underlying bio-chemical mechanisms increased (Henze, 2000). In the meantime, in 1990, the composition of the Task Group changed, when Leslie Grady left and Takashi Mino (Tokyo University) and Mark Wentzel (University of Cape Town) joined the task group. The knowledge acquired on EBPR led to the publication of the Activated Sludge Model No. 2 (ASM2) (Henze *et al.*, 1995), which included the EBPR processes. In particular, ASM2 includes Phosphate Accumulating Organisms (PAO), growing only under aerobic conditions, with the correspondingly associated anaerobic, anoxic, and aerobic reactions. ASM2 was a compromise between complexity and simplicity, and between different points of view on how the correct model should look like to be used as a conceptual platform for further model development (Henze *et al.*, 2000). In 1996, Mark van Loosdrecht (Delft University of Technology) became member of the Task Group, following the departure of Tomonori Matsuo, Mark Wentzel and Gerrit Marais. Because the occurrence of denitrifying EBPR was well-established (e.g. Kuba *et al.*, 1997; Murnleitner *et al.*, 1997) the ASM2 model was expanded in 1999 by the inclusion of denitrifying PAO (DPAO). This version of the model was denoted as ASM2d (Henze *et al.*, 1999). Both ASM2 and ASM2d are similar to ASM1 by assuming the cell to be a black box, as opposed to using the metabolic approach to modelling the processes that take place inside the cell. Parallel to these developments, in 1994, an increasing knowledge of the cell-internal biochemistry of PAO resulted in the development of a metabolic model (TUD bio-P model; Smolders *et al.*, 1994a, b; Murnleitner *et al.*, 1997) describing the anaerobic and aerobic phases of EBPR based on intracellular storage compounds. This model was later fully integrated with ASM by Meijer (2004).

At the same time as the ASM2d model was presented, the Task Group also developed the ASM3 model to correct some of the shortcomings of ASM1. ASM3 was proposed to become the new standard for ASM-based modelling. ASM3 replaced the death-regeneration process for heterotrophic organisms by an endogenous respiration process and also introduced the role of storage of organic substrates (Gujer *et al.*, 1999). In 2000, the Task Group presented the overview of the ASM models 1 to 3 (Henze *et al.*, 2000).

Within the context of model development it is also important to mention the role of hardware. The development of the models and computer capacity (CPU) grew hand to hand (Gujer, 2006). From a technical perspective, it became feasible to work with models that contained a large number of process descriptions and variables. In the 90's, models were increasingly used by researchers, but also mathematical modelling became popular between practitioners. Today, mathematical models are commonly used in the USA, Australia and many countries in Europe (Hauduc *et al.*, 2009). To facilitate its application, software has been developed to assist in design, optimisation, operation and training. Modelling simulators provide a better understanding of wastewater treatment plants since they allow users to view the response of the treatment systems to changes in a number of different variables, and are also used to optimize wastewater treatment plants and to train plant operators. Examples of commercial packages are GPS-X, SIMBA, STOAT, WEST, BioWin etc. For research and training SSSP, ASIM, AQUASIM and even Microsoft Excel are regularly used. Significant benefits are associated with the use of simulators in the analysis, design, and operation of wastewater treatment systems (Meijer and Brdjanovic, 2012). More information on development of activated sludge modelling is provided in Chapter 1 of this book and by Ekama and Takács (2014).

17.2 Recent developments

Several guidelines on how to model a wastewater treatment plant, and protocols on how to characterize the wastewater/sewage and sludge, have been developed during the last years around the world. In 2004, at the 4th IWA World Water Congress in Marrakech, groups that developed various protocols (Hochschulgruppe, STOWA, BIOMATH and WERF) came together to develop plans to synthesize the best modeling practices available. A new IWA Good Modelling Practice Task Group (GMP-TG) was formed on the use of Activated Sludge Models, parallel to and with the full support from the Task group on mathematical modelling for design and operation of biological wastewater treatment. The GMP-TG consists of an international team of modellers collecting experience and knowledge on activated sludge modelling to provide guidance to practitioners (Rieger *et al.,* 2012). One of the aims of GMP-TG was to prepare a scientific and technical report to propose simple and effective procedures for the use of ASM-type models (Rieger *et al.,* 2012). In preparation of this report, the GMP-TG developed and sent out a questionnaire in 2007 to benchmark and collect relevant information on the practical use of modelling. The objectives were to better define the profile of ASM users, to identify the tools/procedures that are used (models, guidelines, protocols) and to highlight the main limitations encountered while building and using ASM-type models (Hauduc *et al.,* 2009). The outcome of the questionnaire, filled in by 96 respondents, showed that models are used by researchers for optimisation purposes, as well as by modellers employed by private companies to carry out design studies. Modelling is seen as an engineering tool, needing relevant training that is often lacking. The most used biokinetic models were ASM1 (57%) and ASM2d (32%), followed by ASM3, other (non-specified), ASDM (BioWin), Mantis (GPSX) and TUD model. The study also revealed that models are sometimes not properly applied, which might be due to a lack of knowledge and standardised procedures. The development of standardised modelling procedures and better knowledge transfer by making available some practical case studies was mentioned as a key instrument to address certain obstacles like the complexity of the model theories and procedures, the time consuming steps and finally the reliability of the models.

Besides sending out the questionnaire several workshops, meetings, and courses on activated sludge modelling were organised, such as the wastewater treatment modelling seminars. The GMP-TG was involved in the development of a new IWA Model Notation System (Corominas *et al.*, 2010), and interviewed several distinguished modellers (Peter Dold, George Ekama, Willi Gujer, Mogens Henze, Mark van Loosdrecht, among others). One of the advices was to suggest typical values for regularly used ratios, variables, and parameters in wastewater modelling. The feedback was compiled in the IWA book "Guidelines for Using Activated Sludge Models", Scientific and Technical Report No. 22 (Rieger *et al.*, 2012).

On the website of the GMP-TG (https://iwa-gmp-tg.irstea.fr/) adjusted Gujer matrices for 7 published models can be downloaded as an MS Excel spreadsheet: (1) ASM1 (Henze *et al.*, 1987, 2000a); (2) ASM2d (Henze *et al.,* 1999); (3) ASM3 (Gujer *et al.,* 1999); (4) ASM3+BioP (Rieger *et al.*, 2001); (5) ASM2d+TUD (Meijer, 2004); (6) Barker & Dold model (Barker and Dold, 1997); and (7) UCTPHO+ (Hu *et al.*, 2007). In the same website, a comparison of different parameter-naming rules including the new IWA Model Notation System (Corominas *et al.*, 2010) can also be found.

Recently, among other initiatives to improve the knowledge transfer by facilitating and making available some practical case-studies (Hauduc *et al.*, 2009), UNESCO-IHE published the book "A Practical Guide to Activated Sludge Modelling" (Meijer and Brdjanovic, 2012) which is used in the modelling course delivered every year at UNESCO-IHE in cooperation with Delft University of Technology (http://www.unesco-ihe.org/modelling-wastewater-treatment-processes-and-plants). With a very practical focus, it presents all the steps performed as a part of a modelling project in Croatia where five WWTPs were subject to upgrade to EU effluent discharge standards. Besides the general modelling protocols and guidelines for wastewater characterization and fractionation, methods for quantitative influent assessment, addressing different components of the urban wastewater chain and introducing a methodology for quantification of sewage components are presented. Guidelines for plant flows and measurement points necessary for the preparation of an additional sampling program are also given, as well as an inventory of all the regular day-to-day sampling to be used in a modelling project, a methodology for activated sludge plant assessment and methods to evaluate raw plant data and to filter out possible errors which affect the model reliability and results. Practical methods for wastewater data evaluation were developed by Meijer *et al.*, 2002. New developments on data evaluation are presented in chapter 15 of this book in relation to a well-documented case study performed on WWTP Houtrust in the Netherlands in 2010. A methodology for secondary settlers' design and assessment is described, including the five most commonly used settler design procedures. The last part of the book elaborates on the methodology applied in model calibration and the main steps thereof.

17.3 Statistics on ASM literature

As a mean to assess the development and historical impact of mathematical modelling on the wastewater treatment field, a statistical analysis of the scientific literature was carried out, excluding the technical reports. Google Scholar was used as web search engine. The Google Scholar index includes most peer-reviewed online journals, plus scholarly books and other non-peer reviewed journals. In addition, such a search engine was selected since, contrary to other databases and scientific search engines, it could provide an estimation of the citations and references of and from documents dated back to the 1980's and even older. Thus, in March 2014, the top 50 most cited papers were identified. It is interesting to notice that an abbreviated report of ASM1 (Henze *et al.*, 1987), ASM2 (Gujer *et al.*, 1995), ASM2d (Henze *et al.*, 1999) and ASM3 (Gujer *et al.*, 1999) were published as papers in addition to their availability as technical reports. The results of the search are presented in Figure 17.1. The papers were classified in five groups concerning their focus and contents (as indicated by the colour of bars): (i) model development and model presentation; (ii) characterization and modelling protocols; (iii) metabolic modelling; (iii) applied modelling; and, (iv) others. As seen in Figure 17.1, the top-50 list is dominated by papers on model development and metabolic modelling, followed by those on protocols for characterization and modelling and thereafter by the applied modelling papers. Interestingly, in the top 5, four papers on model development and presentation and one on metabolic modelling can be found. The two by far most cited papers are the ASM3 paper from Gujer *et al.* (1999) and a paper by Takács *et al.* (1991) on the clarification-thickening process. To find the paper on settling as the second most cited paper is remarkable, as it deals with modelling of the activated sludge system, but unlike the ASM models, not with the biokinetics. The third most cited is the ASM2d paper by Henze *et al.* (1999).

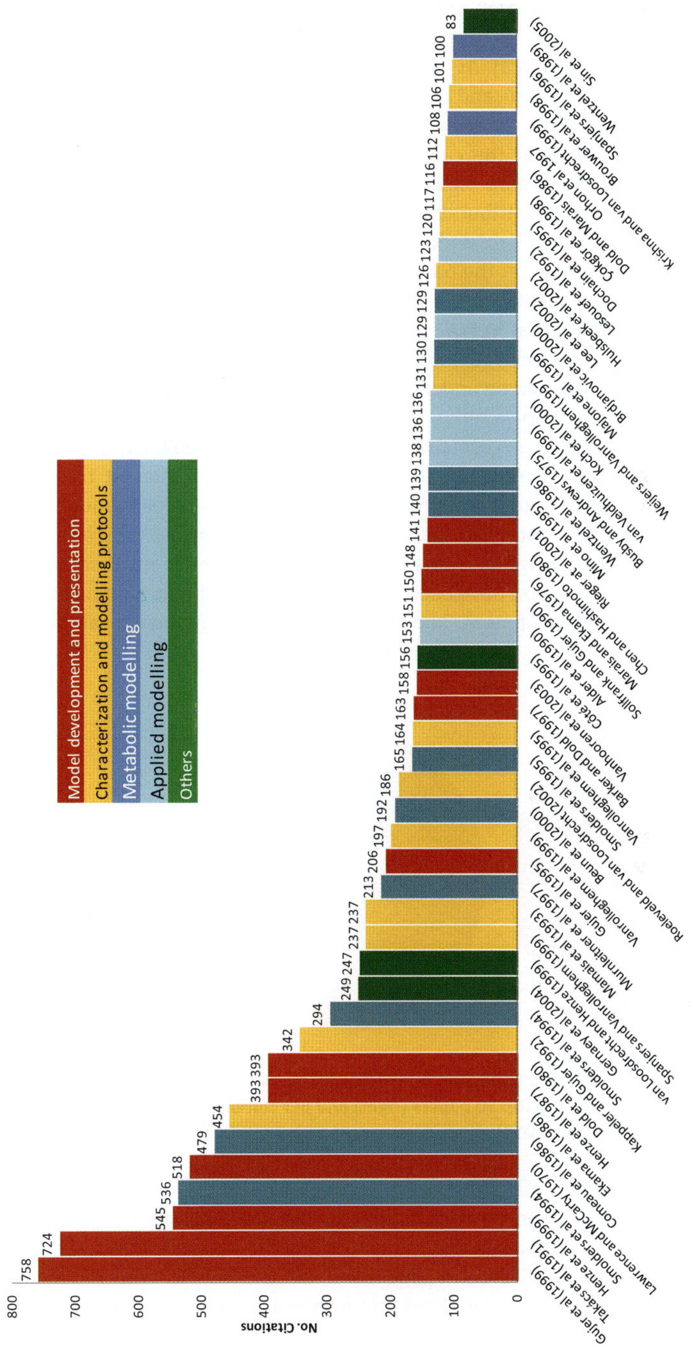

Figure 17.1 The top 50 most cited papers on mathematical modelling (excluding technical reports). Source: Google Scholar (March 2014).

The results of the literature search are also presented in the form of a pie chart, displaying as percentage the number of papers that belong to the top 50 most cited papers when grouped per focus and contents (Figure 17.2a). As observed, most of the top 50 most cited papers deal with protocols for characterization and modelling closely followed by papers on modelling development and metabolic modelling, but by far by the applied modelling papers. Nevertheless, when looking at the cumulative number of citations of the top 50 most cited papers (Figure 17.2b) the papers on model development and presentation have the highest number of citations. Such a distribution can suggest that there has been an increasing need to use well-established protocols to understand and implement the mathematical models but, logically, they cannot outnumber the basic modelling papers since they are indeed the subject of study and application. Remarkably, the papers on metabolic modelling rank third close to the first two groups. This is a clear indication that metabolic modelling has been widely accepted particularly for its contribution to modelling the EBPR process (e.g. Comeau *et al.,* 1986; Smolders *et al.,* 1994). It should be underlined that applied modelling papers received the lowest number of citations. This can be a reflection of the findings of the GMP-TG regarding the need to make available more practical case-studies (Hauduc *et al.,* 2009). In practice, possibly the modelling studies are so case-specific that they offer a limited interest to a broad scientific audience, which limits the potential number of citations that an applied modelling paper could get. But also, it cannot be discarded that the present degree of maturity and reliability of the models reduces the chances to find a novelty in the field. Thereby in any case, the historical links between academia and industry have been a key catalyser and should continue to work hand-to-hand for the future development and establishment of mathematical modelling in the wastewater treatment field.

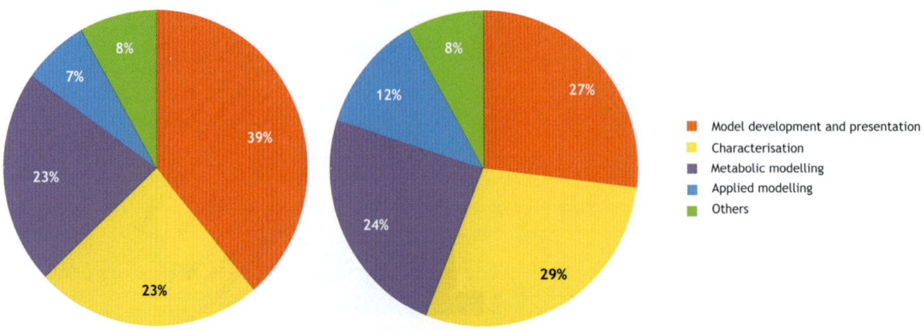

Figure 17.2 The top 50 most cited papers on mathematical modelling (excluding technical reports): percentage per group within the top 50 papers (right) and per cumulative number of citations (left). Source: Google Scholar (March 2014).

In parallel, a search was done on the citations of the five publications on the different ASM models. This included the technical reports on ASM1, ASM2, ASM1-ASM3, and papers on ASM2d and ASM3 (which are only published in papers and not as a technical report). As seen in Figures 17.3a and 17.3b, between 1998 and 2000, more than 10 years after the publication of ASM1, there was a boost in the number of citations to the ASM models. Simulation tools were already available from the early 90's (like GPS-X and SIMBA), indicating that the software capabilities were not a factor that delayed their implementation (and possibly the hardware was not either). Interestingly, the ASM started to have more citations after the publications of the most cited papers on characterization and modelling protocols (like the papers of Kappeler and Gujer, 1992;

Mamais *et al.,* 1993; Spanjers and Vanrolleghem, 1999; Vanrolleghem *et al.,* 1995; Vanrolleghem *et al.,* 1999; and Roeleveld and van Loosdrecht, 2002) (Figure 17.1). Possibly, the lack of understanding and confidence to apply the ASM models delayed their implementation, which was overcome by the availability of reliable characterization and modelling protocols. Furthermore, the publication of the ASM1-ASM3 technical report (Henze *et al.,* 2000), which consolidated in a single publication the contents and features of all ASM models, could be another important driver that helped to promote the ASM models, increasing the number of citations after 2000. Due to the popularity of the ASM1-ASM3 technical report (Figure 17.3a) practically all the latest publications refer to it. This makes difficult to clearly distinguish which model was used for a particular study, possibly overestimating the popularity or application of some ASM models and underestimating the others. Importantly, the yearly number of citations to ASM reports is stabilizing or even decreasing (Figure 17.3). Apparently, the number of citations per year has passed its exponential growth and started to level off since 2009.

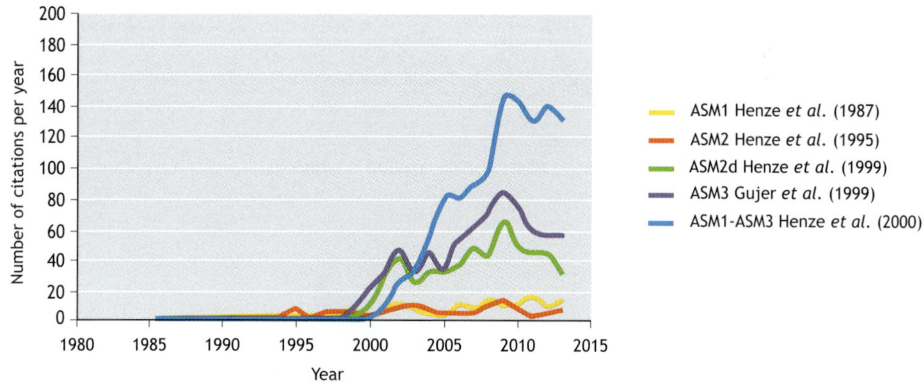

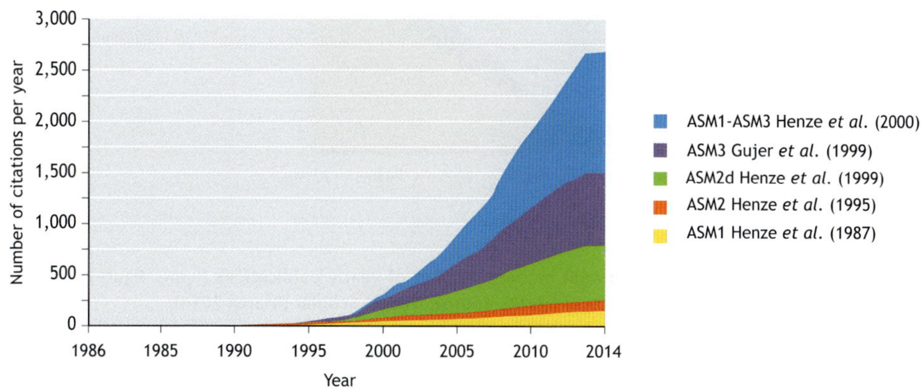

Figure 17.3 ASM reports/papers: number of yearly citations (upper figure), and cumulative number of yearly citations (lower figure). Source: Google Scholar (March 2014).

In an effort to get an estimation of the individual potential application of the ASM models, an additional search was carried out, looking at the yearly number of publications that refer to each model (using as exact phrase in Google Scholar anywhere in the article or document: "Activated sludge model No. ASM X"). Compared to the previous search, the difference is that the focus is not on the technical reports, but on the number of yearly citations to the models themselves. The results are depicted in Figures 17.4a and 17.4b. From Figure 17.4a, it can be seen that ASM1 is still the model of reference, followed by ASM3, ASM2 and ASM2d. Its simplicity and flexibility to apply it, not only to municipal activated sludge systems, but also to industrial conditions (Petersen *et al.,* 2002) and even to other wastewater treatment technologies (e.g. biofilm systems and wetlands) (Wanner *et al.,* 2006; Langergraber *et al.,* 2009) may have contributed to its popularity.

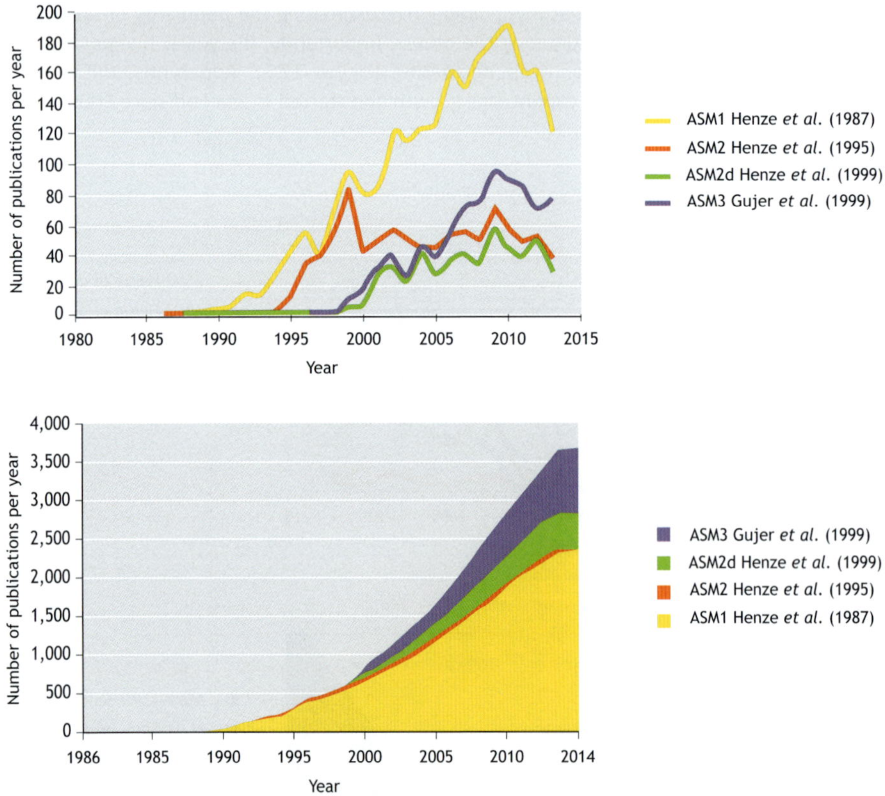

Figure 17.4 Publications referring to ASM models: number of yearly publications (upper figure); and, (b) cumulative number of publications (lower figure). Source: Google Scholar (March 2014).

Similar to Figures 17.3a and 17.3b, the maximum number of citations was reached around the years 2009-2010. Thereafter, a decrease in the number of citations is observed. A similar trend observed by Gujer (2006) when searching the Web of Science. The fact that the number of citations to the ASM models decreases does not automatically mean that the number of studies performed has progressively reduced. As previously discussed (Gujer, 2006), it could likely be a consequence of the fact that mathematical modelling is mature, standardized and well-established and that publications on modelling are not easy to publish in peer-reviewed journals due to a lack or insufficiency of scientific novelty.

17.4 Calibration and model accuracy related to ASM development

A recurring topic in ASM publications is the issue of model calibration and related model accuracy. Early in the ASM development, the need was recognized for the development of better model calibration procedures. This led to the development of practical modelling protocols (Hulsbeek *et al.*, 2002, Rieger *et al.*, 2012). In these protocols, methods are proposed to structure the modelling procedure by addressing specific parameters for calibrating the different biological processes with the goal to obtain more reliable and reproducible model results. Related to model calibration, methods were developed to directly measure kinetic rates of the different bioconversions during full scale practice. They were adopted from lab-scale ASM research and applied in full-scale operation such as the use of activated sludge batch-tests and respirometry (Spanjers *et al.* (1996 and 1999), Vanrolleghem *et al.* (1999)). These tests result in kinetic data which can be directly be applied in the model calibration process.

In 1995, Barker and Dold related model inaccuracy to the general lack of reliable wastewater treatment data. They recognized that in full-scale practice the poor quality of wastewater treatment data affects the model calibration results and reduces the overall accuracy of the model. This important conclusion has led to a new line of research on data accuracy in wastewater treatment, which is developing parallel to the research on ASM models and their practical application. Several publications are available on this topic; Meijer *et al.* (2002), Puig *et al.* (2008), Rieger *et al.* (2011), Spindler *et al.* (2012 and 2014) and more recent work published in chapter 15 of this book by Meijer *et al.*.

Techniques for data evaluation and quality improvement (error detection and reconciliation) are available from process industry, however, not yet widely applied in wastewater treatment. The proposed methods that can be used to improve wastewater treatment data rely on a combination of statistical and mass-balancing techniques. It is possible to detect gross errors in wastewater treatment data and also to correct and improve on the data before using them in a modeling study. Several successful examples are available demonstrating the effectiveness of data evaluation in wastewater treatment and also the usefulness for reliable modeling. Important conclusions of the parallel research on data accuracy are (i) most full-scale wastewater treatment data contain (large) measurement errors and misinterpretations, (ii) when reliable model results are required, also some kind of data evaluation needs to be performed and (iii), the inaccuracy of the measured treatment data is often larger than the inaccuracy of the model and calibrated parameters. This means that effective model calibration is only possible when the plant data are accurate.

17.5 The future

It is too soon to assess whether the efforts of the IWA GMP-TG, in the form of the standardized procedures, available Gujer matrices for all models, and the publication of the Guidelines on Good Modelling Practice in 2012 will lead to an increased work on modelling and increased number of publications. As previously discussed, the number of citations to the ASM models has decreased likely because mathematical modelling is becoming a standardized and mature practice in developed countries. On the opposite, case-studies and publications from developing countries are limited, which might be partly caused by the fact that activated sludge treatment systems are mainly found in developed countries. Likewise, the questionnaire prepared by the GMP-TG (Hauduc *et al.,* 2009) also showed that among the 96 responses, 65% came from European countries and 20% from North America. Other continents (South America, Africa, Asia and Australia) were under-represented. Nevertheless, this book includes some pioneering and (very) recent studies on activated sludge modelling in India (Brdjanovic *et al.*, 2007; Lopez-Vazquez *et al.*, 2013), Bosnia and Herzegovina (Hodzic *et al.*, 2011, Price and Vojinovic, 2010),

Mexico (Fall *et al.,* 2012), Croatia (Meijer and Brdjanovic, 2012) and Uruguay (Betancur, 2014) (Chapters 8, 10, 12, 13, and 15). This demonstrates that efforts are undertaken also in developing countries to apply the models to existing activated sludge systems (mostly for optimization or upgrade).

Concerning the future development of activated sludge modelling, it is important to underline and take into consideration the current and future needs and developments. In this regard, the trends regarding existing wastewater process technologies will likely need to further focus on providing a better description of the nutrient removal processes, not only for the sake of the nutrients removal itself, but also to reduce the associated energy costs and environmental impact. Modelling the description of nitrous oxides emissions in nutrient removal plants (Ni *et al.,* 2013) as well as achieving a satisfactory description of the actual (anaerobic, aerobic and anoxic) metabolic activities of the relevant EBPR populations (e.g. *Accumulibacter* type I and type II) and their interactions with "new" bacterial populations (such as *GAO*) (Lopez-Vazquez *et al.,* 2009; Oehmen *et al.,* 2010) can be some examples of the required developments. Regarding innovative wastewater treatment technologies, a stronger focus can be expected on modelling the bioprocesses and hydrodynamics involved in the aerobic granular sludge (de Kreuk *et al.,* 2007), on the implementation of the Anammox process and related N-removal processes in the mainstream treatment line (Kartal *et al.,* 2010; Wett *et al.,* 2013; Lackner *et al.,* 2014), and even on the intracellular storage processes of organics for bioplastic production (Bengtsson *et al,.* 2008; Jian *et al.,* 2012). While the aerobic granular technology offers significant benefits for WWTP optimization on both resources and footprint, the last two open new doors towards the conversion of existing WWTP from 'removal-type systems' towards 'resource-recovery systems' and 'energy-factories' (Kartal *et al.,* 2010; van Loosdrecht and Brdjanovic, 2014). Also, the application of the sulphur-cycle processes to wastewater treatment will likely continue to attract attention for the treatment of sulphur contaminated waters resulting from saline water intrusion, use of seawater for sanitation to alleviate water scarcity, cooling purposes, and industrial activities (Wang *et al.,* 2009; Hao *et al.*, 2014). Undoubtedly, mathematical modelling will be a strong tool to get a better understanding of the factors affecting these processes and facilitate their implementation. Such efforts may imply the need to incorporate and take into account more complex biokinetic models (Nielsen *et al.,* 2010; Oehmen *et al.*, 2010), elemental balance approaches (Takács *et al.*, 2007; Lu *et al.*, 2009), and likely also to establish stronger links with genomics, molecular techniques and metabolome analyses (Fiehn, 2001) as well as to develop the required experimental methods to determine and understand the microbial activities involved. The incorporation of the new processes (e.g. Anammox, and sulphur conversions) will likely follow the IWA GMP-TG concepts (Rieger *et al.*, 2012) and certainly most of the developments could be expected on the implementation of Anammox processes. The increased complexity of these combined biological processes require development of effective process controls in which application of dynamic ASMs could become crucial for successful full-scale application.

As a consequence of the interaction between the existing but also for the implementation of new technologies, plant-wide modelling will acquire and deserve special attention to: (i) develop an optimal WWTP control strategy, (ii) increase the efficiency of the WWTP removal processes, (iii) reduce the operating costs, (iv) maximize energy recovery through biogas production, and (v) maximize the removal and recovery of nutrients in the side-stream processes.

As a plant-wide modelling starting point, the mathematical description of the separation processes in the primary settling tanks (PST) affecting the COD fractions needs to improve (Vanrolleghem *et al.,* 2014). This will contribute to maximize the recovery of energy via the anaerobic digestion of organics and favour the role of WWTP as energy factories (Kartal *et al.,*

2010). Currently, most of PST separation processes are still modelled as black boxes with lumped and gross removal coefficients assigned to all particulate organics, whereas unbiodegradable particulate organics have shown to be subject to higher removal efficiencies than the biodegradable organics (Ikumi *et al.,* 2014a,b).

Another important aspect in plant-wide modelling is the coupling of the state variables (Volcke *et al.,* 2006) from the activated sludge process tanks and those from secondary settling tanks (SST) (Bürger *et al.,* 2011; Torfs *et al.,* 2013). Models for clarifiers use total suspended solids as a state variable, which is not explicitly used in ASM models and need to be calculated as a composite variable of the activated sludge processes. In addition, due to the different redox conditions created, the bottom of a clarifier needs to be dynamically modelled in a similar way like a bioreactor to take into account the potential redox effects on the active biological processes. Among them, rising sludge due to denitrification under anoxic conditions in nitrogen removal plants, secondary P-release under anaerobic conditions in EBPR systems and the description of the sludge settleability are some examples of the need to secure a satisfactory modelling description of the operation of SST. So far, some of the mentioned processes could be mimicked by the addition of an anoxic (denitrifying) tank in addition to the secondary settling tank. However, this is more an intermediate than the final solution to the problem (e.g. Chapter 3, Figure 3.6). Computational fluid dynamics (CFD) can help to improve the description and operation of SST (Plósz *et al.,* 2012). Furthermore, by studying the influence of diffusion limitations and gradients through CFD, the interactions between bacterial morphology and bacterial competition could be better understood and lead to a better prediction of the sludge bulking phenomenon (Martins *et al.,* 2004a,b). The latter affecting not only the SST performance but also the whole WWTP efficiency and capacity by reducing the sludge settleability. However, besides the added complexity, reliable experimental methods and set-ups are still needed to support model development and cope with the lack of (quality) data that has limited the use of advanced settling models (Plósz *et al.,* 2012). Such an approach requires tightening the collaboration links between practice and research to assess and provide feedback on newly developed models under real case scenarios.

Nevertheless, to apply plant-wide modelling the biggest challenges can be found when coupling ASM models with anaerobic digestion models (like the ADM1) (Batstone *et al.,* 2002). Those challenges are related not only to the description of the sludge digestion processes that take place in sludge thickeners and anaerobic sludge digesters but also to the physicochemical processes occurring within these systems. One of the first challenges is the different sets of state variables used by ASM models and ADM1. Overall, there are two different ways to deal with this issue: (1) the 'super model' approach where a complete set of variables valid for both aerobic and anaerobic environments is defined (Grau *et al.,* 2009), which is also available in e.g. the BioWim simulator developed by EnviroSim (REF) and (2) the use of established interlinked models by applying a set of algebraic transformation equations ('transformers') based on a Gujer matrix description of the two models (Vanrolleghem *et al.,* 2005; Volcke *et al.,* 2006; Nopens *et al.,* 2009). A pioneering and successful attempt of plant-wide modelling and coupling ASM and ADM models using a designed interface/transformer was demonstrated on the WWTP Anjana in India (Brdjanovic *et al.,* 2007), arguably the first published application of activated sludge modelling in a developing country (Chapter 8). Overall, the previous approaches can satisfactorily help out to apply a plant-wide model, the supermodel having the biggest potential. However, they cannot be directly applied to the plant-wide model description of a system performing EBPR. The latter because the anaerobic digestion fate of EBPR sludge (highly rich in phosphorus and intracellular compounds) is not included in ADM1 (Ikumi *et al.,* 2014a,b). Thus, the anaerobic endogenous processes at which an enriched EBPR sludge is exposed in anaerobic digesters cannot be currently described by ADM1. This is also directly linked with the strong need to achieve a satisfactory description of the metabolism of the dominant EBPR populations since it will define the fractions of the different intracellular compounds (from poly-P to glycogen and poly-hydroxy-alkanoates) contained in the sludge.

Concerning the physical-chemical processes occurring within the anaerobic systems, ASM models contain only the alkalinity state, which acts mostly as an indicator of the potential inhibition of a biological process if alkalinity decreases, and a single gas transfer model. Whereas the pH description in ADM1 is only valid for dilute systems and do not include a mechanistic (pH-based) precipitation (Batstone *et al.,* 2012). However, if EBPR sludge is anaerobically digested the related physical-chemical processes that take place in the anaerobic sludge digestion systems are required to be modelled. Of particular importance are the physical-chemical processes affecting the multi-mineral precipitation (with cations such as iron and aluminium and anions like ortho-phosphate, carbonate and even sulphur) and the prevailing pH in anaerobic sludge digesters. In this regard, in recent years, an IWA Task Group was initiated for the development of a Generalized Physical-chemical Framework (Batstone *et al.,* 2012). Although the fundamentals of the physical-chemical reactions are well understood, available from other disciplines, and do not need calibration (since thermodynamics define the end points of the kinetic processes), the organic compounds driving the bioprocesses and the thermodynamics of the precipitation rates are not yet defined/known, implying that the end points of these reactions need to be known, involving considerable calibration efforts (Ekama, *personal communication*, 2014). These plant-wide modelling aspects require further research involving both experimental and modelling development activities to clarify and achieve a satisfactory modelling of the physical-chemical processes. Together with those concerning the implementation of recently developed technologies (such as the implementation of Anammox for the treatment of nitrogen-rich reject waters) it can contribute to reach the objectives of the plant-wide modelling philosophy. Ikumi *et al.* (2011, 2014a) have made one of the first steps towards upgrading ADM1 and account for the potential effects of the anaerobic digestion of EBPR sludge.

Besides the potential energy savings that hydroinformatics tools (such as CFD) could bring (Rieger *et al.*, 2012), with an increasing need and interest in water reuse and integrated modelling the biological and physical-chemical removal processes of micro-pollutants will be another modelling area of major expansion and development (Gujer *et al.*, 2006; Clouzot *et al.,* 2013) where CFD could also be applied (Radu *et al.,* 2010). Moreover, a growing interest in integrated (urban) water modelling will continue to motivate integration of wastewater treatment process models with receiving water quality (RWQM) and sewer models (Gujer, 2006; Vanrolleghem *et al.,* 2014). Until a few years ago, only hydraulics and pollutant transport phenomena in the sewers were taken into account (Hvitved-Jacobsen, 2013). However, recent models start to consider the chemical and biological processes that take place in the sewer system, looking at the sewers as physical, chemical and biological reactors (Rauch *et al.,* 2002). One of the first examples of holistic modelling (combined sewage network, WWTP and the recipient/river) using different models (combining Mike Urban, BioWin and HEC-RAS), although carried out in a sequential mode (as opposite to a better and more realistic but much more complex real-time approach), showed great advantages of such modelling application (Hodzic *et al.,* 2011, Chapter 10; Price and Vojinovic, 2010). This is of major importance for the design, operation and maintenance of sewer networks, not only from a holistic water management perspective but also from a potential future asset management focus (which needs a satisfactory modelling description of the removal of micropollutants). In collaboration with the University of Cape Town, WEST, a hydraulic modelling software developed by Gent University (Vanhooren *et al.,* 2003) and nowadays held by the Danish Hydraulic Institute (DHI), has been upgraded to make the prime efforts to link a wastewater treatment model with a RWQM and a sewer model

(Ikumi *et al.,* 2014b). Together with the plant-wide modelling advances, this can open promising lines towards the development of an integrated urban water model suitable and capable to describe and optimize the entire urban water system. Likely, such a holistic approach will be also of importance and useful when dealing with secondary quality water sources for sanitation to contribute to alleviate water scarcity issues (like the use of saline water for sanitation and the implementation of the SANI process) (Lu *et al.,* 2009; Wang *et al.,* 2009).

Bearing in mind that 2.6 billion people still do not have access to sanitation, that most of population in developing countries is not connected to sewer systems, and that only a small fraction of sewage in developing countries is treated, brings the issue of holistic modelling where the urban drainage and sewerage models and wastewater treatment models (not only ASM and ADM) will be complemented by and integrated with (de)centralized sanitation models in cities which are not fully covered by sewerage. There are several similar examples worldwide, especially in developing countries. Recent advances in this direction include a decentralized sanitation model created by the software developers of the SIMBA simulator (funded by the Bill & Melinda Gates Foundation and therefore expected to be freely accessible) and a decision support tool called WAMEX by UNESCO-IHE and funded by the Asian Development Bank (and also freely available).

Cloud computing has gained in interest lately (Armbrust *et al.,* 2010), by joining efforts and contributing to standardize approaches and notation (Corominas *et al.*, 2010), sharing wastewater treatment models between researchers, software developers, and practitioners, while being in different longitudes and latitudes, may not be far from a dream. This can be a strong tool to facilitate the application of plant-wide and integrated urban water modelling to contribute to optimize the water quality and quantity transported through the aquatic veins and arteries of an urban settlement.

From a commercial and practical perspective, the incorporation of the processes and approaches described previously will considerably increase the model complexity. However, understandably, practitioners feel uncomfortable working with increasingly complex models. So, possibly, vendors with specific modelling skills will appear on the market, since conventional wastewater treatment 'generalists' will not be able to cope with the fast release and development of more complex models for particular applications. Thus, like in other fields, in the near future consultants will outsource their modelling activities to specific vendors (Ekama, *personal communication*, 2014).

It is not impossible to imagine that sooner or later new interfaces and way of interactions between (probably or even likely less specialized) users and models will be created. Maybe, in the form of multi-layer serious gaming and using 3D urban water system simulators with simplified 'surface' user interfaces and more complex expert models running 'invisibly' in the background (Ekama and Brdjanovic, *personal communication*, 2014). An expected future development is the use of models built in data acquisition systems (SCADA) of larger wastewater treatment facilities. Thereby the complex knowledge contained in ASMs is made available for process operators making more efficient and safe plant operation possible on a daily basis. It is also expected that the modelling boundaries will be further extended reaching trans-disciplinary character as other issues will be included, e.g. emergencies, risks, and social aspects (Vojinovic and Abbott, 2011; Brdjanovic *et al.,* 2013). By doing so, modelling will come closer to decision makers and increase and facilitate the use of models by different and currently not involved stakeholders.

Last but not least and despite all the expected developments and release of more complex models for several wastewater treatment applications and further, one must keep in mind that a model is still a mere representation of reality, generally, applied as a tool for improvement and optimization purposes. By no means, a model must be solely used as a substitute of an educational programme or design criterion, but rather as a complement. Particularly, since, for most regular and day-a-day design activities, a single spreadsheet could do a fair enough job once the knowledge is settled.

References

Alder A.C., Siegrist H., Gujer W., Giger W. (1990) Behaviour of NTA and EDTA in biological wastewater treatment. *Water Res*, 24(6): 733–742.

Armburst M., Fox A., Griffith R., Joseph AD., Katz R., Konwinski A., Lee G., Patterson D., Rabkin A., Stoica I., Zaharia M. (2010) A view of cloud computing. Communications of the ACM, 53(4):50-58.

Barker P.S., Dold P.L. (1997) General model for biological nutrient removal activated-sludge systems: model presentation. Water Environ Res, 69(5): 969-984.

Barker, P. S. and Dold, P. L. (1995) Cod and Nitrogen Mass Balances in Activated-Sludge Systems. Water Research, 29(2), 633–643.

Bastone D.J., Keller J., Angelidaki I., Kalyuzhyi S.V., Pavlostathis S.G., Rozzi A., Sanders W.T.M., Siegrist H., Vavillin V.A. (2002) The IWA Anaerobic digestion model No. 1 (ADM1). Wat Sci Tech, 45(10):65-73.

Batstone D.J., Amerlinck Y., Ekama G., Goel R., Grau P., Johnson B., Kaya I., Steyer J.-P., Tait S., Takács I., Vanrolleghem P.A., Brouckaert C.J., Volcke E. (2012) Towards a generalized physicochemical framework. Wat Sci Tech, 66(6):1147-1161.

Bengtsson S., Werker A., Christensson M., Welander T. (2008) Production of polyhydroxyalkanoates by activated sludge treating a paper mill wastewater. Biores Technol, 99(3):509-516.

Bentancur S. (2014) Evaluation and modelling of UPM pulp mill wastewater treatment plant in Fray Bentos, Uruguay. MSc. Thesis. UNESCO-IHE Institute for Water Education. Delft, the Netherlands.

Beun J.J., Paletta P., van Loosdrecht M.C.M., Heijnen J.J. (2000) Stoichiometry and kinetics of poly-β-hydroxybutyrate metabolism in aerobic, slow growing, activated sludge cultures. Biotech Bioeng 67(4): 379–389.

Brdjanovic D., Moussa M.S., Mithaiwala M., Amy G., van Loosdrecht M.C.M. (2007) Use of plant-wide modeling for optimization and upgrade of a tropical wastewater treatment plant in a developing Country. Water Sci Tech, 56(7):21-31.

Brdjanovic D., van Loosdrecht M.C.M., Versteeg P., Hooijmans C.M., Alaerts G.J., Heijnen J.J. (2000) Modeling COD, N and P removal in a full-scale wwtp Haarlem Waarderpolder. Water Res 34(3):846–858.

Brdjanovic D., Zakaria F, Mawioo P.M., Garcia H.A., Hooijmans C.M., Pean T.Y., Setiadi T. (2014) eSOS – Emergency Sanitation Operation System. *Journal WASH Dev* (accepted).

Brouwer H., Klapwijk A., Keesman K.J. (1998) Identification of activated sludge and wastewater characteristics using respirometric batch-experiments. Wat Res. 32(4), 1240–1254.

Bürger R., Diehl S. and Nopens I. (2011) A consistent modelling methodology for secondary settling tanks in wastewater treatment. Water Res, 45:2247-2260.

Busby J.B., Andrews J.F. (1975) Dynamic modeling and control strategies for the activated sludge process. J Water Pollut Control Fed, 47(5), 1055-1080.

Chen R., Hashimoto A.G. (1980) Substrate utilization kinetic model for biological treatment process. Biotech Bioeng 22(10): 2081–2095.

Clouzot L., Choubert J.M., Cloutier F., Goel R., Love N.G., Melcer H., Ort C., Patureau D., Plósz B.G., Pomiès M., Vanrolleghem P.A. (2013) Perspectives on modelling micropollutants in wastewater treatment plants. Water Sci Tech 68(2):448-461.

Çokgör E.U., Sözen S., Orhon D., Henze M.(1998) Respirometric analysis of activated sludge behaviour—I. Assessment of the readily biodegradable substrate. Wat Res, 32(2): 461–475.

Comeau Y., Hall K.J., Hancock R.E.W., Oldham W.K. (1986) Biochemical model for enhanced biological phosphorus removal. Wat Res 20(12), 1511–1521.

Corominas L., Rieger L., Takács I., Ekama G., Hauduc H., Vanrolleghem P.A.,Oehmen A., Gernaey K.V., Comeau Y. 2010. New framework for standardized notation in wastewater treatment modelling. Water Sci Tech 61(4), 841–857.

Côté M., Grandjean B.P.A., Lessard P., Thibault J. (1995) Dynamic modelling of the activated sludge process: Improving prediction using neural networks. Water Res 29(4), 995–1004.
De Kreuk, M. K., Picioreanu, C., Hosseini, M., Xavier, J. B., van Loosdrecht, M. C. M. (2007). Kinetic model of a granular sludge SBR: influences on nutrient removal. Biotechno Bioeng, 97(4), 801-815.
Dochain D., Vanrolleghem P.A., Van Daele M. (1995) Structural identifiability of biokinetic models of activated sludge respiration. Water Res 29(11), 2571–2578.
Dold P.L., Ekama G.A., Marais G.v.R. (1980) A General Model for the Activated Sludge Process. Progress Water Tech 12(6), 47-77.
Dold P.L., Marais G.v.R. (1986) Evaluation of the General Activated Sludge Model Proposed by the IAWPRC Task Group. Wat Sci Tech 18(6), 63–89.
Ekama G.A., Dold P.L., Marais G.v.R. (1986) Procedures for Determining Influent COD Fractions and the Maximum Specific Growth Rate of Heterotrophs in Activated Sludge Systems. Wat Sci Tech 18(6), 91-114.
Ekama G.A., Takács I (2014) Modelling. In: Activated sludge: 100 years and counting. Eds. Wanner J. and Jenkins D. IWA Publishing. London, U.K. ISBN: 9781780404936.
Fall C., Flores-Alamo N., Esparza-Soto M., Hooijmans C.M. (2012) Tracer test and hydraulics modeling of a large WWTP. Water Practice and Technology, 7(1).
Fiehn O. (2001), Combining genomics, metabolome analysis, and biochemical modeling to understand metabolic networks. Comparative and functional genomics, 2(3):155-168.
Gernaey K.V., van Loosdrecht M.C.M., Henze M., Lindd M., Jørgensen S.B. (2004) Activated sludge wastewater treatment plant modelling and simulation: state of the art. Environ Modelling Software 19(9):763-783.
Grau P., Copp J., Vanrolleghem P.A., Takács I. and Ayesa E. (2009) A comparative analysis of different approaches for integrated WWTP modelling. Wat. Sci. Tech., 59(1), 141-147.
Gujer W. (2006). Activated sludge modelling: past, present and future. Water Sci Tech, 53(3), 111-119.
Gujer W., Henze M. (1991) Activated sludge modelling and simulation. Wat Sci Tech 23, 1011-1023.
Gujer W., Henze M., Mino T. and van Loosdrecht M.C.M. (1999) Activated Sludge Model No. 3. Water Sci. Tech., 3, 183-193.
Gujer W., Henze M., Mino T., Matsuo T. Wentzel M.C., Marais G.v.R. (1995) The Activated Sludge Model No. 2: Biological phosphorus removal. Water Sci Tech 31(2), 1-11.
Hao T.W., Xiang P.Y., Mackey H.R., Chi K., Lu H., Chui H.K., van Loosdrecht M.C.M., Chen G.H. (2014) A review of biological sulfate conversions in wastewater treatment. Wat Res, 65:1-21.
Hauduc H., Gillot S., Rieger L., Ohtsuki T., Shaw A., Takács I., Winkler S. (2009) Activated sludge modelling in practice-An international survey. Water Sci Tech 60(8), 1943–1951.
Henze M. (1992) Characterization of wastewater for modeling of activated sludge processes. Water Sci. Tech., 25, 1-15.
Henze M. (Ed.). (2000). Activated sludge models ASM1, ASM2, ASM2d and ASM3 (IWA Scientific and Technical Report No. 9). IWA, London.
Henze M., Grady C. P. L., Gujer W., Marais G.v.R. and Matsuo T. (1987) Activated sludge model No 1. IAWPRC Scientific and Technical Report No. 1. IAWQ, London.
Henze M., Grady J.R.C.P.L., Gujer W. (1987) A general model for single-sludge wastewater treatment systems. Water Res 21(5), 505-515.
Henze M., Gujer W., Mino T., Matsuo T., Wentzel M.C., and Marais G. V. R. (1995) Activated Sludge Model No. 2. IAWQ Scientific and Technical Report No. 3. London, IAWQ.
Henze M., Gujer W., Mino T., Matsuo T., Wentzel M.C., Marais G.VR., van Loosdrecht M.C.M.(1999) Activated sludge model no. 2d, ASM2d. Water Sci Tech 39(1), 165–182.
Henze M., Harremoes P., Jansen J.L.C and Arvin E. (1997) Waste water treatment biological and chemical processes. Springer, Berlin.
Hodzic A., Vojinovic Z., Seyoum S.D., Pathirana A., Meijer S.C.F., Brdjanovic D. (2011) Model-based evaluation of the urban wastewater infrastructure reconstruction options in a developing country: Case study Sarajevo in Bosnia and Herzegovina. 2nd IWA Development Congress. November 21st-24th, 2011. Kuala Lumpur, Malaysia.
Hulsbeek J.J.W., Kruit J., Roeleveld P.J. and van Loosdrecht M.C.M. (2002) A practical protocol for dynamic modelling of activated sludge systems. Water Sci. Tech., 45, 127-136.
Hvitved-Jacobsen, T., Vollertsen, J., & Nielsen, A. H. (2013). Sewer processes: microbial and chemical process engineering of sewer networks. CRC press. ISBN: 978-1-4398-8177-4.

Ikumi D.S., Brouckaert C.J. and Ekama G.A. (2011) A 3 phase anaerobic digestion model. In: Proceedings of the 8th IWA International Symposium on Systems Analysis and Integrated Assessment in Water Management (Watermatex2011). San Sebastian, Spain, June 20-22 2011.

Ikumi D.S., Harding T.H., Ekama G.A. (2014a) Biodegradability of wastewater and activated sludge organics in anaerobic digestion. Water Res, 56:267-279.

Ikumi D.S., Harding T.H., Vogts M., Lakay M.T., Mafungwa H., Brouckaert C.J., Ekama G.A. (2014b) Mass balances modelling over wastewater treatment plants III. Final Report. Water Research Comission. South Africa.

Jiang Y., Marang L., Tamis J., van Loosdrecht M.C.M., Dijkman H., Kleerebezem, R. (2012). Waste to resource: Converting paper mill wastewater to bioplastic. Water Res, 46(17):5517-5530.

Kappeler J., Gujer W. (1992) Estimation of Kinetic Parameters of Heterotrophic Biomass under Aerobic Conditions and Characterization of Wastewater for Activated Sludge Modelling. Water Sci Tech 25(6), 125–139.

Koch G., Kuhni M., Gujer W., Siegrist H. (2000) Calibration and validation of activated sudge model No. 3 for Swiss municipal wastewater. Water Res 34(14), 3580-3590.

Krishna C., Van Loosdrecht M.C.M. (1999) Substrate flux into storage and growth in relation to activated sludge modeling Wat Res 33(14), 3149–3161.

Kuba T., van Loosdrecht M.C.M., Brandse F.A., Heijnen J.J. (1997) Occurrence of denitrifying phosphorus removing bacteria in modified UCT-type wastewater treatment plants. Water Res. 31, 777–786.

Lackner S., Gilbert E. M., Vlaeminck S. E., Joss A., Horn H., van Loosdrecht M. (2014). Full-scale partial nitritation/anammox experiences–An application survey. Water Res, 55:292-303.

Langergraber G., Rousseau D.P.L., Garcia J., Mena J. (2009) CWM1 - A general model to describe biokinetic processes in subsurface flow constructed wetlands. Wat Sci Tech 59(9):1687-1697.

Lawrence A.W., McCarty P.L. (1970) Unified basis for biological treatment design and operation. J Sanitary Eng Division 96 (3), 757-778.

Lee Y., Cho J., Seo Y., Lee J.W., Ahn K.H., (2002) Modeling of submerged membrane bioreactor process for wastewater treatment, Desalination. 146(1–3), 451–457.

Lesouef A., Payraudeau M., Rogalla F., Kleiber B. (1992) Optimizing Nitrogen Removal Reactor Configurations by On-Site Calibration of the IAWPRC Activated Sludge Model. Water Sci Tech 25(6); 105–123.

Loosdrecht van M.C.M., Henze M. (1999) Maintenance, endogeneous respiration, lysis, decay and predation. Water Sci Tech 39(1), 107-117.

Lopez-Vazquez C.M., Mithaiwala M., Moussa M.S., van Loosdrecht M.C.M., Brdjanovic D. (2013) Coupling ASM3 and ADM1 for wastewater treatment process optimisation and biogas production in a developing country: case-study Surat, India. J Water Sanitation Hygiene Develop, 3(1), 12-25.

Lopez-Vazquez C.M., Oehmen A., Hooijmans C.M., Brdjanovic D., Gijzen H.J., Yuan Z., van Loosdrecht M.C.M. (2009) Modeling the PAO–GAO competition: effects of carbon source, pH and temperature. Water Res 43(2), 450-462.

Lu H., Wang J., Li S., Chen G.H., van Loosdrecht M.C.M., Ekama GA (2009) Steady-state model based evaluation of sulfate reduction, autotrophic denitrification and nitrification integrated (SANI) process. Water Res 43(14):3613-3621.

Majone M., Dircks K., Beim J.J. Aerobic storage under dynamic conditions in activated sludge processes. The state of the art. (1999) Water Sci Tech. 39(1), 61–73.

Mamais D., Jenkins D., Pitt P. (1993) A rapid physical-chemical method for the determination of readily biodegradable soluble COD in municipal wastewater. Water Res 27(1), 195-197.

Marais G.v.R., Ekama G.A. (1976) The activated sludge process Part 1 - Steady-state behaviour. Water SA 2(4), 164-200.

Martins A.M.P., Pagilla K., Heijnen J.J., van Loosdrecht M.C.M. (2004a). Filamentous bulking sludge—a critical review. Water Res 38 (4): 793-817.

Martins A.M.P., Picioreanu C., Heijnen J.J., van Loosdrecht M.C.M. (2004b). Three-dimensional dual-morphotype species modeling of activated sludge flocs. Environ Sci Tech, 38(21), 5632-5641.

Meijer S.C.F. (2004) Theoretical and practical aspects of modelling activated sludge processes. PhD dissertation. Delft, University of Technology. 218 p. ISBN: 90-9018027-3.

Meijer S.C.F. and, Brdjanovic D. (2012) A Practical Guide to Activated Sludge Modeling. UNESCO-IHE Lecture Notes, 291 p. ISBN: 9073445264.

Meijer S.C.F., van der Spoel H., Susanti S., Heijnen J.J. and van Loosdrecht M.C.M. (2002) Error diagnostics and data reconciliation for activated sludge modeling using mass balances. Water Sci. Tech. 25(6), 145-156.

Mino T., Liu W.T., Kurisu F., Matsuo T. (1995) Modelling glycogen storage and denitrification capability of microorganisms in enhanced biological phosphate removal processes. Water Sci Tech 31(2), 25–34.

Murnleitner E., Kuba T., van Loosdrecht M.C.M., Heijnen J.J. (1997) An integrated metabolic model for the aerobic and denitrifying biological phosphorus removal. Biotech Bioeng 54(5), 434-450.

Ni B.-J., Yuan Z., Chandran K., Vanrolleghem P.A., Murthy S. (2013) Evaluating four mathematical models for nitrous oxide production by autotrophic ammonia-oxidizing bacteria. Biotechnol Bioeng, 110(1): 153–163.

Nielsen P.H., Mielczarek A.T., Kragelund C., Nielsen J.L., Saunders A.M., Kong Y., Hansen A.A., Vollertsen J. (2010) A conceptual ecosystem model of microbial communities in enhanced biological phosphorus removal plants. Water Res 44(17): 5070-5088.

Nopens I., Batstone D.J., Copp J.B., Jeppsson U., Volcke E., Alex J., Vanrolleghem P.A. (2009) An ASM/ADM model interface for dynamic plant-wide simulation. Water Res, 43(7):1913-1923

Oehmen A., Carvalho G., Lopez-Vazquez C.M., van Loosdrecht M.C.M., Reis M.A.M. (2010) Incorporating microbial ecology into the metabolic modelling of polyphosphate accumulating organisms and glycogen accumulating organisms. Water Res, 44(17): 4992-5004.

Orhon D., Ateş E., Sözen S., Çokgör E.U. (1997) Characterization and COD fractionation of domestic wastewaters. Environmental Pollution, 95(2)2, 191–204.

Petersen B., Vanrolleghem PA., Gernaey K., Henze M. (2002) Evaluation of an ASM1 model calibration procedure on a municipal-industrial wastewater treatment plant. J Hydroinformatics 4:15-38.

Petersen E.E. (1965) Chemical reaction analysis. Prentice-Hall, Englegood Cliffs, N.J.

Plózc B.G., Nopens I., Rieger L., Griborio A., De Clercq J., Vanrolleghem P.A., Daigger G.T., Takács I., Wicks J., Ekama G.A. (2012) A critical review on clarifier modelling: state-of-the-art and enginering practices. In: Proceedings 3rd IWA/WEF Wastewater Modelling Seminar. Mont-Sainte-Anne, Québec, Canada. Feb 26-28th, 2012.

Price R.K. and Vojinovic Z. (2010) Urban Hydroinformatics: Data, Models and Decision Support for Integrated Urban Water Management, IWA Publishing.

Puig, S., van Loosdrecht, M. C. M., Colprim, J., and Meijer, S. C. . (2008) Data evaluation of full-scale wastewater treatment plants by mass balance. Water Research, 42, 4645–4655.

Radu A. I., Vrouwenvelder J. S., van Loosdrecht M.C.M., Picioreanu C. (2010) Modeling the effect of biofilm formation on reverse osmosis performance: Flux, feed channel pressure drop and solute passage. J Membrane Sci 365(1): 1-15.

Rauch W., Bertrand-Krajewski J.L., Krebs P., Mar O., Schilling W., Schutze M., Vanrolleghem P.A. (2002) Deterministic modelling of integrated urban drainage systems. Wat Sci Tech 45(3):81-94.

Rieger L., Gillot S., Langergraber G., Ohtsuki T., Shaw A., Takács I., Winkler S. (2012) Guidelines for using activated sludge models. IWA Task Group on Good Modelling Practice. Scientific and Technical Report No. 22. IWA publishing. London, UK. ISBN: 9781843391746.

Rieger L., Koch G., Kühni M., Gujer W., Siegrist H. (2001) The EAWAG Bio-P module for activated sludge model No. 3 Wat. Res. 35(16), 3887–3903.

Rieger, L., Takács, I., Villez, K., Siegrist, H., Lessard, P., Vanrolleghem, P. A., and Comeau, Y. (2010) Data Reconciliation for Wastewater Treatment Plant Simulation Studies—Planning for High-Quality Data and Typical Sources of Errors. Water Environment Research, 82(5), 426–433.

Roeleveld P.J. and van Loosdrecht M.C.M. (2002) Experience with guidelines for wastewater characterisation in the Netherlands. Water Sci. Tech., 45, 77-87.

Sin G., van Hulle S.W.H., de Pauw D.J.W., van Griensven A., Vanrolleghem P.A. (2005) A critical comparison of systematic calibration protocols for activated sludge models: A SWOT analysis. Wat Res 39(12), 2459–2474.

Smolders G.J.F., van der Meij J., van Loosdrecht M.C.M., Heijnen J.J. (1994a) Model of the anaerobic metabolism of the biological phosphorus removal process: Stoichiometry and pH influence. Biotech Bioeng 43(6), 461-470.

Smolders G.J.F., van der Meij J., van Loosdrecht M.C.M., Heijnen J.J. (1994b) Stoichiometric model of the aerobic metabolism of the biological phosphorus removal process. Biotech Bioeng 44(7):837-848.

Smolders G.J.F., van der Meij J., van Loosdrecht M.C.M., Heijnen J.J. (1995) A structured metabolic model for anaerobic and aerobic stoichiometry and kinetics of the biological phosphorus removal process. Biotech Bioeng 47(3), 277-287.

Sollfrank U., Gujer W. (1990) Characterisation of domestic wastewater for mathematical modelling of the activated sludge process. Water Sci Tech 23(4-6), 1057–1066.

Spanjers H., Vanrolleghem P. (1999) Respirometry as a tool for rapid characterizatiom of wastewater and activated sludge. Water Sci Tech 31(2), 105-114.

Spanjers H., Vanrolleghem P., Olsson G., Dold P. (1996) Respirometry in control of the activated sludge process. Wat Sci Tech 34 (3–4), 117–126.

Spindler, A. (2014) Structural Redundancy of Wastewater Treatment Data. An Algorithm Derived from Data Reconciliation Methods. Water Research (to be published) Institute of Water Quality and Resource Management, Vienna University of Technology, Karlsplatz 13/226-1, 1040 Wien, Austria (E-mail: a.spi@iwag.tuwien.ac.at)

Spindler, A. and Vanrolleghem, P. A. (2012) Dynamic mass balancing for wastewater treatment data quality control using CUSUM charts. Water science and technology: a journal of the International Association on Water Pollution Research, 65(12), 2148–2153.

Takács I., Patry G.G., Nolasco D. (1991) A dynamic model of the clarification-thickening process. Wat Res 25(10), 1263-1271.

Takács I., Vanrolleghem P., Wett B., Murthy S. (2007) Elemental balance based methodology to establish reaction stoichiometry in environmental modeling. Water Sci Tech, 56(9):37-41.

van Loosdrecht M.C.M., Brdjanovic D. (2014) Anticipating the next century of wastewater treatment. Science 344(6191):1452-1453.

van Loosdrecht M.C.M., Hooijmans C.M. Brdjanovic D., Heijnen J.J. (1997) Biological phosphorus removal processes: a mini review. Applied Microbiol Biotechnol 48, 289-296.

Vanhooren H., Meirlaen .J, Amerlinck Y., Claeys F., Vangheluwe H., Vanrolleghem P.A. (2003) WEST: modelling biological wastewater treatment. J Hydroinformatics 5(1):27-50.

Vanrolleghem P., Flores-Alsina X., Guo L., Solon K., Ikumi D., Batstone D., Brouckaert C., Takács I., Grau P., Ekama G.E., Jepsson U., Gernaey K. (2014) Towards BSM2-GPS-X: A plant-wide benchmark simulation model not only for carbon and nitrogen, but also for greenhouse gases (G), phosphorus (P), sulphur (S) and micropollutants (X), all within the fence of WWTPs/WRRFs." In: Procs. 4th IWA/WEF Wastewater Treatment Modelling Seminar (WWTmod2014). March, 2014. Spa, Belgium.

Vanrolleghem P.A., Rosen C., Zaher U., Copp J., Benedetti L., Ayesa E., Jeppsson U. (2005) Continuity-based interfacing of models for wastewater systems described by Petersen matrices. Wat. Sci. Tech., 52(1-2), 493-500.

Vanrolleghem P.A., Spanjers H., Petersen B., Ginestet P., Takács I. (1999) Estimating (combinations of) Activated Sludge Model No. 1 parameters and components by respirometry. Water Sci Tech 39(1):195-214.

Vanrolleghem P.A., Van Daele M., Dochain D. (1995) Practical identifiability of a biokinetic model of activated sludge respiration Water Res 29(11), 2561–2570.

Veldhuizen, van H.M., van Loosdrecht M.C.M., Heijnen J.J. (1999) Modelling biological phosphorus and nitrogen removal in a full scale activated sludge process. Water Res 33(16), 3459–3468.

Vojinovic Z. and Abbott M.B. (2011) Flood Risk and Social Justice: From Hydroinformatics of Quantities to Hydroinformatics of Qualities, IWA Publishing.

Volcke E.I., van Loosdrecht M.C.M., Vanrolleghem P.A. (2006) Continuity-based model interfacing for plant-wide simulation: A general approach. Water Res, 40(15): 2817-2828.

Wang J., Lu H., Chen G.H., Lau G.N., Tsang W.L., van Loosdrecht M.C.M. (2009) A novel sulfate reduction, autotrophic denitrification, nitrification integrated (SANI) process for saline wastewater treatment. Water Res 43(9):2363-2372.

Wanner O., Eberl H.J., Morgenroth E., Noguera D.R., Picioreanu C., Rittman B.E., van Loosdrecht M.C.M. (2006) Mathematical modelling of biofilms. IWA Publishing. London, UK.

Weijers S.R., Vanrolleghem P.A. (1997) A procedure for selecting the best identifiable parameters in calibrating activated sludge model No. 1 to full-scale plant data. Water Sci Tech 36(5), 69-79.

Wentzel M.C., Dold P.L., Ekama G.A., Marais G.v.R. (1989) Enhanced polyphosphate organism cultures in activated sludge systems. Part III: Kinetic model. Water SA 15(2), 89-102.

Wentzel M.C., Ekama G.A., Loewenthal R.E., Dold P.L., Marais G.v.R. (1986) Enhanced polyphosphate organism cultures in activated sludge systems. Part II: Experimental behaviour. Water SA 15(2), 71-88.

Wett B., Omari A., Podmirseg S.M., Han M., Akintayo O., Gómez Brandón M., Murthy S., Bott C., Hell M., Takács I., Nyhuis, O'Shaughnesy M. (2013) Going for mainstream deammonification from bench-to full-scale for maximized resource efficiency. Water Sci Tech 68(2): 283-289.